Bio-Farms for Nutraceuticals

ADVANCES IN EXPERIMENTAL MEDICINE AND BIOLOGY

Recent Volumes in this Series

Volume 690
THE RENIN-ANGIOTENSIN SYSTEM: CURRENT RESEARCH PROGRESS IN THE PANCREAS
Edited by Po Sing Leung

Volume 691
ADVANCES IN TNF FAMILY RESEARCH
Edited by David Wallach

Volume 692
NEUROPEPTIDE SYSTEMS AS TARGETS FOR PARASITE AND PEST CONTROL
Edited by Timothy G. Geary and Aaron G. Maule

Volume 693
POST-TRANSCRIPTIONAL REGULATION BY STAR PROTEINS
Edited by Talila Volk and Karen Artzt

Volume 694
PROTEIN METABOLISM AND HOMEOSTASIS IN AGING
Edited by Nektarios Tavernarakis

Volume 695
THE CELL BIOLOGY OF STEM CELLS
Edited by Eran Meshorer and Kathrin Plath

Volume 696
SOFTWARE TOOLS AND ALGORITHMS FOR BIOLOGICAL SYSTEMS
Edited by Hamid R. Arabnia and Quoc-Nam Tran

Volume 697
HOT TOPICS IN INFECTION AND IMMUNITY IN CHILDREN VII
Edited by Andrew Pollard, Adam Finn, and Nigel Curtis

Volume 698
BIO-FARMS FOR NUTRACEUTICALS: FUNCTIONAL FOOD AND SAFETY CONTROL BY BIOSENSORS
Maria Teresa Giardi, Giuseppina Rea and Bruno Berra

A Continuation Order Plan is available for this series. A continuation order will bring delivery of each new volume immediately upon publication. Volumes are billed only upon actual shipment. For further information please contact the publisher.

Bio-Farms for Nutraceuticals

Functional Food and Safety Control by Biosensors

Edited by

Maria Teresa Giardi, PhD
Institute of Crystallography, Department of Agrofood and Department of Molecular Design, National Council of Research, Area of Research of Rome, Monterotondo Scalo, Rome, Italy

Giuseppina Rea, PhD
Institute of Crystallography, Department of Agrofood and Department of Molecular Design, National Council of Research, Area of Research of Rome, Monterotondo Scalo, Rome, Italy

Bruno Berra, PhD
Institute of General Physiology and Biochemistry, Department of Molecular Sciences Applied to Biosystems, DISMAB, University of Milan, Milan, Italy

Springer Science+Business Media, LLC

Landes Bioscience

Springer Science+Business Media, LLC
Landes Bioscience

Printed in the USA.

Springer Science+Business Media, LLC, 233 Spring Street, New York, New York 10013, USA
http://www.springer.com

Please address all inquiries to the publishers:
Landes Bioscience, 1002 West Avenue, Austin, Texas 78701, USA
Phone: 512/ 637 6050; FAX: 512/ 637 6079
http://www.landesbioscience.com

The chapters in this book are available in the Madame Curie Bioscience Database.
http://www.landesbioscience.com/curie

Bio-Farms for Nutraceuticals: Functional Food and Safety Control by Biosensors, edited by Maria Teresa Giardi, Guiseppina Rea and Bruno Berra. Landes Bioscience / Springer Science+Business Media, LLC dual imprint / Springer series: Advances in Experimental Medicine and Biology.

ISBN: 978-1-4419-7346-7

Library of Congress Cataloging-in-Publication Data

Bio-farms for nutraceuticals : functional food and safety control by biosensors / edited by Maria Teresa Giardi, Giuseppina Rea, Bruno Berra.
p. ; cm. -- (Advances in experimental medicine and biology ; v. 698)
Includes bibliographical references and index.
ISBN 978-1-4419-7346-7
1. Functional foods. 2. Biosensors. 3. Food--Biotechnology. I. Giardi, Maria Teresa. II. Rea, Giuseppina. III. Berra, B. IV. Series: Advances in experimental medicine and biology ; v. 698. 0065-2598
[DNLM: 1. Functional Food. 2. Biosensing Techniques. 3. Food Additives. 4. Food Handling. 5. Food Technology. 6. Health Food. W1 AD559 v.698 2010 / QU 145.5]
QP144.F85B535 2010
615'.3--dc22

2010030182

DEDICATION

This book is dedicated to the nutritionist Dr. Paolo Sorbini, founder of Also-S.p.A., now known as Enervit S.p.A. Enervit S.p.A. is the foremost Italian company specialising in sports nutrition and wellness. It manufactures and distributes products satisfying specific needs for those involved in sports at the professional level and for those merely interested in preserving their own healthy lifestyle. Since the 1980s, Dr. Sorbini, an early contributor of great discoveries in this field, succeeded in completely revolutionizing the idea of nutrition before, after and while performing sports.

Supporting the opinion that good results can be obtained in sporting competition by using a suitable nutrition strategy while avoiding doping, he understood the importance of restoring the right hydro-saline balance in the blood and the necessary supplementation for muscle recovery after intense physical activity. Born in 1926 to an Italian family of Montepulciano 'apothecaries', he dedicated his life to scientific research and sports nutrition, collaborated with universities, sports institutions and placed his experience at the service of events and ventures across diverse disciplines and high level events: from Moser's hour record in the World Cycling competition to the exploits of Valentina Vezzali in fencing, from Messner's primacy in ascending all fourteen mountain peaks over 8,000 metres to Tomba and Di Centa in the Skiing World Cup, from Simeoni's world record in the women's high jump to the successes of Inter-Milan and Juventus in football. The energy and enthusiasm of Paolo Sorbini also opened his entry into the field of medicine concerning obesity. He supported the idea that these worldwide nutritional disorders could be controlled by establishing the right balance of insulin in the blood and collaborated with the famous American biochemist Dr. Barry Sears, creator of the Zone diet. This collaboration brought about the development of a new important brand called EnerZona which combines a series a food supplements to establish the correct hormonal balance in the body. After his death in December 2006, his sons Maurizia, Pino and Alberto founded the 'Paolo Sorbini Foundation for Nutrition Science'. The foundation was established with the aim to promote scientific research and to communicate the knowledge acquired in the field of diet and nutrition through the organization of seminars, congresses, practical courses, work experience and editorial projects.

Finally, particularly close to the foundation's heart is to inform the younger generation of the importance of appropriate nutrition and diet when approching physical activity.

PREFACE

This volume can be said to have been born of the NUTRA-SNACKS project within the Sixth Framework Programme Priority on Food Quality and Safety. One objective of NUTRA-SNACKS was to improve the nutritional and eating properties of ready-to-eat products and semi-prepared foodstuffs through better monitoring of the quality and safety of raw materials and the development of innovative processes along the production chain. Another main objective of the project was the production of ready-to-eat snacks with high nutraceutic activity. Seven research institutes and three companies in six European countries were involved in this effort. The co-operation resulted in the production of food having a high content of natural metabolites with the following beneficial health effects: anticancer, antilipidemic, anticholesterol, antimicrobial, antibacterial, antifungal, antiviral, antihypertensive, anti-inflamatory and antioxidant activities.

Bio-Farms for Nutraceuticals treats comprehensively yet concisely the subjects of phytochemicals in food and feedstuffs, the production of phytochemicals by field crops and tissue cultures as important safety and quality issues in developing phytochemical products.

The early chapters describe the efficacy and safety of some medicinal, nutraceutic herbs and plants, providing a comprehensive review of the most notable phytochemicals being researched today. Information is given about carotenoids, polyphenols, sulfur-containing compounds and secondary plant metabolites showing antioxidant activity. Diet-gene interactions are treated as well as the mechanism and the therapeutic potential against cancer, obesity and aging disorders. Nutrition strategies are also proposed to prevent diseases and improve general well-being. The middle chapters expand the information on phytochemicals and focus on the production of useful secondary products by using molecular biological techniques. The most important methods to enhance the quantity of phytonutrients are discussed as well as the possibility of producing plant compounds on a large scale using industrial bioreactors.

The later chapters deal with testing the safety and quality of nutraceutics using both classical and new research methodologies and technologies. Special attention is given to the currently very active field of research into biosensors and their applications. Different forms of these devices, which are able to detect the presence of active phytochemicals and toxic compounds or evaluate the antioxidant activity of bioactive species, are described. The final chapters add a comprehensive analysis of the main regulations about functional food and nutraceutical products. The most important ethical and safety issues are elaborated to provide a broader understanding of the field of functional foods, their pros and cons and the future prospects for a worldwide nutraceuticals market.

Maria Teresa Giardi, PhD
Department of Agrofood and Department of Molecular Design, CNR-IC National Council of Research, Institute of Crystallography, Rome, Italy

Giuseppina Rea, PhD
Department of Agrofood and Department of Molecular Design, CNR-IC National Council of Research, Institute of Crystallography, Rome, Italy

Bruno Berra, PhD
Faculty of Pharmacy, Department of Molecular Sciences Applied to Biosystems, University of Milan, Milan, Italy

ABOUT THE EDITORS...

MARIA TERESA GIARDI, PhD, is Research Manager at the National Council of Research (CNR) in Rome, Italy. Her background is in organic-industrial chemistry. She has extensive experience in molecular biology; modification of photosystem II to increase its resistance to high temperature and sensitivity to radiation; signal transduction in optical and electrochemical biosensors; protein stabilization and utilization in biosensor design, and analytical procedures in general. She is a referee for the US Department of Agriculture for industrial projects and has worked in various research institutions and research and development companies both in Europe and the USA. She is a supervisor or coordinator of a number of national and international projects in the field of biosensors: European Agency Life V project, 20xx-2004; a European Space Agency project involving space flight in 2005; a NATO "linkage grant", dedicated to sensors based on plant and microrganisms.

ABOUT THE EDITORS...

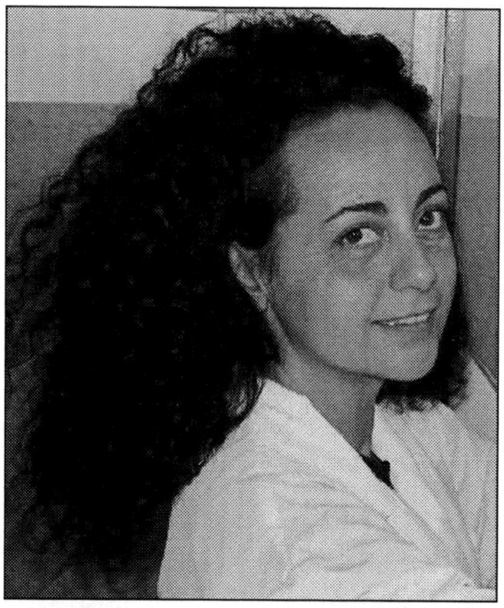

GIUSEPPINA REA, PhD, is a molecular biologist and is on the permanent staff of CNR-IC. She is an expert in the molecular mechanisms involved in plant defense responses and has many years of experience in plant molecular biology, biochemistry, biotechnology and protein engineering techniques. She monitors the CNR project Biotechnological Applications of Oxygenic Photosynthetic Organisms in the Agro-food Field and Space Environment, and the EU projects Multibioplat and Sensbiosyn, aimed at producing bacteria and algae with improved photosynthetic performance to exploit them in biological farms for nutraceutical applications and biosensing elements in biosensor development. Her research activities are mainly focused on *Chlamydomonas reinhardtii* and *Rhodobacter sphaeroides*. She is advisor to several PhD students in Plant Biotechnology, in the fields of exobiology, nutraceutics and biosensors.

ABOUT THE EDITORS...

BRUNO BERRA, PhD, received his academic degree in Chemistry from the University of Pavia, Italy. He is Professor f.r. of Biochemistry at the University of Milan and works at the Department of Molecular Sciences Applied to Biosystems in the Faculty of Pharmacy. During his active teaching career he taught Biochemistry, Molecular Biology, Nutrition Biochemistry and Cosmetology. His most important research areas include: isolation, fractionation, purification and analysis of glycolipids from brain, extranervous tissues and biological fluids; biochemical diagnosis of sphingolipidosis and follow-up of Gaucher's patients receiving enzyme replacement therapy; nutritional biochemistry with reference to fats and antioxidants; membrane component studies in spontaneous or induced degenerative processes, with particular reference to solid and tumor cells; skin biochemistry and molecular biology of the skin; and space biology. He is a member of numerous national and international scientific organisations including the Italian Society of Biochemistry (1957-2004), the Italian Society for Fat Research, the International Conference of Biochemistry of Lipids (ICBL), the Euro Fed Lipid, the American Oil Chemist's Society, and the American Society of Diabetes.

PARTICIPANTS

William W. Adams, III
Department of Ecology
and Evolutionary Biology
University of Colorado
Boulder, Colorado
USA

Cecilia Ambrosi
Institute of Crystallography
Department of Agrofood
and Department of Molecular Design
National Council of Research
Area of Research of Rome
Monterotondo Scalo
Rome
Italy

Amina Antonacci
Institute of Crystallography
Department of Agrofood
and Department of Molecular Design
National Council of Research
Area of Research of Rome
Monterotondo Scalo
Rome
Italy

Lise Barthelmebs
IMAGES Group
Centre de Phytopharmacie
University of Perpignan
Perpignan
France

Giovanni Basile
Biosensor srl
Palombara Sabina
Rome
Italy

Bruno Berra
Institute of General Physiology
and Biochemistry
Department of Molecular Sciences
Applied to Biosystems
DISMAB
University of Milan
Milan
Italy

Patrizia Berselli
Institute of General Physiology
and Biochemistry
Department of Molecular Sciences
Applied to Biosystems
DISMAB
University of Milan
Milan
Italy

Alessandra Bertoli
DCBB-Department of Bioorganic
Chemistry and Biopharmacy
University of Pisa
Pisa
Italy

Iain Brown
Cancer Medicine Research Group
School of Medicine and Dentistry
Aberdeen University
Aberdeen, Scotland
UK

Vittorio Calabrese
Department of Chemistry
Clinical Biochemistry
and Clinical Molecular Biology
University of Catania
Catania
Italy

Carole Calas-Blanchard
BIOMEM Group
Centre of Phytopharmacy
University of Perpignan
Perpignan
France

Monica Campàs
BIOMEM Group
Centre of Phytopharmacy
University of Perpignan
Perpignan
France

Calogero Caruso
Immunosenescence Unit
Department of Pathobiology
and Biomedical Methodologies
University of Palermo
Palermo
Italy

Paola Corsetto
Institute of General Physiology
and Biochemistry
Department of Molecular Sciences
Applied to Biosystems
DISMAB
University of Milan
Milan
Italy

Montserrat Cortina-Puig
BIOMEM Group
Centre of Phytopharmacy
University of Perpignan
Perpignan
France

Anna Czubacka
Institute of Soil Science
and Plant Cultivation
Department of Biochemistry
State Research Institute
Pulawy
Poland

Barbara Demmig-Adams
Department of Ecology
and Evolutionary Biology
University of Colorado
Boulder, Colorado
USA

Sandra Eremia
Centre of Bioanalysis
National Institute for Biological
Sciences
Bucharest
Romania

Tahira Fatima
Department of Biology
University of Western Ontario
London, Ontario
Canada

Simone Ferrari
Department of Plant Biology
University of Rome La Sapienza
Rome
Italy

Dirk Fischer
Institut für Pflanzenphysiologie
Martin-Luther Universität
Halle-Wittenberg
Halle, Saale
Germany

Demetrios F. Ghanotakis
Department of Chemistry
University of Crete
Voutes-Heraklion
Greece

Maria Teresa Giardi
Institute of Crystallography
Department of Agrofood
and Department of Molecular Design
National Council of Research
Area of Research of Rome
Monterotondo Scalo
Rome
Italy

Annalisa Giovannini
Research Unit for Floriculture
and Ornamental Species
CRA-Agricultural Research Council
Sanremo
Italy

Roberto Giunta
Ethical Committee
ASL-Local Health Agency
Milan
Italy

Avtar K. Handa
Department of Horticulture
and Landscape Architecture
Purdue University
West Lafayette, Indiana
USA

Steven D. Heys
Cancer Medicine Research Group
School of Medicine and Dentistry
Aberdeen University
Aberdeen, Scotland
UK

Anja Hohtola
Department of Biology
University of Oulu
Oulu
Finland

Georges Istamboulie
IMAGES Group
Centre de Phytopharmacie
University of Perpignan
Perpignan
France

Udo Johanningmeier
Institut für Pflanzenphysiologie
Martin-Luther Universität
Halle-Wittenberg
Halle, Saale
Germany

Justyna Krzyzanowska
Institute of Soil Science
and Plant Cultivation
Department of Biochemistry
State Research Institute
Pulawy
Poland

Maya Lambreva
Institute of Crystallography
Department of Agrofood
and Department of Molecular Design
National Council of Research
Area of Research of Rome
Monterotondo Scalo
Rome
Italy

Teresa Lavecchia
Biosensor srl
Palombara Sabina
Rome
Italy

Simona Carmen Litescu
Centre of Bioanalysis
National Institute for Biological
Sciences
Bucharest
Romania

Mariella Lucchesini
Department of Crop Plant Biology
University of Pisa
Pisa
Italy

Andrea Margonelli
Institute of Crystallography
Department of Agrofood
and Molecular Design
National Council of Research
Area of Research of Rome
Monterotondo Scalo
Rome
Italy

Jean-Louis Marty
BIOMEM Group
Centre of Phytopharmacy
University of Perpignan
Perpignan
France

Autar K. Mattoo
Sustainable Agricultural Systems
Laboratory
The Henry A. Wallace Beltsville
Agricultural Research Center
United States Department of Agriculture
Agricultural Research Service
Beltsville, Maryland
USA

Anna Mensuali-Sodi
Superior School of Advanced Studies
Sant' Anna
Pisa
Italy

Gigliola Montorfano
Institute of General Physiology
and Biochemistry
Department of Molecular Sciences
Applied to Biosystems
DISMAB
University of Milan
Milan
Italy

Manuela Negroni
Institute of General Physiology
and Biochemistry
Department of Molecular Sciences
Applied to Biosystems
DISMAB
University of Milan
Milan
Italy

Thierry Noguer
BIOMEM Group
Centre of Phytopharmacy
University of Perpignan
Perpignan
France

Wieslaw Oleszek
Institute of Soil Science
and Plant Cultivation
Department of Biochemistry
State Research Institute
Pulawy
Poland

Paolo Petroni
Enervit S.p.A.
Zelbio
Como
Italy

Riccardo Pina
Enervit S.p.A.
Zelbio
Como
Italy

Laura Pistelli
Department of Plant Crop Biology
University of Pisa
Pisa
Italy

Luisa Pistelli
DCBB-Department of Bioorganic Chemistry and Biopharmacy
University of Pisa
Pisa
Italy

Beatriz Prieto-Simón
BIOMEM Group
Centre of Phytopharmacy
University of Perpignan
Perpignan
France

Gabriel Lucian Radu
Centre of Bioanalysis
National Institute for Biological Sciences
Bucharest
Romania

Giuseppina Rea
Institute of Crystallography
Department of Agrofood and Department of Molecular Design
National Council of Research
Area of Research of Rome
Monterotondo Scalo
Rome
Italy

Angela Maria Rizzo
Institute of General Physiology and Biochemistry
Department of Molecular Sciences
Applied to Biosystems
DISMAB
University of Milan
Milan
Italy

Giuseppe Rodio
Enervit S.p.A.
Zelbio
Como
Italy

Dino Rotondo
Strathclyde Institute of Pharmacy and Biomedical Sciences
Strathclyde University
Glasgow, Scotland
UK

Régis Rouillon
IMAGES Group
Centre de Phytopharmacie
University of Perpignan
Perpignan
France

Barbara Ruffoni
Research Unit for Floriculture and Ornamental Species
CRA-Agricultural Research Council
Sanremo
Italy

Giovanni Scapagnini
Department of Health Sciences
University of Molise
Campobasso
Italy

Vijaya Shukla
Sustainable Agricultural Systems Laboratory
The Henry A. Wallace Beltsville Agricultural Research Center
United States Department of Agriculture
Agricultural Research Service
Beltsville, Maryland
USA

Arianna Tibuzzi
Biosensor srl
Palombara Sabina
Rome
Italy

Eleftherios Touloupakis
Department of Chemistry
University of Crete
Voutes-Heraklion
Greece

Francesca Vergari
Institute of Crystallography
Department of Agrofood
and Department of Molecular Design
National Council of Research
Area of Research of Rome
Monterotondo Scalo
Rome
Italy

Klaus W.J. Wahle
Cancer Medicine Research Group
School of Medicine and Dentistry
Aberdeen University
Aberdeen, Scotland
UK

Surender K. Yachha
Department of Pediatric
Gastroenterology
Sanjay Gandhi Postgraduate Institute
of Medical Sciences
Lucknow
India

Stefania Zava
Institute of General Physiology
and Biochemistry
Department of Molecular Sciences
Applied to Biosystems
DISMAB
University of Milan
Milan
Italy

CONTENTS

1. THE NUTRA-SNACKS PROJECT: BASIC RESEARCH AND BIOTECHNOLOGICAL PROGRAMS ON NUTRACEUTICS 1

Giuseppina Rea, Amina Antonacci, Maya Lambreva, Andrea Margonelli, Cecilia Ambrosi and Maria Teresa Giardi

Abstract 1
Introduction 1
Rationale of the Nutra-Snacks Project 1
Technological Innovations 6
Conclusion 13

2. OVERVIEW OF DIET-GENE INTERACTIONS AND THE EXAMPLE OF XANTHOPHYLLS 17

Barbara Demmig-Adams and William W. Adams, III

Abstract 17
Introduction: Overview of Diet-Gene Interaction in Human Disease 17
A Case Study: Xanthophylls and Their Synergism with Other Dietary Factors in Human Health—Protection against Eye Disease and Other Chronic Diseases 20
Dietary Sources of Zeaxanthin and Lutein and Their Potential Enhancement 22
Conclusion 23

3. THERAPEUTIC POTENTIAL OF DIETARY POLYPHENOLS AGAINST BRAIN AGEING AND NEURODEGENERATIVE DISORDERS 27

Giovanni Scapagnini, Calogero Caruso and Vittorio Calabrese

Abstract 27
Introduction 27
Curcumin 28

Caffeic Acid Phenethyl Ester 31
Ethyl Ferulate 31
Epigallocatechin-3-Gallate 31
Conclusion 32

4. PLANT PHENOLICS IN THE PREVENTION AND TREATMENT OF CANCER 36

Klaus W.J. Wahle, Iain Brown, Dino Rotondo, and Steven D. Heys

Abstract 36
Introduction: Epidemiology of Plant Foods and Disease Incidence 36
Anticancer Phytochemicals in Foods, Beverages and Spices 38
Classification and Occurrence of Plant Phenolic Compounds 38
Cellular Mechanisms Modified by Plant Phenolics That Can Reduce Carcinogenesis and Tumour Progression 43
Conclusion 48

5. ENDOGENOUS ANTIOXIDANTS AND RADICAL SCAVENGERS 52

Angela Maria Rizzo, Patrizia Berselli, Stefania Zava, Gigliola Montorfano, Manuela Negroni, Paola Corsetto and Bruno Berra

Abstract 52
Introduction 52
Endogenous Antioxidant Molecules 55
Reactive Oxygen Species: Always Bad? 62
Conclusion 63

6. A NUTRITIONAL STRATEGY FOR REDUCING DISEASE AND OBESITY RISKS 68

Teresa Lavecchia, Paolo Petroni, Giuseppe Rodio and Riccardo Pina

Abstract 68
Introduction 68
The Zone Diet Nutrition Strategy: General Outline 68
Conclusion 72

7. DIETARY PHYTOCHEMICALS AND HUMAN HEALTH 74

Justyna Krzyzanowska, Anna Czubacka and Wieslaw Oleszek

Abstract 74
Introduction 74
Carotenoids 77
Phenolic Compounds 81
Phytoestrogens 86
Polyunsatured Fatty Acids 89
Conjugated Linoleic Acids 90

Tocopherols and Tocotrienols 92
Limonene 93
Allicin and Diallyl Disulfide 93
Glucosinolates 94
Capsacinoids 95
Conclusion 95

8. BIOACTIVE COMPOUNDS FROM NORTHERN PLANTS 99

Anja Hohtola

Abstract 99
Introduction 100
How External Factors Influence the Biosynthesis of Secondary Metabolites 100
Research and Exploitation 101
Examples of Northern Plants Containing Bioactive Compounds 102
Conclusion 106

9. NUTRACEUTICAL USE OF GARLIC SULFUR-CONTAINING COMPOUNDS 110

Eleftherios Touloupakis and Demetrios F. Ghanotakis

Abstract 110
Introduction 110
Garlic Chemical Compounds 111
Garlic Biological Activities 112
Alliin and Allicin 114
Garlic Alliinase 115
Garlic Supplements 116
Layered Double Hydroxides 116
Alginates 117
Conclusion 118

10. GENETIC ENGINEERING TO ENHANCE CROP-BASED PHYTONUTRIENTS (NUTRACEUTICALS) TO ALLEVIATE DIET-RELATED DISEASES 122

Autar K. Mattoo, Vijaya Shukla, Tahira Fatima, Avtar K. Handa and Surender K. Yachha

Abstract 122
Introduction 123
Diet and Human Diseases 124
Phytonutrients and Antiproliferative Activity 126
Genetic Engineering to Improve Nutrient (Nutraceutical) Content in Produce 127
Proteins and Amino Acids 129
Carotenoids 135
Folates 136
Vitamin C (Ascorbate) 137

Polyphenolics 137
Tocopherols 138
Iron 138
Conclusion 138

11. PERSPECTIVE FOR THE USE OF GENETIC TRANSFORMANTS IN ORDER TO ENHANCE THE SYNTHESIS OF THE DESIRED METABOLITES: ENGINEERING CHLOROPLASTS OF MICROALGAE FOR THE PRODUCTION OF BIOACTIVE COMPOUNDS **144**

Udo Johanningmeier and Dirk Fischer

Abstract 144
Introduction 144
Microalgae as Transgenic Bioreactors 145
Genetic Engineering of Plant and Microalgal Chloroplasts 145
Expression of Recombinant Proteins in *C. reinhardtii* Chloroplasts 147
Application to Food Technology 147
Conclusion 149

12. BIOLOGICAL ELICITORS OF PLANT SECONDARY METABOLITES: MODE OF ACTION AND USE IN THE PRODUCTION OF NUTRACEUTICS **152**

Simone Ferrari

Abstract 152
Introduction 152
Biological Elicitors of Defence Responses in Plants 154
Perception and Transduction of General Elicitors 155
Emerging Techniques to Improve Secondary Metabolites Production, Based on Elicitor Signalling Pathways 159
Conclusion 161

13. HAIRY ROOT CULTURES FOR SECONDARY METABOLITES PRODUCTION **167**

Laura Pistelli Annalisa Giovannini, Barbara Ruffoni, Alessandra Bertoli and Luisa Pistelli

Abstract 167
Introduction 167
How *Agrobacterium rhizogenes* Genes Can Affect Plant Tissues Development? 168
***RolC* Mechanism of Action** 170
***RolB* Mechanism of Action** 171
***RolD* Mechanism of Action** 172
***RolA* and ORF13 Mechanism of Action** 173
***A. rhizogenes* Transformed Medicinal Plants as Farm for Aromatic and Nutraceutical Metabolite Production** 173
Conclusion 179

14. PLANT TISSUE CULTURE—AN OPPORTUNITY FOR THE PRODUCTION OF NUTRACEUTICALS 185

Mariella Lucchesini and Anna Mensuali-Sodi

Abstract 185
Introduction 185
Why to Cultivate In Vitro Plants for the Production of Secondary Metabolites? 185
In Vitro Technologies for the Cultivation of Nutraceutical Plants 187
The Micropropagation 189
Conclusion 200

15. PLANT CELL CULTURES: BIOREACTORS FOR INDUSTRIAL PRODUCTION 203

Barbara Ruffoni, Laura Pistelli, Alessandra Bertoli and Luisa Pistelli

Abstract 203
Introduction 203
Historical Background 204
The Plant Bioreactors 206
Bioreactors and Hairy Root Culture 211
Scale-Up of Medicinal and Aromatic Plants for Secondary Metabolite Production 213
Conclusion 218

16. DETERMINATION OF THE ANTIOXIDANTS' ABILITY TO SCAVENGE FREE RADICALS USING BIOSENSORS 222

Montserrat Cortina-Puig, Beatriz Prieto-Simón, Mónica Campàs, Carole Calas-Blanchard and Jean-Louis Marty

Abstract 222
Introduction 222
Monitoring Superoxide Radical ($O_2^{\bullet-}$) 223
Monitoring Hydroxyl Radical ($OH^{\bullet}$) 227
Monitoring Nitric Oxide Radical ($NO^{\bullet}$) 229
Conclusion 230

17. BIOSENSORS FOR THE DETERMINATION OF PHENOLIC METABOLITES 234

Simona Carmen Litescu, Sandra Eremia and Gabriel Lucian Radu

Abstract 234
Introduction 234
Biosensors Used in the Determination of Polyphenols 235
Conclusion 239

18. METHODS FOR THE DETERMINATION OF ANTIOXIDANT CAPACITY IN FOOD AND RAW MATERIALS 241

Simona Carmen Litescu, Sandra Eremia and Gabriel Lucian Radu

Abstract 241
Introduction 241
Models Based on HAT Mechanisms 243
Models Based on SET Mechanisms 244
Assessment of Antioxidant Efficacy Using Significant Biological Substrates 247
Conclusion 248

19. ANALYTICAL METHODS FOR THE EXTRACTION AND IDENTIFICATION OF SECONDARY METABOLITE PRODUCTION IN 'IN VITRO' PLANT CELL CULTURES 250

Alessandra Bertoli, Barbara Ruffoni, Laura Pistelli and Luisa Pistelli

Abstract 250
Introduction 250
Sampling and Storage 251
Sample Preparation and Extraction 253
Chromatographic Separation and Detection of Phytochemicals 257
Conclusion 261

20. BIOSENSORS FOR FUNCTIONAL FOOD SAFETY AND ANALYSIS 267

Teresa Lavecchia, Arianna Tibuzzi and Maria Teresa Giardi

Abstract 267
Introduction to Safety and Functionality Analysis in Food 267
Classical Methods for Food Analysis: Chemical and Microbiological Methods 268
Biosensor Technology 269
Photosynthetic Protein-Based Biosensors 272
Functional Foods 273
Examples of Biosensors for Functional Food Analysis 273
The Market for Biosensors 277
Conclusion 278

21. BIOSENSORS FOR SECONDARY METABOLITES, TWO CASE STUDIES: OCHRATOXIN A AND MICROCYSTIN 282

Monica Campàs, Beatriz Prieto-Simón and Régis Rouillon

Abstract 282
Introduction 282
A Mycotoxin: Ochratoxin A 282
A Cyanotoxin: Microcystin 286
Conclusion 290

22. BIOSENSORS AS ANALYTICAL TOOLS IN FOOD FERMENTATION INDUSTRY 293

Lise Barthelmebs, Carole Calas-Blanchard, Georges Istamboulie, Jean-Louis Marty and Thierry Noguer

Abstract 293
Introduction 293
Biosensors, General Aspects 294
Main Food Fermentation and Compounds of Interest 294
Potential Applications 295
Commercial Devices 299
Conclusion 302

23. AN OVERVIEW OF THE FUNCTIONAL FOOD MARKET: FROM MARKETING ISSUES AND COMMERCIAL PLAYERS TO FUTURE DEMAND FROM LIFE IN SPACE 308

Francesca Vergari, Arianna Tibuzzi and Giovanni Basile

Abstract 308
Introduction 308
Marketing of Functional Food 309
The Nutraceutical and Functional Food Market 310
Functional Food Market Segments in Europe 312
Suppliers of Functional Food 317
A Market Focus on Carotenoids 318
Conclusion 319

24. LEGISLATION ON NUTRACEUTICALS AND FOOD SUPPLEMENTS A COMPARISON BETWEEN REGULATIONS IN USA AND EU 322

Roberto Giunta, Giovanni Basile and Arianna Tibuzzi

Abstract 322
Introduction 322
Regulations on Dietary Supplements in USA 323
Regulation on Food Supplements in the European Union 324
Conclusion 326

INDEX 329

ACKNOWLEDGEMENTS

We are grateful to Dr. Teresa Lavecchia—Biosensor srl, Palombara Sabina, Rome, Italy—for her participation to the work of proofreading and editing; to Prof. Rodolfo Federico, Prof. Riccardo Angelini, Dr. Alessandra Cona, and Dr. Paraskevi Tavladoraki—Department of Biology, University Roma Tre, Rome, Italy—for their scientific contributions.

Chapter 1

The NUTRA-SNACKS Project:
Basic Research and Biotechnological Programs on Nutraceutics

Giuseppina Rea, Amina Antonacci, Maya Lambreva, Andrea Margonelli, Cecilia Ambrosi and Maria Teresa Giardi*

Abstract

The Nutra-Snacks project aims at creating novel high quality ready-to-eat foods with functional activity, useful for promoting public health. The team is composed of seven research institutes and three SMEs from different countries whose activities span from basic to applied research providing the right technological transfer to small and medium industries involved in the novel food production chain. Strategic objectives include the application of plant cell and in vitro culture systems to create very large amounts of high-value plant secondary metabolites with recognized anticancer, antilipidemic, anticholesterol, antimicrobial, antiviral, antihypertensive and anti-inflammatory properties and to include them in specific food products. To this end, the screening of a vast number of working organisms capable of accumulating the desired compounds and the characterization of their expression profiles represent fundamental steps in the research program. The information allows the identification of plant species hyper-producing metabolites and selection of those metabolites capable of specifically counteracting the oxidative stress that underlies the development of important pathologies and diseases. In addition, devising safe metabolite extraction procedures is also crucial in order to provide nutraceutical-enriched extracts compatible with human health. New biotechnological approaches are also undertaken including the exploitation of photosynthetic algal strains in bio-farms to enhance the synthesis of antioxidant compounds and the design of novel bioreactors for small and large scale biomass production. Further outstanding objectives include the development of (i) safety and quality control protocols (ii) biosensor techniques for the analysis of the emerging ready-to-eat food and (iii) a contribution to define a standard for new regulations on nutraceutics.

Introduction: Rationale of the Nutra-Snacks Project

In recent years, food quality and safety have become a fundamental issue in public opinion and the media. Compelling evidence indicates inappropriate dietary habits as a leading cause of poor health that results in a worldwide increase in health care costs. These concerns prompted policymakers to declare a thematic priority to Food Quality and Safety Programmes, in an attempt to improve the health and well-being of European citizens through higher quality food and improved control of food production and related environmental factors. The supported program gave priority to identifying

*Corresponding Author: Maria Teresa Giardi—Institute of Crystallography, Department of Agrofood, CNR, 00016 Monterotondo Scalo, Rome, Italy. Email: giardi@mlib.cnr.it

Bio-Farms for Nutraceuticals: Functional Food and Safety Control by Biosensors, edited by Maria Teresa Giardi, Giuseppina Rea and Bruno Berra.

the major concerns for consumers along the food production chain, outlining issues associated with primary production, animal feeds, processing, distribution, consumption and environmental health risks (http://ec.europa.eu/research/research-eu).

The Nutra-Snack project found application in this social research context. The project aims at the realization of ready-to-eat food for breakfast and sport with a high content of compounds endowed with functional activity useful for promoting public health. There is already a wide range of foods available to today's consumer but now the impetus is to identify functional foods that have the potential to improve health and well-being, reduce the risk from—or delay the onset of major diseases such as cardiovascular disease, cancer, osteoporosis and so on. Combined with a healthy lifestyle, functional food can make a positive contribution to health and well being.

The Influence of Various Life Style and Environmental Factors on Human Health

Human health is determined by the interaction of genetic, nutritional, socio-cultural, economic, physical infrastructure and ecosystem factors that, in turn, are all influenced by the environment and by its changes. Currently, several disciplines, including geography, gender studies, health and social sciences, are bringing to light the complex pathways by which factors in the environment influence patterns of health. Environmental changes arising from urbanization, population increases, industrial and agricultural activities have resulted in thermal fluctuations, reduced quality of air, water and soil, a greater exposure to radiation and persistent chemical pollutants that can trigger disease processes. In particular, environmental degradation and chemical overload coming from agriculture practices have been linked to diseases such as respiratory and cardiovascular disease, neurological and physiological disorders, and increased incidence of many cancers.[1] In addition to the built environment, several life style choices such as smoking habits, decrease of physical activity and consumption of alcohol and drugs can have profound impacts on our health and have been associated with the abovementioned pathologies.

Among lifestyles, dietary factors also affect health and well-being. The growing epidemic of obesity is one of the major public health issues in the world, and is often a consequence of high calorie intake compared to energy expenditure. Poor eating-habits leading to inadequate intake of calories can negatively affect human health and have been unequivocally and causally associated with the risk of obesity, cardiovascular disease, type 2 diabetes, stroke, cancers and neurodegenerative disorders.[2]

Oxidative Stress Is an Important Health Risk Factor

The term oxidative stress describes a harmful process resulting from an imbalance between the excessive production of reactive oxygen (ROS) and/or nitrogen species (RNS) and limited action of antioxidant defenses. Oxidative stress is implicated in the development of many important pathologies including atherosclerosis, hypertension, ischemia-reperfusion injury, inflammation, cystic fibrosis, diabetes, cancer, Parkinson's and Alzheimer's diseases. [3-9]

Due to its bi-radical nature, molecular oxygen can accept unpaired electrons giving rise to a series of dangerous partially reduced species that include superoxide ($O^{\cdot-}_2$), hydrogen peroxide, hydroxyl radical and peroxyl and alkoxyl radicals. In addition, $O^{\cdot-}_2$ can react with other radicals including nitric oxide yielding products, the dangerous peroxynitrites, which have a very potent oxidant activity. These compounds may be involved in the initiation and propagation of free radical-mediated chain reactions that can involve proteins, lipids, polysaccharides and DNA, leading to biological damage. [10]

ROS and RNS can be derived either from such essential metabolic processes in the human body as aerobic respiration, peroxisomal β-oxidation of fatty acids, microsomal cytochrome P450 metabolism of xenobiotic compounds, stimulation of phagocytosis by pathogens or lipopolysaccharides, arginine metabolism and activity of tissue specific enzymes, or from external sources such as exposure to ozone, cigarette smoking, air pollutants and chemicals. In physiological conditions, ROS are efficiently scavenged from the cell mainly by the activity of detoxifying enzymes like superoxide dismutase, catalase, or glutathione peroxidase or electron carriers like ubiquinol.[11]

In addition to deleterious effects, ROS and RNS also perform several useful functions.[12] Moreover, sub-toxic levels of ROS and RNS serve as important intracellular signalling molecules that can influence cellular survival.[13] In fact, it has been shown that numerous cellular processes in several cell types can be regulated by slight changes in redox balance. Examples of these processes include the activation of certain nuclear transcription factors, determination of cellular fate by apoptosis or necrosis, modulation of protein and lipid kinases and phosphatases, membrane receptors and ion channels.[14,15]

As oxidative stress is believed to play a major role in the ageing process and in several diseases there is considerable public interest in the anti-oxidative effects of dietary factors.[2,16] Indeed, with proper nourishment, the body can, on its own, make sufficient quantities of antioxidant enzymes and substrates for those enzymes. These can facilitate the quenching of excess free radicals by antioxidants. An enhancement in dietary intake of antioxidants and phytochemicals with related functions can counteract oxidizing species and potentially restore a healthy cellular redox balance.[17] Apart from traditional components with potential health benefits, new food is being developed to enhance or incorporate these beneficial nutraceutics for their health benefits or desirable physiological effects.

Why Novel Nutra-Snacks?

In this context, the research developed in the Nutra-Snack project is focused on the creation of new foods capable of acting in concert with human defense responses to provide protection against oxidative stress. The idea was the development of a new generation of healthy dietary formulations containing natural but bolstered ingredients able to meet and satisfy the consumer's taste and eating-habits.

Faced with a highly diverse food supply, consumers can take advantage of a wide variety of dietary intake of micro- and macro-nutrients. Nowadays, the consumption of midday and/or midnight snack food is becoming one of the most popular eating-habit trends, despite its close correlation with several metabolic disorders. Eating small amounts of food on a regular basis is considered healthy, but often available snacks in the market provide excess calories and fats and little or no nutrient value to the diet. Moreover, most snacks contain preservatives, sweeteners and flavoring that can have negative effects on health.

Healthy diets include factors such as making proper food, keeping track of meal times and regulating the amount of food intake. There are many different healthy eating diet plans available today. Among these, the Zone diet is one of the most popular focused on food management rather than weight loss. The Zone diet suggests the 40:30:30 ratio for carbohydrates, proteins and fats that a person should consume. With this ratio of components, the body is able to balance insulin and glucagon, as well as provide a more effective internal metabolism. One of the key partners in the Nutra-Snack project is Enervit®, a well-known firm in the development of nutrition programs to enhance the performances of athletes and sportsmen. As the manufacturer of several products for breakfast, everyday meals, snacks and food supplements, the company is a leading supporter of the Zone diet. Currently, experts from the Enervit® team are involved in the development of modern alimentary strategies for those who believe in health through correct nutrition and food supplementation (http://www.enervitwellness.it).

Plant Secondary Metabolites as High-Value Compounds for Human Nutrition

In essence, healthy eating refers to the idea of maintaining a well-balanced food intake comprised of food types from different food sources. Much epidemiological and experimental evidence supports the hypothesis that vegetables and fruits in the diet can reduce risk of diseases; this has led to the use of a number of phytometabolites as preventive and/or protective agents, promoting a dramatic increase in their consumption as dietary supplements. The therapeutic potential of plant-based formulations has been ascribed to various plant secondary metabolites whose biological activities are beneficial at the cellular and molecular levels. In addition, plant-based extracts are the most favorable choice, although a large dosage of a single compound is not considered advantageous to deliver it to many tissues.[18,19]

Plants are basic ingredients of our diet whose nutritional and medicinal value have been recognized since ancient times. In addition to essential primary metabolites, like carbohydrates, lipids and amino acids, plants are capable of synthesizing an amazing variety of low-molecular-weight organic compounds, called secondary metabolites, usually with unique and complex structures. Compared to the main and most abundant molecules found in plants, these secondary metabolites are defined by their low abundance, often less than 1-5% of the dry weight. Secondary metabolites have no recognized role in the maintenance of fundamental life processes; nevertheless, these molecules contribute largely to plant wellbeing by interacting with the ecosystems. Their production, in fact, is often enhanced or induced in response to biotic and abiotic stress conditions, after that, they can be stored in specific cells and/or organs of the plant.

Some of these secondary metabolites are endowed with functional activity useful for promoting public health as they provide health benefits beyond basic nutrition.[18] Some of the aforementioned compounds and their functional roles are briefly described below and will be treated in detail later in this book.

Carotenoids

Photosynthetic carotenoids are red, orange, and yellow lipid-soluble pigments found embedded in the membranes of chloroplasts and chromoplasts. They are polyisoprenoid compounds containing 40 carbon atoms formed by the condensation of eight isoprene units. They possess a long chain of conjugated double bonds whose linkage order is reversed in the central part of the molecule giving rise to a symmetrical molecule. This set of conjugated double bonds is responsible for the absorption of light in the visible region of the spectrum. Left and right end groups are variable and different levels of hydrogenation and introduction of oxygen-containing functional groups create a large family of over 600 natural compounds.[20]

In photosynthetic organisms, carotenoids play a vital role in the photosynthetic reaction center. They either participate in the energy-transfer process, or protect the reaction center from auto-oxidation. In non-photosynthesizing organisms, carotenoids have been linked to oxidation-preventing mechanisms. Animals are incapable of synthesizing carotenoids and must obtain them through their diet. Carotenoids absorbed through the diet, and often metabolized into other compounds, are responsible for the color of familiar animals such as lobster, flamingo, and red fish.

However, carotenoids may have many physiological functions. Apart from the carotenoid, β-carotene, the primary vitamin A source, carotenoids are of physiological interest in human nutrition, since they exhibit radical or singlet oxygen trapping activity,[21] and as such have potential antioxidant effects in vivo.

Given their structure, carotenoids are efficient free-radical scavengers, capable of enhancing the vertebrate immune system. Indeed, epidemiological studies have shown that people with high β-carotene intake and high plasma levels of β-carotene have a significantly reduced risk of lung cancer. However, studies of supplementation with large doses of β-carotene in smokers have shown the opposite effect—an increase in cancer risk—possibly because excessive beta-carotene results in breakdown products that reduce plasma vitamin A and induce the lung cell proliferation caused by smoke.[19]

Moreover, there is evidence supporting a protective role for lutein and zeaxanthin in delaying chronic diseases, including age-related vision loss via macular degeneration and cataract formation, cancer and heart disease. Lutein and zeaxanthin are the only carotenoids present in both the macula and lens of the human eye and are also referred to as macular pigment. Functions of these pigments include improving visual function, quenching free radicals and thereby acting as an antioxidant to protect the macula from oxidative damage. The involvement of lutein and zeaxanthin in protection against photo-induced damage is as blue light filters, antioxidants and shielding potentially harmful short-wave radiation. The retina is susceptible to oxidative stress because of its high demand for oxygen, the high proportion of polyunsaturated fatty acids, and aerobic metabolism. Age-related macular degeneration is the leading cause of blindness in the elderly in several countries. Many researchers have found significant associations between lutein and zeaxanthin concentrations in ocular tissues, serum, and plasma, with a possible reduced risk of macular degeneration.[22]

Flavonoids

Flavonoids are widespread constituents of fruit, vegetables, cereals, olives, dry legumes, chocolate and beverages, such as tea, coffee and red wine. Structurally, they are polyphenolic compounds, produced by the phenylpropanoid biosynthetic pathway, having a 15 carbon atoms skeleton and two benzene rings, joined by a linear three-carbon chain, which may form a closed pyran ring with one of the benzene rings. Six subclasses can be also identified, depending on the oxidation state of the central pyran ring: flavonols, flavones, flavanones, isoflavones, anthocyanidins and flavanols (catechins and proanthocyanidins).

Flavonoids, exclusively produced in plants, are known as health promoting substances in the human diet. As antioxidants, flavonoids may protect cell constituents against oxidative damage and, therefore, limit the risk of various degenerative diseases associated with oxidative stress. Experimental studies strongly support their role in the prevention of cardiovascular disease, cancer, osteoporosis, diabetes mellitus and neurodegenerative disease.[23,24]

Alliin and Allicin from Garlic

Garlic (*Allium sativum*) contains the highest levels of the two important organosulfur compounds, cysteine sulfoxide (alliin) and cysteine thiosulfinate (allicin), respectively. They constitute the active principles of the garlic phytotherapy that has been used in folk medicine for thousands of years. Nowadays there has been renewed interest in the therapeutic properties of these compounds which are the objects of an increasing numbers of biochemical and clinical studies.

In intact cells, alliin is stored as an odorless and stable compound; when the cells are chopped, chewed or crushed, the enzyme alliinase promptly converts the exposed alliin into allicin which gives off the typical fresh garlic aroma. Allicin accounts for about 70-80% of total thiosulfinates in garlic. Once formed, the fairly unstable allicin further reacts with itself producing more stable sulfur compounds within a few hours to a few days. Allicin and many of its derivatives have been reported to exert diverse biological effects such as: (a) induction of carcinogen detoxification, (b) inhibition of tumor cell proliferation, (c) antimicrobial effect, (d) free radical scavenging, (e) inhibition of DNA adduct formation, (f) induction of cell cycle arrest and apoptosis etc. Epidemiological and experimental carcinogenesis provides overwhelming evidence to support the claim that people consuming a diet rich in organosulfur are less susceptible to different types of cancers.[25-27]

Rosmarinic Acid

Rosmarinic acid is widespread in nature in a great number of aromatic, spicy and medicinal plants such as rosemary, sage, oregano, basil, which resulted in a development of natural antioxidant formulations for food, cosmetic and other applications.[28,29] Rosmarinic acid belongs to the class of polyphenols and, structurally, is a dimer of caffeic acid (α-O-caffeoyl-3,4-dihydroxyphenyl-lactic acid). Rosmarinic acid exhibits various pharmacological activities, such as antibacterial, antiviral, antiallergic and anti-inflammatory effects.[30] Moreover, it possesses anti-apoptotic and anti-oxidant effects in astrocytes[31] as well as neuroprotective effects in the neurotoxicity of β-amyloid, indicating that this compound may be useful for the treatment of dysfunction not only in the immune system but also in the nervous system.[32,33] The main propriety, however, of rosmarinic acid is its antioxidant activity in parallel with a very low toxicity,[34] that makes it interesting for commercial production.

Resveratrol

Resveratrol (3,5,4'-trihydroxystilbene) is a polyphenolic phytoalexin occurring as two geometric isomers, *cis-* and *trans-*; the *trans-* form can undergo isomerisation to the cis- form when exposed to ultraviolet irradiation.[35] It is naturally produced by grapes, certain vines, pine trees, peanuts and other plants as a defense response against pathogen attacks. Nowadays, resveratrol is extensively used as a nutraceutical supplement being an antioxidant chemically stable compound at room temperature.[36] Resveratrol is known to be cardioprotective and to have beneficial circulatory effects. Resveratrol protects the cardiovascular system by mechanisms that include defense against ischemic-reperfusion injury, promotion of vasorelaxation, protection and maintenance of intact endothelium, anti-atherosclerotic properties, inhibition of low-density lipoprotein oxidation, suppression of platelet aggregation, and estrogen-like actions.[37,38]

Curcumin

Curcumin is a yellow pigment derived from the plant *Curcuma longa* L., whose powdered rhizome, called turmeric, is commonly used in the preparation of curries. Curcumin, a polyphenol with a diarylheptanoid structure that contains two α,β-unsaturated ketones, is considered to be the major active constituent of turmeric. In addition to its preservative, flavoring, or coloring properties in the diet, turmeric has noteworthy antioxidative activities.[39] Moreover, for generations it has been used for the treatment of inflammation, skin wounds, hepatic and biliary disorders, cough, as well as certain tumors. Epidemiological studies suggest that turmeric consumption may reduce the risk of some forms of cancer and accumulating preclinical studies have shown that curcumin can interfere with a large number of molecular processes involved in cancer. Recently, the chemopreventive and chemotherapeutic activities of curcumin have been associated with its strong anti-oxidant and pro-oxidant effects. Even at low concentrations, the activity of curcumin reduces and keeps the cellular content of ROS within the physiological levels, mediating the cancer chemopreventive properties of curcumin. At higher concentrations, the pro-oxidant activity of curcumin would increase the accumulation of ROS to cytotoxic levels, acting as a chemotherapeutic agent.[40]

Antioxidant Peptides

Some low molecular weight peptides derived from plant and animal sources, have been found to possess antioxidant properties.[41] In the human body, sulfur-containing amino acids like cysteine and methionine, and small molecules like glutathione, play a crucial role in the oxidative damage protection.[42] However, other amino acids behave as quenchers of free radicals. Several studies reported that peptides derived from natural sources like fermented mussel sauce, fish skin gelatin hydrolysate and digests of soybean proteins can act as potential antioxidants. In vitro experiments carried out on human lung fibroblasts, revealed that they are good scavengers for free oxygen radicals, carbon-centered radicals and lipidic radicals. Moreover, they showed a higher antioxidant power compared to the natural antioxidant α-tocopherol. In the bioactive peptides important roles are carried out by the hydrophobic amino acids: e.g. the sulfidrilic groups and the imidazolic ring are good scavengers of free radicals;[43-46] moreover, the imidazolic ring of the histidines allows the chelation of metal ions, which can generate other radicals. Amino acids like proline, lysine or arginine seem also to be important for antioxidant activity,[47] because of their three-dimensional structure in the peptide.

Technological Innovations

The Nutra-Snacks project is developing the application of plant cell and in vitro culture systems together with biotechnological approaches to enable the production of new high quality ready-to-eat food with functional activity promoting properties. The realization of the strategic objectives have been structured into subtasks depicted in Figure 1 and described as follows:

Selection of Organisms and Biologically Functional Metabolites

Plant secondary metabolites are a very important industrial and economic source of many drugs, flavors, insecticides, fragrances and dyes. Today, about a quarter of all pharmaceuticals are of plant origin, either as a pure compound or derived from a natural synthon.[48,49] The technology of large-scale plant cell culture is feasible for the industrial production of plant-derived fine chemicals, but still has many limitations. The natural content of the compound in a plant is often very low, or production is often impossible due to a lack of raw material. Some compounds can only be isolated from rare plants. On the other hand, chemical synthesis of these compounds is often not technically or commercially feasible because of their highly complex structures.[50]

The project aims at improving plant cell culture systems in such a way that the production of high-value compounds is industrially rational. Various approaches are being studied to increase yields that encompass screening and selection of high producing cell lines, media optimization,

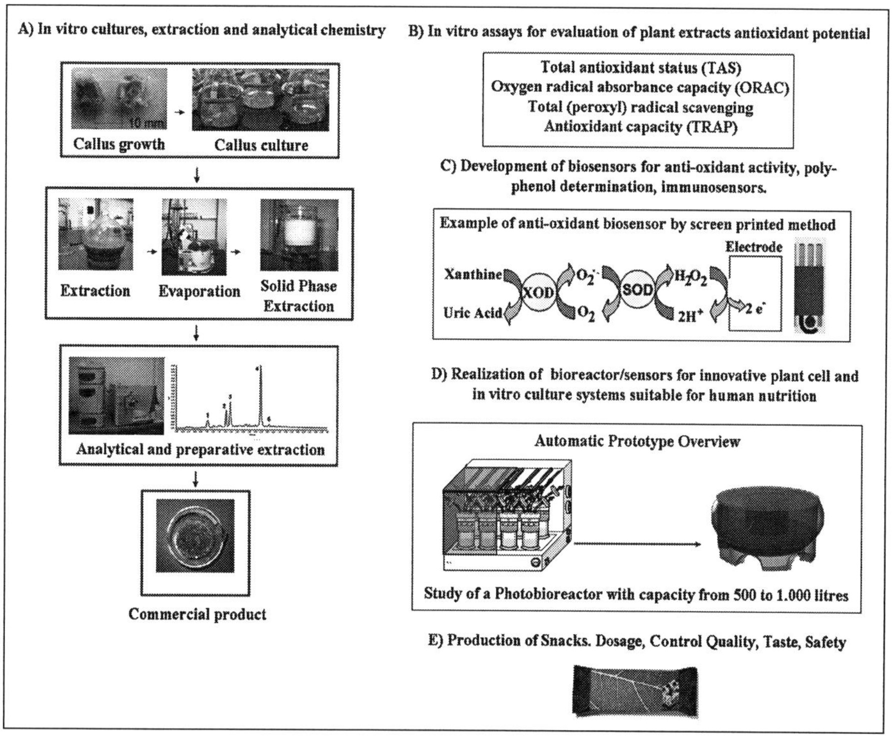

Figure 1. Basic research and precompetitive commercial approach for Nutra-Snack project.

culturing of organ cultures and elicitation.[51] Furthermore, as in recent years metabolic engineering has opened new promising perspectives for improved production in a plant or plant cell culture,[52,53] a molecular biological approach is being undertaken to study the expression profile of genes involved in biosynthetic pathways.[54] Most activities are dedicated to the screening, either the literature or in the laboratory, of a wide number of photosynthetic organisms such as algae, cyanobacteria, and higher plants for their ability to produce functional metabolite enriched cultures, and optimization of protocols for their extraction from the best candidate organisms.[55] Cultures of the unicellular green alga *Chlamydomonas reinhardtii* are used as model systems to study the accumulation and over-expression of photosynthetic anti-oxidant compounds like carotenoids and plastoquinones in both physiological and pathological conditions, and, as an innovation in this field, as biological farms for the production of the compounds of interest. In a later phase, after the completion of analytical metabolite profiling and establishment of growth environmental conditions, selected cell cultures and growth environmental conditions will be applied to scale up the procedure and set up a nutritional safe extraction protocols. It is of interest that we explore whether lutein and zeaxanthin protect the retina from damage in people subjected to strong oxidative stress, as it is the case of astronauts subjected to ionizing radiation emitted by solar flares that give the visual phenomenon of light flashes (Giardi and Rea, unpublished).

In addition, a collection of medical plants and strains, already available in our laboratories, has been tested for their potential to growth. Many of these species have long been used in ancient traditional medicine systems, associated in prevention of some diseases as they contain very high levels of phenolics, quinones and flavonoids. These species encompass Leguminosae, *Acacia catechu L.*, Punicaceae, *Punica granatum L.*, Palmae, *Areca catechu L.*, Rosaceae, *Sanguisorba officinalis L*, Theaceae, *Camellia sinensis L.;* Combretaceae, *Terminalia chebula*, Anacardiaceae, *Rhus chinensis*.

Table 1. Phytochemicals, sources and health-promoting activities proposed for new functional food design

Active Metabolite Classes	Species Botanical Family Common Name	Main Nutraceutic Activity	Culture System
Alliin, Allicin[57]	*Allium sativum* L.	Lipid-lowering Anti cholesterol Antibiotic effects	Callus
Flavonones[58]	*Mentha spp* Labiatae (Peppermint, spearmint)	Antibacterial Anticolitis	Shoots
Isoflavones[59]	*Glicine max* Leguminosae Soybeans	Anti-arterosclerosis	Cell culture
Phenolic[60]	*Salvia spp* Labiatae (Sage) *Taraxacum officinale* Compositae (Dandelion)	Cardio-protective Anti-hypertensive Antioxidants	Callus
Plastoquinones and protein metabolites[61]	*Spinacia Oleracea* *Passifloraceae* (Passion flower)	Anti-cancer Antioxidants	Cell culture
Xanthophylls [62]	*Chlamydomonas reinhardtii*	Antioxidants	Cell culture
Rosmarinic acid[34]	*Salvia spp* (Sage) *Ocymum Basilicum* Labiatae	Antimicrobial Antioxidants	Callus
Resveratrol[36]	Vitis Vinifera Vitaceae (Grape)	Antiarterioschlerosis Circulatory effects Cardioprotective	Callus Cell culture
Ferulic acids and derivatives (curcumin)[63]	*Echinacea* spp Compositae Curcuma longa	Immunostimulation and cardioprotective	Callus

In total, we selected more than ten classes of metabolites which, in traditional medicine, have been recognized as having a positive effect against a number of major diseases. The benefits of these substances for human nutrition are well documented and lately significant progress has been made at the molecular level regarding the synthesis and accumulation of these compounds in plants.[56] A selection of target bioactive compounds and sources which have been screened are reported in Table 1.

As mentioned above, this work allows the preliminary screening of plants hyper-producing the most valuable secondary metabolites. However, due to the excessive metabolic energy required for their synthesis, secondary metabolites usually occur at low levels in a plant. Nevertheless, being effective defense compounds, their accumulation could be induced at high levels in response to biotic and abiotic stress.[51] Physical elicitations involving high light intensities, high temperatures, combinations of high light and high intensities and UVC irradiation have been successfully applied to both plants and plant cell cultures to induce the synthesis of metabolites. Chemical elicitation is applied mainly by treating cell cultures with jasmonates at safe concentrations. Biological elicitors have also been exploited but, as they may, under some circumstances, pose a threat to human health, particular attention has been focused on the selection of safe sources for human consumption.

Research carried out in the Nutra-Snacks project employs oligogalacturonides (OGs) to increase the production of secondary metabolites in plant cell cultures.[64,65] The advantage of OGs over other chemical elicitors is that these molecules are derived from plant pectins and therefore are completely safe both for human consumption and the environment. Furthermore, they are produced after the digestion of pectin, which is easily obtained from fruit peels and other waste products. Another challenge, successfully addressed by the project, is the optimization of a protocol for cost-effective large-scale production of OGs on an industrial scale.

Enhanced production of secondary metabolites can also be obtained through plant cell immobilization procedures.[66,67] In this case, plant cells or micro aggregates are encapsulated in polymers like agar, agarose, calcium alginate, and carageenan. This procedure allows cell viability in the stationary producing phase to be extended enabling maintenance of biomass over a prolonged time period, but necessarily requires the extra-cellular secretion of the products (Touloupakis, Giardi, Ghanotakis, unpublished).

Concerning the safety issues, it is important to highlight that an important task/innovation proposed by the project is the establishment of culture conditions and chemical compounds and/ or elicitation compatible with human nutrition. This aspect considers not only the selection of biological elicitors, but also the composition of the growth and maintaining media utilized in most plant cell and in vitro cultures. This means that the presence of heavy metal salts, pathogenic fungi, high salts and generally poisonous and/or unhealthy substances is excluded, and healthy elicitors having a good taste able to improve the production of the metabolites are selected.

Cell and Organ Growth In Vitro Techniques

After the selection of the most promising individual plants, the activities are directed towards the set up of in vitro cultures. The rationale of this approach relies on the evidence that the production of phytochemicals in field cultivation is, in many instances, environmentally and seasonally dependent.

Plant cell, tissue, and organ culture in vitro techniques are used for the growth and multiplication of cells and tissues using nutrient solutions in an aseptic and controlled environment. This technology explores conditions that promote cell division and genetic reprogramming with in vitro conditions. The hard work involved in this activity provides the formulation of the best adapted media useful for callus initiation, with special attention to the mineral and organic constituent composition and balance of hormones that govern dedifferentiation/differentiation mechanisms. Once calli are obtained, they have to be stabilized. It is well known that calli can undergo somaclonal variation usually during several subculture cycles. This is a critical period where, due to this in vitro variation, secondary metabolite production is often variable from one subculture cycle to another. After a period of time, genetic stability may occur and each callus can be considered as homogeneous cell aggregate, just as if it was derived from single cell cloning. In our work we assume that a cell line is not stabilized until growth parameters could be repeated during three consecutive subculture cycles in stable culture conditions. When genetic stability is reached, each callus cell line is separately assessed for its growth speed as well as metabolite accumulation. This allows an evaluation of the productivity of each cell line, so that only the best one is taken for cell suspension preparation and subsequent studies.

Another biotechnological tool exploited for our purposes utilizes hairy roots. Hairy roots are obtained after the successful transformation of a plant with *Agrobacterium rhizogenes*. They have received considerable attention from plant biotechnologists for the production of secondary compounds; since can be sub-cultured and indefinitely propagated on a synthetic medium without phyto-hormones and usually display interesting growth capacities owing to the profusion of lateral roots. This growth can be simulated by an exponential growth model, when the number of generations of lateral roots becomes very large. The specific growth rates are generally comparable, if not superior, to the parameters observed with undifferentiated cells. Biomass doubling times ranging from less than one day to almost one week has been reported. A major characteristic of hairy roots is that they are able to produce secondary metabolites concomitantly with growth. Hence, it is possible to generate a continuous supply of secondary compounds from actively growing hairy roots. In addition, it is also possible to cultivate roots under light, in order to get green tissues. The same elicitation strategies developed for cell cultures, are also used to increase the metabolite production from hairy roots.[68]

Efforts are directed towards the creation of a sample metabolite profile with the aid of analytical techniques. The composition of plant extracts is usually evaluated by liquid chromatography combined to mass spectrometry LCDAD-ESI/MSn that allows the identification of major constituents when appropriate standards are available. Initially it must be decided the developmental stage at which the characterization should be performed. Compared to cell growth kinetics, which usually has an exponential curve, most secondary metabolites are produced during the plateau phase. This lack of production during the early stages can be explained by carbon allocation mainly distributed for primary metabolism (building of cell structures and respiration) when growth is very active. On the other hand, when growth stops, carbon is no longer needed in large quantities for primary metabolism and secondary compounds are more actively synthesized. It has been frequently observed that many new enzymatic activities, absent during lag or log phases, instead appear during this plateau phase. This has led many authors to talk of a possible biochemical differentiation of the cells when growth stops. However, some secondary plant products like carotenoids are known to be growth-associated with undifferentiated cells. These considerations are taken in carefully consideration for the subsequent extraction procedures and for the bioreactor design.[69]

Concerning the extraction procedures, we set up different protocols depending on the particular metabolite under study, but in every case the use of solvents not compatible with human health are avoided.

Genetic Engineering for a Biological Farm

The object of these pioneering studies is the exploitation of the unicellular green alga *Chlamydomonas reinhardtii* as a potential source of valuable compounds useful in basic and applied research. The former includes elucidation of the molecular mechanisms of the biosynthesis of photosynthetic antioxidant compounds (e.g. carotenoids and plastoquinones) with the aim of developing a rational metabolic engineering approach to overproduce them. Recent advances in the metabolic engineering of microorganisms have enabled the production of both native and heterologous secondary metabolites, high levels of production and the synthesis of targeted molecules. One recent example is the production of astaxanthin by *Haematococcus pluvialis*.[70]

Applied research activities are mainly directed to the over-production of proteins or peptides with functional proprieties. Physiologically active peptides derived from plant and animal proteins are potential health enhancing components for food applications.[71-73] Hydrolysis of proteins in vitro or in vivo from these sources and subsequent analyses of peptide fragments for bioactivity revealed that certain peptide fragments exert a multitude of healthy effects like antioxidant or antithrombotic activities, cholesterol-lowering abilities or antimicrobial properties, to mention only a few. However, the industrial production of bioactive peptides with specific functions is often complex and expensive. Eukaryotic microalgae have recently gained interest as bioreactors because they provide attractive alternatives to the bacterial, yeast, plant and other systems currently in use.[74-77] Over the last decades there has been considerable progress in genetic engineering technologies of algae to alter metabolic pathways and express valuable proteins in different cell compartments. Particularly the eukaryotic green alga *Chlamydomonas reinhardtii* has been widely used as a model organism since its complete genome sequence is known, databases are available for any searching or cloning requirements and successful high-level expression of recombinant proteins in the chloroplast compartment has been obtained recently.[77] *C. reinhardtii* belongs to the group of green algae many of which belong to the category of organisms with a GRAS status (Generally Regarded As Safe) granted by the FDA.[54] Apart from being easily transformable, transgenic strains can be obtained in 14 days and production volumes within a few weeks in full containment. In this context we used an innovative strategy for the overproduction of bioactive peptides through the expression of a fusion cassette encompasses the useful peptide and the so-called D1-subunit of photosystem II. However, as the use of transformants for food nutrition has not yet gained public acceptance, the metabolites will be safely extracted from the transformants without their direct use in the nutritional chain.

Production of Snacks, Protocols for Dosage, Control Quality, Taste and Safety

This project is aimed at defining suitable controls for establishing the functionality of the new produced nutraceutical food. To this end the main objectives are the setup of control analyses, to develop standards and new regulations on nutraceutics, to help the assessment of protocols for production of ready-to-eat food containing a range of nutraceuticals embracing the main health risks; to analyze changes induced on specific cellular targets by the new produced functional foods using in vitro systems and to analyze the anti-oxidative protective effect derived by the use of ready-to-eat food in specific populations (i.e., sportsmen and astronauts, both of whom are exposed to oxidative stress).

As already stated, oxidative stress is a dangerous condition caused by an imbalance between the production of free radicals and the capability of organism to cope with their excess. Being asymptomatic, if not promptly recognized and treated, it can accelerate the ageing process and favor the appearance of chronic diseases. In this context, the nutritional countermeasures against oxidative stress represent an important issue in human health, and for these reasons, for control trials and risk assessment, it has been proposed to investigate the antioxidant metabolism of in vitro cell culture and in vivo systems analyzing human blood prior and after utilization of the novel metabolite enriched food. Hence, we determined the redox state of the cell and activities of enzymes involved in anti-oxidative responses after treatments with specific substances and/or physiological agents capable of inducing stress. The program included analyses on cell morphology and vitality by staining with haematoxylin and immuno-fluorescence methods using specific antibodies against cyto-scheletal and specific cell type proteins; the detection of apoptotic cells using the DNA-fragmentation marker and the determination of heat shock proteins expression which are induced also in response to non-thermal stress (cold shock, cytoxicity, UVA and UVB).

Ready to eat food will be tested in specific populations such as athletes in which physical activity raises the level of free radicals in blood.

The assays to be performed on human blood serum include the d-ROMs (Reactive Oxygen Metabolites) test that allows the determination of the hydroperoxides concentration generated in the cells by the oxidative attack of ROS on different organic substrates. This is considered to be an important parameter to evaluate the metabolism in patients affected by different pathologies and it can be applied in any field of medicine either in the prevention or therapy of oxidative stress-related diseases. The determination of total fatty acid composition is also included to quantify polyunsaturated fatty acids by Gas-liquid Chromatography, as well as the ratio arachidonic acid/eicosapentaenoic acid that is measured as an index of omega-3 fatty acid anti-inflammatory activity. In addition, the quantification of alfa-tocopherol content is performed by HPLC, while for the monitoring of plasma susceptibility to lipid peroxidation a fluorescence method in which peroxidation is induced in vitro by $CuSO_4$ is used. The resulting kinetics can be used to calculate the lag-time and the propagation velocity of plasma peroxidation. Experiments on isolated erythrocytes allow the assessment of glutathione, catalase, superoxide dismutase, glutathione peroxidase and reductase activities. Subsequently, after purification of the cell membrane the total content of protein, cholesterol and phospholipidic Pi is determined. The lipid peroxidation and the sensitivity to oxidative stress is measured on lipid extracts determining endogenous and iron/ascorbate induced level of thiobarbituric reactive substances.

It is important to emphasize that in this project only nutraceutics already well characterized and exploited in human nutrition are utilized. Nutraceutics derived by molecular biological means are not considered at this stage. Moreover, the selected nutraceutics are analyzed to establish their active metabolite content and to ensure the absence of any interfering compound(s).

Finally, particular attention is dedicated to the formulation of the snacks. The choice of the proper flavoring materials is a critical point vis-à-vis consumer acceptance and in any case it should be a healthy choice as it will also determine the quality of the food products (low fat, low saccharose, low salt).

Building a Bioreactor for Industrial Plant Cell and/or Organ Growth

Bioreactor studies are the final step for a possible commercial production of secondary metabolites and recombinant protein/peptides from plant cell cultures. Compared to traditionally grown whole plants, the exploitation of bioreactors guarantees defined controlled process conditions and minimizes or even prevents variations in product yield and quality. In addition, the possible cultivation of transformed cells or organs could find more public acceptance as they are, by definition, cultivated under closed conditions and therefore do not present environmental risk.

The research in this sector aims at developing a laboratory, miniaturized pilot bioreactor and designing special automatic systems useful as an industrial bioreactor. Depending on cell type and organ culture, different prototypes are proposed attempting to avoid specific bottlenecks deriving from large scale production of phytochemicals. This is an important phase as numerous problems arise in culture scaling up. In cell suspension cultures, for example, growth is considerably modified when cells are cultivated in large tanks and the production of cell biomass remains a critical point for bioreactor productivity. This is mainly due to limitations of oxygen and nutrients transfer to biomass, as well as shaking stress or inhomogeneous culture systems which cause cell sedimentation and death.[69]

Compared to cell cultures, organ cultures generally display a lower sensitivity to shear stress although some hairy root lines have also been described to be susceptible. Immobilization of hairy roots into a polymer matrix is a well-known technique for improving their growth, nevertheless immobilization can also be reached spontaneously, as hairy roots usually grow actively and rapidly colonize the entire reactor. In bioreactors, it is also possible to protect the roots from agitation by using screens or wire meshes.

Finally, a set of methods for the extraction of secondary metabolites will be evaluated as well as extraction efficiency and HPLC metabolite profiling of the extracts produced by the plant material in the bioreactor. This work is supervised by a robotics expert in the field of automation in collaboration with experts in cell culture systems and growth in bioreactors.[69]

Building Biosensors to Control Metabolite Content

Biosensors are integrated devices combining a bio-sensing element with a physico-chemical transducer which are capable of providing specific quantitative or semi-quantitative analytical information. The biological sensing element is responsible for the specific recognition of the analyte, while the transducer supplies an electric output signal which is processed by the electronic component. Biosensors find applications in several industrial bioreactor processes allowing, for example, the control of pH, temperature, CO_2 and O_2 parameters which, it is hoped, will lead to optimization of the procedure.

In the context of the Nutra-Snack project, the interest was mainly focused on the development of biosensors capable of revealing the antioxidant capacity in extracts enriched in secondary metabolites, food products containing extracts enriched in secondary metabolites, large-scale in vitro cultures in bioreactor during secondary metabolites synthesis and blood serum of subjects nourished with the new functional food.

The common methods of antioxidant capacity evaluation rely on the measurement of the inhibition of the radical chain reaction by a compound presumed to act like an antioxidant. For plant extracts the most widely applied methods are spectrophotochemical ones, based on measurement of the decrease of specific absorbency of a long-lived free radical in the presence of the plant extract. The 2,2'-diphenyl-1-picrylhydrazyl and 2,2'-azino-bis(3-ethylbenzthiazoline-6-sulphonic acid) cation radical are the two mainly used free radicals when the efficacy of plant extracts is to be assessed. These procedures are applied to validate and compare data obtained from the newly developed biosensors. Enzymes like polyphenoloxidases (PPOs). Tyrosinase and Laccase are used as immobilized biomediators on different supports for their high structure-enzyme specificity. Subsequently, the characterization of the developed biosensor is performed on compounds and secondary metabolites like rosmarinic acid, caffeic acid, resveratrol, resorcinol, chlorogenic acid, and classical substrates of PPOs like pyrocatechol, cresol and gallic acid. In order to obtain a response of the biosensor related to the content of the metabolites of interest, the first step was the study of the enzymatic reaction in solution, mainly to evaluate the pH influence on the redox potential of the investigated compounds using various types of electrode materials such as glassy carbon and gold. The performance of biosensors is assessed in terms of sensitivity, linearity range and reproducibility, stability and lifetime.[78,79]

Conclusion

The activities carried out in this project are intended to help establishing the integrated scientific and technological bases needed to develop an environmentally friendly production of safer, healthy food and control food-related risks as well as health risks associated with environmental changes.

In addition, much effort is focused on attempts to define a standard for new regulations on Nutraceutics. At present this is lacking since functional food has not yet been clearly defined by legislation.[80] As interest in nutraceutics has grown, new products have appeared and interest has turned to the development of standards and guidelines for the development and promotion of such foods. The subject of health claims is becoming increasingly important and there is broad consensus on the needs of a regulatory framework that will protect consumers, promote fair trade and encourage product innovation in the food industry. The communication of health benefits to consumers is also of critical importance so that they have the knowledge to make informed choices about the foods they eat and enjoy. The future of food will increasingly be concerned with how it affects our health and well-being. Everyday food products will be adapted to become a part of disease prevention.

Acknowledgements

We are grateful to European Community for support (EU N. FOOD-CT-2005-023044). Fai-MoMa was financed by ASI for Nutraceutics in support to astronauts in space.

References

1. Valko M, Leibfritz D, Moncol J et al. Free radicals and antioxidants in normal physiological functions and human disease. Int J Biochem Cell Biol 2007; 39:44-84.
2. Mattson MP. Awareness of hormesis will enhance future research in basic and applied neuroscience. Crit Rev Toxicol 2008; 38:633-639.
3. Singh U, Jialal I. Oxidative stress and atherosclerosis. Pathophysiology 2006; 13:129-42.
4. Ward NC, Croft KD. Hypertension and oxidative stress. Clin Exp Pharmacol Physiol 2006; 33:872-876.
5. Pacher P, Szabo C. Role of the peroxynitrite-poly(ADP-ribose) polymerase pathway in human disease. Am J Pathol 2008; 173:2-13.
6. Benz CC, Yau CA. Oxidative stress and cancer: paradigms in parallax. Nat Rev Cancer 2008; 8:875-879.
7. Manos J, Arthur J, Rose B, Bell S et al. Gene expression characteristics of a cystic fibrosis epidemic strain of Pseudomonas aeruginosa during biofilm and planktonic growth. FEMS Microbiol Lett 2009; 292:107-114.
8. Maritim AC, Sanders RA, Watkins JB 3rd. Diabetes, oxidative stress, and antioxidants: a review. J Biochem Mol Toxicol 2003; 17:24-38.
9. Andersen JK. Oxidative stress in neurodegeneration: cause or consequence? Nat Rev Neurosci 2004; 5:S18-S25.
10. Trachootham D, Lu W, Ogasawara MA et al. Redox regulation of cell survival. Antioxid Redox Signal 2008;10:1343-74.
11. Spitz DR, Azzam EI, Li JJ, Gius D. Metabolic oxidation/reduction reactions and cellular responses to ionizing radiation: A unifying concept in stress response biology. Cancer Metastasis Rev 2004; 23:311-322.
12. Fialkow L, Wang Y, Downey GP. Reactive oxygen and nitrogen species as signaling molecules regulating neutrophil function. Free Radic Biol Med 2007; 42:153-164.
13. D'Autréaux B, Toledano MB. ROS as signalling molecules: mechanisms that generate specificity in ROS homeostasis. Nat Rev Mol Cell Biol 2007; 8:813-24.

14. Baran CP, Zeigler MM, Tridandapani SM et al. The role of ROS and RNS in regulating life and death of blood monocytes. Curr Pharm Des 2004; 10:855-866.
15. Moran LK, Gutteridge JMC, Quinlan GJ. Thiols in cellular redox signalling and control. Curr Med Chem 2001; 8:763-772.
16. Lesgards J-F, Durand P, Lassarre M et al. Assessment of lifestyle effects on the overall antioxidant capacity of healthy subjects. Environ Health Perspect 2002; 110:479-486.
17. Vattem DA, Ghaedian R, Shetty K. Enhancing health benefits of berries through phenolic antioxidant enrichment: focus on cranberry. Asia Pac J Clin Nutr 2005; 14:120-30.
18. Singh B, Bhat TK, Singh B. Potential therapeutic applications of some antinutritional plant secondary metabolites. J Agric Food Chem 2003; 51:5579-97.
19. Tran E, Demmig-Adams B. Vitamins and minerals: Powerful medicine or potent toxins? Nutr Food Sci 2007; 37:50-60.
20. Bouvier F, Rahier A, Camara B. Biogenesis, molecular regulation and function of plant isoprenoids. Prog Lipid Res 2005; 44:357-429.
21. Miller NJ, Sampson J, Candeias LP et al. Antioxidant activities of carotenes and xanthophylls. FEBS Lett 1996; 384:240-242.
22. Mares-Perlman JA, Millen AE, Ficek TL et al. The body of evidence to support a protective role for lutein and zeaxanthin in delaying chronic disease. Overview. J Nutr 2002; 132:518S-524S.
23. Scalbert A, Manach C, Morand C et al. Dietary polyphenols and the prevention of diseases. Crit Rev Food Sci Nutr 2005; 45:287-306.
24. Wahle KWJ, Rotondo D, Heys SD. Plant phenolics in the prevention and treatment of cancer, this volume.
25. Moriarty RM, Naithani R, Surve B. Organosulfur compounds in cancer chemoprevention. Mini Rev Med Chem 2007; 7:827-38.
26. Nagini S. Cancer chemoprevention by garlic and its organosulfur compounds-panacea or promise? Anticancer Agents Med Chem 2008; 8:313-21.
27. Touloupakis E, Ghanotakis D F. Nutraceutical use of garlic, this volume.
28. Del Bano MJ, Lorente J, Castillo J. Phenolic diterpenes, flavones, and rosmarinic acid distribution during the development of leaves, flowers, stems, and roots of Rosmarinus officinalis. J Agric Food Chem 2003; 51:4247-4253.
29. Kintzios S, Makri O, Panagiotopoulos E et al. In vivo rosmarinic acid accumulation in sweet basil (Ocimum Basilicum l.). Biotechnol Lett 2003; 25:405-408.
30. Lee BJ, Hendricks DG. Antioxidant effects of L-carnosine on liposomes and beef homogenates. J Food Sci 1997; 62:931-934.
31. Gao LP, Wei HL, Zhao HS et al. Antiapoptotic and antioxidant effects of rosmarinic acid in astrocytes. Pharmazie 2005; 60:62-5.
32. Estrada LD, Soto C. Disrupting β-amyloid aggregation for alzheimer disease treatment. Curr Topics Medl Chem 2007; 7:115-126.
33. Rivière C, Richard T, Vitrac X et al. New polyphenols active on β-amyloid aggregation. Bioorg Med Chem Lett 2008; 18:828-831.
34. Tepe B. Antioxidant potentials and rosmarinic acid levels of the methanolic extracts of Salvia virgata (Jacq), Salvia staminea (Montbret & Aucher ex Bentham) and Salvia verbenaca (L.) from Turkey. Bioresour Technol 2008; 99:1584-1588.
35. Lo C, Le Blanc JCY, Yu CKY et al. Detection, characterization, and quantification of resveratrol glycosides in transgenic arabidopsis over-expressing a sorghum stilbene synthase gene by liquid chromatography/tandem mass spectrometry. Rapid Commun Mass Spectrom 2007; 21:4101-4108.
36. Prokop J, Abrman P, Seligson AL et al. Resveratrol and its glycon piceid are stable polyphenols. J Med Food 2006; 9:11-4.
37. Hao HD, He LR. Mechanisms of cardiovascular protection by resveratrol. J Med Food 2004; 7:290-298.
38. Klinge CM, Wickramasinghe NS, Ivanova MM, Dougherty SM. Resveratrol stimulates nitric oxide production by increasing estrogen receptor -Src-caveolin-1 interaction and phosphorylation in human umbilical vein endothelial cells. FASEB 2008; 22:2185-2197.
39. Maheshwari RK, Singh AK, Gaddipati J, Srimal RC. Multiple biological activities of curcumin: a short review. Life Sci 2006; 78:2081-2087.
40. Lopez-Lazaro M. Anticancer and carcinogenic properties of curcumin: considerations for its clinical development as a cancer chemopreventive and chemotherapeutic agent. Mol Nutr Food Res 2008; 52:S103-S127.
41. Elias RJ, Kellerby SS, Decker EA. Antioxidant activity of proteins and peptides. Crit Rev Food Sci Nutr 2008; 48:430-441.
42. Atmaca G. Antioxidant effects of sulfur-containing amino acids. Yonsei Med J 2004; 45:776-788.

43. Rajapakse N, Mendis E, Jung WK et al. Purification of a radical scavenging peptide from fermented mussel sauce and its antioxidant properties. Food Res Int 2005; 38:175-182.
44. Rajapakse N, Mendis E, Byun H et al. Purification and in vitro antioxidative effects of giant squid muscle peptides on free radical-mediated oxidative systems. J Nutr Biochem 2005; 16:562-569.
45. Mendis E, Rajapakse N, Byun H-G et al. Investigation of jumbo squid (Dosidicus gigas) skin gelatine peptides for their in vitro antioxidant effects. Life Sci 2005; 77:2166-2178.
46. Chen HM, Muramoto K, Yamauchi F, et al. Antioxidative properties of histidine-containing peptides designed from peptide fragments found in the digests of a soybean protein. J Agric Food Chem 1998; 46:49-53.
47. Kitts DD, Weiler KA. Bioactive proteins and peptides from food sources: applications of bioprocesses used in isolation and recovery. Curr Pharm Des 2003; 9:1309-1323.
48. Davies KM. Plant Colour and Fragrance. In: Verpoorte R, Alfermann AW, eds. Metabolic Engineering of Plant Secondary Metabolism. Dortrecht: Kluwer Academic Publishers, 2000:127-163.
49. Oksman-Caldentey KM, Arroo R. Regulation of tropane alkaloid metabolism in plants an plant cell cultures. In: Verpoorte R, Alfermann AW, eds. Metabolic Engineering of Plant Secondary Metabolism. Dortrecht: Kluwer Academic Publishers, 2000:253-281.
50. Verpoorte R, van der Heijden R, ten Hoopen HJG et al. Metabolic engineering of plant secondary metabolite pathways for the production of fine chemicals. Biotechnol Lett 1999; 21:467-479.
51. Ferrari S. Biological elicitors of plant secondary metabolites: mode of action and use in the production of Nutraceutics, this volume.
52. Hain R, Grimmig B. Modification of plant secondary metabolism by genetic engineering. In: Verpoorte R, Alfermann AW, eds. Metabolic Engineering of Plant Secondary Metabolism. Dortrecht: Kluwer Academic Publishers, 2000:217-231.
53. De Luca V. Metabolic engineering of crops with the tryptophan decarboxylase of catharanthus roseus. In: Verpoorte R, Alfermann AW, eds. Metabolic Engineering of Plant Secondary Metabolism. Dortrecht: Kluwer Academic Publishers, 2000:179-194.
54. Johanningmeier U, Fischer D. Perspective for the use of genetic transformants in order to enhance the synthesis of the desired metabolites. Engineering chloroplasts of microalgae: application to nutraceutical technology, this volume.
55. Krzyzanowska J, Czubacka A, Oleszek W. Dietary phytochemicals and human health, this volume.
56. Cahoon EB, Hall SE, Ripp KG et al. Metabolic redesign of vitamin E biosynthesis in plants for tocotrienol production and increased antioxidant content. Nature Biotechnol 2003; 21:1082-1087.
57. Burba JL,Portela JA, Lanzavechia S. Argentine garlic I: a wide offer of clonal cultivars. Acta Horticulturae 2005; 688:291-296.
58. Fiamegos CG N, Vervoort J, Stalikas CD. Analytical procedure for the in-vial derivatization-extraction of phenolic acids and flavonoids in methanolic and aqueous plant extracts followed by gas chromatography with mass-selective detection. J Chromatogr A 2004; 1041:11-18.
59. Romani A, Vignolini P, Galardi C et al. Polyphenolic content in different plant parts of soy cultivars grown under natural conditions. J Agric Food Chem 2003; 51:5301–5306.
60. Vagi E, Rapavi E, Hadolin M, Peredi KV et al. Phenolic and triterpenoid antioxidants from Origanum majorana L. herb and extracts obtained with different solvents. J Agric Food Chem 2005; 53:17-21.
61. Kruk J, Myśliwa-Kurdziel B, Jemioła-Rzemińska M et al. Fluorescence lifetimes study of α-tocopherol and biological prenylquinols in organic solvents and model membranes. Photochem Photobiol 2006; 82:1309-1314.
62. Jahns P, Depka B, Trebst A. Xanthophyll cycle mutants from Chlamydomonas reinhardtii indicate a role for zeaxanthin in the D1 protein turnover. Plant Physiol Biochem 2000; 38:371-376.
63. Yongyue S, Shufen L, Can Q. Solubility of ferulic acid and tetramethylpyrazine in supercritical carbon dioxide. J Chem Eng Data 2005; 50:1125-1128.
64. Aziz A, Heyraud A, Lambert B. Oligogalacturonide signal transduction, induction of defense-related responses and protection of grapevine against Botrytis cinerea. Planta 2004; 218:767-774.
65. Ferrari S, Denoux C, Galletti R et al. Resistance to Botrytis cinerea induced in Arabidopsis thaliana by elicitors is independent of salicylic acid, ethylene or jasmonate signaling but requires PAD3. Plant Physiol 2007; 144:367-379.
66. Ramachandra Rao S, Ravishankar GA. Plant cell cultures: chemical factories of secondary metabolites. Biotechnol Adv 2002; 20:101-153.
67. Savitha BC, Thimmaraju R, Bhagyalakshmi N et al. Different biotic and abiotic elicitors influence betalain production in hairy root cultures of Beta vulgaris in shake-flask and bioreactor. Process Biochem 2006; 41:50-60.
68. Pistelli L, Giovannini A, Bertoli A et al. Hairy root cultures for secondary metabolites production, this volume.

69. Ruffoni B, Pistelli L, Bertoli A et al. Plant cell cultures: bioreactors for industrial production, this volume.
70. He P, Duncan J, Barber J. Astaxanthin accumulation in the green alga Haematococcus pluvialis: effects of cultivation parameters. J Integr Plant Biol 2007; 49:447-451.
71. Korhonen H, Pihlanto A. Food-derived bioactive peptides-opportunities for designing future foods. Curr Pharm Des 2003; 9:1297-1308.
72. Korhonen H, Pihlanto A. Bioactive peptides: production and functionality. Int Dairy J 2006; 16:945-960.
73. Hartmann R, Meisel H. Food-derived peptides with biological activity: from research to food applications. Curr Opin Biotechnol 2007;18:163-9.
74. Pulz O, Gross W. Valuable products from biotechnology of microalgae. Appl Microbiol Biotechnol 2004; 65:635-648.
75. Walker TL, Purton S, Becker DK et al. Microalgae as bioreactors. Plant Cell Rep 2005; 24:629-641.
76. Spolaore P, Joannis-Cassan C, Duran E et al. Commercial applications of microalgae. J Biosci Bioeng 2006; 101:87-96.
77. Manuell AL, Beligni MV, Elder JH et al. Robust expression of a bioactive mammalian protein in Chlamydomonas chloroplast. Plant Biotechnol J 2007; 5:402-412.
78. Litescu SC, Eremia S. Methods for antioxidant capacity determination from food and raw materials, this volume.
79. Lavecchia T, Tibuzzi A, Giardi MT. Biosensors for functional food safety and analysis, this volume.
80. Giunta R, Basile G and Tibuzzi A. Legislation on nutraceuticals and food supplements: a comparison between regulations in United States and European Union, this volume.

CHAPTER 2

Overview of Diet-Gene Interactions and the Example of Xanthophylls

Barbara Demmig-Adams* and William W. Adams, III

Abstract

This chapter provides an overview of diet-gene interaction and the role of dietary factors in human health and disease. Human master control genes that regulate processes of fundamental importance, such as cell proliferation and the immune response, are introduced and their modulation by nutraceuticals, produced by plants and photosynthetic microbes, is reviewed. Emphasis is placed on antioxidants and polyunsaturated fatty acids as regulators of master control genes. Furthermore, a case study is presented on xanthophylls, a group of carotenoids with multiple health benefits in the protection against eye disease and other chronic diseases, as well as the synergism between xanthophylls and other dietary factors. Lastly, dietary sources of the xanthophylls zeaxanthin and lutein are reviewed and their enhancement via genetic engineering is discussed.

Introduction: Overview of Diet-Gene Interaction in Human Disease

The role of food components—like vitamins—in human physiology has long been recognized. However, it is only now being realized that a multitude of additional dietary factors has profound effects on human health and the risk for disease. All of these dietary factors possess the remarkable ability to modulate the expression of key regulatory genes in humans. Most of these factors are synthesized by plants or algae (with the latter being the ultimate source of the much-touted fish oils; see below) and are termed phytochemicals (from *phyto* = plant), phytonutrients, or nutraceuticals. These phytochemicals modulate the expression of human master control genes that regulate processes of fundamental importance, such as cell proliferation and the immune response. These central processes, and especially imbalances in their regulation, play a key role in all major human diseases.

Disturbances in cellular signal transduction are being identified as underlying mechanisms for a plethora of human diseases. A major example is disturbances in the cellular redox homeostasis. The cellular redox state, or balance between oxidants and antioxidants, regulates "life-and-death" decisions of cells (such as cell division and programmed cell death), as well as vital defenses (such as the immune response and inflammation). Why? In eukaryotic cells, the presence of oxygen/oxidants acts as (i) a sign of plentiful energy (available through oxidative respiration), thus stimulating cell division and growth and/or (ii) as a sign for the presence of invaders (that have caused immune cells to fight back with strong oxidants), thus triggering a full immune response/inflammation[1-4]. Any disturbances in these essential processes can lead to run-away cell division and either insufficient or excessive elimination of cells, as well as excessive or chronic inflammation. It is now recognized that all major chronic human diseases, including cancers, diabetes, heart

*Corresponding Author: Barbara Demmig-Adams—Department of Ecology and Evolutionary Biology, University of Colorado, Boulder, Colorado 80309-0334, USA.
Email: barbara.demmig-adams@colorado.edu

Bio-Farms for Nutraceuticals: Functional Food and Safety Control by Biosensors
edited by Maria Teresa Giardi, Giuseppina Rea and Bruno Berra.

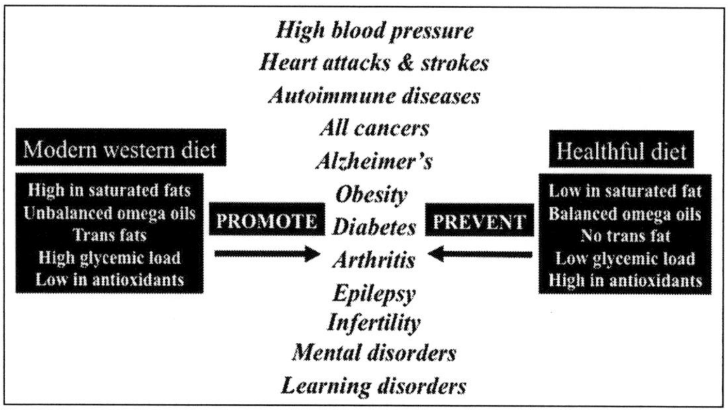

Figure 1. Schematic depiction of the effects of the modern western diet, compared to a more healthful diet, in either promoting or preventing chronic human diseases and other conditions. Diseases and conditions listed are those for which a wealth of information exists on the role of diet in disease promotion or prevention. See text for a detailed description of omega oils (polyunsaturated fats with omega-6 and omega-3 fatty acids). Glycemic load is the product of the amount of a food consumed and its glycemic index, i.e., the speed with which its carbohydrate content is converted to free glucose. Trans fats are those containing hydrogenated fatty acids. Foods high in antioxidants include all fruits, vegetables, herbs and spices.

disease, neurodegenerative diseases like Alzheimer's, autoimmune diseases, mental and learning disorders and many others, involve these same types of disturbances (Fig. 1). In addition, the susceptibility to infectious diseases is affected by the cellular redox state and any disturbances in it (Fig. 1; see also below).

Redox-regulated proteins include key signaling proteins that serve as master control proteins high up in the hierarchy of control pathways for metabolism. Two key redox-regulated proteins (nuclear factor kappa B, NF-κB and activator protein 1, AP-1) are gene regulatory proteins, or transcription factors, that orchestrate the immune response as well as cell division and cell death.[5,6] It is now thought that inheritance of overly or insufficiently active versions of these proteins (or rather their associated genes) is the basis for the genetic propensity for, and family history of, an enhanced risk for any of the above-mentioned chronic diseases. Furthermore, both NF-κB and AP-1 interact with infectious diseases in multiple ways; both transcription factors activate genes necessary for defense and both are also the target of several viruses. Astoundingly, several human viruses, including human immunodeficiency virus (HIV; associated with AIDS) and human papilloma virus (HPV; associated with cervical cancer) actually employ these human redox-activated gene regulatory proteins, activated in response to invaders, for their own replication.[7-11]

It is of crucial importance to note that these same vital master control processes are highly susceptible to modulation by a wide range of environmental factors such as diet, pollutants (heavy metals, asbestos, silica and others), radiation (UV and ionizing radiation) and, as discussed above, infectious agents.[12-15] By virtue of their redox effects, these environmental factors thus also have a profound effect on the risk for chronic disease as well as infectious diseases. In general, dietary antioxidants and other phytochemicals ameliorate chronic disease (Fig. 1). Furthermore, the human viruses HIV and HPV are negatively impacted by antioxidants.[16-20] Plant-based foods contain a myriad of redox-active compounds, mostly oxidant-removing antioxidants, that directly and indirectly modulate redox-controlled master control proteins.

Another group of key dietary factors involved in the regulation of the immune response and cell division/cell death are essential fats, i.e., essential polyunsaturated fatty acids.[21,22] These essential fats cannot be manufactured in the human body, yet are vital precursors for major groups of hormone-like messengers such as prostaglandins, leukotrienes and others. These essential fatty acids furthermore interact with the above-described antioxidants, since the conversion of all of the essential fatty acids to active messengers is an oxidative process that is slowed by antioxidants. Plant-based foods are a major source of these essential unsaturated fatty acids and it is especially in cooler climate zones that plants manufacture polyunsaturated fatty acids. The other major source is cold-water fish (that consume cold-water algae rich in polyunsaturated fatty acids).

The polyunsaturated fatty acids fall into two groups, omega-6 fatty acids that become converted to typically highly pro-inflammatory (and often proliferation-promoting) messengers and omega-3 fatty acids that, in turn, are converted to messengers ending inflammation and slowing down cell division. As a rule of thumb, many processed foods are low in antioxidants and rich in omega-6 fatty acids and thus promote inflammation and proliferation. In particular, certain plant oils, such as corn, soybean, sunflower, safflower, and cottonseed oil, that do contain some omega-3 fats are also very high in omega-6 fatty acids and it is these plant oils that are frequently used to produce highly processed foods. It is the ratio of omega-6 to omega-3 fats that determines the effect of any particular oil on human health. Both omega-6 and omega-3 fatty acids have to be consumed by humans in their diet and a balanced ratio of their consumption is crucial to avoid metabolic imbalances.

It is now thought that a balanced ratio of omega-6 to omega-3 fats is around 1-2 to 1, while the modern western diet, with much highly processed food, typically exhibits an unbalanced ratio of 10-20 to 1.[23] Such a high ratio of omega-6 to omega-3 fatty acids is thought to overstimulate inflammation (and a host of pro-inflammatory diseases) as well as cell division (and cancers).[24,25] It is thought that conditions ranging from diabetes and heart disease[26,27] to cancers[24] and a host of neuropsychiatric and neurodevelopmental disorders[28,29,30] are promoted by an excessively high omega-6 to omega-3 fatty acid ratio.

Current recommendations are that the sum of all polyunsaturated fats consumed in the human diet should not exceed 20-30% of total fat consumption, with the lion's share of the remainder recommended to stem from monounsaturated fats like oleic acid that is also found in many plant foods. Utilization of oils derived from oil-rich seeds such as sunflower, corn, soybean, or cotton that have been genetically engineered to contain high levels of oleic acid and much reduced levels of omega-6 linoleic acid should thus have many benefits.[31-35] The messengers derived from omega-3 fats act as strong physiological antagonists of highly pro-inflammatory omega-6-derived messengers and omega-3 fats therefore act as anti-inflammatory agents. Whole plant-based foods (with balanced omega-6 to -3 ratios and high antioxidant levels) as well as cold-water fish (with high levels of omega-3 fats and of the antioxidant vitamin E, as well as the antioxidant mineral selenium in certain fishes) are thus generally anti-inflammatory foods.

Plants and other photosynthetic organisms, as the primary producers of energy-rich food molecules, thus also produce the majority of (metabolism-regulating) antioxidants and essential fatty acids. As stated above, even the high levels of omega-3 fats in fish are derived from the consumption (by the fish) of photosynthetic algae that synthesize and accumulate high levels of these fats. These trophic relationships establish food produced by plants or photosynthetic microbes as a key modulator of human metabolism and health (in addition to being the ultimate source of energy and carbon-based materials for humans and other consumers). It is therefore important to recognize the many roles of these foods and to promote an improved availability of nutrient-rich food from plant and algal sources to all of humankind. In view of the finite area available for agricultural food production, further enhancement of the nutritional quality of foods produced by plants and photosynthetic microbes must be considered—via enhancement of the levels of antioxidant and essential fatty acids (the latter at a balanced, low ratio of omega-6 to omega-3 fatty acids).

It should be noted that manipulation of plant or algal antioxidant or polyunsaturated fatty acid content is likely to also affect critical regulatory processes within the photosynthetic organism itself, since the above-described key regulatory role of the cellular redox state applies to all lifeforms. As a generalization, over-expression of antioxidants in plants and algae may (i) make these organisms more resistant to abiotic stresses, while (ii) possibly increasing their susceptibility to some pests/pathogens and (iii) likely also affecting growth, development and reproduction. It may be feasible to simultaneously decrease crop losses due to environmental stress (like drought, heat, or frost), while enhancing the nutritional quality of food plants. However, potential costs in terms of possible reductions in growth rates and/or increased pest/pathogen susceptibility, remain to be fully established.

Furthermore, the synergistic effects of antioxidant and omega-3 fatty acid overexpression in plants should be investigated. While plants, like animals, use oxidation products of polyunsaturated fatty acids as messengers that induce critical developmental and stress responses, plants use omega-3-derived messengers (like the plant stress hormone jasmonic acid) to trigger defenses against pathogens and environmental stress, while animals/humans use omega-6-derived messengers to trigger e.g., inflammation as a pathogen defense. Overexpression of omega-3 fatty acid production in plants may be expected to increase e.g., pathogen fighting potential and may thus counterbalance an overexpression of antioxidants. However, little experimental evidence is available to date on these issues, making these questions an important target for research. The ability to produce improved food crops and ameliorate human disease clearly requires an improved understanding of the underlying molecular processes at the intersection between physiology, disease and stress for both plants and humans.

Why should one aim for an enhancement of the nutritional quality of whole plant foods as opposed to simply taking supplements? There have now been over 15 years of clinical trials with high dose single supplements, e.g., the antioxidant vitamin E or the carotenoid beta-carotene, all with disappointing outcomes. Rather than demonstrating health benefits, these high dose supplements often had adverse effects.[36] It has therefore been concluded that phytochemicals work best when consumed in the 'whole package', i.e., as a whole food with its multiple, synergistically acting ingredients. The underlying mechanisms why whole foods rich in e.g., antioxidants protect against disease, while high-dose antioxidant supplements can have the opposite effects are currently largely unknown.

It is attractive to speculate that these contrasting effects of supplements and whole foods may be explained by the complex nature of redox-regulated signaling pathways. For example, while the transcription factors NF-κB and AP-1 are activated by a signaling (MAP kinase) pathway stimulated by oxidants, DNA binding of these transcription factors is activated by antioxidants. It is possible that the cocktail of antioxidants, present at moderate concentrations for each component in whole foods, mainly acts to dampen MAP kinase signaling towards NF-κB and AP-1, while high dose antioxidant supplements may promote DNA binding of one or both of these transcription factors directly. Since multi-factor cocktails of nutraceuticals are needed rather than single high dose antioxidants/phytochemicals, the aim of plant and algal engineering efforts should also be directed at an upregulation of multiple nutraceuticals in appropriate proportions. These multiple nutraceuticals to consider should include the key classes of thiol antioxidants, phenolic antioxidants/phytochemicals, terpenoids (terpenes and carotenoids), antioxidant minerals (e.g., zinc and selenium, the accumulation of which by plants could be targeted), as well as polyunsaturated omega-3 and monounsaturated fatty acids.

A Case Study: Xanthophylls and Their Synergism with Other Dietary Factors in Human Health—Protection against Eye Disease and Other Chronic Diseases

Carotenoids have received considerable attention as potential modulators of human health. Initially, most of this attention was directed towards beta-carotene, also known as provitamin A, the precursor of vitamin A. Vitamin A serves in the vision process, but also has vital roles in the regulation of the immune system. Additional health-promoting effects of beta-carotene were proposed, but have remained elusive. In the meantime, carotenoids other than beta-carotene have received increasing attention, particularly lycopene (the red pigment of tomatoes) as well as zeaxanthin and lutein (the yellow pigments responsible for the color of corn and eggs; see below). In the case of lycopene, there is some evidence for a protective role against prostate cancer.[37] For lutein and zeaxanthin, considerable evidence[38] is available for a role of these xanthophylls in the protection against age-related vision loss, such as age-related macular degeneration or AMD[39] and cataracts.[40,41] In addition, evidence is also accumulating for a role of zeaxanthin and lutein in the protection against cancer and heart disease.[38]

Zeaxanthin and lutein are synthesized by plants (as well as photosynthetic microbes), particularly in the green, photosynthetic parts of plants and especially when these plants are grown in high light or under otherwise stressful conditions.[42,43] Zeaxanthin protects against the formation of potentially destructive reactive oxygen species in leaves exposed to intense sunlight alone or moderate levels of sunlight in the presence of environmental conditions unfavorable for plant growth.[44] This photoprotective process is necessary for plant survival and reproductive success;[45] oxygenic photosynthesis could not exist without it. Zeaxanthin facilitates the harmless dissipation of excess energy absorbed by chlorophyll[44] via a mechanism that involves an exchange of energy and/or a reversible electron exchange between chlorophyll and zeaxanthin.[46,47] Zeaxanthin's close isomer lutein also plays a role, albeit a more minor one, in this dissipation process.[48] In addition, zeaxanthin serves in photoprotection via a second, not fully understood mechanism that involves a direct inhibition of the oxidation of fatty acids of biological membranes (lipid peroxidation).[49,50]

Zeaxanthin and lutein—neither of which can be synthesized by animals—have a host of beneficial effects when consumed by humans. Epidemiological studies have identified inverse links between zeaxanthin/lutein and a wide range of human diseases, including age-related eye disease, various cancers and other conditions.[38,44,51] However, the underlying mechanisms for these apparent protective effects remain poorly understood. Although zeaxanthin is clearly the primary protective xanthophyll in plants, greater attention has been paid to lutein in human studies. This is, however, not based on actual evidence and may simply be due to the fact that lutein has long been commercially available, whereas zeaxanthin has not.

In the human eye, zeaxanthin and lutein form the yellow spot of the retinal macula, with higher ratios of zeaxanthin to lutein in the areas of highest light exposure. The zeaxanthin/lutein ratio also increases from the content of the intestinal tract to the blood plasma and, again, to the retina of the eye (reviewed in ref. 38). Furthermore, individuals suffering from age-related eye disease (e.g., AMD) have lesser xanthophyll densities throughout their retinas and there is an inverse correlation between dietary zeaxanthin and lutein levels and the risk for AMD as well as cataracts (reviewed in ref. 38). Just like in plants (and photosynthetic microbes), zeaxanthin and lutein protect the human eye from damage by intense light. A breakthrough in the understanding of the function of retinal zeaxanthin was made when it was shown that zeaxanthin prevents programmed cell death of retinal photoreceptor cells in an animal model.[52,53] On the other hand, studies with human cancer cell lines provided evidence that lutein can stimulate programmed cell death of human breast cancer cells[54] and leukemia cells.[55] Lutein furthermore selectively induces programmed cell death in mouse tumor cells, but decreases programmed cell death in cancer-fighting immune cells (blood leukocytes) of tumor-bearing mice.[56] It is presently not known how xanthophylls exert these remarkable and beneficial roles, including opposite effects on programmed cell death in different cell types.

What is known is that zeaxanthin can modulate the oxidation of fatty acids (lipid peroxidation) in plants[49,50] as well as in humans (e.g., in epithelial cells of the eye's lens[57]). Zeaxanthin protects lipids against destructive reactive oxygen in vitro and while this ability of zeaxanthin is enhanced by the presence of the antioxidant vitamin E (tocopherol), zeaxanthin has the primary, more potent effect in this synergism.[58,59]

As hinted above, age-related blindness (with photoreceptor death) and cancer (with run-away cell proliferation) involve seemingly opposite problems, i.e., either too little or too much programmed cell death. While runaway cell division and/or insufficient cell death is a key feature of cancer, excessive cell death is involved in other diseases such as heart disease, diabetes, age-related blindness and neurodegenerative diseases. One might be concerned that compounds that trigger the elimination of unwanted cells might possibly exacerbate conditions involving excess cell death. However, this concern is apparently unfounded. Indeed, the available evidence suggests that several dietary factors possess the capacity to ameliorate both of these contrasting conditions. Thus far, phenolics,[60,61] the polyunsaturated fatty acids of fish oil[62] and zeaxanthin and lutein[63] have all been shown to possess the notable ability to 'work both ways', i.e., triggering programmed cell death of unwanted cells while aiding in the survival of needed cells. This makes these food-derived compounds potentially highly desirable nutraceuticals.

Dietary Sources of Zeaxanthin and Lutein and Their Potential Enhancement

Plant Sources and Engineering Approaches

As reviewed above, zeaxanthin and lutein have roles in protecting human vision and other aspects of human health. However, there is great disparity in how much zeaxanthin versus lutein can be obtained from leafy green plant foods. While the green parts of plants after harvest typically contain high levels of lutein, they retain mere traces of zeaxanthin. This is so because plants carefully modulate the level of zeaxanthin in response to the amount of light to which they are exposed. The level of zeaxanthin is finely controlled by a set of biochemical reactions,[64] the xanthophyll cycle. Only under direct exposure to high light does the xanthophyll cycle produce zeaxanthin (from a biochemical precursor, the xanthophyll violaxanthin); whenever direct high light exposure ceases, zeaxanthin is relatively quickly reconverted to violaxanthin. To control its zeaxanthin levels this way is important to maintain the plant's ability to dissipate potentially harmful excess light one minute (via zeaxanthin, acting as a dissipator of excess light) and to quickly return to a high light use efficiency the next, by converting the dissipator zeaxanthin to the nondissipator violaxanthin.[42,43,65]

Human consumption of high levels of zeaxanthin is highly desirable, because zeaxanthin needs to be preferentially accumulated and incorporated into the parts of the mammalian retina exposed to high irradiance[66] (see also above) and thus appears to be even more important in human diets than the more readily available lutein. For this reason, arrest of the xanthophyll cycle in green leaves in the state of zeaxanthin may be a desirable trait to incorporate into crops that provide green leafy foods. Such a retention of zeaxanthin can be accomplished e.g., by knocking out or silencing the enzyme/gene (zeaxanthin epoxidase of the xanthophyll cycle) responsible for zeaxanthin conversion to violaxanthin and/or by overexpressing enzymes in earlier portions of the carotenoid biosynthetic pathway (see below).

Xanthophyll cycle mutants that accumulate zeaxanthin have already been produced in model plants and algae[67] and these traits can be transferred to crop plants. However, since constantly elevated zeaxanthin levels can cut into the ability of the green leaf to efficiently collect solar energy during the parts of the day when light levels are low and limiting to photosynthesis, e.g., early morning and late afternoon, the effect of zeaxanthin retention on the productivity of crop plants needs to be further examined. Due to the need for leaves to photosynthesize efficiently, overexpression of zeaxanthin in fruit, rather than leaves, is attractive. Tomato fruit with an increased zeaxanthin content has recently been engineered via two other manipulations (overexpression of lycopene-cyclase and beta-carotene hydroxylase[68]). Zeaxanthin-rich potato has also been produced[69] as well as zeaxanthin-accumulating *E. coli*.[70] Moreover, corn is a food naturally high in zeaxanthin and lutein: all of its yellow color stems directly from these two pigments (see ref. 71 for analysis of various corn products). White corn, on the other hand, does not provide these xanthophylls. Intensely yellow corn, produced by traditional breeding and/or engineering, is the preferable type of corn, as long as the ratio of zeaxanthin to lutein is no less than 1 to 1.

A natural condition that can lead to long-term retention of zeaxanthin in perennial plants is environmental stress, such as drought or cold temperatures.[43] However, since environmental stress can severely diminish crop productivity, this is not the most attractive option. Furthermore, additional plant defense compounds, some of which can have adverse effects on human health, may also be formed under the influence of environmental stress. Selective engineering of plants with enhanced zeaxanthin and/or antioxidant content should be the safer route. The production of zeaxanthin by microbes should also be further explored. However, currently available blue-green algal (cyanobacterial) supplements cannot be recommended, due to potential damage to the eye by retinal accumulation of ketocarotenoids also contained in these.[72]

A Source Other Than Plants: Eggs

In addition to corn, the other naturally available food with high zeaxanthin and lutein levels (and a favorable zeaxanthin to lutein ratio) is chicken eggs. However, the chickens are no less capable of synthesizing their own xanthophylls than humans and the coloration (and zeaxanthin/lutein content) of their eggs thus depends on their feed. As long as a food source of xanthophyll is available (from e.g., alfalfa or corn or other sources containing zeaxanthin/lutein), chickens deposit zeaxanthin and lutein into the their eggs. Higher yolk content of xanthophylls leads to a more intense yellow to orange color (cf. 73). These xanthophylls are also completely transferred from yolk to chick where an important portion accumulates in the retina.[74] Some of the dietary lutein is also moved into birds' feathers,[75] possibly to signal the possession of a strong immune system to potential mates[76] since lutein has been reported to stimulate aspects of the immune response.[77,78] For some time, chicken eggs were considered a food that, due to their cholesterol content, should be consumed in moderation or not at all by individuals at risk for e.g., heart disease. However, this recommendation is currently being revised and all but reverted, since (i) the cholesterol found in arterial plaques is predominantly derived from dietary saturated fats converted to cholesterol in the human body itself (rather than from the direct consumption of dietary cholesterol) and (ii) due to the beneficial health effects of the yellow xanthophylls conveniently provided by eggs.

Conclusion

A myriad of phytochemicals in plant-based foods, including antioxidants and essential omega-3 and omega-6 fatty acids, directly and indirectly modulate the human master control proteins that regulate processes of fundamental importance, such as cell proliferation and the immune response. By virtue of these effects, dietary antioxidants and other phytochemicals in plant foods can ameliorate many chronic diseases. Whole plant foods appear to be a better option than high-dose supplements. The effects of engineering or breeding plants for enhanced antioxidant and/or essential fatty acid content on the productivity and stress/pest/disease resistance of the plants themselves remains to be explored. As an example for a group of potent regulatory phytochemicals with multiple health benefits, the yellow carotenoids zeaxanthin and lutein serve in the protection against eye disease and other chronic human diseases. One emerging mechanism of action is the modulation of programmed cell death by zeaxanthin/lutein. Existing foods rich in zeaxanthin/lutein include yellow corn and (the yellow yolks of) eggs.

References

1. Dalton TD, Shertzer HG, Puga A. Regulation of gene expression by reactive oxygen. Annu Rev Pharmacol Toxicol 1999; 39:67-101.
2. Allen RG, Tresini M. Oxidative stress and gene regulation. Free Radic Biol Med 2000; 28:463-499.
3. Maher P, Schubert D. Signaling by reactive oxygen species in the nervous system. Cell Mol Life Sci 2000; 57:1287-1305.
4. Shackelford RE, Kaufmann WK, Paules RS. Oxidative stress and cell cycle checkpoint function. Free Radic Biol Med 2000; 28:1387-1404.
5. Lavrovsky Y, Chatterjee B, Clark RA et al. Role of redox-regulated transcription factors in inflammation, aging and age-related diseases. Exp Gerontol 2000; 35:521-532.
6. Janssen-Heiniger YMW, Poynter ME, Baeuerle PA. Recent advances towards understanding redox mechanisms in the activation of nuclear factor kappa B. Free Radic Biol Med 2000; 28:1317-1327.
7. Nees M, Geoghegan JM, Hyman T et al. Papillomavirus type 16 oncogenes downregulate expression of interferon-responsive genes and upregulate proliferation-associated and NF-kappa B-responsive genes in cervical keratinocytes. J Virol 2001; 75:4283-4296.

8. Pande V, Ramos MJ. Nuclear Factor Kappa B: a potential target for anti-HIV chemotherapy. Curr Med Chem 2003; 10:1603-1615.
9. Devadas K, Hardegen NJ, Wahl LM et al. Mechanisms for macrophage-mediated HIV-1 induction. J Immunol 2004; 173:6735-6744.
10. Gasparian AV, Fedorova MD, Kisseljove FL. Regulation of matrix metalloproteinase-9 transcription in squamous cell carcinoma of uterine cervix: the role of human papillomavirus gene E2 expression and activation of transcription factor NF-kappa B. Biochemistry-Moscow 2007; 72:848-853.
11. Mukerjee R, Sawaya BE, Khalili K et al. Association of p65 and C/EBP beta with HIV-1 LTR modulates transcription of the viral promoter. J Cell Biochem 2007; 100:1210-1216.
12. Schreck R, Albermann K, Baeuerle PA. Nuclear factor kappa-B—an oxidative stress-responsive transcription factor of eukaryotic cells (a review). Free Radic Res Comm 1992; 17:221-237.
13. Chen F, Shi XL. NF-kappa B, A pivotal transcription factor in silica-induced diseases. Mol Cell Biochem 2002; 234:169-176.
14. Flaherty DM, Monick MM, Carter AB et al. Oxidant-mediated increases in redox factor-1 nuclear protein and activator protein-1 DNA binding in asbestos-treated macrophages. J Immunol 2002; 168:5675-5681.
15. Valko M, Morris H, Cronin MTD. Metals, toxicity and oxidative stress. Curr Med Chem 2005; 12:1161-1208.
16. Hirano F, Tanaka H, Miura T et al. Inhibition of NF-kappa B-dependent transcription of human immunodeficiency virus 1 promoter by a phosphodiester compound of vitamin C and vitamin E, EPC-K1. Immunopharmacology 1998; 39:31-38.
17. Garland M, Fawzi W. Antioxidants and progression of human immunodeficiency virus (HIV) disease. Nutr Res 1999; 19:1259-1276.
18. Guiliano A. The Role of nutrients in the prevention of cervical dysplasia and cancer. Nutrition 2000; 16:570-573.
19. Beniston RG, Campo MS. Quercetin elevates p27 (Kip1) and arrests both primary and HPV16 E6/E7 transformed human keratinocytes in G1. Oncogene 2003; 22:5504-5514.
20. Kaiser JD, Campa AM, Ondercin JP et al. Micronutrient supplementation increases CD4 count in HIV-infected individuals on highly active antiretroviral therapy: A prospective, double-blinded, placebo-controlled trial. J Acquir Immune Defic Syndr 2006; 42:523-528.
21. Yaqoob P. Fatty acids as gatekeepers of immune cell regulation. Trends Immunol 2003; 24:639-645.
22. Lapillonne A, Clarke SD, Heird WC. Polyunsaturated fatty acids and gene expression. Curr Opin Clin Nutr Metab Care 2004; 7:151-156.
23. Simopoulos AP. Omega-6/omega-3 essential fatty acid ratio and chronic disease. Food Rev Int 2004; 20:77-90.
24. Simopoulos AP. Omega-3 fatty acids and cancer. Indoor Built Environ 2003; 12:405-412.
25. Simopoulos AP. The omega-6/omega-3 fatty acid ratio, genetic variation and cardiovascular disease. Asia Pac J Clin Nutr 2008; 17:131-134.
26. Haag M, Dippenaar NG. Dietary fats, fatty acids and insulin resistance: short review of a multifaceted connection. Med Sci Mon 2005; 11:RA359-367.
27. Blaschke F, Takata Y, Caglayan E et al. Obesity, peroxisome proliferator-activated receptor and atherosclerosis in type 2 diabetes. Arteroscler Thromb Vasc Biol 2006; 26:28-40.
28. Richardson AJ, Ross MA. Fatty acid metabolism in neurodevelopmental disorder: a new perspective on associations between attention-deficit/hyperactivity disorder, dyslexia, dyspraxia and the autistic spectrum. Prostaglandins Leukot Essent Fatty Acids 2000; 63:1-9.
29. Young G, Conquer J. Omega-3 fatty acids and neuropsychiatric disorders. Reprod Nutr Dev 2004; 45:1-28.
30. Wainwright PE. Dietary essential fatty acids and brain function: a developmental perspective on mechanisms. Proc Nutr Soc 2002; 61:61-69.
31. Warner K, Knowlton S. Frying quality and oxidative stability of high-oleic corn oils. J Am Oil Chem Soc 1997; 74:1317-1322.
32. Forster VA. Genetically modified crop approvals and planted acreages. Crop Biotechnol 2002; 829:17-22.
33. Liu Q, Singh SP, Green AG. High-stearic and high-oleic cottonseed oils produced by hairpin RNA-mediated posttranscriptional gene silencing. Plant Physiol 2002; 129:1732-1743.
34. Liu Q, Singh S, Green A. High-oleic and high-stearic cottonseed oils: Nutritionally improved cooking oils developed using gene silencing. J Am Coll Nutr 2002; 21:205S-211S.
35. Smith SA, King RE, Min DB. Oxidative and thermal stabilities of genetically modified high oleic sunflower oil. Food Chem 2007; 102:1208-1213.
36. Tran E, Demmig-Adams B. Vitamins and minerals: Powerful medicine or potent toxins? Nutr Food Sci 2007; 37:50-60.

37. Ellinger S, Ellinger J, Stehle P. Tomatoes, tomato products and lycopene in the prevention and treatment of prostate cancer: do we have the evidence from intervention studies? Curr Opin Clin Nutr Metab Care 2006; 9:722-727.
38. Mares-Perlman JA, Millen AE, Ficek TL et al. The body of evidence to support a protective role for lutein and zeaxanthin in delaying chronic disease. Overview. J Nutr 2002; 132:518S-524S.
39. Seddon JM, Ajani UA, Sperduto RD et al. Dietary carotenoids, vitamin A, vitamin C and vitamin E and advanced age-related macular degeneration. J Am Med Assoc 1994; 272:1413-1420.
40. Chasan-Taber L Willett WC, Seddon JM et al. A prospective study of carotenoid and vitamin A intakes and risk of cataract extraction in US women. Am J Clin Nutr 1999; 70:509-516.
41. Brown L, Rimm EB, Seddon JM et al. A prospective study of carotenoid intake and risk of cataract extraction in US men. Am J Clin Nutr 1999; 70:517-524.
42. Demmig-Adams B, Adams WW III. The role of xanthophyll cycle carotenoids in the protection of photosynthesis. Trends Plant Sci 1996; 1:21-26.
43. Demmig-Adams B, Adams WW III. Photoprotection in an ecological context: the remarkable complexity of thermal dissipation. New Phytol 2006; 172:11-21. <doi: 10.1111/j.1469-8137.2006.01835.x>
44. Demmig-Adams B, Adams WW III. Antioxidants in photosynthesis and human nutrition. Science 2002; 298:2149-2153.
45. Külheim C, Agren J, Jansson S. Rapid regulation of light harvesting and plant fitness in the field. Science 2002; 297:91-93.
46. Holt NE, Zigmantas D, Valkunas L et al. Carotenoid cation formation and the regulation of photosynthetic light harvesting. Science 2005; 307:433-436.
47. Ahn TK, Avenson TJ, Ballottari M et al. Architecture of a charge-transfer state regulating light harvesting in a plant antenna protein. Science 2008; 320:794-797.
48. Pogson BJ, Niyogi KK, Björkman O et al. Altered xanthophyll compositions adversely affect chlorophyll accumulation and nonphotochemical quenching in Arabidopsis mutants. Proc Natl Acad Sci USA 1998; 95:13324-13329.
49. Havaux M, Niyogi KK. The violaxanthin cycle protects plants from photooxidative damage by more than one mechanism. Proc Natl Acad Sci USA 1999; 96:8762-8767.
50. Havaux M, Dall'Osto L, Cuine S et al. The effect of zeaxanthin as the only xanthophyll on the structure and function of the photosynthetic apparatus in Arabidopsis thaliana. J Biol Chem 2004; 279:13878-13888.
51. Sajilata MG, Singhal RS, Kamat MY. The carotenoid pigment zeaxanthin—A review. Compr Rev Food Sci Food Saf 2008; 7:29-49.
52. Thomson LR, Toyoda Y, Langner A et al. Elevated retinal zeaxanthin and prevention of light-induced photoreceptor cell death in quail. Investig Ophthalmol Vis Sci 2002; 43:3538-3549.
53. Thomson LR, Toyoda Y, Delori FC et al. Long term dietary supplementation with zeaxanthin reduces photoreceptor death in light-damaged Japanese quail. Exp Eye Res 2002; 75:529-542.
54. Sumatran VN, Zhang R, Lee DS et al. Differential regulation of apoptosis in normal versus transformed mammary epithelium by lutein and retinoic acid. Canc Epidemiol Biomarkers Prev 2000; 9:257-263.
55. Müller K, Carpenter KLH, Challis IR et al. Carotenoids induce apoptosis in the T-lymphoblast cell line Jurkat E6.1. Free Radic Res 2002; 36:791-802.
56. Chew BP, Brown CM, Park JS et al. Dietary lutein inhibits mouse mammary tumor growth by regulating angiogenesis and apoptosis. Anticancer Res 2003; 23:3333-3339.
57. Chitchumroonchokchai C, Bomser JA, Glamm JE et al. Xanthophylls and α-tocopherol decrease UVB-induced lipid peroxidation and stress signalling in human lens epithelial cells. J Nutr 2004; 134:3225-3232.
58. Wrona M, Korytowksi W, Ròzanowska M et al. Cooperation of antioxidants in protection against photosensitized oxidation. Free Radic Biol Med 2003; 35:1319-1329.
59. Wrona M, Ròzanowska M, Sarna T. Zeaxanthin in combination with ascorbic acid or alpha-tocopherol protects APRE-19 cells against photosensitized peroxidation of lipids. Free Radic Biol Med 2004; 36:1094-1101.
60. Youdim KA, Spencer JPE, Schroeter H et al. Dietary flavonoids as potential neuroprotectants. Biol Chem 2002; 383:503-519.
61. Surh YJ. Cancer chemoprevention with dietary phytochemicals. Nat Rev Canc 2003; 3:768-780.
62. Seo T, Blaner WS, Deckelbaum RJ. Omega-3 fatty acids: molecular approaches to optimal biological outcomes. Curr Opin Lipidol 2005; 16:11-18.
63. Maccarrone M, Bari M, Gasperi V et al. The photoreceptor protector zeaxanthin induces cell death in neuroblastoma cells. Anticancer Res 2005; 25:3871-3876.
64. Yamamoto HY. Biochemistry of the violaxanthin cycle in higher plants. Pure Appl Chem 1979; 51:639-648.

65. Demmig-Adams B. Linking the xanthophyll cycle with photoprotective energy dissipation. Photosynth Res 2003; 76:73-80.
66. Landrum JT, Bone RA. Lutein, zeaxanthin and the macular pigment. Arch Biochem Biophys 2001; 385:28-40.
67. Niyogi KK. Safety valves for photosynthesis. Curr Opin Plant Biol 2000; 3:455-460.
68. Dharmapuri S, Rosati C, Pallara P et al. Metabolic engineering of xanthophyll content in tomato fruits. FEBS Lett 2002; 519:30-34.
69. Romer S, Lubeck J, Kauder F et al. Genetic engineering of a zeaxanthin-rich potato by antisense inactivation and cosuppression of carotenoid epoxidation. Metab Eng 2002; 4:263-272.
70. Albrecht M, Misawa N, Sandmann G. Metabolic engineering of the terpenoid biosynthetic pathway of Escherichia coli for production of the carotenoids beta-carotene and zeaxanthin. Biotechnol Lett 1999; 21:791-795.
71. de Oliveira GPR, Rodriguez-Amaya DB. Processed and prepared corn products as sources of lutein and zeaxanthin: Compositional variation in the food chain. J Food Sci 2007; 72:S079-085.
72. Daicker B, Schiedt K, Adnet JJ et al. Canthaxanthin retinophathy—an investigation by light and electron-microscopy and physicochemical analysis. Graefes Arch Clin Exp Ophthalmol 1987; 225:189-197.
73. Stewart G. Investigating the effect of diet on nutrient concentration in eggs: How your breakfast might be healthier than you think. Inquiry (The University of New Hampshire) 2007: <http://www.unh.edu/inquiryjournal/07/articles/stewart.html>
74. Wang YM, Conner SL, Wang W et al. The selective retention of lutein, meso-zeaxanthin and zeaxanthin in the retina of chicks fed a xanthophyll-free diet. Exp Eye Res 2007; 84:591-598.
75. McGraw KJ, Beebee MD, Hill GE et al. Lutein-based plumage coloration in songbirds is a consequence of selective pigment incorporation into feathers. Comp Biochem Physiol B Biochem Mol Biol 2003; 135:689-696.
76. Moller AP, Biard C, Blount JD et al. Carotenoid-dependent signals: Indicators of foraging efficiency, immunocompetence or detoxification ability? Avian Poultry Biol Rev 2000; 11:137-159.
77. Kim HW, Chew BP, Wong TS et al. Dietary lutein stimulates immune response in the canine. Vet Immunol Immunopathol 2000; 74:315-327.
78. Kim HW, Chew BP, Wong TS et al. Modulation of humoral and cell-mediated immune responses by dietary lutein in cats. Vet Immunol Immunopathol 2000; 74:331-341.

Chapter 3

Therapeutic Potential of Dietary Polyphenols against Brain Ageing and Neurodegenerative Disorders

Giovanni Scapagnini,* Calogero Caruso and Vittorio Calabrese

Abstract

In recent years there has been a growing interest, supported by a large number of experimental and epidemiological studies, in the beneficial effects of some commonly used food-derived products in preventing various age-related pathologic conditions, ranging from cancer to neurodegenerative diseases. Spices and herbs often contain active phenolic substances endowed with potent antioxidative and chemopreventive properties. Curcumin is a phytochemical compound extracted from the rhizome of *Curcuma Longa*. It is the pigment responsible for the characteristic yellow color of Indian curry. Data from our and other laboratories demonstrated that curcumin, as well as some other polyphenols, strongly induce heme oxygenase 1 and Phase II detoxification enzymes in neurons and, by this activation, protect neurons against different modes of oxidative challenge. The potential role of curcumin as a preventive agent against brain aging and neurodegenerative disorders has been recently reinforced by epidemiological studies showing that in India, where this spice is widely used in the daily diet, there is a lower incidence of Alzheimer's disease than in the USA. These studies identify a novel class of compounds that could be used for therapeutic purposes as preventive agents against the acute neurodegenerative conditions that affect many in the world's increasingly ageing population.

Introduction

Nowadays, people live much longer than they used to do, however they are not free of disabilities and diseases, which still represent the dark side of aging and longevity. Age related diseases, such as cardiovascular diseases, osteoporosis, cancer and neurodegenerative disorders, represent the major cause of morbidity and mortality in Western countries. Alzheimer's disease (AD) is the fifth leading cause of death in Americans aged 65 and older, but whereas other major causes of death have been on the decrease, deaths attributable to AD have been rising dramatically in the last decade (47% increase between 2000 and 2006).[1] With the increasingly aging population of the United States, the number of AD patients is predicted to reach 14 million in 2050, with an expected incidence of close to a million people per year. Similar considerations apply worldwide.[1,2] This situation will pose immense economic and personal burdens on current and future generations, creating a critical emergency for effective therapeutic and preventive interventions. Neurodegeneration in AD appears to be multifactorial, whereby several biochemical processes operate sequentially and/or in parallel.

*Corresponding Author: Giovanni Scapagnini—Department of Health Sciences, University of Molise, Campobasso, Italy. Email: g.scapagnini@gmail.com

Bio-Farms for Nutraceuticals: Functional Food and Safety Control by Biosensors
edited by Maria Teresa Giardi, Giuseppina Rea and Bruno Berra.

Neuro-pathological hallmarks are senile plaques, resulting from the accumulation of several proteins and an inflammatory reaction around deposits of amyloid, a fibrillar protein, Abeta, a product of the cleavage of a much larger protein, the beta-amyloid precursor protein (APP) and neurofibrillary tangles. Amyloid deposition, due to the accumulation of Abeta peptide, is the main pathogenetic mechanism. AD ethiogenesis and progression has been linked to inflammation and oxidative stress.[3] Free radicals are produced from a number of sources, among which are enzymatic, mitochondrial and redox metal ion-derived sources.[4] On the other hand, irrespective of the source and mechanisms that lead to the generation of intracellular toxic oxidants, mammalian cells have developed highly refined inducible systems to counteract stressful conditions.[5] When properly activated, each one of these cell systems has the possibility to restore cellular homeostasis and resume the ability to fight off oxidation. Activation of antioxidant pathways is particularly important for tissue with relatively weak antioxidant defences, such as the brain. Increasing evidence points to the notion that reduced cellular expression and activity of antioxidant proteins and the consequent oxidative stress are fundamental causes for brain aging processes and neurodegenerative diseases.[6,7] In fact, compared to other organs, the brain is more susceptible to oxidative stress for the following reasons: (a) its high content of peroxidizable unsaturated fatty acids; (b) high oxygen consumption per unit weight; (c) high content of lipid peroxidation key ingredients (iron and ascorbate); and (d) the scarcity of antioxidant defense systems. Aging, the major risk factor for AD, leads to loss of free radical scavenging ability by endogenous mechanisms and it is widely held that free radical-induced oxidative stress increases in brain aging.[8,9] Oxidative damage to key intracellular targets such as DNA or proteins by free radicals has been shown to be a major cause of the neuronal cell death related to AD. Since oxidative stress may underlie some, if not all, aspects of AD neurodegeneration and since to date most of the available treatments are merely symptomatic,[10] considerable research has been aimed at reducing the effects of oxidative stress in order to prevent AD progress, by use of free radical scavengers.[11] Thus, one therapeutic strategy is to delay the onset of AD dementia sufficiently long as to slow the neuronal damage associated with Aβ-induced oxidative stress, particularly Aβ-induced lipid peroxidation. Brain-accessible antioxidants potentially may provide the means of implementing this therapeutic strategy of delaying the onset of AD, acting as neuro-protective agents. By definition, neuroprotection is an effect that may result in salvage, recovery or regeneration of the nervous system, its cells, structure and function. Although there are several lines of evidence supporting the hypothesis that neuroprotection may be a practical and achievable pharmacological target, few effective compounds have been developed for clinical application. To date the use of antiapoptotic therapies for neurodegenerative disorders has not been successful, particularly because many of the compounds have high levels of toxicity. Moreover interfering with apoptotic pathways has often resulted in an augmented risk of cancerogenesis. On the other side the use of safe antioxidant compounds, such as α-tocopherol, in the treatment of neurodegenerative disorders is strongly limited by the difficulty to reach an active concentration of these molecules in the brain.[12] Thus, in the field of the pharmacological treatment of neurodegenerative disorders, there remains the need and desire for safe, nontoxic and orally effective pharmacological agents. In recent years there has been a growing interest, supported by a large number of experimental and epidemiological studies, in the beneficial effects of some commonly used food-derived products in preventing various age-related pathologic conditions, included brain ageing.[13] Spices and herbs often contain active phenolic substances endowed with potent antioxidative and chemopreventive properties[14] and recently a series of papers focused on specific neuroprotective effects of some of those polyphenols derived from nutritional sources.[15,16]

In this chapter we will provide insight into the possible therapeutic significance of curcumin and a closely related group of polyphenols, in brain ageing and their possible utility for AD therapy.

Curcumin

Curcumin (1,7-bis[4-Hydroxy-3-methoxyphenyl]-1,6-heptadiene-3,5-dione), a coloring agent and food additive commonly used in Indian culinary and traditional medical preparations from time immemorial, is extracted from the rhizome of *Curcuma Longa*.[17] It is a polyphenolic substance that has the potential to inhibit lipid peroxidation and to effectively intercept and neutralize ROS (superoxide, peroxyl, hydroxyl radicals)[18,19] and NO-based free radicals (nitric oxide and peroxynitrite).[20] In this regard, curcumin has been demonstrated to be several times more potent than vitamin E.[21] It is generally assumed that the phenol moiety is responsible for the antioxidant properties of any plant phenolic compound. Consequently, the free radical chemistry of curcumin has focused on its phenol ring. The possible involvement of the β-diketone moiety in the antioxidant action of curcumin has been considered[22] and H-atom donation from the β-diketone moiety to a lipid alkyl or a lipid peroxyl radical has been reported as the potentially more important mechanism underlying its antioxidant action.[23] Of particular interest is the ability of curcumin to inhibit COX-1 and COX-2 enzymes[24] and to reduce the activation of nuclear transcription factor NF-κB.[25] Its anti-inflammatory properties and cancer-preventive activities have been consistently reported using in vitro and in vivo models of tumor initiation and promotion.[26] In addition to its ability to scavenge carcinogenic free radicals,[27] curcumin also interferes with cell growth through inhibition of protein kinases. Although the exact mechanisms by which curcumin promotes these effects remains to be elucidated, the electrophilic properties of this yellow pigment appear to be an essential component underlying its pleiotropic biological activities. Curcumin contains two electrophilic α,β-unsaturated carbonyl groups, which can react with nucleophiles such as glutathione.[28] By virtue of its Michael reaction acceptor function and its electrophilic characteristics, curcumin and other compounds derived from closely related plants have been recently demonstrated to induce the activities of the Phase I and Phase II detox system.[29,30] This finding is in agreement with other studies demonstrating that curcumin can increase the activity of γ-glutamyl-cysteinyl synthetase and other GSH-linked detoxifying enzymes.[31] Recent and unprecedented data from our group revealed that low concentrations of curcumin, potently induces heme oxygenase-1 (HO-1) expression and activity in vascular endothelial cells,[32] in rat astrocytes[33] and in cultured hippocampal neurons.[34] In our experiments we also demonstrated that preincubation (12 h) of cultured neurons with low concentration of curcumin resulted in an enhanced cellular resistance to glucose oxidase mediated oxidative damage; this cytoprotective effect was considerably attenuated by zinc protoporphyrin IX, a specific inhibitor of HO activity. In other experiments we demonstrated the efficacy of curcumin to protect cortical neurons against apoptotic cell death induced by β-amyloid peptide (1-40).[35] The ability of curcumin to induce HO-1 can explain, at least in part, its strong anti-oxidant and anti-inflammatory properties, which depend more on its ability to activate cellular signals than on its radical scavenger effect.[36]

In the last decade the stress proteins system has been strongly emphasized for its potential significance in maintaining cellular homeostasis. Among the molecules belonging to the stress protein family, HO-1 has been the object of intensive studies in the brain for its potential role in protecting neurons against cell death. HO-1 is a ubiquitous and redox-sensitive inducible stress protein and, together with the constitutive isozyme HO-2, provides the first and rate-limiting step in heme degradation.[37] HO cleaves the heme ring via oxidation at the alpha methene bridge to give biliverdin, gaseous carbon monoxide and free iron. Biliverdin is subsequently converted to bilirubin by biliverdin reductase and both these molecules can act as intracellular antioxidants. A substantial body of evidence demonstrates that both bilirubin and CO effectively contribute to modulate important physiological processes within the cardiovascular, immune and nervous systems. These include the regulation of vessel tone, inhibition of platelet aggregation and prevention of cell death and tissue injury. In the CNS, the HO system has been reported to be very active and its modulation seems to play a crucial role in the pathogenesis of neurodegenerative disorders. Deregulation of the HO system has been associated with the pathogenesis of Alzheimer's disease,[38] multiple sclerosis and brain aging.[39,40] Many studies clearly demonstrate that activation of HO-1 in neurons is strongly protective against oxidative damage and cell death.[41,42] In fact, the activation of HO-1 seems to represent an important defensive mechanism for neurons exposed to oxidative stress. Thus, modulation of HO-1 should represent a potential pharmaceutical strategy for the treatment of neurodegenerative disorders. Although several challenges are able to regulate HO-1 expression and activity, to date no safe and specific inducers have been developed. The activation of "classic" detoxifying enzymes and the induction of HO-1 by phenolic natural substances involves common transcription mechanisms that are sensitive to the distinctive chemistry of these compounds. The HO-1 gene has a heat shock consensus sequence as well as AP1, AP2 and NF-κB binding sites in its promoter region and it is rapidly up-regulated by oxidative and nitrosative stresses as well as by many toxic compounds. It has been recently demonstrated that HO-1 response is also mediated by a cis-acting element, the antioxidant response element (ARE, also called EpRE or OSRE). The ARE is also found in other cytoprotective proteins related to the so-called cellular stress response, such as the gluthatione-S-transferases, ferritin, gamma glutamyl cysteine synthetase, NAD(P)H quinone oxidoreductase and others. Curcumin has been demonstrated to stimulate the mitogen-activated protein kinase pathway and to activate *heterodimers of NF-E2-related factors 2* (Nrf2), leading to induction of the antioxidant responsive element activated reporter genes.[43] Nrf2, a member of the Cap'n'Collar family of transcription factors, is sequestered in the cytoplasm by binding to protein Keap1 in nonstimulated conditions. However, several stimuli, including oxidative stress, lead to the disruption of this complex, freeing Nrf2 for translocation to the nucleus and dimerization with basic leucine zipper transcription factors such as Maf and Jun family members. A corollary of our studies is that, by these pathways (ARE/Nrf2) curcumin and other similar polyphenols strongly induce the expression of cellular stress response genes (phase II detoxification enzymes and HO-1), resulting in cell protection and enhancing cell survival.[34]

The involvement of curcumin in restoring cellular homeostasis and rebalancing redox equilibrium by the activation of defensive genes, suggests that it might be a useful adjunct also in AD treatment. Neuroprotective effects of curcumin have been demonstrated by Rajakrishnan[44] in ethanol-induced brain damage, in which oral administration of curcumin to rats caused a significant reversal in lipid peroxidation, brain lipid modifications, as well as increase in glutathione levels. Epidemiological studies suggested that curcumin, as one of the most prevalent nutritional and medicinal compounds used by the Indian population, is responsible for the significantly reduced (4.4-fold) prevalence of AD in India compared to United States.[45] Furthermore, elderly Singaporeans who ate curry with turmeric had higher MMSE scores than those who did not.[46] However, the relatively short duration of follow-up, cultural factors and other potential confounders suggest caution in interpreting these findings. Consistent with these data, Lim and colleagues have provided convincing evidence that dietary curcumin, given to an Alzheimer's transgenic mouse model (Tg2576) for 6 months, resulted in a suppression of indices of inflammation and oxidative damage in the brain of these mice and to reverse Aβ-induced cognitive deficits.[47,48] The same group demonstrated in a continuation of the research that curcumin was a better Aβ40 aggregation inhibitor than ibuprofen and naproxen and prevented Aβ42 oligomer formation and toxicity at very low concentration (between 0.1 and 1.0 microM).[49] They also showed that curcumin readily entered the brain to label plaques in vivo inhibiting the formation of Aβ oligomers and their toxicity. Among the several mechanisms by which curcumin is able to clear amyloid is the induction of HSPs, that function as molecular chaperones to block protein aggregate formation.[50] Recently curcumin has been evaluated in a pilot clinical trial in AD patients—preliminary results were encouraging.[51] Curcumin is highly lipophilic and might cross the BBB and reach the brain and although its bioavailability is very low, since the drug is rapidly metabolized by conjugation, curcumin may reach brain concentrations sufficient to activate signal transduction events and to decrease Aβ aggregation.[52]

Other plant-derived phenolic agents with analogous chemical structures to curcumin have been demonstrated to strongly activate HO-1 expression and to defend cells against oxidative stress. In particular carnosol,[53] zerumbone,[54] resveratrol,[55] rosolic acid[56] and sulphoraphanes.[57] Furthermore in our labs we have demonstrated that other phenolics, such as caffeic acid phenethyl ester (CAPE), ethyl ferulate (EFE) and epigallocatechin-3-gallate (EGCG), are able to protect neurons via HO-1 induction.

Caffeic Acid Phenethyl Ester

Caffeic acid phenethyl ester is a phenolic structurally related to curcumin, originating from plants. CAPE is, in fact, an active component of propolis derived from the bark of conifer trees and carried by honeybees to their hives.

The similarity to curcumin is striking because CAPE is also a Michael reaction acceptor that has a broad spectrum of biological activities, including anti-inflammatory,[58,59] antioxidant[60] and anti-cancer[61,62] effects. We have reported that CAPE is a potent inducer of HO-1 in astroglial cells and in neurons. CAPE is capable of transcriptionally activating a gene battery that also includes NAD(P)H:quinone oxidoreductase, aldo-keto reductases, glutathione S-transferases, γ-glutamylcysteine synthetase and glutathione synthetase, through the ARE/Nrf2 pathway. Cape demonstrated a neuroprotective capacity comparable to curcumin.[33,35]

Ethyl Ferulate

Ferulic acid, which is the precursor of lignin biosynthesis, has long been recognized for its antioxidative and anti-inflammatory activities.[63] Cinnamic acid derivatives, including ferulic acid, are abundant in plants, playing important roles in the cross-linking of the cell walls of various grasses. Ferulic acid, is found in many fruits and vegetables such as the tomato.[64] Tomato consumption has been demonstrated to result in absorption and excretion of ferulic acid by humans.[65] Ferulic acid has been demonstrated to have antioxidant activity against peroxynitrite[66] and against lipid peroxidation.[67,68] Recently, we demonstrated that ferulic acid was protective against protein oxidation and lipid peroxidation in synaptosomal membranes and against cell death and protein carbonyl formation in neuronal cell culture induced by the peroxyl radical initiator AAPH.[69] Although it has been demonstrated to be effective in in vitro experiments, its low lipophilicity impairs its in vivo efficiency, bioavailability and stability. Ethyl ferulate, the naturally occurring ester derivative of ferulic acid, is present in various systems of many plants, such as the Solanaceae family, as a trace constituent.[70,71] In comparison with the corresponding acid form, EFE is more lipophilic and has been shown to possess better scavenging properties toward both hydroxyl radicals and superoxide anions. In a recent study of the inhibitory effects of antioxidants on lipid oxidation, EFE was shown to prevent autooxidation of model substrates by extending the induction time of this process.[72] Data from our lab clearly indicate that EFE is able, at low concentrations, to induce HO-1 protein expression and activity in cell lines of rat astrocytes and hippocampal neurons.[73] In our study, we also demonstrated the cytoprotective effects of EFE against oxidative damage in neuronal cells. Accordingly, other studies[74] have shown that EFE exerts strong neuroprotection against the oxidative stress and neurotoxicity induced by amyloid β1–42.

Epigallocatechin-3-Gallate

Green tea, one of the most widely consumed beverages, has recently attracted scientific attention as a potential nutritional strategy to prevent a broad range of age related chronic disorders, including cardiovascular diseases,[75,76] cancer[77], obesity,[78] diabetes[79] and neurodegenerative pathologies.[80] Moreover, a number of epidemiological studies have suggested that consuming green tea on a daily basis, as part of a lifestyle, might reduce the onset of all cause mortality and improve longevity.[81] The health-promoting effects of green tea consumption are mainly attributed to its polyphenol content, which represents 35% of the dry weight.[82] Compared to black tea, green tea is particularly rich in catechins, that include: (-)- epigallocatechin-3-gallate (EGCG?), (-)-epicatechin-3-gallate (ECG), (-)-epigallocatechin (EGC) and epicatechin (EC). EGCG is the most active and abundant compound in green tea, representing approximately 43% of the total phenols. Many of the aforementioned beneficial effects of green tea on age related diseases have been linked to its EGCG content. EGCG possesses antioxidant and anti-inflammatory properties which include the capacity to inhibit over-expression of cyclooxygenase-2[83] and nitric oxide synthase.[84] It also induces apoptosis in several types of cancer cells by inactivating some transcription factors, such as NF-κB,[85] AP-1[86] and STAT-1.[87] EGCG prevents cancer cell invasion, angiogenesis and metastasis by down-regulating the expression of matrix metalloproteinases and by inhibiting the cell adhesion function.[88-90] Several reports have also shown the ability of EGCG to induce a general xenobiotic response in the target cells, activating multiple defence genes. In particular, some studies have recently demonstrated that the neuroprotective mechanisms of EGCG are partly due to increasing activities of antioxidant enzymes and decreasing advanced glycation end products (AGE)-induced damage in rat brain or neuronal cells.[91,92] In a recent study, we attempted to determine the molecular mechanisms underlying the antioxidant effects of EGCG in neurons exposed to oxidative stress, by focusing on the ability of this compound to up-regulate HO-1 expression and other adaptive enzymes involved in cellular stress response. In accordance with other studies, [93] we demonstrated the ability of EGCG to activate HO-1 by the ARE/Nrf2 pathway and by the induction of HO-1 to protect neurons against oxidative damage (unpublished data).

Conclusion

The majority of in vitro and in vivo studies conducted so far have attributed the protective effect of bioactive polyphenols to their chemical reactivity toward free radicals and their capacity to prevent the oxidation of important intracellular components. However, observations from our and other laboratories, reveal a potential novel aspect in the mode of action of curcumin and curcumin/like compounds; that is, the ultimate stimulation of the HO-1 pathway is likely to account for the established and powerful antioxidant/anti-inflammatory properties of these plant-derived compounds. Because the HO-1 gene can be stimulated at transcriptional levels by a plethora of noxious stimuli, the use of plant-derived natural substances to trigger HO-1 expression and other intracellular defensive systems would clearly offer a greater advantage for therapeutic and preventive purposes. It needs to be emphasized that curcumin and other plant constituents eventually become part of the human diet and can be consumed daily as supplements.

Our studies identify a novel class of natural substances that could be used for therapeutic purposes as potent inducers of HO-1 in the protection of tissues against inflammatory and neurodegenerative conditions. Further in vitro and in vivo studies using curcumin-like molecules will give important information on the feasibility of developing new pharmacological strategies for maximizing heme oxygenase activity in targeted tissues as an alternative to or in combination with HO-1 gene therapy.

Acknowledgments

This work was supported, in part, by grants from 'Paolo Sorbini Foundation for Nutrition Science'.

References

1. Alzheimer's Association. 2009 Alzheimer's disease facts and figures. Alzheimers Dement 2009; 5:234-270.
2. Katzman M. The aging brain. Limitations in our knowledge and future approaches. Arch Neurol 1997; 54:1201-1205.
3. Butterfield DA, Drake J, Pocernich C et al. Evidence of oxidative damage in Alzheimer's disease brain: central role for amyloid beta-peptide. Trends Mol Med 2001; 7:548-554.
4. Butterfield DA, Stadtman ER. Protein oxidation processes in aging brain. Adv Cell Aging Gerontol 1997; 2:161-191.
5. Calabrese V, Boyd-Kimball D, Scapagnini G et al. Nitric oxide and cellular stress response in brain aging and neurodegenerative disorders: the role of vitagenes. In Vivo 2004; 18:245-267.
6. Halliwell B. Biochemistry of oxidative stress. Biochem Soc Trans 2007; 35:1147-1150.
7. Finkel T, Holbrook NJ. Oxidants, oxidative stress and the biology of aging. Nature 2000; 408:239-247.
8. Katzman R, Saitoh T. Advances in Alzheimer's disease. FASEB J 1991; 5:278-286.
9. Butterfield DA, Howard BJ, Yatin S et al. Free radical oxidation of brain proteins in accelerated senescence and its modulation by N-tert-butyl-alpha-phenylnitrone. Proc Natl Acad Sci USA 1997; 94:674-678.
10. Skovronsky DM, Lee VM, Trojanowski JQ. Neurodegenerative diseases: new concepts of pathogenesis and their therapeutic implications. Annu Rev Pathol 2006; 1:151-170.
11. Racchi M, Uberti D, Govoni S et al. Alzheimer's disease; new diagnostic and therapeutic tools. Immun Ageing 2008; 5:7.
12. Vatassery GT, Fahn S, Kuskowski MA. The Parkinson Study Group. Alpha tocopherol in CSF of subjects taking high-dose vitamin E in the DATATOP study. Neurology 1998; 50:1900-1902.

13. Gómez-Pinilla F. Brain foods: the effects of nutrients on brain function. Nat Rev Neurosci 2008; 9:568-578.
14. Nakatani N. Phenolic antioxidants from herbs and spices. Biofactors 2000; 13:141-146.
15. Butterfield D, Castegna A, Pocernich C et al. Nutritional approaches to combat oxidative stress in Alzheimer's disease. J Nutr Biochem 2002; 13:444-461.
16. Sun AY, Wang Q, Simonyi A et al. Botanical phenolics and brain health. Neuromolecular Med 2008; 10:259-274.
17. Ammon HPT, Wahl MA. Pharmacology of Curcuma Longa. Planta Med 1991; 57:1-7.
18. Priyadarsini KI, Guha SN, Rao MN. Physicochemical properties and antioxidant activities of methoxy phenols. Free Radic Biol Med 1998; 24:933-941.
19. Martin-Aragon S, Benedi JM, Villar AM. Modifications on antioxidant capacity and lipid peroxidation in mice under fraxetin treatment. J Pharm Pharmacol 1997; 49:49-52.
20. Sreejayan A, Rao MN. Nitric oxide scavenging by curcuminoids. J Pharm Pharmacol 1997; 49:105-107.
21. Zhao BL, Li XJ, He RG et al. Scavenging effect of extracts of green tea and natural antioxidants on active oxygen radicals. Cell Biophys 1989; 14:175-185.
22. Masuda T, Hidaka K, Shinohara A et al. Chemical studies on antioxidant mechanism of curcuminoid: analysis of radical reaction products from curcumin. J Agric Food Chem 1999; 47:71-77.
23. Jovanovic SV, Boone CW, Steenken S et al. How curcumin works preferentially with soluble antioxidants. J Am Chem Soc 2001; 123:3064-3068.
24. Ramos-Gomez M, Kwak MK, Dolan PM et al. Sensitivity to carcinogenesis is increased and chemoprotective efficacy of enzyme inducers is lost in nrf2 transcription factor-deficient mice. Proc Natl Acad Sci USA 2001; 98:3410-3415.
25. Singh S, Aggarwal BB. Activation of transcription factor NF-κB is suppressed by curcumin (diferuloylmethane). J Biol Chem 1995; 270:24995-25000.
26. Huang MT, Newmark HL, Frenkel K. Inhibitory effects of curcumin on tumorigenesis in mice. J Cell Biochem Suppl 1997; 27:26-34.
27. Abe Y, Hashimoto S, Horie T. Curcumin inhibition of inflammatory cytokine production by human peripheral blood monocytes and alveolar macrophages. Pharmacol Res 1999; 39:41-47.
28. Awasthi S, Pandya U, Singhal SS et al. Curcumin-glutathione interactions and the role of human glutathione S-transferase P1-1. Chem Biol Interact 2000; 128:19-38.
29. Dinkova-Kostova AT, Talalay P. Relation of structure of curcumin analogs to their potencies as inducers of Phase 2 detoxification enzymes. Carcinogenesis 1999; 20:911-914.
30. Dinkova-Kostova AT, Massiah MA, Bozak RE et al. Potency of Michael reaction acceptors as inducers of enzymes that protect against carcinogenesis depends on their reactivity with sulfhydryl groups. Proc Natl Acad Sci USA 2001; 98:3404-3409.
31. Singhal SS, Awasthi S, Pandya U et al. The effect of curcumin on glutathione-linked enzymes in K562 human leukemia cells. Toxicol Lett 1999; 109:87-95.
32. Motterlini R, Foresti R, Bassi R et al. Curcumin, an antioxidant and anti-inflammatory agent, induces heme oxygenase-1 and protects endothelial cells against oxidative stress. Free Radic Biol Med 2000; 28:1303-1312.
33. Scapagnini G, Foresti R, Calabrese V et al. Caffeic acid phenethyl ester and curcumin: a novel class of heme oxygenase-1 inducers. Mol Pharmacol 2002; 61:554-561.
34. Scapagnini G, Colombrita C, Amadio M et al. Curcumin activates defensive genes and protects neurons against oxidative stress. Antioxid Redox Signal 2006; 8:395-403.
35. Scapagnini G, Calabrese V, Motterlini R et al. Use of curcumin derivatives or CAPE in the manufacture of a medicament for the treatment of neuroprotective disorders. World Patent Number: WO 2004/075883 A1, Publication date: 2004-09-10.
36. Goel A, Kunnumakkara AB, Aggarwal BB. Curcumin as "Curecumin": from kitchen to clinic. Biochem Pharmacol 2008; 75:787-809.
37. Calabrese V, Signorile A, Cornelius C et al. Practical approaches to investigate redox regulation of heat shock protein expression and intracellular glutathione redox state. Methods Enzymol 2008; 441:83-110.
38. Takahashi M, Dore S, Ferris CD et al. Amyloid precursor proteins inhibit heme oxygenase activity and augment neurotoxicity in Alzheimer's disease. Neuron 2002; 28:461-473.
39. Schipper HM. Heme oxygenase-1: role in brain aging and neurodegeneration. Exp Gerontol 2000; 35:821-830.
40. Colombrita C, Calabrese V, Stella AM et al. Regional rat brain distribution of heme oxygenase-1 and manganese superoxide dismutase mRNA: relevance of redox homeostasis in the aging processes. Exp Biol Med 2003; 228:517-24.
41. Chen K, Gunter K, Maines MD. Neurons overexpressing heme oxygenase-1 resist oxidative stress-mediated cell death. J Neurochem 2000; 75:304-312.

42. Le WD, Xie WJ, Appel SH. Protective role of heme oxygenase-1 in oxidative stress-induced neuronal injury. J Neurosci Res 1999; 56:652-658.
43. Balogun E, Hoque M, Gong P et al. Curcumin activates the haem oxygenase-1 gene via regulation of Nrf2 and the antioxidant-responsive element. Biochem J 2003; 371:887-895
44. Rajakrishnan V, Viswanathan P, Rajasekharan KN et al. Neuroprotective role of curcumin from curcuma longa on ethanol induced brain damage. Phytother Res 1999; 13:571-574.
45. Chandra V, Pandav R, Dodge HH et al. Incidence of Alzheimer's disease in a rural community in India: the Indo-US study. Neurology 2001; 57:985-989.
46. Ng TP, Chiam PC, Lee T et al. Curry consumption and cognitive function in the elderly. Am J Epidemiol 2006; 164:898-906.
47. Lim GP, Chu T, Yang F et al. The curry spice curcumin reduces oxidative damage and amyloid pathology in an Alzheimer transgenic mouse. J Neurosci 2001; 21:8370-8377.
48. Frautschy SA, Hu W, Miller SA et al. Phenolic anti-inflammatory antioxidant reversal of Aβ-induced cognitive deficits and neuropathology. Neurobiol Aging 2001; 22:993-1005.
49. Yang F, Lim GP, Begum AN et al. Curcumin inhibits formation of amyloid beta oligomers and fibrils, binds plaques and reduces amyloid in vivo. J Biol Chem 2005; 280:5892-5901.
50. Cole GM, Teter B, Frautschy SA. Neuroprotective effects of curcumin. Adv Exp Med Biol 2007; 595:197-212.
51. Baum L, Lam CW, Cheung SK et al. Six-month randomized, placebo-controlled, double-blind, pilot clinical trial of curcumin in patients with Alzheimer disease. J Clin Psychopharmacol 2008; 28:110-113.
52. Begum AN, Jones MR, Lim GP et al. Curcumin structure-function, bioavailability and efficacy in models of neuroinflammation and Alzheimer's disease. J Pharmacol Exp Ther 2008; 326:196-208.
53. Martin D, Rojo AI, Salinas M et al. Regulation of heme oxygenase-1 expression through the phosphatidylinositol 3-kinase/Akt pathway and the Nrf2 transcription factor in response to the antioxidant phytochemical carnosol. J Biol Chem 2004; 279: 8919-8929.
54. Surh YJ, Kundu JK, Na HK. Nrf2 as a master redox switch in turning on the cellular signaling involved in the induction of cytoprotective genes by some chemopreventive phytochemicals. Planta Med 2008; 74(13):1526-39.
55. Zhuang H, Kim YS, Koehler RC et al. Potential mechanism by which resveratrol, a red wine constituent, protects neurons. Ann N Y Acad Sci 2003; 993:276-286.
56. Foresti R, Hoque M, Monti D et al. Differential activation of heme oxygenase-1 by chalcones and rosolic acid in endothelial cells. J Pharmacol Exp Ther 2005; 312:686-693.
57. Talalay P, Fahey JW, Healy ZR et al. Sulforaphane mobilizes cellular defenses that protect skin against damage by UV radiation. Proc Natl Acad Sci USA 2007; 104:17500-5.
58. Michaluart P, Masferrer JL, Carothers AM et al. Inhibitory effects of caffeic acid phenethyl ester on the activity and expression of cyclooxygenase-2 in human oral epithelial cells and in a rat model of inflammation. Cancer Res 1999; 59:2347-2352.
59. Natarajan K, Singh S, Burke TR Jr et al. Caffeic acid phenethyl ester is a potent and specific inhibitor of activation of nuclear transcription factor NF-kappa B. Proc Natl Acad Sci USA 1996; 93:9090-9095.
60. Chen YJ, Shiao MS, Wang SY. The antioxidant caffeic acid phenethyl ester induces apoptosis associated with selective scavenging of hydrogen peroxide in human leukemic HL-60 cells. Anticancer Drugs 2001; 12:143-149.
61. Huang MT, Ma W, Yen P et al. Inhibitory effects of caffeic acid phenethyl ester (CAPE) on 12-O-tetradecanoylphorbol-13-acetate-induced tumor promotion in mouse skin and the synthesis of DNA, RNA and protein in HeLa cells. Carcinogenesis 1996; 17:761-765.
62. Frenkel K, Wei H, Bhimani R et al. Inhibition of tumor promoter-mediated processes in mouse skin and bovine lens by caffeic acid phenethyl ester. Cancer Res 1993; 53:1255-1261.
63. Graf E. Antioxidant potential of ferulic acid. Free Rad Biol Med 1992; 13:435-448.
64. Qureshi MJ, Blain JA. Antioxidant activity in tomato extracts. Nucleus (Karachi) 1976; 13:29-33.
65. Bourne LC, Rice-Evans C. Biovailability of ferulic acid. Biochem Biophys Res Comm 1998; 253:222-227.
66. Pannala R, Razaq B, Halliwell S et al. Inhibition of peroxynitrite dependent tyrosine nitration by hydroxycinnamates: nitration or electron donation. Free Rad Biol Med 1998; 24:594-606.
67. Castelluccio C, Paganga G, Melikian N et al. Antioxidant potential of intermediates in phenylpropanoid metabolism in higher plants. FEBS Lett 1995; 368:188-192.
68. Bourne L, Rice-Evans C. The effect of the phenolic antioxidant ferulic acid on the oxidation of low density lipoprotein depends on the pro-oxidant used. Free Rad Res 1997; 27:337-344.
69. Kanski J, Aksenova M, Stoyanova A et al. Ferulic acid antioxidant protection against hydroxyl and peroxyl radical oxidation in synaptosomal and neuronal cell culture systems in vitro: structure—activity studies. J Nutr Biochem 2002; 13:273-281.

70. Clifford MN. Chlorogenic acids and other cinnamates—nature, occurrence and dietary burden. J Sci Food Agric 1999; 79:362-372.
71. Kroon PA, Williamson G. Hydroxycinnamates in plants and food: current and future perspectives. J Sci Food Agric 1999; 79:355-361.
72. Kikuzaki H, Hisamoto M, Hirose K et al. Antioxidant properties of ferulic acid and its related compounds. J Agric Food Chem 2002; 50:2161-2168.
73. Scapagnini G, Butterfield DA, Colombrita C et al. Ethyl ferulate, a lipophilic polyphenol, induces HO-1 and protects rat neurons against oxidative stress. Antioxid Redox Signal 2004; 6:811-818.
74. Perluigi M, Joshi G, Sultana R et al. In vivo protective effects of ferulic acid ethyl ester against amyloid-beta peptide 1-42-induced oxidative stress. J Neurosci Res 2009; 84:418-26.
75. Sano J, Inami S, Seimiya K et al. Effects of green tea intake on the development of coronary artery disease. Circ J 2004; 68:665-670.
76. Wolfram S. Effects of green tea and EGCG on cardiovascular and metabolic health. J Am Coll Nutr 2007; 26:373-388.
77. Moyers SB, Kumar NB. Green tea polyphenols and cancer chemoprevention: multiple mechanisms and endpoints for phase II trials. Nutr Rev 2004; 62:204-211.
78. Boschmann M, Thielecke F. The effects of epigallocatechin-3-gallate on thermogenesis and fat oxidation in obese men: a pilot study. J Am Coll Nutr 2007; 26:389-395.
79. Potenza MA, Marasciulo FL, Tarquinio M et al. Epigallocatechin gallate, a green tea polyphenol, improves endothelial function and insulin sensitivity, reduces bood pressure and protects against myocardial ischemia/reperfusion injury in spontaneously hypertensive rats (SHR). Am J Physiol Endocrinol Metab 2007; 292:1378-1387.
80. Mandel S, Weinreb O, Amit T et al. Cell signaling pathways in the neuroprotective actions of the green tea polyphenol (-)-epigallocatechin-3-gallate: implications for neurodegenerative diseases. J Neurochem 2004; 88:1555-1569.
81. Khan N, Mukhtar H. Tea polyphenols for health promotion. Life Sci 2007; 81:519-533.
82. Yang CS, Landau JM. Effects of tea consumption on nutrition and health. J Nutr 2000; 130:2409-2412.
83. Hussain T, Gupta S, Adhami VM et al. Green tea constituent epigallocatechin-3-gallate selectively inhibits COX-2 without affecting COX-1 expression in human prostate carcinoma cells. Int J Cancer 2005; 113:660-669.
84. Ahmed S, Rahman A, Hasnain A et al. Green tea polyphenol epigallocatechin-3-gallate inhibits the IL-1 beta-induced activity and expression of cyclooxygenase-2 and nitric oxide synthase-2 in human chondrocytes. Free Radic Biol Med 2002; 33:1097-1105.
85. Kim SJ, Jeong HJ, Lee KM et al. Epigallocatechin-3-gallate suppresses NF-kappaB activation and phosphorylation of p38 MAPK and JNK in human astrocytoma U373MG cells. J Nutr Biochem 2007; 18:587-596.
86. Dong Z, Ma W, Huang C et al. Inhibition of tumor promoter-induced activator protein 1 activation and cell transformation by tea polyphenols, (-)-epigallocatechin gallate and theaflavins. Cancer Res 1997; 57:4414-4419.
87. Townsend PA, Scarabelli TM, Pasini E et al. Epigallocatechin-3-gallate inhibits STAT-1 activation and protects cardiac myocytes from ischemia/reperfusion-induced apoptosis. FASEB J 2004; 18:1621-1623.
88. Khan N, Mukhtar H. Multitargeted therapy of cancer by green tea polyphenols. Cancer Lett 2008; 269:269-280.
89. Sartippour MR, Shao ZM, Heber D et al. Green tea inhibits vascular endothelial growth factor (VEGF) induction in human breast cancer cells. J Nutr 2002; 132:2307-2311.
90. Ahmed S, Wang N, Lalonde M et al. Green tea polyphenol epigallocatechin-3-gallate (EGCG) differentially inhibits interleukin-1 beta-induced expression of matrix metalloproteinase-1 and -13 in human chondrocytes. J Pharmacol Exp Ther 2004; 308:767-773.
91. Srividhya R, Jyothilakshmi V, Arulmathi K et al. Attenuation of senescence-induced oxidative exacerbations in aged rat brain by (-)-epigallocatechin-3-gallate. Int J Dev Neurosci 2008; 26:217-223.
92. Lee SJ, Lee KW. Protective effect of (-)-epigallocatechin gallate against advanced glycation endproducts-induced injury in neuronal cells. Biol Pharm Bull 2007; 30:1369-1373.
93. Kweon MH, Adhami VM, Lee JS et al. Constitutive overexpression of Nrf2-dependent heme oxygenase-1 in A549 cells contributes to resistance to apoptosis induced by epigallocatechin 3-gallate. J Biol Chem 2006; 281:33761-33772.

CHAPTER 4

Plant Phenolics in the Prevention and Treatment of Cancer

Klaus W.J. Wahle,* Iain Brown, Dino Rotondo, and Steven D. Heys

Abstract

Epidemiological studies indicate that populations consuming high levels of plant derived foods have low incidence rates of various cancers. Recent findings implicate a variety of phytochemicals, including phenolics, in these anticancer properties. Both monophenolic and polyphenolic compounds from a large variety of plant foods, spices and beverages have been shown to inhibit or attenuate the initiation, progression and spread of cancers in cells in vitro and in animals in vivo. The cellular mechanisms that phenolics modulate to elicit these anticancer effects are multi-faceted and include regulation of growth factor-receptor interactions and cell signaling cascades, including kinases and transcription factors, that determine the expression of genes involved in cell cycle arrest, cell survival and apoptosis or programmed cell death. A major focus has been the inhibitory effects of phenolics on the stress-activated NF-κB and AP-1 signal cascades in cancer cells which are regarded as major therapeutic targets. Phenolics can enhance the body's immune system to recognize and destroy cancer cells as well as inhibiting the development of new blood vessels (angiogenesis) that is necessary for tumour growth. They also attenuate adhesiveness and invasiveness of cancer cells thereby reducing their metastatic potential.

Augmentation of the efficacy of standard chemo- and radiotherapeutic treatment regimes and the prevention of resistance to these agents is another important effect of plant phenolics that warrants further research.

Plant phenolics appear to have both preventative and treatment potential in combating cancer and warrant further, in-depth research. It is interesting that these effects of plant phenolics on cancer inhibition resemble effects reported for specific fatty acids (omega-3 PUFA, conjugated linoleic acids).

Although phenolic effects in cells in vitro and in animal models are generally positive, observations from the less numerous human interventions are less clear. This is surprising given the positive epidemiological data and may relate to mixed diets and synergistic interactions between compounds or the bioavailability of individual compounds. Much of the work in vitro with phenolic compounds has utilized concentrations higher than the amount that can be obtained from the diet suggesting a role of fortified, functional foods in cancer suppression.

Introduction: Epidemiology of Plant Foods and Disease Incidence

The role of nutrition in the prevention of disease has been recognised for centuries. Hippocrates, some 25 centuries ago, stated "Let food be thy medicine and medicine be thy food", thereby recognizing the importance of food in determining the health of an individual. He also highlighted

*Corresponding Author: Klaus W.J. Wahle—Cancer Medicine Research Group, School of Medicine and Dentistry, Aberdeen University, Aberdeen. UK. Email: k.wahle@abdn.ac.uk

Bio-Farms for Nutraceuticals: Functional Food and Safety Control by Biosensors
edited by Maria Teresa Giardi, Giuseppina Rea and Bruno Berra.

the fact that, since time immemorial, plants and plant extracts have been used as treatments of the various ailments known to man. This tenet is further supported by numerous population-based studies that show clear differences in incidence rates of various diseases, including cancers, in different countries throughout the world. For example, there is a distinct North-South divide in Europe with the incidence rates of most types of cancer being significantly lower in populations from Southern European and particularly Mediterranean countries, compared with those from Northern European countries.[1,2] Similarly, populations in South East Asian countries have a much lower risk of developing numerous cancers, including colon, gastrointestinal, prostate, breast compared to their more industrialized, Western counterparts.[1-6] Epidemiological studies have also revealed that certain, specific cancers appear to be more prevalent in people from some cultures than from others.[3-6] As already stated, cancers of the lung, colon, breast and prostate are very common in populations from Western compared to Eastern countries whereas cancers of the head, neck and cervix are more common in Indian populations and stomach cancer is significantly more prevalent in Japanese populations.

That these variations in cancer risk were probably not the result of genetic variations between populations from different countries was also evident from a number of migration studies where people migrating from their native country to an adopted country (mainly the USA) developed the risk profile and cancer incidence of their adopted country within one or two generations.[3-6] Similarly, the heritability of breast cancer was investigated in a study with identical twins (100% homology in DNA) in Sweden and showed that if one twin developed the disease the likelihood of the second twin also developing the disease was only 15-20%, suggesting that the inheritance of faulty genes only made a small contribution to the overall pathogenesis of the disease within a population.[7] This agrees with the known percentage distribution of inherited/mutated breast cancer genes like BRCA1, BRCA2, Her-2/neu.[8]

Since the genetic composition of individuals and populations can only account for a small proportion of their cancer risk, other factors external to the individual must play a major role. It has been estimated that 75%-85% of chronic illnesses and diseases appear to have significant lifestyle factors in their aetiology and chief amongst these are smoking, lack of exercise and poor nutrition.[9] Doll and Peto[10] suggested that a majority of cancers diagnosed in the USA in 1970 might have been prevented if lifestyle factors like diet and smoking had been modulated. A large number of studies have since supported these suggestions and it has been estimated that circa 30% of cancer deaths could be attributed to inadequate diet, which is on a par with cancer deaths attributable to smoking; i.e., poor diet and smoking can account for circa 60% of cancer deaths.[9-11] Clearly, a combination of heavy smoking, high alcohol intake and inadequate diet, a common occurrence in lower socio-economic groups in Western societies, would be expected to elicit an even greater risk profile.[9-11]

Epidemiological studies have also shown that, those populations with significantly lower incidence of various types of cancer common in industrialized Western populations tended to consume diets that were high in foods of plant origin, namely fruits-including berries, nuts, vegetables, whole-grains, legumes, seeds, various types of tea and a bewildering array of spices, but also low in red meat and animal fats. Such diets are exemplified by the classical Mediterranean diet of Southern Europe and the highly vegetarian diets of South East Asian populations.[11-18] An analysis of results from 206 human epidemiological studies and 22 laboratory animal studies also supported an inverse relationship between the consumption of fruit and vegetables and the risk of developing various cancers.[13] Such compelling evidence of cancer prevention by dietary plants has resulted in great deal of research over the past two decades to identify the specific components of dietary plants responsible for these beneficial effects and the mechanisms by which they elicit these effects. It should be noted here that the quest for natural, plant-derived phytochemicals that express anticancer properties is not restricted to food plants. Pharmaceutical companies are currently isolating and assessing the anticancer potential of numerous phytochemicals from all regions of the World in order to discover more efficient treatments for various types of cancer. The validity of such an approach is evident from the widely used taxanes, first extracted from the bark and leaves of the Pacific and European yew tree, in chemotherapy of various cancers, particularly breast and prostate cancer.[19] Consideration must be given to the possibility that although individual components of plants may have significant, specific anti-cancer effects, these effects may be even greater when these components are consumed in various combinations, as occur naturally in foods. The interactions and synergisms between compounds occurring within a food matrix should be given greater consideration than at present.

The catalogue of plants, their specific derivatives, their health benefits and the cellular mechanisms by which these benefits are elicited is far too extensive for the somewhat eclectic remit of the present review which will focus mainly on specific phenolic phytochemicals derived mainly from food, beverage and spice plants, although many of these compounds are ubiquitous in the plant kingdom. For a more comprehensive treatise the reader is directed to the many excellent reviews on the different aspects of phytochemicals and human health, some of which are referenced in the current text below.

Anticancer Phytochemicals in Foods, Beverages and Spices

Fruits, vegetables, nuts and whole grains are major sources of fibre, trace metals, essential oils and vitamins all of which are important for the maintenance of human health, including the prevention of cancer, through their effects on numerous cell mechanisms.[1,2,10-18] They also contain a bewildering array of secondary metabolites (micronutrients) including flavonoids and allied monophenolic and polyphenolic compounds, terpenoids, nitrogen-containing alkaloids, phytosterols and sulphur-containing compounds, to name a few, that have been implicated in a reduced cancer incidence. The compounds that have been most cited as being cancer protective include the various catechins from green and, to a lesser extent, from black tea, diallyl sulphides and allicin from garlic and onions; sulforaphanes and indole-3-carbanols from brassica vegetables; genistein from soya; delphinidine and ellagic acid from soft fruits and various nuts; curcumin and resveratrol from turmeric and red grapes respectively; anthocyanins in berries, grapes, cherries and many other fruits.[1,2,10-18,20] This short review will be limited to considering certain aspects of the reported anticancer benefits derived from plant phenolic compounds and the putative cellular mechanisms by which these benefits are derived.

Classification and Occurrence of Plant Phenolic Compounds

The phenolics under consideration are secondary plant metabolites characterized by having one (monophenolic) or more (polyphenolic) aromatic rings with one or more hydroxyl groups attached. Over 8000 different plant phenolics have been identified and they are widely distributed throughout the plant kingdom.[1,2,20-26] The nature, synthesis and distribution of plant phenolics can vary significantly and depends on the plant species, growing conditions and tissue type. The major route of synthesis of mono- and polyphenolic compounds is from carbohydrates by way of the shikimic acid, phenylpropanoid and flavonoid biosynthetic pathways.

Their structure ranges from simple, low molecular weight compounds with a single aromatic ring to the large, complex, multi-ringed tannins. They can be classified by the number and arrangement of their carbon atoms (see Fig. 1 and structures), are generally conjugated with sugars and organic acids and can be grouped into flavonoids and nonflavonoids.

For a more complete description and classification of the major plant phenolics the reader is referred to references.[2,24-26] The following is a brief outline of the salient aspects of phenolic classification and occurrence.

Flavonoids

The polyphenolic flavonoids are the most abundant and numerous plant phenolics with a ubiquitous distribution. They can be categorized into 13 classes comprising more than 5,000 compounds. They are found in high concentrations in leaf epidermis and fruit skins and have important roles to play in the plants as secondary metabolites in such diverse processes as pigmentation, protection against UV radiation and disease resistance.[2,24-26] The major sub-classes

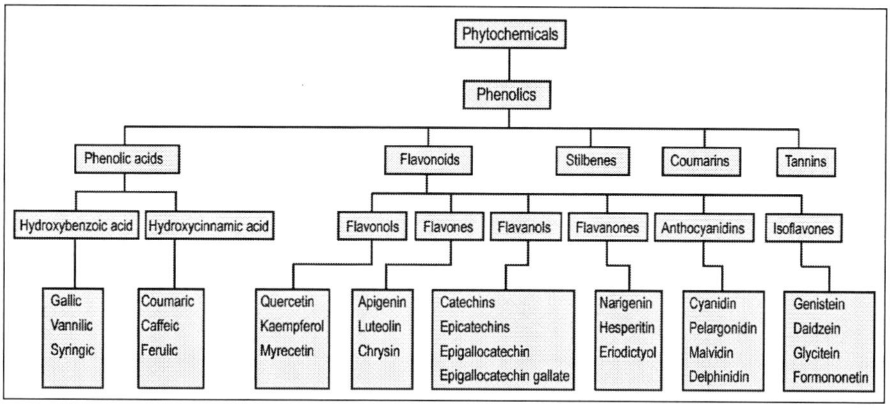

Figure 1. A brief classification of some phenolic phytochemicals reported to have health benefits.

of flavonoids are the flavones, flavonols, flavan-3-ols, isoflavones, flavanones and anthocyanidins (Fig. 1). Other sub-groups are of less importance in nutrition such as dihydroflavones, flavan-3,4,-diols, coumarins, chalcones, dihydrochalcones and aurones but they may still have anticancer effects.[2,24-26]

The bioactivity of flavonoids is also dependent on the various structural sub-constituents such as hydroxyl groups, usually present at the 4-, 5- and 7-positions in the ring. Sugar attachments are common and the majority of flavonoids exist naturally as glycosides. Methyl and isopentyl groups are also present, thereby conferring lipophilicity (lipid solubility) to these compounds; sugars and hydroxyl groups increase the water solubility of the compounds.[2,24-26]

Flavonols

Flavonols are the most widespread of all the flavonoids and are found throughout the plant kingdom with the exception of fungi and algae. Compounds such as myrecetin, quercetin, kaempferol and isorhamnetin most commonly occur as O-glycosides. Numerous flavonol conjugates exist with over 200 different sugar conjugates of kaempferol alone. This emphasizes the complexity of the flavonol profile occurring in fruit, vegetables and beverages as well as their activity levels in relation to health benefits in mammalian systems. This complexity is compounded further by the marked differences in concentration of these compounds that can occur in apparently similar produce which is possibly due to seasonal changes, growing regions and different varieties of the same produce.[2,24-26]

Flavones

Flavones are structurally very similar to flavonols (Fig. 1, ia) but they lack the hydroxyl group at position 3 of the C-ring. As with the flavonols, a large number of substitutions are possible including hydroxylation, methylation, as well as alkylation and glycosylation. Flavones, unlike flavonols, are not widely distributed in nature and occur mainly in parsley, celery and certain herbs. Polymethoxylated flavones like nobiletin and tangeretin occur in citrus fruits.[2,24-26]

Flavan-3-Ols (Flavanols)(Catechins)

Flavanols are structurally the most complex of the flavonoids. They occur as simple monomers of (+)-catechin and (−)-epicatechin and range in complexity to oligomeric and polymeric proanthocyanidins which constitute the condensed tannins. Proanthocyanidins can occur as polymers of circa 50 catechin units. Flavanols can be hydroxylated to form gallocatechins and can be esterified to gallic acid. These compounds are abundant in black grapes (oligomeric proanthocyanidins) and consequently in red wine. Green tea is a rich source of flavanols, mainly

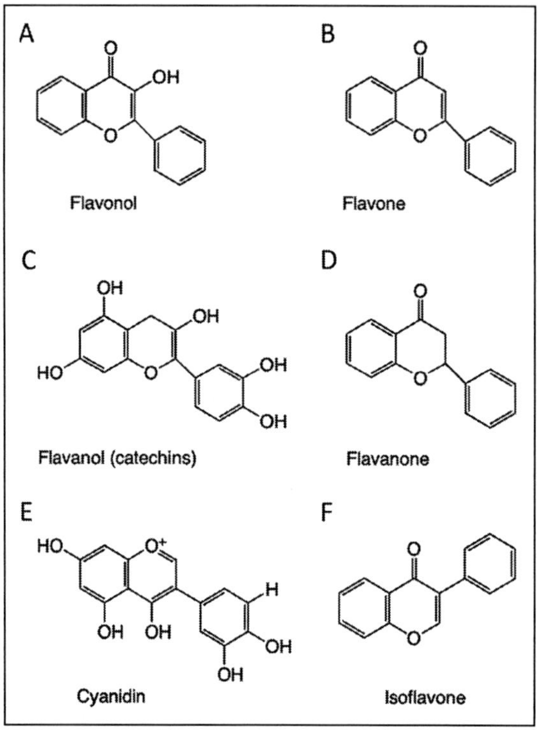

Figure 2. General structures of common flavonoids. In order A) Flavonols; B) Flavones; C) Flavanols; D) Flavanones; E) Anthocyanidins; F) Isoflavones.

(−)-epigallocatechin, (−)-epigallocatechin galate and (−)-epicatechin gallate. Fermentation of the green tea leaves results in a decline in the levels of catechins and consequently black tea contains mainly high molecular weight thearubigens, which are also flavanoids, but their structures have not been clearly elucidated.[2,24-26] A number of intervention and epidemiological studies have shown that the consumption of flavanols can confer health benefits in animals and man, including anticancer benefits (vide infra).

Flavanones

Flavanones are highly reactive compounds and can readily undergo hydroxylation, glycosylation and methylation reactions. They are present in high concentrations in citrus fruits, particularly in the peel (i.e., hesperidin-a flavanone rutiside; naringin from grapefruit peel). For detailed structures see reference.[2,24-26]

Anthocyanidins

Anthocyanidins occur widely in the plant kingdom, principally as their sugar-conjugated derivatives, anthocyanins, where they are responsible for the red, blue and purple colours in a variety of fruits and flowers. They are also found in leaves, stems, seeds and roots of numerous plant species where they are involved in protecting sensitive plant tissues from excessive light and in attracting insects for pollination. The most common anthocyanidins are pelargonidin, cyanidin, delphindin, peonidin, petunidin and malvidin.[2,24-26] Anthocyanins have also been reported to have anticancer effects (see below).

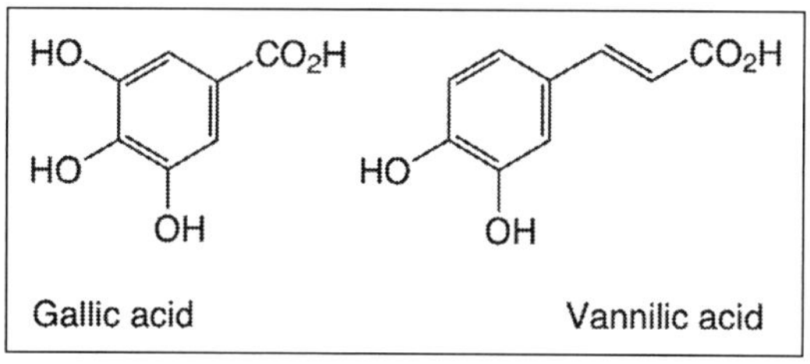

Figure 3. Structure of common phenolic acids; gallic and vannilic acids.

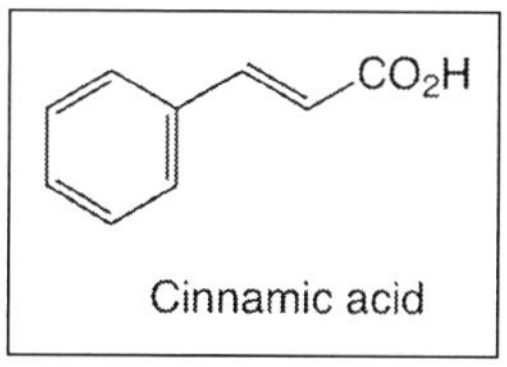

Figure 4. Structure of cinnamic acid (see also Figure 6, coumaric acid).

Isoflavones

Isoflavones are derived from the flavonoid synthetic pathway. Naringenin is converted to genistein via 2-hydroxyisoflavanone and removal of the hydroxy group on carbon 5- of the A ring yields daidzein. Isoflavones such as genistein and daidzein, like other members of the flavonoid family, can readily undergo hydroxylation, methylation and prenylation reactions to yield a range of isoflavonoids including coumestans, rotenoids and pterocarpins.[2,24-26] The isoflavones derived from soya and clovers (genistein, daidzein and the coumestan-coumesterol respectively), are known to have high oestrogenic activity and are termed phytoestrogens. Their structure resembles that of oestradiol, a mammalian steroid hormone that blocks ovulation and can seriously affect reproduction in grazing animals.

Dietary intake of genistein and daidzein in soya products has also been implicated in a reduction of prostate and breast cancer incidence rates in humans. The initiation and progression of these cancers involve androgen and oestrogen dependent mechanisms respectively and the isoflavones compete with the natural hormonal receptor mechanisms and suppress the growth and progression of the tumours.[2,24-27]

Nonflavonoid Phenolic Compounds

The major dietary nonflavonoid phenolics are: (a) the C_6-C_1 phenolic acids, namely gallic and ellagic acids and derivatives (hydrolysable tannins); (b) the C_3-C_3 hydroxycinnamates and their conjugated derivatives and (c) the polyphenolic C_6-C_2-C_6 stilbenes.[2,24-26]

Phenolic Hydroxybenzoic Acids and their Analogues

Gallic acid, one of the main phenolic acids, is formed primarily via the shikimic acid pathway. It is further converted to ellagic acid, another major phenolic acid. These compounds can then be further modified (sugar conjugations) to yield pentagalloylglucose or pentaellagoylglucose and a range of gallotannins and ellagitannins which together form the hydrolysable tannins which are more readily hydrolysed by dilute acids than the condensed tannins.[2,24-26]

Hydroxycinnamates

Cinnamic acid, a C_6-C_3 compound, is produced by deamination of the amino acid phenylalanine which is common to all plants. Hydroxylation of cinnamic acid then results in formation of

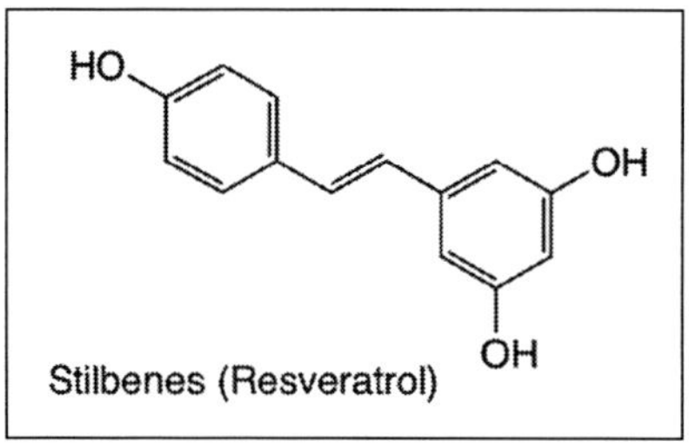

Figure 5. Structure of a common stilbene, resveratrol.

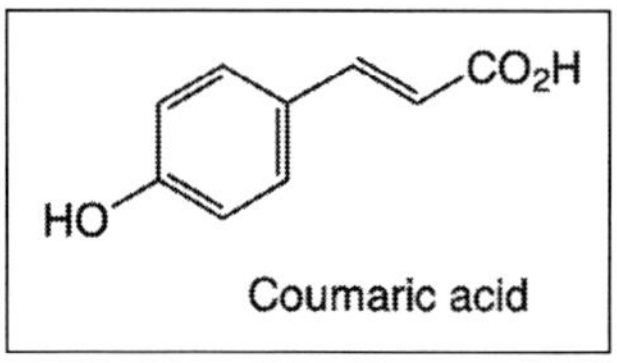

Figure 6. Structure of coumaric acid.

coumaric acid. The most common hydroxycinnamates are caffeic, 4-coumaric, ferulic and sinapic acids which are formed by a series of hydroxylation and methylation reactions. The quinic acid conjugate of caffeic acid, chlorogenic acid, is a common constituent of fruit and vegetables.[2,24-26]

Stilbenes

The stilbenes are polyphenolic compounds with a C_6-C_2-C_6 structure; two aromatic rings linked by an ethene bridge. They are produced by plants in response to various traumas like UV radiation and attack by fungal, bacterial and viral pathogens (infections). Resveratrol is a member of the stilbene sub-group of plant phenolics. It is present in many plant tissues in both the cis- and trans-configuration, often as the glucoside and it is the main phenolic in red wine that has been attributed with many health benefits in man and animals (vide infra).[2,24-26]

Coumarins

Coumarins are lactones derived by cyclisation of cis-ortho-hydroxycinnamic acid. They are characterized by their chemical/structural diversity due to the differing oxygenation in their benzopyrene ring. The majority of coumarins in nature are C7-hydroxylated derivatives. They occur in fruits, vegetables, olive oil, wine and beverages like tea and coffee. They have also been attributed with anti-oxidant and anti-cancer effects in cells and animal models.[2,24-26]

Tannins

Tannins are polyphenolic compounds in two classes characterized by their ability to be hydrolysed; nonhydrolysable tannins are the condensed forms. They are complex polyphenols that can be degraded to sugars and phenolic acids by both enzymatic and non-enzymatic hydrolytic processes. The basic units of hydrolysable, polyester tannins are gallic acid and its derivatives.

Tannins are high in many unripe fruits and levels decline as the fruit ripens. They also occur in nuts, food flavourings and red wine. Their astringent taste and adverse effects on gastric digestive enzymes makes them unpalatable to many mammalian species unless they are adapted to detoxify them. They have also been ascribed important antioxidant and anti-inflammatory effects in human cancer cells.[2,24-26]

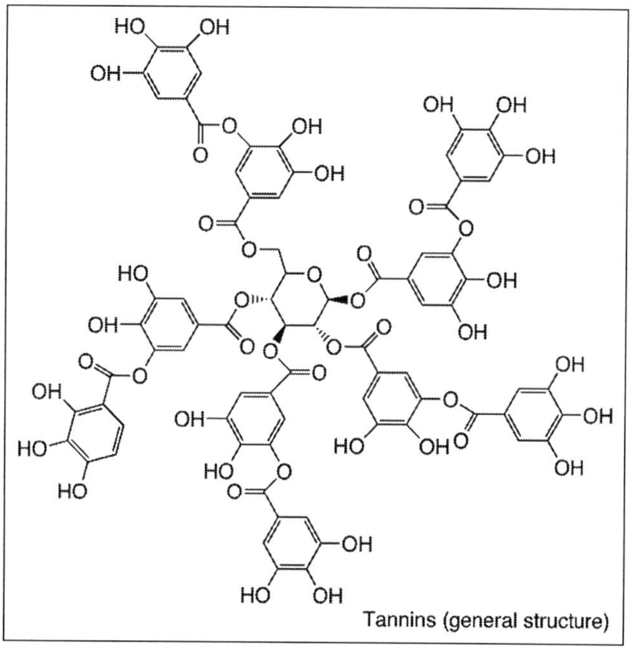

Figure 7. General structure of a tannin.

Other Important Dietary Mono- and Polyphenolic Compounds

Tyrosol and Hydroxyryrosol

These two monophenolic compounds are simple phenolic acid analogues and are important constituents of olive oil. Along with the high content of oleic acid (18:1), they have been implicated as major factors in the general health benefits of the "Mediterranean diet" including anti-cancer and anti-cardiovascular disease effects. They are also present in other edible oils and in wine. They are reported to have antioxidant and pro-apoptotic effects in various human cancer cells.[2,21-26]

Oleuropein

Another important phenolic compound found in olive oil, where it is present mainly as the glycoside esterified to a sugar moiety. As with the tyrosols, oleuropein is attributed with antioxidant and anti-inflammatory effects as well as inducing cell cycle arrest and apoptosis in various human cancer cells.[2,21-23]

Cellular Mechanisms Modified by Plant Phenolics That Can Reduce Carcinogenesis and Tumour Progression

Conventional therapeutic and surgical modalities have been responsible for the marked improvement in recent years in the survival and enhanced quality of life of patients with a variety of cancers. There is now a great need to prevent or attenuate the increasing incidence rate of various cancers in Western and industrialised populations. A judicious modification of diet with increased intakes of fruit and vegetables, with its increased availability of constituent phytochemicals, appears to be a strong lifestyle change that meets such a requirement.[1,2,10-18,20-27]

A wealth of epidemiological evidence suggests that one-third of cancers can be prevented through the consumption of appropriate diets containing adequate levels of fruit, vegetables and whole grains. Nutrition experts suggest that 5 or more helpings of fruit and vegetables would confer significant benefit to the individual although it is difficult for people to assess what constitutes a helping and consequently such loose terminology needs to be clarified.[1,2,10-18,20-27]

The complex, multi-step processes leading to carcinogenesis can be activated by a number of environmental factors such as cigarette smoke, industrial pollutants, oxidative and inflammatory agents. The process of carcinogenesis can be broadly categorized into three distinct phases: tumour initiation, promotion and progression. Phytochemicals, including specific plant phenolic compounds, appear to play a significant role in suppressing all three stages of tumour formation and metastasis, including the transformative, hyperproliferative and inflammatory processes involved in initiation, the angiogenic processes required for tumour growth as well as the vascular adhesive properties necessary for metastasis or tumour dispersion.[1,2,16-18,20-27] In recent years, following the positive epidemiological evidence of phytochemical intake preventing cancer, a great deal of research has identified a variety of cellular signaling mechanisms that support this hypothesis and shows that numerous phytochemicals can modulate distinct cell receptors and signal transduction pathways to suppress the carcinogenic process, both in vitro and in vivo in animals and to a lesser extent in man.[1,2,20,22-25,28-30]

Effects of Phenolics on Growth Factors and Receptors (GFRs) Implicated in Cancer Initiation and Progression

Growth factors are proteins that bind to specific receptors on cell surfaces to elicit a signaling cascade responsible for the normal activation of cell proliferation/differentiation required for tissue growth and repair. Aberrant growth factor expression/availability results in a signaling cascade leading to uncontrolled cell proliferation and differentiation, suppression of apoptotic signals and ultimately in carcinogenesis, tumour growth/progression and metastasis.[2,20,22,24,25,29]

Major growth factors implicated in carcinogenesis are epidermal growth factor (EGF), plate-derived growth factor (PDGF), fibroblast growth factors (FGFs), transforming growth factors-α and -β (TGFs-α and -β), insulin-like growth factor (IGF), erythropoietin (EPO), as well as the inflammation-related cytokines interleukin1,-2,-6,-8 (IL-1,-2,-6,-8), tumour necrosis factor-α (TNF-α), interferon-γ (IFN-γ) and colony stimulating factors (CSFs).[24,25,29]

The binding of these factors to their specific receptors elicits powerful cell proliferation signals through the activation of signal cascades involving various receptor-regulated and cytosolic kinases and transcription factors which forms the basis of growth factor/receptor driven carcinogenesis and tumour progression.[24,25,29] Compounds that can attenuate GF binding and the attendant signal cascade are generally regarded as excellent chemo-preventive agents.

Plant phenolics such as curcumin, genistein, resveratrol and catechins are potent inhibitors of a number of growth factor binding and signaling pathways implicated in cancer. Thus, curcumin inhibits EGFR action and reduces the invasive potential of cancer cells.[31,32] Curcumin also inhibits EGF kinase activity in A431 cells[32] and EGF expression in Ishikawa endometrial cancer cells probably by inhibiting the tyrosine kinase activity.[33] This phenolic attenuates the Her2/neu receptor expression that is often over-expressed in breast, prostate, ovarian and lung cancer, indicating poor prognosis, by inhibiting the tyrosine kinase activity (e.g., c-Src) involved in activation of the G-protein-coupled receptor.[34] Similarly, the flavonoids apigenin and quercetin were shown to induce apoptosis by increasing the phosphorylation of Her2/neu via the PI3/Akt kinase pathway and subsequent proteosomal degradation in breast cancer cells over-expressing this receptor/transcription factor and suppress tumour growth in DMBA-induced rat mammary tumours in vivo by inhibiting the PTK pathway.[35,36] Consumption of olive oil, with its high oleic acid and phenolic content, has long been implicated in the reduced cancer incidence in Mediterranean compared with Northern European countries. The phenolics oleuropein aglycone and hydroxytyrosol have also been shown to deplete the over-expressed HER2/neu receptor expression (mRNA and protein) on breast cancer cells. Furthermore, oleuropein resulted in a synergistic augmentation of the Herceptin-induced down-regulation of Her2/neu expression indicating that such phenolics could be used as adjunct therapeutic agents to the standard chemotherapy in cancers expressing high HER2/neu.[36,37]

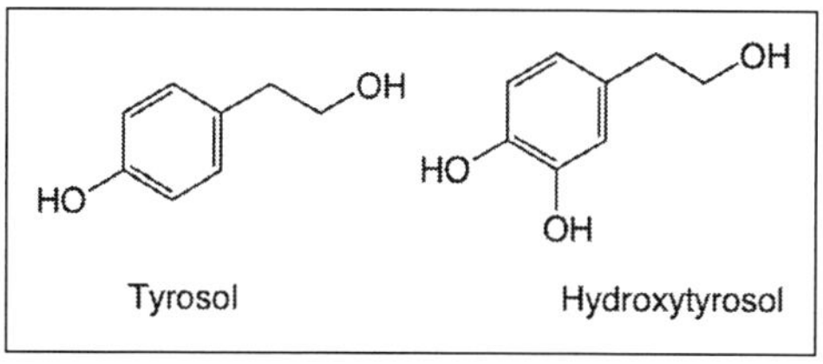

Figure 8. Structure of tyrosol and hydroxytyrosol from olive oil.

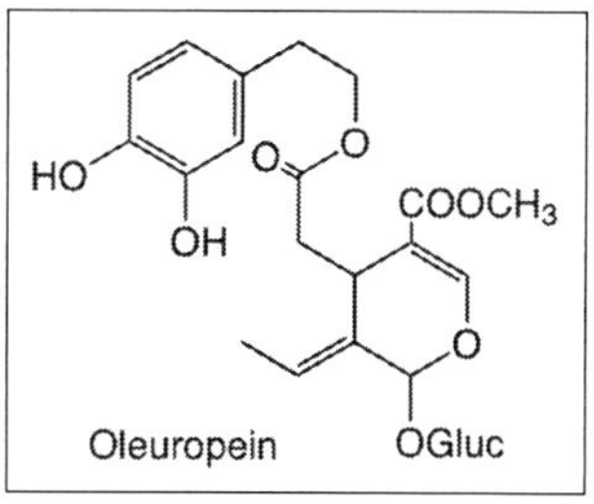

Figure 9. Structure of oleuropein from olive oil.

Resveratrol inhibited IL-6 and IL-8 expression in stimulated animal and human monocytic cells and suppressed proliferation of Ishikawa cells by down-regulating of EGF effects.[25]

The major green tea phenolic EGCG attenuates IL-6 and IL-8 expression, reduces VEGF production in breast cancer cells and inhibits VEGF-induced angiogenesis by suppressing VE-cadherin (an adhesin molecule) phosphorylation and inactivating the Akt pathway.[24,25] EGCG is known to inhibit activation of EGFR and human EGFR-2 signaling pathways in human colon cancer cells and this is thought to be through direct inhibition of the ERK1/2 and Akt kinases. Green tea also inhibits expression of other factors including FGF, VCAM-1 (an adhesion molecule) and HER-2/neu.[24,25] Flavonoids can also attenuate the expression of ICAM-1, E-selectin and E-cadherin, which possibly explains their attenuation of metastasis.[24,25,29,38]

Plant phenolics are regarded as potent antioxidants and their major beneficial effects on health, including anti-atherogenic and anti-cancer effects, appear to relate, in part, to their ability to down-regulate the oxidative and inflammatory signal cascades through such redox-sensitive transcription factors as NF-κB and AP-1 and subsequent modulation of down-stream gene expression.[20,24,25,29,39,40]

Attenuation by Phenolics of Oxidation and Inflammatory Processes

Oxidative and inflammatory stresses are now recognized as being closely involved in the initiation and progression of carcinogenesis. Excessive formation of reactive oxygen species (ROS) in cells can damage proteins, DNA and RNA and can oxidise membrane polyunsaturated fatty acids, thereby increasing the likelihood of deleterious mutations occurring in the genome.[20,24,25,29] ROS are regarded as major activators of the NF-κB and AP-1 transcription factor pathways.[20,24,25,29,41] Numerous dietary phenolics are potent antioxidants, capable of scavenging deleterious reactive species such as superoxide anions, singlet oxygen hydroxy radicals, nitric oxide and peroxynitrite.[20,24,25,29,42] Various phenolics are also able to attenuate ROS generation through inhibition of redox sensitive transcription factors such as NF-κB and AP-1 responsible for the expression of the ROS-induced inflammatory enzyme cascade. Xanthine oxidase, COX-II and LOX were shown to be reduced by dietary phenolics like curcumin, silymarin and resveratrol.[20,24,25,38,39,43,44]

Polyphenols can also induce detoxifying enzymes such as glutathione-S-transferase (GST) and quinone reductase (QR) which can protect cells from carcinogenic intermediates, exogenous or endogenous.[20,40,45]

Effects of Phenolics on Cell Cycle and Apoptotic Mechanisms

Disruption of normal cell cycling mechanisms and the over-expression of growth promoting mechanisms such as cyclin D1 and the cyclin-dependent kinases (CDKs) are major events in carcinogenesis.[20,24,25,46,47]

Phenolic compounds such as resveratrol have been shown to inhibit a variety of cells at different stages of the cell cycle, i.e., at G1, S, S/G2 and G2, both in vitro and in vivo in animal models of cancer.[20,24,25,47,48] ECGC elicits its effects by directly inhibiting CDKs or indirectly by inducing p21 and p27 gene expression and inhibiting cyclin d1 and Rb phosphorylation. Recently, resveratrol was shown to arrest HL-60 cells in the S/G2 transition phase, increase cell numbers in G1/S phase and induce apoptosis, possibly as a result of decreased Bcl-2 expression, a major anti-apoptotic oncogene.[20,24,25,29,47] Apigenin (celery, parsley), genistein (soya) and silymarin (milk thistle) have also been shown to modulate the effects of deregulated cell cycle checkpoints and are thereby believed to contribute to the prevention of cancer. They appear to elicit their beneficial effects not only through cell cycle arrest and the induction of cyclin-dependent kinase inhibitors (p15, Cip-1/p21 and Kip-1/p27) but also through down-regulation of anti-apoptotic gene products (Bcl-2, Bcl-xL) and induction of pro-apoptotic p53 and Bax.[24,25,29,47,49] Apoptosis, or programmed cell death, is an important mechanism in normal development and in anticancer surveillance. The process is regulated by various oncogenes/proteins, including the important pro-apoptotic p53, the anti-apoptotic and cell survival Bcl-2 and the caspase cascade.[24,25,29,47]

Effects of Phenolics on Chemokines and Cytokines

Chemokines are small chemotactic cytokines responsible for leucocyte targeting and activation of inflammatory responses and they are involved in the regulation of tumour growth. Chemokines exert their migration- inducing effects on leukocytes through binding to specific chemokine receptors such IL-8/CXCL8 which stimulates endothelial chemotaxis, proliferation and angiogenesis in vivo and have been detected in high concentrations in a variety of tumours.[24,25,50]

Dietary phenolics like curcumin, resveratrol, quercetin, green tea polyphenols, black tea theaflavin, soya genistein and capsaicin from peppers have all been shown to attenuate chemokine and cytokine expression.[25,50-54]

Curcumin is a potent anticancer agent. It inhibits expression of inflammatory cytokines and chemokines (mRNA, protein) and their signal cascades in tumour cells but effects are reversible within 24hrs after removal of the phenolic.[25] This suggests that a constant intake of these compounds is necessary in order to inhibit carcinogenesis. Similarly, resveratrol inhibits phorbol ester-induced IL-8 expression in U937 cells, partly through inhibition of AP-1 activation.[51] Suppression of NF-κB activation by quercetin is also responsible, in part, for the inhibition of IL-1 induced expression of MCP-1.[52] Green tea EGCG inhibits the expression of chemokines IL-8, MIP3 and of prostaglandin E2 (PGE2), a down stream product of NF-κB transcriptional activity, in TNF-stimulated colon epithelial cells.[53] Theaflavin, from black tea, can inhibit TNF-stimulated IL-8 gene expression, probably at the transcriptional level (mRNA) and through inhibition of both IKK/NF-κB and AP-1 pathways.[54] Genistein and capsaicin have also been shown to inhibit expression of various growth factors/chemokines/cytokines in different, stimulated, cell types including melanoma cells through the suppression of the NF-κB pathway.[24,25,40]

Phenolic Suppression of Angiogenesis

As stated previously, angiogenesis, or new blood vessel formation, is vital for supplying nutrients and oxygen to the tumour and ensuring its growth and progression, its invasiveness and spread to other tissues. Prevention of angiogenesis would "starve" the tumour and prevent metastasis.[20,24,25,49,50] Consequently, angiogenesis inhibition is an active objective in current anticancer research.

Quercetin, genistein, resveratrol and various phenolic acids are capable of inhibiting angiogenesis in in vitro cell-based systems as well as in animal models. These effects appear to be due to attenuation of matrixmetalloproteinases (MMPs) inhibition of VEGF and their Src kinases.[24,49] These phenolics have also been shown to attenuate the expression of vascular adhesion molecules which would reduce the metastatic process.[25,38]

Epigenetic Modulation as a Novel Chemopreventive Role for Phenolics

DNA methylation state is regulated through the activity of DNA-methyltransferases (DNMTs) and demethylating reactions. Hypermethylation of reporter regions of specific genes results in their transcriptional silencing. Although hypermethylated genes can be inherited, the process of methylation is reversible and genes can be reactivated by removal of methyl groups.[24,25,29]

Numerous genes have been shown to be hypermethylated and consequently inactivated or hypomethylated and activated, in cancer cells.[55] These include genes involved in cell cycle regulation (p16, p21waf1/cip1, p151, Rb), genes associated with DNA repair (BRCA1 and 2, MGMT), redox enzymes (GPx1 and 4, MnSOD), apoptosis/tumour suppression (p53), drug resistance, detoxification, angiogenesis and metastasis, all of which are susceptible to hypermethylation and silencing. Various phenolics, including green tea polyphenols (epicatechin, catechin, EGCG) and flavonoids (quercetin, myricetin, fisetin) have been shown to reactivate silenced genes in cancer cells by inhibiting DNMT activity and reversing hypermethylation of their promoter DNA in a time and concentration manner.[55-57] Phenolics (e.g., EGCG) are also able to reverse hypomethylation that is observed in many cancers, presumably silencing pro-carcinogenic gene expression.[55]

Cancer Cell Sensitization and Reversal of Drug Resistance by Phenolics

Most chemotherapeutic agents and radiotherapy used in cancer treatment activate the NF-κB survival pathway which can eventually result in resistance to the treatment. Co-administration of phenolic chemopreventive agents such curcumin will tend to upregulate proapoptotic pathways (p53, p21waf1/cip1) whilst also downregulating the cell survival mechanisms (PI3K, AKT, NF-κB, AP-1) and changing the survival Bcl-2:Bax ratio to proapoptotic one.[29] Resveratrol was shown to enhance chemo-sensitisation by downregulating another cell survival gene, survivin.[58,59] EGCG from green tea also inhibits growth in human breast cancer cells by suppressing surviving expression; a member of the inhibitor of apoptosis proteins (IAP) highly expressed in cancer cells/tissues.[60]

Immune Response-Mediated, Anti-Cancer Mechanisms of Phenolics

The removal of cancer cells by the host immune response has long been regarded as the most desirable anti-cancer strategy. This is primarily because the immune system, due to its highly specific recognition mechanisms, has the potential to eradicate cancerous cells with the least damage to normal cells. The enhancement of immune cell responses that can specifically target cancer cells could prove to be a very effective preventive and treatment strategy. Thus, either nutrients or drug treatment modalities that enhance anti-tumour effector cell activities have naturally been perceived as an ideal mechanism in cancer treatment/prevention. The most simplistic perspective of enhanced immunity toward cancer cells is the direct enhancement of specific cell activities, particularly those that can kill tumour cells. T-cells are a potential target for any immunotherapeutic approach and enhancement of T-cell activity could enhance anti-tumour immunity. $CD8^+$ cytotoxic T-cells directly engage and kill appropriate target kills and one of the first T-cell responses to stimulation is the rapid clonal proliferative expansion of the T-cell population. The study of Gao et al,[61] showed that resveratrol at low concentrations (<12.5 μM) enhanced concanavalin-induced proliferation with no effect on IL-2-induced proliferation. This difference demonstrated that resveratrol does not enhance the proliferative response per se, rather it enhances the upstream mechanisms which signal for proliferation, most likely enhancing the T-cell receptor-mediated induction of IL-2, the major T-cell growth factor. However, resveratrol did not enhance the killing capability of these T-cells against lymphoma target cells. This study also showed that very high resveratrol concentrations (>20 μM) can suppress the effects, indicating a biphasic response to resveratrol which can have implications for the use of this phenolic in human studies, i.e., high doses can have the opposite effect on immunity than intended.

A comparable study had also demonstrated similar results in human immune cells. Resveratrol was shown to increase the proportion of $CD8^+$ cells that produced interferon-gamma and IL-2 in response to activation of the T-cell receptor via CD3 and also costimulation via CD28.[62] These are cytokines that are important in the activation of T-cell receptor-mediated, cell killing capability; this was confirmed in the study when the authors also showed that specific cell killing (cytotoxicity) was enhanced. These effects were shown at low concentrations of resveratrol (≤10 μM) whereas higher concentrations (≥20 μM) had a suppressive action.[62] This study further showed that resveratrol enhanced natural killer (NK) cell cytotoxic activity against target human cancer cells at even lower concentrations (≤1 μM). This is an extremely important observation since NK cells are a major mechanism whereby mutated cells can be eliminated in the earliest stages of cancer development. An example of this, with respect to specific cancers, is the control of melanoma by NK cells, where it has been confirmed that NK cells can directly lyse melanoma cells when the absence of self-antigens is detected.[63] These studies clearly indicate that low concentrations of resveratrol can enhance the profile of immune responses that are capable of suppressing/eliminating cancer cells in a specific and coordinated manner. Similarly, the flavonoid quercetin can also upregulate human antitumour mechanisms by selectively increasing the production of interferon-gamma thereby augmenting immune-cell mediated cell killing actions and downregulating suppressive cytokines such as IL-4.[64]

By contrast the flavonoid kaempferol has been shown to be suppressive toward $CD8^+$ cell activities including interferon-gamma production and may be counterproductive in cell-mediated cancer immunotherapy.[65] This polyphenol does attenuate adhesion molecule expression in vascular cells[25,38] and may thus be more effective in preventing metastatic spread of cancers.

A more recent consideration in our understanding of how immune responses control anticancer activity is the regulation of suppressor mechanisms in the ex vivo inhibition of suppressor/regulatory T-cell (Treg) which when switched-off can be permissive for a series of diverse immune cell responses. A recent study by Yang et al,[66] showed that resveratrol administration can regulate activities in tumour-derived cells, especially the release of cytokines (including transforming growth factor-beta), that inhibit the activation of many immune cells. Furthermore, the authors confirmed that this inhibition of cells also resulted in increased immune stimulatory activities, such as interferon-gamma production by $CD8^+$ cells, resulting in an unfavourable microenvironment for tumour cells to grow and proliferate by enhancing cell cytotoxic mechanisms. This gives an important insight into the possible mechanisms by which polyphenolic compounds such as resveratrol can modulate immune responses directly relevant to the elimination of cancer cells. They also show the importance of appropriate doses and that low doses, such as the ones that can be obtained from the diet, could be extremely beneficial in the immunotherapy of cancer.

Conclusion

This brief, eclectic review highlights the fact that plant phenolic compounds, consumed as part of a balanced diet, are able, not only to prevent cancer initiation, progression and metastasis through a variety of cell mechanisms, including immune cell functions and angiogenesis, but are also capable of enhancing standard chemo- and radiotherapeutic modalities by reversing the cell mechanisms that lead to desensitization. Such adjunct treatment with natural, dietary components would be expected to reduce the known debilitating side effects of standard treatments and should reduce the likelihood of cells/tissues becoming desensitized. The anticancer effectiveness of multi-matrices, as found in foods, should not be underestimated. Tumour cells utilize a multitude of survival mechanisms to remain viable and consequently preventative and treatment modalities that are capable of suppressing a number of pro-carcinogenic/pro-tumorigenic pathways, such as the ingestion of plant phenolics from our diet, have great potential in the future prevention and treatment of cancer.

Observations that omega-3 polyunsaturated fatty acids and conjugated linoleic acids appear to elicit similar anticancer effects on various cancer cells and animal models and that the same signaling cascades are affected suggests a common aetiology and warrants investigation.[67]

References

1. World cancer research fund and American institute for cancer research, In food, nutrition and prevention of cancer; a global perspective, 1997.
2. Plants: Diet and health. British nutrition foundation. Edit. G. Goldberg; Blackwell publ 2003.
3. Ziegler RG, Hoover RN, Pike MC et al. Migration patterns and breast cancer risk in Asian-American women. J Natl Cancer Inst 1993; 85:1819-1827.
4. Haenszel W, Kurihara M. Studies of Japanese migrants. I. Mortality from cancer and other diseases among Japanese in the United States. J Natl Cancer Inst 1968; 40:43-68.
5. Kolonel LN, Altshuler D, Henderson BE. The multi-ethnic cohort study: exploring genes, lifestyle and cancer risk. Nat Rev Cancer 2004; 4:519-527.
6. Wiencke JK. Impact of race/ethnicity on molecular pathways in human cancers. Nat Rev Cancer 2004; 4:9-84.
7. Locatelli I, Lichtenstein P, Yashin AI. The heritability of breast cancer: a Bayesian correlated frailty model applied to Swedish twins. Twin Res 2004; 7:182-191.
8. Oldenburg RA, Meijers-Heijboer H, Cornelisse CJ et al. Genetic susceptibility for breast cancer: how many more genes to be found? Crit Rev Oncol Hematol 2007; 63:125-149.
9. Jemal A, Siegel R, Ward E et al. Cancer statistics, 2007. CA Cancer Journal for Clinicians. 2007; 56:106-130.
10. Doll R, Peto R. Avoidable risks of cancer in the United States. J Natl Canc Inst 1981; 66:1197-1265.
11. Block G, Patterson B, Subar A. Fruit, vegetables and cancer prevention: a review of the epidemiological evidence. Nutr Cancer 1992; 18:1-29.
12. Mathew A, Peters U, Chatterjee N et al. Fat, fibre, fruits, vegetables and risk of colorectal adenomas. Int J Cancer 2004; 108:287-292.
13. Gandini S, Merzenich H, Robertson C et al. Meta-analysis of studies on breast cancer risk and diet: the role of fruit and vegetable consumption and the intake of associated micronutrients. Eur J Cancer 2000; 36:636-646.
14. Reddy L, Odhav B, Bhoola KD. Natural products for cancer prevention: a global perspective. Pharmacol Ther 2003; 99:1-13.
15. Willett WC. Diet and health: what should we eat? Science 1994; 264:532-537.
16. Steinmetz KA, Potter JD. Vegetables, fruit and cancer prevention: a review. J Am Diet Assoc 1996; 96:1027-1039.
17. Imai K, Suga K, Nakachi K. Cancer-preventive effects of drinking green tea among a Japanese population. Prev Med 1997; 26:769-775.
18. Messina MJ, Persky V, Setchell KDR et al. Soy intake and cancer risk: a review of the in vitro and in vivo data. Nutr Cancer 1994; 21:113-131.
19. Nabholtz JM, Tonkin K, Smylie M et al. Chemotherapy of breast cancer: are taxanes going to change the natural history of breast cancer? Exert Opin Pharmacother 2000; 1:187-206.
20. Duthie SJ. Berry phytochemicals, genomic stability and cancer: evidence for chemoprotection at several stages in the carcinogenic process. Mol Nutr Food Res 2007; 51:665-674.
21. La Vecchia C. Mediterranean diet and cancer. Public Health Nutr 2004; 7:965-968.
22. Wahle KWJ, Caruso D, Ochoa J et al. Olive oil and modulation of cell signaling in disease prevention. Lipids 2004; 39:1223-1231.
23. Colomer R, Menendez JA. Mediterranean diet, olive oil and cancer. Clin Transl Oncol 2006; 8:15-21.
24. Fresco P, Borges F, Diniz C et al. New insights into the anticancer properties of polyphenols. Med Res Rev 2006; 26:747-766.
25. Aggarwal BB, Shishodia S. Molecular targets of dietary agents for prevention and therapy of cancer. Biochem Pharmacol 2006; 71:1397-1421.
26. Liu RH. Potential synergism of phytochemicals in cancer prevention: mechanisms of action. J Nutr 2004; 134:3479S-3485S.
27. Rosenberg-Zand RS, Jenkins DJA, Diamandis EP. Favonoids and steroid dependent cancers. J Chromat 2002; 777:219-232.
28. Kelloff GF. Perspectives on cancer chemoprevention research and drug development. Adv Canc Res 2000; 78:199-334.
29. Dorai T, Aggarwal BB. Role of chemopreventive agents in cancer therapy. Canc Lett 2004; 25:129-140.
30. Hahn WC, Weinberg RA. Rules for making human tumour cells. N Engl J Med 2002; 347:1593-1603.

31. Korutla L, Cheung JY, Mendelsohn J et al. Inhibition of ligand induced activation of epidermal growth factor receptor tyrosine phosphotylation by curcumin. Carcinogenesis 1995; 16:1741-1745.
32. Korutla L, Kumar R. Inhibitory effect of curcumin on epidermal growth factor receptor kinase activity in A431 cells. Biochim Biophys Acta 1994; 1224:597-600.
33. Kaneuchi M, Sasaki M, Tanaka Y et al. Resveratrol suppresses growth of Ishikawa cells through down regulation of EGF. Int J Oncol 2003; 23:1167-1172.
34. Hong RL, Spohn WH, Hung MC. Curcumin inhibits tyrosine kinase activity of p185neu and also depletes p185neu. Clin Cancer Res 1999; 5:1884-1891.
35. Levy J, Teuerstein I, Marbach M et al. Tyrosine protein kinase activity in the DMBA-induced rat mammary tumor: inhibition by quercetin. Biochem Biophys Res Commun 1984; 123:1227-1233.
36. Way TD, Kao MC, Lin JK. Apigenein induces apoptosis through proteosomal degradation of HER2/neu in HER2/neu-overexpressing breast cancer cells via the phosphoinositol 3-kinase/Akt-dependent pathway. 2004.
37. Menendez JA, Vazquez-Martin A, Colomer R et al. Olive oil's bitter principle reverses acquired autoresistance to trastuzumab (Herceptin TM) in HER2-overexpressing breast cancer cells. BMC Cancer 2007; 7:80-99.
38. Ferrero ME, Bertelli AA, Pelegatta F. Phytoalexin resveratrol (3,4,5,-trihydroxystilbene) modulates granulocyte and monocyte endothelial cell adhesion. Transplant Proc 1998; 30:4191-4193.
39. Suhr Y-J. Cancer chemoprevention with dietary phytochemicals. Nature Rev Cancer 2003; 3:768-780.
40. Nair S, Wenge LI, Kong A-N. T. Natural dietary anticancer chemopreventive compounds: redox-mediated differential signaling mechanisms in cytoprotection of normal cells versus cytotoxicity in tumour cells. Acta Pharmacol Sin 2007; 28:459-472.
41. Kovacic P, Jacintho JD. Mechanisms of carcinogensis: focus on oxidative stress and electron transfer. Curr Med Chem 2001; 8:773-796.
42. D'Alessandro T, Prasain J, Benton MR et al. Polyphenols, inflammatory response and cancer prevention: chlorination of isoflavones by human neutrophils. J Nutr 2003; 133:3773S-3777S.
43. Kundu JK, Suhr YJ. Molecular basis of chemoprevention by resveratrol: NF-kB and AP-1 as potential targets. Mutat Res 2004; 555:65-80.
44. Le Core L, Chalabi N, Ho CT et al. Resveratrol and breast cancer chemoprevention: Molecular mechanisms. Mol Nutr Food Res 2005; 49:462-471.
45. Fiander H, Schneider H. Dietary ortho-polyphenols that induce glutathione-S-transferase and increase the resistance of cells to hydrogen peroxide are potential cancer chemopreventives that act by two mechanisms: the alleviation of oxidative stress and the detoxification of mutagenic xenobiotics. Cancer Lett 2000; 156:117-124.
46. Semczuk A, Jacowicki JA. Alterations of pRb1-cyclinD1-cdk4/6-p16 (INK4A) pathway in endometrial carcinogenesis. Cancer Lett 2004; 203:1-12.
47. Meeran SM, Katiyar SK. Cell cycle control as a basis for cancer chemoprevention through dietary agents. Front Biosci 2008; 13:2191-2202.
48. Gusman J, Malonne H, Atassi GH. A reappraisal of the potential chemopreventive and chemotherapeutic properties of resveratrol. Carcinogenesis 2001; 22:1111-1117.
49. Park OJ, Suhr Y-J. Chemopreventive potential of epigallocatechin gallate and genistein: evidence from epidemiological and laboratory studies. Toxicol Lett 2004; 150:43-56.
50. Koch AE, Polverini PJ, Kunkel SL et al. Interleukin-8 as a macrophage-derived mediator of angiogenesis. Science 1992; 258:1798-1801.
51. Shen F, Chen SJ, Dong XJ et al. Suppression of IL-8 gene transcription by resveratrol in phorbol ester treated human monocyte cells. J Asian Natl Prod Res 2003; 5:151-157.
52. Ishikawa Y, Sugiyama H, Stylianou E et al. Bioflavonoid quercetin inhibits interleukin-1 induced transcriptional expression of monocyte chemoattractant protein-1 in glomerula cells via suppression of nuclear factor kB. J Am Soc Nephrol 1999; 10:2290-2296.
53. Porath D, Riegger C, Drew J et al. Epigallocatechin-3- gallate impairs chemokine production in human colon epithelial cell lines. J Pharmacol Exp Ther 2005; 315:1172-1180.
54. Aneja R, Odoms K, Denenberg AG et al. Theaflavin, a black tea extract, is a novel anti-inflammatory compound. Crit Care Med 2004; 32:2097-2103.
55. Das PM, Singal R. DNA methylation and cancer. J Clin Oncol 2004; 22:4632-4642.
56. Fang MZ, Wang Y, Ai N et al. Tea polyphenol (−)-epigallocatechin-3-gallate inhibits DNA methyltransferase and reactivates methylation silenced genes in cancer cell lines. Cancer Res 2003; 63:7563-7570.
57. Lee WJ, Shim JY, Zhu BT. Mechanisms for the inhibition of DNA methyltransferases by tea catechins and bioflavonoids. Mol Pharmacol 2005; 68:1018-1030.
58. Fulda S, Debatin KM. Sensitisation for tumour necrosis factor-related apoptosis-inducing ligand induced apoptosis by the chemopreventive agent resveratrol. Cancer Res 2004; 64:337-346.

59. Chendil D, Ranga RS, Meigooni D et al. Curcumin confers radiosensitising effect in prostate cancer cell line PC-3. Oncogene 2004; 23:1599-1607.
60. Tang Y, Zhao DY, Elliot S et al. Epgallocatechin-3-gallate induces growth inhibition and apoptosis in human breast cancer cells through surviving suppression. Int J Oncol 2007; 31:705-711.
61. Gao X, Deeb D, Media J et al. Immunomodulatory activity of resveratrol: discrepant in vitro and in vivo immunological effects. Biochem Pharmacol 2003; 66:2427-35.
62. Falchetti R, Fuggetta MP, Lanzilli G et al. Effects of resveratrol on human immune cell function. Life Sci 2001; 70:81-96.
63. Solana R, Casado JG, Delgado E et al. Lymphocyte activation in response to melanoma: interaction of NK-associated receptors and their ligands. Cancer Immunol Immunother 2007; 56:101-9.
64. Nair MP, Kandaswami C, Mahajan S et al. The flavonoid, quercetin, differentially regulates Th-1 (IF-Ngamma) and Th-2 (IL4) cytokine gene expression by normal peripheral blood mononuclear cells. Biochim Biophys Acta 2002; 1593:29-36.
65. Okamoto I, Iwaki K, Koya-Miyata S et al. The flavonoid Kaempferol suppresses the graft-versus-host reaction by inhibiting type 1 cytokine production and CD8+ T-cell engraftment. Clin Immunol 2002; 103:132-44.
66. Yang Y, Paik JH, Cho D et al. Resveratrol induces the suppression of tumor-derived CD4+CD25+ regulatory T-cells. Int Immunopharmacol 2008; 8:542-7.
67. Wahle KWJ, Heys SD, Rotondo D. Conjugated linoleic acids (CLAs): Are they beneficial or detrimental to health? Progr Lipid Res 2004; 43(6):553-587.

Chapter 5

Endogenous Antioxidants and Radical Scavengers

Angela Maria Rizzo, Patrizia Berselli, Stefania Zava, Gigliola Montorfano, Manuela Negroni, Paola Corsetto and Bruno Berra*

Abstract

All living organisms are constantly exposed to oxidant agents deriving from both endogenous and exogenous sources capable to modify biomolecules and induce damages. Free radicals generated by oxidative stress exert an important role in the development of tissue damage and aging. Reactive species (RS) derived from oxygen (ROS) and nitrogen (RNS) pertain to free radicals family and are constituted by various forms of activated oxygen or nitrogen. RS are continuosly produced during normal physiological events but can be removed by antioxidant defence mechanism: the imbalance between RS and antioxidant defence mechanism leads to modifications in cellular membrane or intracellular molecules. In this chapter only endogenous antioxidant molecules will be critically discussed, such as Glutathione, Alpha-lipoic acid, Coenzyme Q, Ferritin, Uric acid, Bilirubin, Metallothioneine, L-carnitine and Melatonin.

Introduction

Living organisms are constantly exposed to oxidative stress capable to modify biomolecules (nucleic acids, proteins, carbohydrates and polyunsaturated lipids) and to exert an important role in the development of tissue damage.[1] The most studied reactive species are the radicals derived from molecular oxygen (ROS) and nitrogen (RNS), but recently chlorine (RCS), bromine (RBS) and sulphur-derived (RSS) radical species have also been identified (listed in Table 1).

The term ROS is a collective term that includes both oxygen radicals, such as superoxide anion radicals ($O_2^{\cdot-}$) and hydroxyl radicals ($OH\cdot$) and some nonradicals that are oxidizing agents and/or are easily converted into radicals, such as H_2O_2 and singlet oxygen ($^1O_2$). This means that all oxygen radicals are ROS, but not all ROS are oxygen radicals. Also the term "reactive" is relative: some species are capable of attacking every molecule they come in contact with, for example $OH\cdot$, while others (e.g., $O_2^{\cdot-}$ and H_2O_2) are highly selective in their reactions.

ROS are exacerbating factors in cellular injury and aging process, as schematically shown in Figures 1 and 2.

ROS induce oxidative damage of DNA, including strand breaks, base and nucleotide modifications, particularly in sequences with high guanosine content. ROS-induced DNA damage leads to p53 activation, growth arrest and apoptosis.[2]

The oxidation of amino acid residues, the subsequent formation of protein aggregates by cross-linking and the production of protein fragments may result in the loss of activity and inactivation of enzymes and metabolic pathways, finally ending up with cell death.[3]

*Corresponding Author: Bruno Berra—Department of Molecular Sciences Applied to Biosystems, DISMAB, University of Milan, Italy. Email: bruno.berra@unimi.it

Bio-Farms for Nutraceuticals: Functional Food and Safety Control by Biosensors
edited by Maria Teresa Giardi, Giuseppina Rea and Bruno Berra.

Table 1. Examples of ROS, RNS, RCS, RBS and RSS

Free Radicals	Non-Radicals
Superoxide $O_2\cdot-$	Hydrogen peroxide H_2O^2
Hydroxyl $OH\cdot$	Hypobromous acid HOBr
Hydroperoxyl $HO_2\cdot$	Hypochlorous acid HOCl
Carbonate $CO_3\cdot-$	Ozone O_3
Peroxyl $RO_2\cdot$	Singlet $O_2{}^1\Delta g$
Alkoxyl $RO\cdot-$	Organic peroxides ROOH
Carbon dioxide radical $CO_2\cdot-$	Peroxynitrite ONOO-
Singlet $O_2{}^1\Sigma g+$	Peroxynitrate O_2NOO-
	Peroxynitrous acid, ONOOH
	Peroxomonocarbonate, $HOOCO_2-$
	Nitrosoperoxycarbonate $ONOOCO_2-$

Carbohydrates oxidation leads to Advanced Glycation End-products (AGEs) accumulation. These molecules are usually more reactive than the initial sugars they were formed from and present a range of pathological effects, including increasing vascular permeability, Low Density Lipoprotein (LDL) oxidation and enhancing oxidative stress.[4]

Free radical oxidation is responsible for the degradation of fatty acids and their esters not only in foods but also in biological membranes and lipoproteins and plays an important role in a wide range of pathological events.[5-18]

The damage on cell membrane architecture induced by lipid peroxidation of membranes fatty acid is schematically depicted in Figure 3.

Lipid hydroperoxides are important nonradical intermediates of lipid peroxidation. They present several biological activities (e.g., cytoxicity) and are relatively easily decomposed to reactive radical species that can propagate the oxidative stress (see Fig. 4).[19]

The oxidative damage to lipoproteins and especially LDL, is known to play a role in a number of diseases, such as cardiovascular diseases, arthritis, dementia and the metabolic syndrome.[20] High circulating oxidized LDL concentrations have been linked to an increased cardiovascular disease risk.[21] Oxidized LDLs promote endothelial cell damage, are chemotactic for leukocytes and may be endocytosed in an uncontrolled manner by macrophages. Furthermore, they can also induce

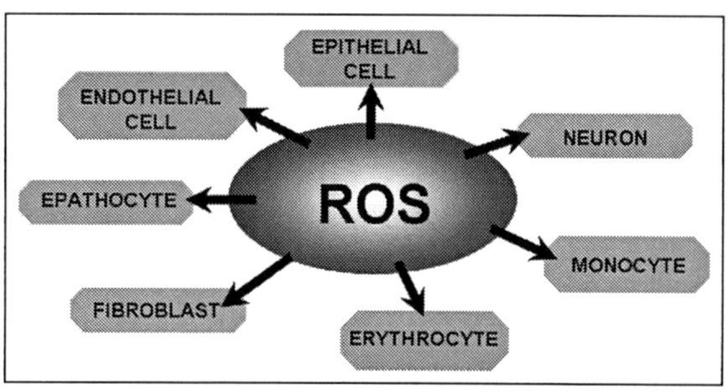

Figure 1. Main cell types affected by ROS. The picture shows as a great number of human cell types are "sensitive" to ROS: this explains why all cells have developed a variety of antioxidant countermeasures.

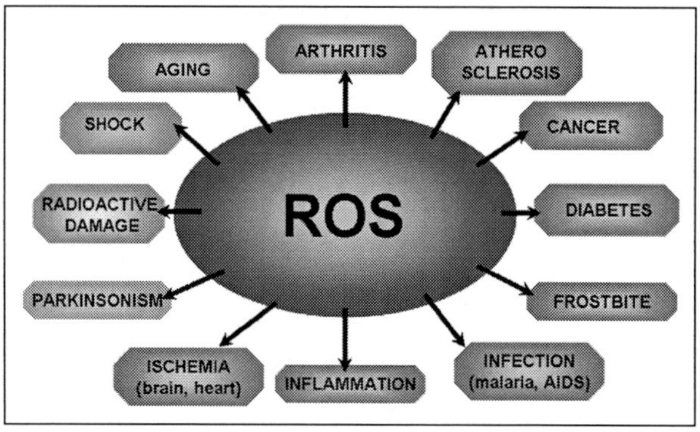

Figure 2. Oxidative stress has been correlated with a continuously increasing number of diseases and disorders: up to now at least one hundred diseases present ROS involvement. The great attention directed to ROS themselves and to their effects on cellular homeostasis is therefore well justified.

smooth muscle cells proliferation.[19] LDL oxidative process involves the degradation of unsaturated fatty acids and cholesterol on the surface and in the core of the lipoprotein and produces a variety of products. In particular, 7-oxygenated oxysterols are the major and more stable oxysterols formed during lipid peroxidation in LDL.[20] The oxidative modification of LDL in the human body is inhibited by endogenous antioxidative defence systems as well as by dietary antioxidants.[21] Many studies have examined the protective effect of antioxidant-rich foods (fresh fruits, vegetables, red wine, olive oil and beverage plants) and phenolic compounds (flavonoids, polyphenols, α-tocopherol) on LDL particle oxidizability.[21,22]

There is a physiological balance between the generation of ROS and their inactivation by the antioxidant system in organisms. Under pathological conditions, ROS are overproduced and result in oxidative stress. In fact, when ROS accumulate, the endogenous antioxidant defences result

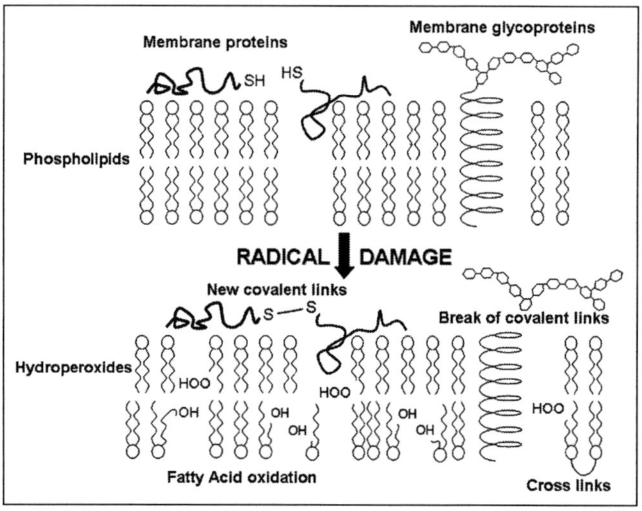

Figure 3. Radical damage occurring at cell membrane: among the different targets, phospholipids fatty acids and membrane proteins/glicoproteins are mainly involved. The consequence of the radical damage is a drastic perturbation of the well organized structure of cell membrane, that can result in altered cell functions, leading even to cell death.

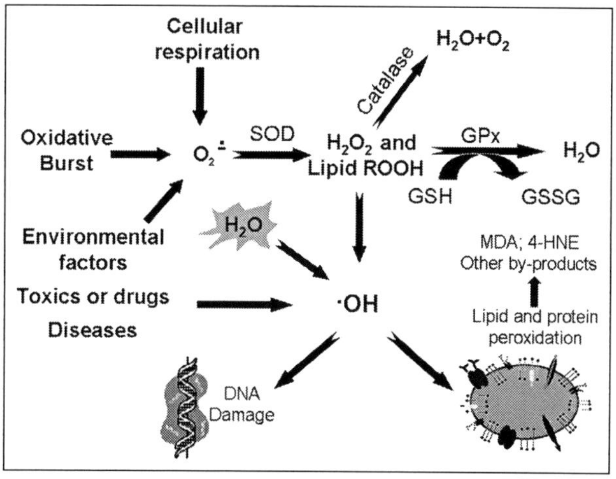

Figure 4. Reactive oxygen species production and resultant disruption of cellular homeostasis. The production of ROS occurs both physiologically (by cellular respiration) and pathologically (by oxidative burst, environmental factors, toxics or drug and diseases). Abbreviations: GSH: reduced glutathione; GSSG: oxidized glutathione; GPx: reduced glutathione peroxidase; SOD: superoxide dismutase; MDA: malondialdehyde; 4-HNE: 4-hydroxynonenal.

inadequate. The imbalance between ROS and antioxidant defence mechanisms leads to oxidative modifications in cellular membrane or intracellular molecules.[23]

An example of the ROS mechanism for inducing cell damage at molecular level is shown in Figure 5. This pathway involves the activation of Nuclear Factor Kappa B (NF-κB), a family of highly regulated dimeric transcription factors, involved in regulating many aspects of cellular growth, differentiation and proliferation. Consistent with these roles, incorrect regulation of NF-κB has been linked to cancer, inflammatory and autoimmune diseases, septic shock, viral infection and improper immune development.[24]

Besides ROS also RNS, which include some radicals such as nitric oxide and nitrogen dioxide in addition to nonradical molecules such as nitrous acid, dinitrogen tetroxide, peroxinitrite and the nitronium cation, can damage proteins, lipids and DNA.

Endogenous Antioxidant Molecules

Antioxidant compounds protect against oxidative damage using four main mechanisms: (i) sequestration of transition metal ions into complexes, (ii) scavenging or quenching free radicals and other ROS and RNS, (iii) breaking chain reactions initiated by free radicals and (iv) repairing damaged molecules, as schematically shown in Figure 6.

Among the endogenous molecules protecting cells from ROS and RNS damage, the following can be included: glutathione, alpha-lipoic acid (thioctic acid), coenzyme Q, ferritin, uric acid, bilirubin, metallothionein, L-carnitine and melatonin.

Reduced Glutathione

Reduced glutathione (GSH) is a water-soluble tripeptide (γ-L-Glutamyl-L-cysteinyl-glycine) mainly present inside cells. High levels of GSH are found in the lower respiratory tract: this is of particular importance, as air may contain ROS and RNS such as ozone and nitrogen dioxide. GSH is also responsible for providing antioxidant protection to red blood cell. In fact, GSH has numerous antioxidant properties: (i) it can react directly with ROS and RNS by virtue of its thiol group; (ii) it can break the disulfide bridges formed inside and between proteins by the action of oxidants.

In its antioxidant actions GSH forms an intermolecular disulfide nonradical end product, glutathione disulfide (GSSG), which can either be exported from the cells or transformed back

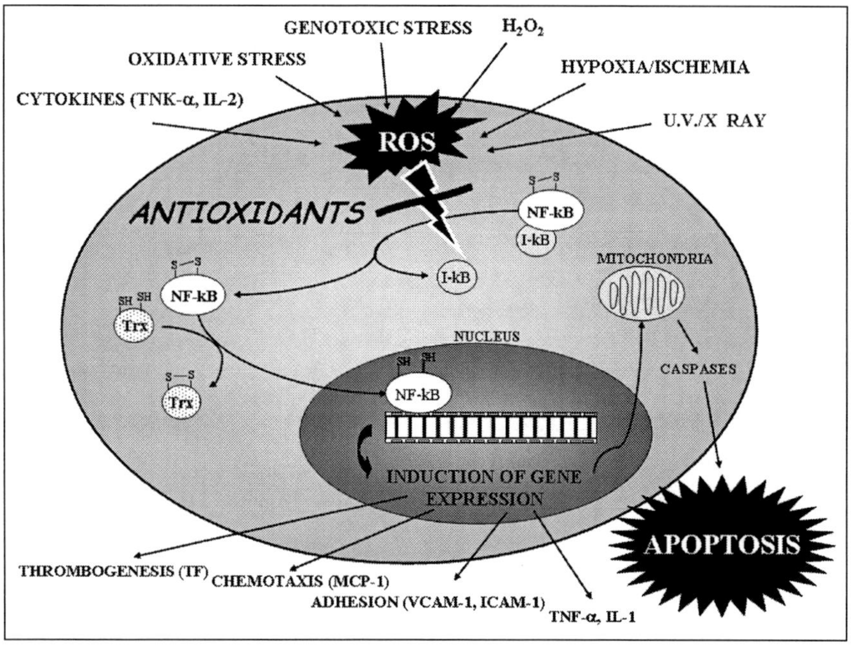

Figure 5. NF-κB activation by ROS. The most common NF-κB form is a heterodimer bound to the inhibitory subunit (I-κB). In the classical activation pathway, in response to extracellular inducers, such as ROS, I-κB becomes phosphorylated and this leads to its degradation and to the release of NF-κB, that needs to be in a reduced form, at a cysteine residue. Therefore Thioredoxin (Trx), an ubiquitous protein with oxidoreductase activity, reduces NF-κB and activates it. NF-κB is therefore free to translocate from the cytoplasm to the nucleus, where it can bind to its κB motif and interact with DNA-associated factors as well as the general transcriptional apparatus in order to mediate an effective transcriptional activation. This pathway terminates with the increased transcription of target genes encoding chemokines, cytokines, adhesion molecules.[25] Mitochondrial release of cytochrome c can be triggered by ROS and leads to the activation of the caspase cascade, which ends with DNA cleavage and apoptosis.[26]

to GSH by the combined action of GSH reductase and the Nicotinamide Adenine Dinucleotide Phosphate (NADPH) cofactor.

GSH also acts as an antioxidant indirectly by regenerating ascorbate from dehydroascorbate and alpha-tocopherol from tocopheroxyl radical.[27-29]

The endogenous protection exerted by GSH versus radicals and oxidant species is further substantiated by the identification of a group of nutrients, including GSH itself, that can directly or indirectly protect mitochondria from oxidative damage and improve mitochondrial function, named "mitochondrial nutrients".[30] The direct protection offered by these molecules includes preventing the generation of oxidants, scavenging free radicals or inhibiting oxidant reactivity and elevating cofactors of defective mitochondrial enzymes with increased Michaelis-Menten constant to stimulate enzyme activity and also protect enzymes from further oxidation; the indirect protection includes repairing oxidative damage by enhancing antioxidant defense systems either through activation of enzymes, such as hemeoxygenase 1, NAD(P)H:quinone oxidoreductase 1, which neutralize ROS, or through increase in mitochondrial biogenesis.[31]

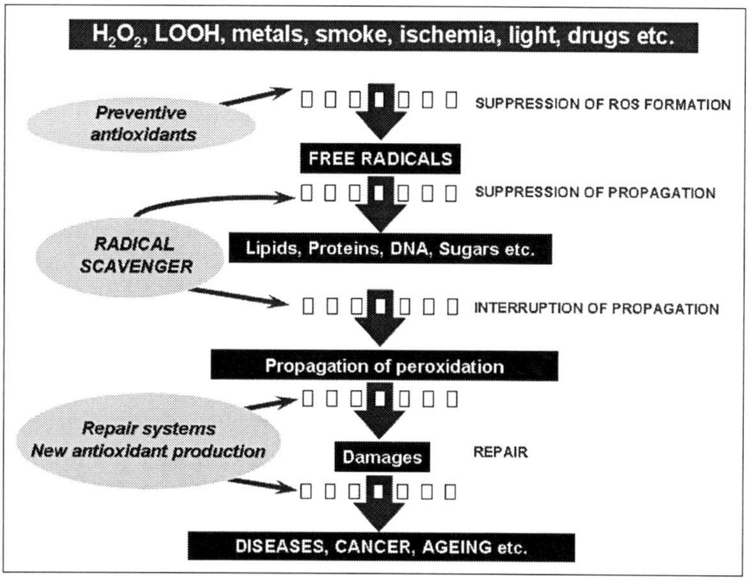

Figure 6. In vivo defences against ROS. Cells have developed many different conteracting measures that act at the different stages of ROS formation and propagation. These mechanisms range from preventive antioxidants to radical scavengers and repair systems.

Alpha-Lipoic Acid (LA)

The lipoic acid amide is the coenzyme of the dihydrolipoate acyltransferase, subunit of multi-enzymatic mitochondrial complexes catalyzing oxidative decarboxilation of pyruvate, alpha-ketoglutarate and branched-chain alpha-keto acid. At present, the majority of researchers believe that LA is not a vitamin. In fact, it is supposed that it is synthesized in human and animal body in mitochondria; and being octanoic acid and cysteine direct precursors of lipoic acid.[32] LA and its derivatives improve the age-associated decline of memory[33,34] and mitochondrial structure and function.[35] Since now authors attribute efficiency of therapy with lipoic acid in diabetes mellitus and neurodegenerative diseases to the antioxidant properties of lipoate/dihydrolipoate system influencing the tissue concentration of the reduced forms of other antioxidants, including GSH (lipoate therefore acting as antioxidant of antioxidants). However, new evidences indicate that lipoic acid might also counteract NF-κB activation trigged by oxygen shock.[35]

Coenzyme Q

For Coenzyme Q (CoQ or CoQ_{10}) a number of functions have been established during the years but its role as an effective antioxidant of cellular membranes remains of dominating interest. This compound is our only endogenously synthesized lipid soluble antioxidant, present in all membranes and exceeding both in amount and efficiency the other antioxidants. The protective effect is extended to lipids, proteins and DNA mainly because of its close localization to the oxidative events and the effective regeneration by continuous reduction at all locations. Its biosynthesis is influenced by nuclear receptors which may give the possibility, in the future, by using agonists or antagonists, of reestablishing the normal level in deficiencies caused by genetic mutations, aging or cardiomyopathy.[36]

The high concentration in different organelles may reflect a compartmentalization for absolving different functions. For example, in the inner mitochondrial membrane CoQ has at least four different functions: redox carrier, antioxidant, activator of uncoupling proteins and factor influencing the permeability transition pore (PTP). In addition, it was suggested that lysosome contains a NADH-dependent CoQ reductase involved in traslocation of protons into the lysosomal lumen.[36]

The effectiveness of CoQ as inhibitor of lipid peroxidation is based on its complex interaction during the peroxidation process. The primary action is the prevention of lipid peroxyl radicals ($LOO^{\cdot}$) production during initiation. The reaction of perferryl radical (Fe^{3+}–$O_2^{\cdot-}$) with a polyunsaturated lipid to produce lipid radicals and its subsequent reaction with O_2 to form lipid peroxyl radicals represent the well-known initiating steps for lipid peroxidation. $CoQH_2$ reduces the perferryl radical with the formation of ubisemiquinone ($CoQH^{-}$) and H_2O_2, in presence of NADPH. Moreover, it is possible that $CoQH_2$ eliminates $LOO^{\cdot}$ directly, therefore acting on the propagation steps of lipid peroxidation as well. It is also established that the reduced lipid effectively regenerates vitamin E from gamma-tocopheroxyl radical, a favored reaction in comparison with the regeneration with ascorbate.[37]

Echtay et al[38] showed that CoQ is an obligatory cofactor for Uncoupling Proteins (UCPs). UCPs are situated in the inner mitochondrial membrane and can traslocate H^+ from the outside to the inside of the mitochondria. Thus, the proton gradient built by the respiratory chain is uncoupled from oxidative phosphorilation. Five UCPs are known and UCP1 is the most well characterized, present in brown adipose tissue where it participates in thermogenesis. In addition to thermogenesis UCPs could be involved in suppression of oxygen radicals and there are also suggestions that altered UCPs expression could be related to human diseases such as obesity and diabetes.[39]

The inner membrane of mitochondria has a low permeability to ions and solutes in order to permit energy conservation in the form of a proton electrochemical gradient over the membrane. In order to facilitate transmembrane transport the inner membrane contains a number of macromolecules transporters and ion channels (e.g., PTP). Several analogues of CoQ have been shown to affect mitochondrial PTPs,[40] suggesting CoQ as a structural element and modulator. In fact, CoQ seems to prevent PTP opening-dependent apoptosis related events, i.e., mitochondrial membrane depolarization, ATP hydrolysis, cytochrome *c* release, caspase-9 activation and DNA fragmentation, via the inhibition of PTP opening.[41]

Finally, a protective effect of CoQ on decreased oxidation stress resistance induced by simvastatin was recently reported.[42]

The different functions of CoQ can be summarized as follows:

Partaking as electron carrier in the mitochondrial respiratory chain;
Participation in extra-mitochondrial electron transport (PLM, lysosomes);
Participation in the endogenous synthesis of lipid-soluble antioxidant;
Regulation of mitochondrial permeability transition pore;
Participation in the activation of mitochondrial uncoupling proteins;
Regulation of the physicochemical properties of membranes;
Modulation of the amount of beta 2 integrins of the surface of blood monocytes;
Improvement of endothelia disfunction (probably by increasing Nitric Oxide (NO)).

Ferritin

Ferritin is a globular protein complex consisting of 24 protein subunits and is the main intracellular iron storage protein in both prokaryotes and eukaryotes, keeping it in a soluble and nontoxic form. Ferritin is present in every cell type. In vertebrates, these subunits are both of the light (L) and the heavy (H) type with an apparent molecular weight of 19 kDA or 21 kDA, respectively.[43] Iron is essential for cellular functions. Excessive cellular iron can be dangerous to the cell because of the increased oxidation of DNA, proteins and cell membranes resulting from the generation of free radicals by iron-driven Fenton reactions. Therefore it is necessary to maintain a tight control on the availability of iron. Cellular iron balance is regulated by an elegant post transcriptional mechanism involving Iron Regulatory Protein-one and Iron Regulatory Protein-two (IRP-1 and IRP-2) which allow the regulation of ferritin translation and in the mean time the regulation of transferrin receptor. IRP-1 and IRP-2 are involved in the receptor mediated endocytosis of ferric transferrin in liver and intestinal crypt cell; while transferrin is a plasma and extracellular fluid iron binding transport protein. When a transferrin protein loaded with iron encounters a transferrin receptor on the surface of a cell (importantly, to erythroid precursors in the bone marrow), it binds to it and is consequently transported into the cell in a vesicle. The H^+ ATPase of the cell will decrease the pH of the vesicle, causing transferrin to release its iron ions. The receptor is then transported through the endocytic cycle back to the cell surface, ready for another round of iron uptake. Each transferrin molecule has the ability to carry two iron ions in the ferric form (Fe^{3+}).

Uric Acid

Uric acid is a nitrogenous waste product of purine metabolism and is ubiquitous in body fluids. It is also very important as a free radical scavenger: according to one estimate, uric acid and albumine are the two major antioxidants in human plasma, contributing 24% and 33% respectively of the total antioxidant activity.[44]

Although previous epidemiological studies have suggested that hyperuricemia may be a risk factor for cardiovascular diseases, new data suggest that the elevation of uric acid, a natural antioxidant, might be a defence mechanism against advanced atherosclerosis.[45] Along this line some authors have hypothesized that the cardio-protective effect of red wine may be mediated by urate.[46]

Uric acid appears to be elevated transiently only following both endurance and short duration high-intensity exercise. Increased plasma urate may arise from the increased demand of ATP during high-intensity exercise. ATP can be regenerated via the adenylate chinase catalyzed reaction and subsequent break down of nucleotide by-products via xantine oxidase. On the other hand exercise training appears to reduce both plasma urate concentration at rest and its increase following high-intensity exercise. These findings might indicate that exercise training reduces the contribution of urate to the body's antioxidant capacity; however, it might also mean that adaptation from exercise training reduces oxidant production via the xantine oxidase pathway.[47]

Bilirubin

Bilirubin: Heme oxygenase-1 (HO-1) is the inducible isoform of heme oxygenase, the enzyme that generates carbon monoxide, iron and biliverdin using heme as a substrate.[48] Biliverdin is rapidly reduced to bilirubin by biliverdin reductase, so that the actual end product of the heme degradation pathway is bilirubin. The discovery that bile pigments are potent scavengers of free radicals in vitro[49] has helped to reinforce the concept that HO-1 is a crucial inducible antioxidant system engaged to combat oxidative injury and other forms of cellular stress. Recent findings have shown important roles for bilirubin in the prevention of ischemic injury in isolated hearts,[50] attenuation of oxidative damage in cultured cells[51] and modulation of airway smooth muscle contractility.[52] In addition, bilirubin effectively defends against hydrogen peroxide-mediated damage in neurons,[53] where a redox cycle between biliverdin and bilirubin appears to amplify this protective effect.[54] Recently, administration of biliverdin in vivo was also reported to protect rat kidney, liver and gut from ischemia-reperfusion injury.[55-57] Similarly, in vitro experiments produced evidence on the ability of bile pigments to scavenge NO and NO-related species.[58] Of great interest are also epidemiological studies sustaining a beneficial action of bilirubin against the development of cardiovascular disease and cancer.[59,60]

Lanone S et al demonstrated that bilirubin can counteract hypothension elicited by endotoxin through a mechanism mediated by direct inhibition of NAD(P)H oxidase, involved in inducible NO synthase (iNOS) induction, that consequently leads to a decrease of iNOS expression and activity.[61]

The protection of bilirubin against classic coronary heart diseases risk factor was also demonstrated by Troughton J et al[62] and increased serum total bilirubin level is associated with reduced peripheral artherial disease.[63] Moreover elevated serum bilirubin concentration protects from coronary flow reserve impairment, coronary microvascular disfunction and possibly coronary atherosclerosis.[64] Other predictors of atherosclerosis, i.e., endothelian disfunction and increased carotid intima-media thickness, are decreased in healthy subjects by lower serum bilirubin concentration.[65]

Recent studies suggest that, due to its in vivo antioxidant and free radical scavenging properties at physiological concentration, bilirubin constitutes one of the defence mechanisms in neurotrauma patients and in full term normal neonates.[66-67]

A summary of the studies that examined the relationship of bilirubin as an antioxidant and radical scavenger both in neonates and adults is reported in the paper by Sedlac et al.[68]

Metallothioneins

Metallothioneins (MTs) are ubiquitous low molecular weight proteins and polypeptides which present extremely high amount of heavy metals bound to cysteine residues that give rise to metal-thiolate clusters.[69] MTs constitute a protein superfamily of 15 families comprising many sequences inferred from both amino acid and polynucleotide sequences obtained from all animal phyla examined to date and also from certain fungi, plants and cyanobacteria.[70]

The MT gene family in mammals consists of four subfamilies designated MT-1 through MT-4. Increasing evidence shows that mammalian MT-1/MT-2 isoforms are involved in zinc homeostasis and protection against heavy metal toxicity and oxidative stress. MT-3 is expressed mainly in neurons but also in glia; while MT-4 is mostly present in differentiating stratified squamous epithelial cells.

The Metal responsive element-binding Transcription Factor-1 (MTF-1) plays an important role in MT transcription. Several lines of evidence suggest that the highly conserved six-zinc finger DNA-binding domain of MTF1 functions as a zinc-sensing domain and the linkers between the six different fingers can actively participate in modulating MTF1 translocation to the nucleus and its binding to the MT1 gene promoter.[71]

MT-3 is also known as a growth inhibitory factor and is a brain specific isomer of metallothionein.[72] In previous studies, MT-3 exhibited free radical scavenging activity and regulatory control of zinc metabolism in the brain to protect neurons against toxic metals.[73] Several reports suggest that MT-3 is related to neuronal loss in neurodegenerative diseases. In fact, MT-3 induces HO-1 expression in dopaminergic SH-SY5Y neuronal cells and this expression confers neuroprotection against oxidative injury. MT-3 also induces the Nuclear factor-E2 Related Factor (Nrf2) nuclear translocation, which is upstream of MT-3-induced HO-1 expression and activates Akt and ERK1/2 phosphorylation. The PI3K/Akt and ERK1/2 pathways are involved in MT-3-induced Nrf2 nuclear translocation, HO-1 expression and neuroprotection.[74]

L-Carnitine

L-carnitine (4-*N*-trimethylammonium-3-hydroxybutyric acid) is a natural biomolecule that plays important physiological roles shuttling the long-chain fatty acids across the inner mitochondrial membrane for β-oxidation and ATP production in peripheral tissues. However, in neural function a major modulatory role for L-carnitine may be played by the transfer of acetyl groups for acetylcholine synthesis, as well as by influencing signal transduction pathway and gene expression.[75,76]

L-carnitine prevents oxidative stress, regulates NO and the activity of enzymes involved in defence against oxidative damage.[77] Also, L-carnitine has a protective effect on the activity of mitochondrial enzyme succinate dehydrogenase as well as on the activity of the antioxidant enzymes CAT and SOD against the 3-Nitro Propionic Acid (NPA)-induced neurotoxicity.[76] As previously indicated, the antioxidant defense system is composed of mainly three enzymes: GPx, CAT and SOD. Since L-carnitine can protect these enzymes from further peroxidative damage, it results very effective in normalizing age-associated alterations.

According to data of the study by Gülçin I.,[75] L-carnitine was found to be an effective antioxidant in different in vitro assays, such as the determination of the total reducing power by potassium ferricyanide reduction, by diphenylpicrylhydrazyl (DPPH) radical and superoxide anion radical scavenging, by hydrogen peroxide scavenging and metal chelating activities, when it is compared to standard antioxidant compounds such as alpha-tocopherol and trolox, which is a water-soluble analogue of tocopherol.

Dokmeci and coworkers concluded that a L-carnitine pretreatment may have a protective effect in esperimental testicular torsion-detorsion model in rats by its well-known antioxidant potential.[78]

Finally, lipoic acid and acetyl-L-carnitine, the acetylated form of L-carnitine, can reverse iron-induced oxidative stress in human fibroblast and, via the reduction of oxidative stress in mitochondria, may contribute to the regulation of blood pressure and vascular tone.[79,80]

In line with these data are the conclusions of a study by Xie J. and coworkers, who found that L-carnitine has a protective effect on ischemia reperfusion injury, which is partly due to its prevention of energy loss and to its antioxidant activity.[81]

According to Thangasamy T. and coworkers, defects in immunological functions are responsible for many manifestations of aging and form the basis for many age-related diseases. Since L-carnitine demonstrates free radicals quenching properties in lymphocytes of aged animals, the authors conclude that the antioxidant role of L-carnitine could contribute to its cell protective functions.[82]

In order to survive to different types of injuries, brain cells have evolved several responses to control stress and its related damages. Among activated gene products, heat shock proteins (HSP) are involved and form a highly conserved system responsible for repairing the ROS damages. The heat shock response contributes to establish a cytoprotective state in a wide variety of human diseases. There is now a strong interest in discovering and developing pharmacological agents capable of inducing the heat shock response. Acetyl-L-carnitine is proposed as a therapeutic agent for several neurodegenerative disorders and there is now evidence that it may play a critical role as modulator of cellular stress response in healthy and disease states.[83]

Melatonin

Melatonin (N-acetyl-5-methoxytryptamine) is a naturally occurring hormone present in bacteria, plants, eukaryotes and fungi as well as in all phyla of multicellular animals and also in most animals, including humans; probably its original evolutionary role was to act as an antioxidant[84] and its antioxidant properties have been shown in both tissue culture and in intact animals. However, the discovery of different targets for melatonin in the cell suggests a variety of mechanisms of action for this compound. At the present, melatonin functions seem to fall into three categories: receptor-mediated, protein-mediated and nonreceptor-mediated effects.

Receptor-mediated melatonin events involve both membrane and nuclear receptors. Although membrane melatonin receptors are well-characterized in humans,[85] some of the receptor-related antioxidant effects of melatonin seem also to be related to its nuclear receptors.[86]

One problem still opened and much discussed is whether its antioxidant action is direct or mediated by the activation of critical antioxidant enzymes.[87] The evidence for a direct effect resides on the fact that melatonin acts as a powerful free radical scavenger in isolated cell-free-systems.[88] On the contrary there are also reports indicating that melatonin can behave as a pro-oxidant in such systems.[89]

A wide range of antioxidant enzymes are induced by melatonin: GPx, Glutathione Reductase (GRd), CAT and SOD.[90-91] Moreover, the levels of some prooxidant enzymes such as lipoxygenase and iNOS are depressed after melatonin treatment.[92,93] The dual mechanism of action of melathonin, i.e., influencing antioxidant enzymes activity and induction of their expression, in a single molecule is unique for an antioxidant. In this view the interaction between membrane and nuclear melatonin signalling has been proposed.[86]

Melatonin is also effective in protecting nuclear DNA, membrane lipids and cytosolic proteins from oxidative damage and from increased membrane rigidity[94] induced by a variety of free radical generating agents, including the carcinogen safrole, Fenton reagents, glutathione depletion, carbon tetrachloride and ionizing radiation.[95]

In brain, melatonin is present at a concentration that is only 5% of that found in serum[97]; therefore it likely can give little scavenging contribution in comparison to predominant antioxidant species such as glutathione and alpha-tocopherol.[98] However, melatonin is present in cerebrospinal fluid at concentrations higher than in serum, suggesting that the pineal gland may be a source of blood-born melatonin.[99]

Within the brain there are three major plasma member receptors for melatonin. On the other side the presence of additional melatonin binding sites in the nuclei of many cell types suggests the existence of mechanisms different from that mediated by the interaction with plasma membrane receptors.[100] The specificity of melatonin may reside in its properties as a neurohormone which affects transcriptional events in the CNS.[101] A review on these new aspects of melatonin as a "new" antioxidant have been recently published.[102]

The protective effects of melatonin were further analyzed also in isolated mitochondria and compared with those of other known antioxidants, such as N-acetylcysteine (NAC) and vitamins C and E.[103] Two main considerations suggest a role for melatonin also in mitochondrial homeostasis. First, mitochondria produce high amounts of ROS and RNS, being superoxide anion the main radical produced. Second, mitochondria depend on the GSH uptake from the cytoplasm, although they have GPx and GRd to maintain GSH redox cycling; thus, the antioxidant effect of melatonin and its ability to increase GSH levels may be of great importance for mitochondrial physiology.[96] In fact, after oxidative stress in mitochondria virtually all GSH is oxidized to GSSG and the activity of both GPx and GRd are reduced to practically zero. In this situation, melatonin counteracts these effects, restores basal levels of GSH and the normal activities of both GPx and GRd.

As already indicated, aerobic cells use oxygen for the production of 90-95% of the total amount of ATP they use and the antioxidant systems must avoid the oxidative damage to the mitochondrial membrane which leads to an impaired ATP production.

During aging and some neurodegenerative diseases, oxidatively damaged mitochondria are unable to maintain the energy demand of the cell leading to a further increased production of free radicals.

Both processes, i.e., defective ATP production and increased oxygen radicals, may induce mitochondrial-dependent apoptotic cell death.

One of the most recent advances in basic gerontology is the free radical theory of aging were ROS, as previously indicated, damage cellular macromolecules resulting in mutations and genome instability; all these abnormalities can lead to the development of age-related pathological phenomena.[104] There is great deal of evidence indicating also that aging is characterized by a progressive deterioration of circadian time keeping.[105]

Reduced melatonin concentration during aging, especially in the nocturnal levels, has been reported in pineal gland, plasma, CSF and in urine,[106] although in some studies the difference in age-related melatonin level was not statistically significant.[107] Besides the age-related decline of melatonin production, age-related changes in the timing of the melatonin level rhythm have also been reported.[108] The possible mechanism of age-related melatonin level changes was associated with changes in the pineal gland morphology and its calcification.

In addition to the changes in melatonin levels rhythms during aging, a decrease of core body temperature, cortisol, vasopressin, blood pressure, pulsatile LH, testosterone secretion, β-endorphine levels and other parameters have also been reported.[106] These findings suggest that the changes observed in the melatonin level rhythm may be part of a general effect of aging, in particularly on the central clock of SCN and/or its regulation.

Melatonin has been reported to exert neuroprotective effects in several experimental and clinical situations involving neurotoxicity and/or excitotoxicity. Additionally, in a series of pathologies in which high production of free radicals is the primary cause of the disease, melatonin is also protective. In fact the common feature of these diseases is the existence of mitochondrial damage due to oxidative stress.

Reactive Oxygen Species: Always Bad?

While it is widely accepted that increased levels of ROS contribute to develop neurodegenerative disease including atherosclerosis, Alzheimer's, diabetes, metabolic syndrome and carcinogenesis, a growing number of studies clearly demonstrate that ROS are part of normal cellular signals and induce cell differentiation and apoptosis. Thus, it has been hypothesized that decreased levels of ROS may lead to cancer development, on the basis of the following proposed mechanism: (i) inactivation of nuclear genes governing mitochondrial DNA replication interferes with mitochondrial biogenesis; (ii) reduced mitochondrial capacity leads to a low level of oxidative phosphorylation and a decreased ROS production, which in turn causes (iii) low levels of manganese SOD activity and (iv) reduced lipid peroxidation, that (v) de-inhibits glucose-6-phosphate dehydrogenase (G6PD), a key enzyme in pentose phosphate pathway that provides riboses and NAD(P)H for uncontrolled DNA synthesis and cell proliferation, well known characteristics of cancer cells. A clarification of the real role of ROS in cancer may shed light on the understanding of how impairment of mitochondria leads to malignant transformation of normal cells and offers new types of strategies for cancer prevention and therapy.[17,109-111]

In this view, it has become increasingly evident that ROS also play a role in the regulation of many intracellular signaling pathways that are important for normal cell growth and inflammatory responses for host defense.

Although there is a large number of evidences indicating that ROS and the redox state have a signaling role in bacteria and plants, until recently there was less evidence in mammalian cells. For example, in bacteria the transcription factor OxyR is redox sensitive.[112] In animals, ROS-based signaling of Protein Tyrosine Phosphatase 1B (PTP-1B), thioredoxin, SERCA2 and Ras is well established.[113-115] Another radical that plays a major role in normal physiological function is nitric oxide ($NO^{\bullet}$): it takes part in the regulation of vascular tone, nerve function and immune regulation.

The role of ROS in the signaling of a number of growth factors has also recently been well established. An excellent example is the role of ROS in angiotensin signaling as studied by Griendling and coworkers.[116] They showed that exposure of vascular smooth muscle to angiotensin II results in a smooth muscle growth that is dependent upon increased production of $O_2^{\cdot-}$ by NAD(P)H oxidase and its subsequent dismutation to H_2O_2. H_2O_2 then activates downstream prosurvival pathways and, in vivo, this results in vascular hypertrophy. Other growth factors, such as the platelet derived growth factor, have been shown to have similar signaling pathways.[117]

Various mechanisms have been explored recently that can explain how ROS can influence intracellular events. These generally involve the oxidation of cysteine residues and formation of –S–S– bonds.[118-121] These bonds can be within a molecule and result in a conformational change or between protein strands, in which case they result in dimerization of proteins. The creation of –S–S– bonds can also result in the release of an inhibitory molecule.

In conclusion "a key concept in dealing with ROS must be to regulate but not eradicate them since turning off the production of ROS is equivalent to turning of the engine that powers us".[122]

Conclusion

It is widely accepted that increased level of ROS contributes to neurodegenerative diseases including atherosclerosis, Alzheimer's disease, diabetes, metabolic syndrome and carcinogenesis. Fortunately, in our bodies, antioxidant compounds protect against oxidative damage by the sequestration of transitional metal ions into complexes scavenging or quenching free radicals and other ROS and RNS, breaking chain reactions and reparing damaged molecules.

Among these endogenous molecules with antioxidant activity, the following can be included which are discussed in this chapter: glutathione, alfa-lipoic acid, coenzyme Q, ferritin, uric acid, bilirubin, metallothioneine, L-carnitine, melatonin. However, a growing number of studies clearly demonstrates that reactive oxygen species are part of normal cellular signals and induce cell differentiation and apoptosis. It has become increasingly evident that ROS also play a role in the regulation of inflammatory response of host defence.

In conclusion, reactive oxygen species are not always bad but they can play a role in the signalling of a number of growth factors. In this view, it has been hypothesized that decreased levels of ROS may lead to degenerative diseases, generating a key concept that ROS must be regulated but not eradicated.

References

1. Halliwell B, Whiteman M. Measuring reactive species and oxidative damage in vivo and in cell culture: how should you do it and what do the results mean? Br J Pharmacol 2004; 142:231-255.
2. Halliwell B. Oxidative stress and cancer: have we moved forward? Biochem J 2007; 401:1-11.
3. Stadman ER, Levine RL. Free radical-mediated oxidation of free amino acids and amino acid residues in proteins. Amino Acids 2003; 25:207-218.
4. Ogino K, Wang DH. Biomarkers of oxidative/nitrosative stress: an approach to disease prevention. Acta Med Okayama 2007; 61(4):181-189.
5. Bruckdorfer KR. Antioxidant and CVD. Proc Nutr Soc 2008; 67(2):214-222.
6. Riccioni G, Bucciarelli T, Mancini B et al. The role of the antioxidant vitamin supplementation in the prevention of cardiovascular diseases. Expert Opinion on Investigational Drugs 2007; 16(1):25-32.
7. Nishikawa M, Inoue M. Oxidative stress and tissue injury. Masui 2008; 57(3):321-326.
8. Tremellen K. Oxidative stress and male infertility—a clinical perspective. Hum Reprod Update 2008; 14(3):243-258.
9. Gao L, Laude K, Cai H. Mitochondrial pathophysiology, reactive oxygen species and cardiovascular diseases. Vet Clin North Am Small Anim Pract 2008; 38(1):137-155.
10. Nunomura A, Moreira PI, Takeda A et al. Oxidative RNA damage and neurodegeneration. Curr Med Chem 2007; 14(28):2968-2975.
11. Rohr-Udilova NV, Stolze K, Sagmeister S et al. Lipid hydroperoxides from processed dietary oils enhance growth of hepatocarcinoma cells. Mol Nutr Food Res 2008; 52(3):352-359.
12. Esme H, Cemek M, Sezer M et al. High levels of oxidative stress in patients with advanced lung cancer. Respirology 2008; 13(1):112-116.
13. Riedl MA, Nel AE. Importance of oxidative stress in the pathogenesis and treatment of asthma. Curr Opin Allergy Clin Immunol 2008; 8(1):49-56.
14. Shi Q, Gibson GE. Oxidative stress and transcriptional regulation in Alzheimer disease. Alzheimer Dis Assoc Disord 2007; 21(4):276-291.
15. Bonomini F, Tengattini S, Fabiano A et al. Atherosclerosis and oxidative stress. Histol Histopathol 2008; 23(3):381-390.
16. Walters DM, Cho HY, Kleeberger SR. Oxidative stress and antioxidants in the pathogenesis of pulmonary fibrosis: a potential role for Nrf2. Antioxid Redox Signal 2008; 10(2):321-332.
17. Wittgen HG, van Kempen LC. Reactive oxygen species in melanoma and its therapeutic implications. Melanoma Res 2007; 17(6):400-409.
18. Hung JH. Oxidative stress and antioxidants in preeclampsia. J Chin Med Assoc 2007; 70(10):430-432.
19. Gerry AB, Satchell L, Leake DS. A novel method for production of lipid hydroperoxide—or oxysterol-rich low-density lipoprotein. Atherosclerosis 2008; 197:579-587.
20. Aldred S. Oxidative and nitrative changes seen in lipoproteins following exercise. Atherosclerosis 2007; 192:1-8.
21. Lapointe A, Couillard C, Lemieux S. Effect of dietary factors on oxidation of low-density lipoprotein particles. J Nutr Biochem 2006; 17:645-658.
22. Saura-Calixto F, Goni I. Antioxidant capacity of the Spanish Mediterranean diet. Food Chem 2006; 94:442-447.
23. Ott M, Gogvadze V, Orrenius S et al. Mitochondria, oxidative stress and cell death. Apoptosis 2007; 12(5):913-922.
24. Nishikori M. Classical and alternative NF-kB activation pathways and their roles in lymphoid malignancies. J Clin Hematopathol 2005; 45:15-24.
25. Raha S, Robinson BH. Mitochondria, oxygen free radical and apoptosis. Am J Med Genet 2001; 106(1):62-70.
26. May MJ, Ghosh S. Signal transduction through NF-kB. Immunol Today 1998; 19(2):80-88.
27. Franco R, Schoneveld OJ, Pappa A et al. The central role of glutathione in the pathophysiology of human diseases. Arch Physiol Biochem 2007; 113(4-5):234-258.
28. Devasagayam TP, Tilak JC, Boloor KK et al. Free radicals and antioxidants in human health: current status and future prospects. J Assoc Physicians India 2004; 52:794-804.
29. Botta D, White CC, Vliet-Gregg P et al. Modulating GSH synthesis using glutamate cysteine ligase transgenic and gene-targeted mice. Drug Metab Rev 2008; 40(3):465-477.
30. Liu J, Ames BN. Reducing mitochondrial decay with mitochondrial nutrients to delay and treat cognitive dysfunction, Alzheimer's disease and Parkinson's disease. Nutr Neurosci 2005; 8(2):67-89.
31. Liu J. The effects and mechanisms of mitochondrial nutrient alpha-lipoic acid on improving age-associated mitochondrial and cognitive dysfunction: an overview. Neurochem Res 2008; 33(1):194-203.
32. Bilska A, Włodek L. Lipoic acid—the drug of the future? Pharmacol Rep 2005; 57(5):570-577.
33. Manda K, Ueno M, Anzai K. Memory impairment, oxidative damage and apoptosis induced by space radiation: ameliorative potential of alpha-lipoic acid. J Behav Brain Res 2008; 187(2):387-395.

34. Makeeva AV, Popova TN, Matasova LV et al. Effects of lipoic acid on citrate content, aconitate hydratase activity and oxidative status during myocardial ischemia in rats. Biochemistry (Mosc) 2008; 73(1):76-79.
35. Lee CK, Lee EY, Kim YG et al. Alpha-lipoic acid inhibits TNF-alpha induced NF-kappa B activation through blocking of MEKK1-MKK4-IKK signaling cascades. Int Immunopharmacol 2008; 8(2):362-370.
36. Gille L, Nohl H. The existence of a lysosomal redox chain and the role of ubiquinone. Arch Biochem Biophys 2000; 375:347-354.
37. Bentinger M, Brismar K, Dallner G. The antioxidant role of coenzyme Q. Mitochondrion 2007; 7(l):S41-50.
38. Echtay KS, Winkler E, Klingenberg M. Coenzyme Q is an obligatory cofactor for uncoupling protein function. Nature 2000; 408:609-613.
39. Echtay KS, Winkler E, Frischmuth K et al. Uncoupling proteins 2 and 3 are highly active H+ transporters and highly nucleotide sensitive when activated by coenzyme Q (ubiquinone). Proc Natl Acad Sci USA 2001; 98:1416-1421.
40. Walter L, Miyoshi H, Leverve X et al. Regulation of the mitochondrial permeability transition pore by ubiquinone analogs, A progress report. Free Radic Res 2002; 36:405-412.
41. Papucci L, Schiavone N, Witort E et al. Coenzyme Q10 prevents apoptosis by inhibiting mitochondrial depolarization independently of its free radical scavenging property. J Biol Chem 2003; 278:28220-28228.
42. Kettawan A, Takahashi T, Kongkachuichai R et al. Protective effects of coenzyme q(10) on decreased oxidative stress resistance induced by simvastatin. J Clin Biochem Nutr 2007; 40(3):194-202.
43. Wood R, Ronnenberg AG. Iron 2006; 248-270. In: Shills MA et al. eds. Modern nutrition in health and disease. X edition. Lipincott 2006.
44. Miller NJ, Evans CA. Spectrophotometric determination of antioxidant activity. Redox Report 1996; 2:161-171.
45. Ishizaka N, Ishizaka Y, Toda E et al. Association between serum uric acid, metabolic syndrome and carotid atherosclerosis in Japanese individuals. Arterioscler Thromb Vasc Biol 2005; 25(5):1038-1044.
46. Rodrigo R, Castello R, Carrasco R et al. Diminution of tissue lipid peroxidation in rats is related to the in vitro antioxidant capacity of wine. Life Sci 2005; 76:889-900.
47. Gomez-Cabrera MC, Borras C, Pallardò FV et al. Decreasing xanthine oxidase-mediated oxidative stress prevents useful cellular adaptations to exercise in rats. J Physiol 2005:113-120.
48. Otterbein LE, Soares MP, Yamashita K et al. Heme oxygenase-1:unleashing the protective properties of heme. Trends Immunol 2004; 24:449-455.
49. Stocker R, Yamamoto Y, McDonagh A et al. Bilirubin is antioxidant of possible physiological importance. Science 1987; 235:1043-1046.
50. Clark JE, Foresti R, Sarathchandra P et al. Heme oxygenase-1-derived bilirubin ameliorates postischemic myocardial dysfunction. Am J Physiol Heart Circ Physiol 2000; 278:H643-H651.
51. Foresti R, Sarathchandra P, Clark JE et al. Peroxynitrite induces haem oxygenase-1 in vascular endothelial cells: a link to apoptosis. Biochem J 1999; 339:729-736.
52. Samb A, Taille C, Almolki A et al. Heme oxygenase modulates oxidant-signaled airway smooth muscle contractility: role of bilirubin. Am J Physiol Lung Cell Mol Physiol 2002; 283:L596-L603.
53. Dore S, Snyder SH. Neuroprotective action of bilirubin against oxidative stress in primary hippocampal cultures. Ann NY Acad Sci 1999; 890:167-172.
54. Baranano DE, Rao M, Ferris CD et al. Biliverdin reductase: a major physiologic cytoprotectant. Proc Natl Acad Sci USA 2002; 99:16093-16098.
55. Kato Y, Shimazu M, Kondo M et al. Bilirubin rinse: A simple protectant against the rat liver graft injury mimicking heme oxygenase-1 preconditioning. Hepatology 2003; 38:364-373.
56. Adin CA, Croker BP, Agarwal A. Protective effects of exogenous bilirubin on ischemia-reperfusion injury in the isolated, perfused rat kidney. Am J Physiol Renal Physiol 2005; 288:F778-F784.
57. Nakao A, Otterbein LE, Overhaus M et al. Biliverdin protects the functional integrity of a transplanted syngeneic small bowel. Gastroenterology 2004; 127:595-606.
58. Kaur H, Hughes MN, Green CJ et al. Interaction of bilirubin and biliverdin with reactive nitrogen species. FEBS Lett 2003; 543:113-119.
59. Hopkins PN, Wu LL, Hunt SC et al. Higher serum bilirubin is associated with decreased risk for early familial coronary artery disease. Arterioscler Thromb Vasc Biol 1996; 16:250-255.
60. Temme EH, Zhang J, Schouten EG et al. Serum bilirubin and 10-year mortality risk in a Belgian population. Cancer Causes Control 2001; 12:887-894.
61. Lanone S, Bloc S, Foresti R et al. Bilirubin decreases NOS2 expression via inhibition of NAD(P)H oxidase: implications for protection against endotoxic shocks in rats. FASEB J 2005; 19(13):1890-2.
62. Troughton J, Woodside JV, Young IA et al. Bilirubin and coronary heart disease risk in the Prospective Epidemiological Study of Myocardial Infraction (PRIME). Eur J Cardiovasc Prev Rehabil 2007; 14(1):79-84.

63. Perlstein TS, Pande RL, Beckman JA et al. Serum total bilirubin level and prevalent lower-extremity peripheral arterial disease: National Health and Nutrition Examination Survey (NHANES) 1999 to 2004. Arterioscler Thromb Vasc Biol 2008; 28(1):166-172.
64. Gullu H, Erdogan D, Tok D et al. High serum bilirubin concentrations preserve coronary flow reserve and coronary microvascular functions. Arterioscler Thromb Vasc Biol 2005; 25(11):2289-2294.
65. Erdogan D, Gullu H, Yildirim E et al. Low serum bilirubin levels are independently and inversely related to impaired flow-mediated vasodilation and increased carotid intima-media thickness in both men and women. Atherosclerosis 2006; 184(2):431-437.
66. Dohi K, Satoh K, Ohtaki H et al. Elevated plasma levels of bilirubin in patients with neurotrauma reflect its pathophysiological role in free radical scavenging in vivo. 2005; 19(5):855-860.
67. Shekeeb Shahab M, Kumar P, Sharma N et al. Evaluation of oxidant and antioxidant status in term neonates: a plausible protective role of bilirubin. Mol Cell Biochem 2008; 317(1-2):51-59.
68. Sedlak TW, Snyder SH. Bilirubin benefits: cellular protection by a biliverdin reductase antioxidant cycle. Pediatrics 2004; 113(6):1776-1782.
69. Carpenè E, Andreani G, Isani G. Metallothionein functions and structural characteristics. J Trace Elem Med Biol 2007; 21(Suppl 1):35-39.
70. http://www.expasy.org/cgi-bin/lists?metallo.txt.
71. Li Y, Kimura T, Laity JH et al. The zinc-sensing mechanism of mouse MTF-1 involves linker peptides between the zinc fingers. Mol Cell Biol 2006; 26:5580-5587.
72. Uchida Y. Growth inhibitory factor, metallothionein-like protein and neurodegenerative diseases. Biol Signals 1994; 3:211-215.
73. Hussain W et al. Role of metallothionein and other antioxidants in scavenging superoxide radicals and their possible role in neuroprotection. Neurochem Int 1996; 29:145-152.
74. Hwang YP, Kim HG, Han EH et al. Metallothionein-III protects against 6-hydroxydopamine-induced oxidative stress by increasing expression of heme oxygenase-1 in a PI3K and ERK/Nrf2-dependent manner. Toxicol Appl Pharmacol 2008; 231(3):318-327.
75. Gülçin I. Antioxidant and antiradical activities of L-carnitine. Life Sci 2006; 18: 78(8):803-811.
76. Binienda ZK, Ali SF. Neuroprotective role of l-carnitine in the 3-nitropropionic acid induced neurotoxicity. Toxicology Letters 2001; 125:67-73.
77. Kremser K, Stangl H, Pahan K et al. Nitric oxide regulates peroxisomal enzyme activities. Europane Journal of Clinic Chemisty and Clinic Biochemistry 1995; 33:763-774.
78. Dokmeci D, Inan M, Basaran UN et al. Protective effect of L-carnitine on testicular ischaemia-reperfusion injury in rats. Cell Biochem Funct 2007; 25(6):611-618.
79. Lal A, Atamna W, Killilea DW et al. Lipoic acid and acetyl-carnitine reverse iron-induced oxidative stress in human fibroblasts. Redox Rep 2008; 13(1):2-10.
80. McMackin CJ, Widlansky ME, Hamburg NM et al. Effect of combined treatment with alpha-Lipoic acid and acetyl-L-carnitine on vascular function and blood pressure in patients with coronary artery disease. J Clin Hypertens (Greenwich) 2007; 9(4):249-255.
81. Xie J, Zeng Q, Wang L. The protective effect of L-carnitine on ischemia-reperfusion heart. J Huazhong Univ Sci Technolog Med Sci 2006; 26(2):188-191.
82. Thangasamy T, Jeyakumar P, Sittadjody S et al. L-Carnitine mediates protection against DNA damage in lymphocytes of aged rats. Biogerontology 2009; 10(2):163-172.
83. Calabrese V, Giuffrida Stella AM, Calvani M et al. Acetylcarnitine and cellular stress response: roles in nutritional redox homeostasis and regulation of longevity genes. J Nutr Biochem 2006; 17(2):73-88.
84. Hardeland R, Poeggeler B. Nonvertebrate melatonin. J Pineal Res 2003; 34:233-234.
85. Conway S, Drew JE, Mowat P et al. Chimeric melatonin mtl and melatonin-related receptors. Identification of domains and residues participating in ligand binding and receptor activation of the melatonin mtl receptor. J Biol Chem 2000; 275:20602-20609.
86. Garcia-Maurino S, Pozo D, Calvo JR et al. Correlation between nuclear melatonin receptor expression and enhanced cytokine production in human lymphocytic and monocytic cell lines. J Pineal Res 2000; 29:129-137.
87. Srinivasan V, Pandi-Perumal SR, Cardinali DP et al. Melatonin in Alzheimer's disease and other neurodegenerative disorders. Behav Brain Funct 2006; 2-15.
88. Tan DX, Reiter RJ, Manchester LC et al. Chemical and physical properties and potential mechanism: melatonin as a broad spectrum antioxidant and free radical scavenger. Curr Top Med Chem 2002; 2:181-197.
89. Buyukavci M, Ozdemir O, Buck S et al. Melatonin cytotoxicity in human leukaemia cells: relation with its prooxidant effect. Fundam Clin Pharmacol 2006; 20:73-79.
90. Rodriguez C, Mayo JC, Sainz RM et al. Regulation of antioxidant enzymes: a significant role for melatonin. J Pineal Res 2004; 36:1-9.
91. Anisimov SV, Popovic N. Genetic aspects of melatonin biology. Rev Neurosci 2004; 15:209-230.

92. Reiter RJ, Acuna-Castroviejo D, Tan DX et al. Free radical-mediated molecular damage. Mechanisms for the protective actions of melatonin in the central nervous system. Ann NY Acad Sci 2001; 939:200-215.
93. Koh PO. Melatonin regulates nitric oxide synthase expression in ischemic brain injury. J Vet Med Sci 2008; 70(7):747-750.
94. Garcia JJ, Reiter JR, Pie J et al. Role of pinoline and melatonin in stabilizing hepatic microsomal membranes against oxidative damage. J Bioenerg Biomen 1999; 31:609-616.
95. Karbownik MA, Reiter RJ. Antioxidative effects of melatonin in protection against cellular damage caused by ionizing radiation. Proc Soc Exp Biol Med 2000; 225:9-22.
96. Lahiri DK, Ge YW, Sharman EH et al. Age-related changes in serum melatonin in mice: higher levels of combined melatonin and 6-hydroxymelatonin sulphate in the cerebral cortex than serum, heart, liver and kidney tissues. J Pineal Res 2004; 36:217-223.
97. Sanchez-Moreno C, Dorfmann SE, Lichtestein AH et al. Dietary fat type affects vitamins C and E and biomarkers of oxidative status in peripheral and brain tissue of golden Syrian hamsters. J Nutr 2004; 134:655-660.
98. Rousseau A, Petren S, Plannthin J et al. Serum and cerebrospinal fluid concentration of melatonin: a pilot study in healthy male volunteers. J Neural Transm 1999; 106:883-888.
99. Filadelfi AM, Castrucci AM. Comparative aspects of the pineal melatonin system of poikilothermic vertebrates. J Pineal Res 1996; 20:175-186.
100. Kotler M, Rodriguez C, Sainz RM et al. Melatonin increases gene expression for antioxidant enzymes in rat brain cortex. J Pineal Res 1998; 24:83-89.
101. Berra B, Rizzo AM. Melatonin: circadian rhythm regulator, chronobiotic, antioxidant and beyond. Clinics in Dermatology 2008; 27(2):202-209.
102. Martin M, Macias G, Escames G et al. Melatonin-induced increased activity of the respiratory chain complexes I and IV can prevent mitochondrial damage induced by ruthenium red in vivo. J Pineal Res 2000; 28:242-248.
103. Urata Y, Honma S, Goto S et al. Melatonin induces glutamylcysteine synthase mediated by activator protein-I in human vascular endothelial cells. Free Rad Biol Med 1999; 27:838-847.
104. Yu BP, Chung HV. Adaptive mechanism to oxidative stress during aging. Mech Ageing Dev 2006; 127:436-443.
105. Skene DJ, Swaab DF. Melatonin rhythmicity: effect of age and Alzheimer's disease. Exp Gerontol 2003; 38:199-206.
106. Swaab DF. The human hypothalamus basic and clinical aspects. Handbook of Clinical Neurology. Aminoff MJ, FrancoisB, Swaab DF series eds. Elsevier, Amsterdam 2003; 79:63-123.
107. Fourtillian JB, Brisson AM, Fourtillian M et al. Melatonin secretion occurs at a costant rate in both young and older men and women. Am J Physiol Endocrinol Metab 2001; 280:E11-E22.
108. Duffy JF, Zeitzer JM, Rimmer DW et al. Peak of circadian melatonin rhythm occurs later within the sleep of older subjects. Am J Physiol Endocrinol Metab 2002; 282:E297-E303.
109. Lu F. Reactive oxygen species in cancer, too much or too little? Med Hypotheses 2007; 69(6):1293-1298.
110. Fruehauf JP, Meyskens FFL. Reactive oxygen species: A breath of life or death? Clin Cancer Res 2007; 13(3):789-794.
111. Tuma S. Reactive oxygen species may have antitumor activity in metastatic melanoma. JNCI 2008; 100:11-12.
112. Droge W. Free radicals in the physiological control of cell function. Physiol Rev 2002; 82:47-95.
113. Lee SR, Yang KS, Kwon J et al. Reversible inactivation of the tumor suppressor PTEN by H_2O_2. J Biol Chem 2002; 277:20336-20342.
114. Adachi T, Weisbrod RM, Pimentel DR et al. S-Glutathiolation by peroxynitrite activates SERCA during arterial relaxation by nitric oxide. Nat Med 2004; 10:1200-1207.
115. Adachi T, Pimentel DR, Heibeck T et al. S-glutathiolation of Ras mediates redox-sensitive signaling by angiotensin II in vascular smooth muscle cells. J Biol Chem 2004; 27:29857-29862.
116. Griendling KK, Sorescu D, Ushio-Fukai M. NAD(P)H oxidase. Role in cardiovascular biology and disease. Circ Res 2000; 86:494-501.
117. Ushio-Fukai M. Alexander RW, Akers M et al. Reactive oxygen species mediate the activation of Akt/ protein kinase B by angiotensin II in vascular smooth muscle cells. J Biol Chem 1999; 274:22699-22704.
118. Forman HJ, Torres M, Fukuto J. Redox signaling. Mol Cell Biochem 2002; 234-235:49-62.
119. Schafer FQ, Buettner GR. Redox environment of the cell as viewed through the redox state of the glutathione disulfide/glutathione couple. Free Radic Biol Med 2001; 30:1191-1212.
120. Xu D, Rovira II, Finkel T. Oxidants painting the cysteine chapel: redox regulation of PTPs. Dev Cell 2002; 2:251-252.
121. Filomeni G, Rotilio G, Ciriolo MR. Cell signalling and the glutathione redox system. Biochem Pharmacol 2002; 64:1057-1064.
122. Magder S. Reactive oxygen species: toxic molecules or spark of life? Critical Care 2006; 10:208-216.

Chapter 6

A Nutritional Strategy for Reducing Disease and Obesity Risks

Teresa Lavecchia, Paolo Petroni, Giuseppe Rodio and Riccardo Pina*

Abstract

Overweight and obesity are the most common nutritional disorders in our age and are becoming more and more common worldwide. The most harmful consequences of an incorrect diet leading to overweight or obesity are a series of cardio-vascular diseases often leading to disability and death. In recent years various studies have shown that a reduction in caloric intake is the main factor involved in reduction of pathology risk. In this article, a nutrition strategy, based on the Zone diet by US biochemist Dr. Barry Sears is proposed. It underlines the importance of choosing certain types of foods over others, their beneficial physiological effects on the human body and how they can reduce the risk of heart disease and cancer.

Introduction

Overweight and obesity are very common nutritional disorders. The foremost physical consequence of overweight and obesity is the formation of fatty plaques on the inner surface of the arteries which harden and narrow, sometimes so much that scarcely any blood can pass. This phenomenon is known in medicine as atherosclerotic cardiovascular disease[1] and is the leading cause of both death and disability in Europe and North America.[2-4] The absence of blood circulation in some organs can cause grave and harmful consequences such as angina pectoris, heart attack, cardiac arrest, transient ischemic attack and stroke. In addition, obesity is accompanied by other medical complications such as diabetes, fatty liver, cholesterol gallstones, sleep apnea, osteoarthritis and polycystic ovary disease. In the 1980s-1990s Dr. Sears noticed that a class of hormones known as eicosanoids play an important role in the protection of human body against certain diseases,[5-7] including heart disease.[8-12] His research suggested that the origin of many diseases could, in fact, be traced back to an imbalance of eicosanoids and the best way to regulate them and thus prevent many diseases, was through food. The proposed dietary strategy to establish the right hormonal balance in our bodies was recognized and launched as 'The Zone Diet'. It proposes a combination of foods so that 40% of the calorie intake is derived from carbohydrates, 30% from proteins and 30% from fats.[13-15] The dietary aspect is considered to be fundamental, but other elements are also very important, in particular physical activity and mental relaxation.

The Zone Diet Nutrition Strategy: General Outline

The Zone dietary scheme teaches us how to choose and combine foods in order to establish the correct hormonal balance in our bodies. A fixed rule in the diet is the percentage of carbohydrates, proteins and fats to be eaten at every main meal and for every snack: 40% of the calories should come from carbohydrates, 30% from proteins and 30% from fats. This ratio among the

*Corresponding Author: Riccardo Pina—Enervit, Zelbio (CO) Italy. Email: r.pina@enervit.it

Bio-Farms for Nutraceuticals: Functional Food and Safety Control by Biosensors
edited by Maria Teresa Giardi, Giuseppina Rea and Bruno Berra.

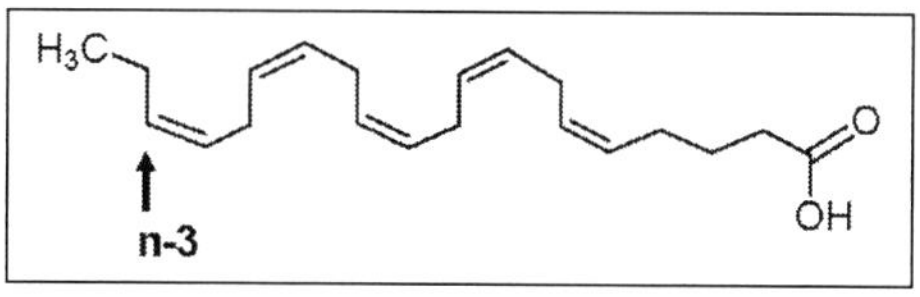

Figure 1. Eicosapentaenoic acid (EPA).

mentioned nutrients is necessary to establish the correct balance of the various hormones in the blood. Each nutrient has effects on the release of hormones: carbohydrates are responsible for the increase of the insulin level in the blood, proteins raise the level of glucagons while some fats can increase "bad" eicosanoids. Eicosanoids are molecules arising from the oxygenation of essential fatty acids (EFAs). EFAs are long–chain unsaturated carboxylic acids having a carbon–carbon double bond in the n–3, n–6 or n–9 position.

Eicosanoids (eicosa-, Greek for "twenty") come from twenty-carbon carboxylic acids, the simplest being the icosapentaenoic (or eicosapentaenoic acid) (Fig. 1).

They represent the first hormones developed by living organisms 500 million years ago and are also the most powerful hormones, since they affect the synthesis of every other hormone in the body. Eicosanoids are capable of great health benefits ("good" eicosanoids) or great harm ("bad" eicosanoids) depending on which eicosanoid a cell produces. "Good" eicosanoids prevent blood clots caused by platelet aggregation, cause vasodilation of blood vessels, reduce pain and cell division and enhance the immune system and brain function. Bad "eicosanoids" favor clots caused by platelet aggregation, cause vasoconstriction of blood vessels, promote pain and cell division and depress the immune system and brain function.

Since different foods contain different types of nutrients, the intake of some foods over others is also important. The general rules of nutrition strategy are few and simple: have at least five meals per day, never go for more than five hours without eating, have at least two snacks and choose the meals and snacks according to individual lifestyle.

Carbohydrate Intake in the Zone Diet

Carbohydrates are responsible for the glucose level in the blood commonly indicated as glycemia. Whenever glycemia is raised above-normal levels the insulin hormone is released into the blood. According to the Zone diet the optimal level of insulin hormone in the blood is maintained when foods containing large amounts of carbohydrates are not consumed excessively. These foods are referred as high glycemic index foods (high GI) and are characterized by a rapid digestion and absorption causing significant increases of insulin in the blood. High GI foods consist of sugar-rich beverages (such as colas, orange soda and other soft drinks), sweets, cookies, processed snack foods, pastries, potatoes, bread, rice and various others. According to the Zone diet other foods that also provide carbohydrates can replace high GI foods. In this category are included most vegetables (with the exception of potatoes, squash, beets and carrots) and fruits (except for bananas, persimmons, some types of exotic fruits and dried fruits such as raisins and dried figs). Table 1 shows types of carbohydrates that should be eaten according to Zone Diet nutrition strategy.

This suggestion is supported by different studies on GI and its correlation with diabetes and overweight people.[16-18]

Protein Intake in Zone Diet

Intake of proteins is responsible for the blood level of glucagon which, like insulin is a hormone produced by the pancreas. The action of glucagon is opposite to that of insulin, in fact, glucagon is released when the glucose level in the blood lowers, causing the liver to convert stored glycogen into glucose. Since the optimal level of glucagon in the blood is maintained by the correct amount of amino acids in the blood plasma, a daily proteins ration is predicted by the Zone diet. Frequently, foods providing amino acids contain also large amounts of fats, some of

Table 1. List of the suggested carbohydrates in Zone Diet nutrition strategy

Preferred Carbohydrates	Carbohydrates To Be Consumed in Limited Quantities	Carbohydrates To Be Consumed Rarely
1. All types of vegetables (except for potatoes, squash, beets and carrots) 2. All types of fruits (except for bananas, persimmons, watermelon, cantaloupe, mangoes and papayas and some types of dried fruits, such as raisins, dates and dried figs) 3. Barley and oats	1. Bread, breadsticks, crackers, melba toast, focaccia, pizza 2. Cookies 3. Polenta 4. Rice, pasta 5. Fruit juices 6. The types of fruit and vegetables indicated above	1. Sweet beverages such as colas, orange sodas and soft drinks 2. Almond nougat, processed snack foods, pastries 3. Alcoholic beverages (which the Zone Diet also places in this category)

which are healthy and others unhealthy. The Zone diet nutrition strategy suggests which types of proteins can be consumed to avoid unhealthy fats or substances. In the 'permitted' category of protein foods all those deriving from seafood products (such as fish, crustaceans and mollusks), less fatty dairy products (such as reduced fat milk, low fat yogurt and certain low fat cheeses) and some types of meat (such as chicken and turkey) are enclosed. In the 'forbidden' category we find giblets and fatty meats, fatty cheeses, whole milk, whole yoghurt. Table 2 shows which types of protein food should be eaten according to Zone Diet nutrition strategy.

Fats Intake in Zone Diet

Fat intake is very important for the good functioning of human body because fats are responsible for various functions. They play a role in maintaining healthy skin and hair, insulating body organs against shock, maintaining body temperature and promoting healthy cell function etc. They also serve as energy stores for the body since their degradation releases free fatty acids and glycerol. The latter can be converted to glucose by the liver and used as a source of energy. Also, some vitamins need to work in conjunction with fats. A, D, E and K vitamins, for example, are fat-soluble molecules and can only be digested, absorbed and transported in the presence of fats. Popular wisdom about fats considers them very harmful for the body. This

Table 2. List of the suggested protein food in Zone Diet nutrition strategy

Preferred Proteins From	Proteins To Be Consumed in Limited Quantities From	Proteins in Very Limited Quantities From
1. Fish and other seafood products such as crustaceans and mollusks 2. Low fat meats such as chicken, turkey, guinea hen and bresaola 3. Low fat dairy products such as reduced fat milk and low fat yogurt 4. Fat cheeses such as parmesan-type, primosale, feta, fresh ricotta, certain types of scamorza and Asiago 5. Soy cheese and meat such as tofu and seitan	1. Speck, prosciutto and cooked ham by removing mechanically the fat 2. Fatty meats such as lean, beef and pork 3. Rabbit 4. Lean canned meat and drained canned tuna	1. Fatty land mammals meats such as goat, lamb, pork, beef 2. Giblets 3. Bacon, salami and sausages 4. Eggs yolk 5. Fatty cheeses, whole milk, whole yogurt

Table 3. List of the suggested fats in Zone Diet nutrition strategy

Preferred Fats Derived From	Fats To Be Consumed in Limited Quantities Derived From
1. Extra virgin olive oil 2. Oily fruit such as almonds, pine nuts, walnuts, hazelnuts and avocados 3. Fish oil (in capsules or liquid form supplements)	1. Dairy products (especially cream, cheese, butter and ghee) 2. Suet, tallow, lard, fatty meat 3. Coconut oil, cottonseed oil, palm kernel oil, margarine 4. Chocolate

is only partially true. Some fats known as saturated fats are correlated with increased incidence of atherosclerosis and coronary heart disease according to a number of studies, in African green monkeys[19] and humans, such as a study of infant diets[20] and hypercholesterolemic men[21-23] and should always be consumed in limited quantities. Some fats are essential nutrients, meaning that they cannot be produced in the body from other compounds and need to be consumed in small amounts. These species, already mentioned above, are the EFAs which are divided in two large families: omega-3 (ω-3) and omega-6 (ω-6). They are named in this way, because of the presence of a double bond in their chemical structure at position 3 and 6 respectively. In 2004, the U.S. Food and Drug Administration gave a "qualified health claim" status to eicosapentaenoic acid (EPA) and docosahexaenoic acid (DHA) n–3 fatty acids, stating that "supportive but not conclusive research shows that consumption of EPA and DHA [n–3] fatty acids may reduce the risk of coronary heart disease".[24] When added to a proper dietary strategy, they help to restore the body's balance and contribute to the regulation of cholesterol and triglyceride levels in the blood. Table 3 shows which types of fat food should be eaten according to Zone Diet nutrition strategy.

Next to a balanced diet, Zone nutrition strategy supports the importance of functional foods to modulate eicosanoids.[25] According to the European Commission's Concerted Action on Functional Food Science in Europe: "a food product can only be considered functional if together with the basic nutritional impact it has beneficial effects on one or more functions of the human organism thus either improving the general and physical conditions or/and decreasing the risk of the evolution of diseases".[26]

The most important compound for functional food is pharmaceutical-grade fish oil since it rapidly lowers the ratio between 'bad' and 'good' ecoisanoids although other nutraceutical compounds are considered fundamental[27] for their synergistic effect. Among these are soybean oil, canola oil, tocopherol and citric acid: soybean and canola oil are particularly rich in polyphenols, which are antioxidant molecules contained in legumes, fresh fruits, vegetables, tea and red wines; tocopherol and citric acid, commonly known as vitamins E and C, are antioxidant molecules mainly contained in plant oils and nuts, peppers, broccoli and citrus fruits.

A molecule is defined as an 'antioxidant' when present in low concentration compared to that of an oxidizable substrate, it significantly delays or prevents the oxidation of that substrate. Antioxidants are normally contained in the human body; they act mainly as a defence against free radicals, which are highly reactive species produced during cellular metabolism.

An overproduction of free radicals inside the human body can cause serious damage to all cell structures including lipids, membranes, proteins and DNA. Various studies[28-30] indicate that atherogenesis, which is the process of forming plaques on the inner lining of arteries is initiated by oxidation of the lipids in low-density lipoprotein, also termed lipid peroxidation. As a consequence of this hypothesis, antioxidants that inhibit lipid peroxidation should limit atherosclerosis, cardiovascular disorders, myocardial infarction and stroke.[31-33]

Conclusion

A correct nutrition strategy is fundamental to improve general well-being and to prevent many diseases. An optimal ratio among the various nutrients in our body determines the right balance of hormones in the blood and is a basic condition for the correct metabolic functioning. Today, the great variety of foods available to consumers, as well as broad socio-economic changes have caused people to turn to inexpensive and fast foods such as pizza, sandwiches, ready meals and soft drinks. These foods are generally rich in fats and carbohydrates but very poor in vitamins, fibres and other components such as nutraceutics with potential health benefits. Imbalanced nutrition in conjunction with increased sedentary work is the main cause for the growth of many diseases. A prevention strategy is possible by properly combining the intake of carbohydrates, proteins, fats and other healthy nutrients and preferring certain types of foods over others.

Acknowledgement

This work was supported by the NUTRA-SNACKS project funded within the EU Framework Programme 6.

References

1. Grundy SM, Hansen B, Smith SC et al. Clinical management of metabolic syndrome: report of the American Heart Association/National Heart, Lung and Blood Institute/American Diabetes. Circulation 2004; 109:551-556.
2. Yoshino M, Kuhlmann MK, Kotanko P et al. International differences in dialysis mortality reflect background general population atherosclerotic cardiovascular mortality. J Am Soc Nephrol 2006; 17:3510-3519.
3. Conroy RM, Pyorala K, Fitzgerald AP et al. Estimation of ten-year risk of fatal cardiovascular disease in Europe: the SCORE project. Eur Heart J 2003; 24(11):987-1003.
4. Parekh RS, Zhang L, Fivush BA et al. Incidence of atherosclerosis by race in the dialysis morbidity and mortality study: A sample of the US ESRD population. J Am Soc Nephrol 2005; 16:1420-1426.
5. Kremer JM, Lawrence DA, Petrillo GF et al. Effects of high-dose fish oil on rheumatoid arthritis after stopping nonsteroidal antiinflammatory drugs. Clinical and immune correlates. Arthritis Rheum 1995; 38:1107-1114.
6. Belluzzi A, Boschi S, Brignola C et al. Effect of an enteric-coated fish-oil preparation on relapses in Crohn's disease. N Engl J Med 1996; 354:1557-1560.
7. Tang K and Honn KV. 12(S)-HETE in cancer metastasis. Adv Exp Med Biol 1999; 447:181-191.
8. Austin MA, Breslow JL, Hennekens CH et al. Low density lipoprotein subclass patterns and risk of myocardinal infarction. JAMA 1988; 260:1917-1920.
9. Jeppesen J, Hein HO, Suadicani P et al. "Low triglycerides-high high-density lipoprotein cholesterol and risk of ischemic heart disease." Arch Intern Med 2001; 161:361-366.
10. Lamarche B, Lemieux I, Despres JP. "The small dense phenotype and the risk of coronary heart disease epidemiology, pathophysiology and therapeutic aspects." Diabetes Metab 1999; 25:199-211.
11. Gaziano JM, Hennekens CH, O'Donnell CJ et al. "Fasting triglycerides, high-density lipoproteins and risk of myocardial infarction." Circulation 1997; 96:2520-2525.
12. GISSI-Prevenzione Investigators. "Dietary supplementation with n-3 polyunsaturated fatty acids and vitamin E after myocardial infarction: results of the GISSI-Prevenzione trial." Lancet 1999; 354:447-455.
13. Sears B. The Zone. New York: Regan Books, 1995.
14. Sears B. The Anti-Aging Zone. New York: Regan Books, 1999.
15. Sears B. The Omega Rx Zone. New York: Regan Books, 2002.
16. Jenkins DJ, Wolever TM, Taylor RH et al. The glycaemic index of foods tested in diabetic patients: a new basis for carbohydrate exchange favouring the use of legumes. Diabetologia 1983; 24(4):257-264.
17. Jenkins DJ, Wolever TM, Taylor RH et al. Glycemic index of foods: a physiological basis for carbohydrate exchange. Am J Clin Nutr 1981; 34(3):362-366.
18. Takeda E, Arai H, Muto K et al. Gene expression in low glycemic index diet—impact on metabolic control. Forum Nutr 2007; 60:127-139.
19. Wolfe MS, Sawyer JK, Morgan TM et al. Dietary polyunsaturated fat decreases coronary artery atherosclerosis in a pediatric-aged population of African green monkeys. Arterioscler Thromb 1994; 14:587-597.
20. Lapinleimu H. "Prospective randomised trial in 1062 infants of diet low in saturated fat and cholesterol". Lancet 1995; 345(8948):471-476.
21. Fuentes F, Lopez-Miranda J, Sanchez E et al. Mediterranean and low-fat diets improve endothelial function in hypercholesterolemic men. Ann Intern Med 2001; 134(12):1115-1119.

22. Rivellese AA, Maffettone A, Vessby B et al. "Effects of dietary saturated, monounsaturated and n-3 fatty acids on fasting lipoproteins, LDL size and postprandial lipid metabolism in healthy subjects". Atherosclerosis 2003; 167(1):149-158.
23. Hu FB, Stampfer MJ, Manson JAE et al. "Dietary fat intake and the risk of coronary heart disease in women". N Engl J Med 1997; 337(21):1491-1499.
24. United States Food and Drug Administration (September 8, 2004). FDA announces qualified health claims for omega-3 fatty acids. Press release.
25. Sears B. The OmegaZone Dietary Program and Eicosanoid Modulation. The Role of Eicosanoids and Chronic Disease. www.zonelabsinc.com
26. Diplock AT, Aggett PJ, Ashwell M et al. Scientific concepts of functional foods in Europe—consensus document. British Journal of Nutrition 1999; 81(1):1-27.
27. Sears B. Omega/Rx ingredients. http://www.zonediet.com/OMEGA3/tabid/64/Default.aspx
28. Siegel-Axel D, Daub k, Seizer P et al. Platelet lipoprotein interplay: trigger of foam cell formation and driver of atherosclerosis. Cardiovasc Res 2008; 78(1):8-17.
29. Soininen P, Öörni K, Maaheimo H et al. ^{1}H NMR at 800 MHz facilitates detailed phospholipid follow-up during atherogenic modifications in low density lipoproteins. Biochem Biophys Res Commun 2007; 360(1,17):290-294.
30. Matsuura E, Kobayashi K, Tabuchi M et al. Oxidative modification of low-density lipoprotein and immune regulation of atherosclerosis. Prog Lipid Res 2006; 45(6):466-486.
31. Kaliora AC, Dedoussis GVZ, Schmidt H. Dietary antioxidants in preventing atherogenesis. Atherosclerosis 2006; 187(1):1-17.
32. Turner R, Etienne N, Alonso MG et al. Antioxidant and anti-atherogenic activities of olive oil phenolics. Int J Vitam Nutr Res 2005; 75(1):61-70.
33. Akila M, Devaraj H. Synergistic effect of tincture of Crataegus and Mangifera indica L. extract on hyperlipidemic and antioxidant status in atherogenic rats. Vascular Pharmacology 2008; 49(4-6):173-177.

CHAPTER 7

Dietary Phytochemicals and Human Health

Justyna Krzyzanowska,* Anna Czubacka and Wieslaw Oleszek

Abstract

This chapter is a comprehensive review of the health promoting phytochemicals commonly found in our daily food. These include carotenoids, phenolics, phytoestrogenes, polyunsaturated fatty acids, conjugated linoleic acids, tocols, allicin, glucosinolates, limonene and capsaicinoids. The review encompasses the main food sources of these chemicals in the diet, the possible mechanisms of their activity, evidence for potential health promoting activity and possible harmful effects. The newly emerged interest in these phytochemicals in animal nutrition as substitutes for synthetic antibiotic growth promoters has also been addressed.

Introduction

Living plants produce a great number of chemicals that are crucial for their function and development. Some of these chemicals are primary metabolites, which include proteins (aminoacids), carbohydrates, fats, nucleic acids etc. However, besides these primary chemicals, the plants also produce so-called secondary metabolites, which are specific to some taxonomic groups (families, genera). Their physiological function was questioned already in earlier work, but recent research has shown that they are important constituents of plants. They are formed under environmental pressure and play a crucial function in protecting plants against some environmental stresses. This group includes classes of compounds such as phenolics, carotenoids, alkaloids, saponins, glucosinolates, cyanogenic glycosides, terpenes etc. Each of these groups contains compounds with different biological activities, which in traditional medicine and ethno-pharmacology was used for centuries to cure or to protect from diseases.

A correlation between diet and health has been known since Hippocrates (460-377 BC), who recognised that "differences of diseases depends on the nutriment".[1]

During the centuries, numerous epidemiological studies have been performed, confirming Hippocrates statement, especially regarding the frequency of some diseases in relation to the intake of particular nutrients. One example might be the study of amine-based substances discovered by Funk in rice bran and which for their importance were called "Vitamin" (Vita = life).[2] Some epidemiological studies, in fact, clearly correlate the frequency of some diseases with diet, especially those with a low consumption of fruit, vegetables and whole grain, which are rich sources of particular classes of phytochemicals.

Many natural products are thought to have health promoting activities. The most promising and extensively studied compounds over recent years are listed in Table 1. This chapter provides an overview of data available on health benefits of some classes of listed phytochemicals.

*Corresponding Author: Justyna Krzyzanowska—Institute of Soil Science and Plant Cultivation, Pulawy, Poland. Email: jkrzyzanowska@iung.pulawy.pl

Bio-Farms for Nutraceuticals: Functional Food and Safety Control by Biosensors
edited by Maria Teresa Giardi, Giuseppina Rea and Bruno Berra.

Table 1. Bioactive food constituents that may prevent diseases (data modified from ref. 3)

Active Compounds	Food Source	Potential Health Benefit	Possible Mechanism and Function
Carotenoids (-carotene, lycopene, lutein)	Tomatoes, carrots, yams, cantaloupe, spinach, sweet potatoes, citrus fruits	Reduce coronary heart disease and cancer	Antioxidants, singlet oxygen and free radical scavengers, inhibit proliferation of acute myeloblastic leukemia
Catechins (epigallo- and epigallocatechin gallate)	Green tea, grapes/wine	Reduce cancer and heart disease	Inhibits initiation, promotion and progression of cancer, antioxidant, reduces free radical/oxidative damage
Phytoestrogens: Isoflavones (daidzein, genistein), Coumestans, Lignans (enterodiol, enterolactone)	Soybeans, soybeanfood Pea (dried) Rapeseed, garlic, wheat bran, lentils, brown rice	Prevent menopausal symptoms, prevents osteoporosis, reduce cancer (breast, prostate)	Inhibit the growth of human breast cancer cell lines, decrease cholesterol, LDL cholesterol and triglicerides, stimulate calcium absorption and bone deposition
Favones, flavonols and flavanols (quercetin, kaempferol, rutin, myricetin, anthocyanidin), procyanidins	Apples, onion, tea, grapefruit and orange juice, broccoli, kale	Reduce coronary heart disease and cancer	Scavengers of reactive oxygen and nitrogen species, antioxidants, transition metal chelation
Stilbenes (resveratrol)	Red wine, yucca bark	Reduce coronary heart disease and cancer, stimulate apoptosis, prevent blood clustering	Antioxidants and free radical scavengers
Tocopherols, tocotrienols	Vegetable oils	Antioxidant, lowers serum cholesterol, inhibits cancer, decreases heart diseases	Inhibits cancer cell poliferation, inhibits HMG-CoA reductase
PUFA (omega-3 fatty acids)	Fish oil, algae, flaxseed	Reduce serum cholesterol and triacylglicerol, reduce heart diseases, immunosuppresants	Lower the total and LDL/HDL ratios, increase serum HDL, inhibit arachadonic acid-derived products such as PGE and leukotrienes

continued on next page

Table 1. Continued

Active Compounds	Food Source	Potential Health Benefit	Possible Mechanism and Function
Conjugated linoleic acids	Dairy products, processed vegetable oils	Anticancer, antiatherosclerosis	Inhibit cancer cell growth by interfering with the hormone regulated mitogenic pathway, reduce LDL/HDL ratio and total/HDL cholesterol in rabbits
Diallyl disulfide and allicin	Garlic, onions	Anticancers, stimulate immune function, free radical scavengers, reduce serum cholesterol and triacylglycerols	Inhibit the proliferation of human tumor cells in culture, inhibit metabolic activation of the toxicants and carcinogens, inhibit cholesterol biosynthesis
Nitrogen and sulphur-containing amino acid derivaties (glucosinolates, alkaloids, capsaicinoids)	Cruciferous vegetables, Green and red pepper	Chemoprevention	Chemopreventive activity, modulation of drug metabolizing enzymes, neuroactivity
Limonene	Citrus fruits	Anticancer	Regulator of malignant cell proliferation, inhibit posttranslational isoprenylation of cell growth-regulatory proteins
Lipoic acid (lipoyllysine)	Spinach, broccoli, tomato, green pea, Brussel sprouts, rice bran	Plays fundamental role in energy metabolism	Antioxidant and redox regulatory activity, potentiates metabolism serving as cofactor in enzyme complexes
Coumarins	Vegetables, citrus fruits	Prevent blood clotting, anticarcinogenic activity	Anticoagulants, inhibitors and inactivators of carcinogen and mutagen, scavengers of superoxide anion radicals
Nondigestable fermentable oligosaccharides, fructans	Garlic, asparagus, chicory	Intestinal fortification, stimulate immune function, inhibit tumorigenesis, reduce serum cholesterol	Prebiotics-effective substrates for bifidobacteria, modulate lipid metabolism

Some natural products found in plants have recently also found application in animal nutrition. For some time, a number of antibiotics have been used as growth promoting factors in animal production, especially in poultry and pigs. However, this practice has had serious consequences in increasing the resistance of some microorganisms dangerous to human health. Thus, some countries have introduced regulations prohibiting the use of antibiotics in animal production. In addition, starting from 2006, the European Community has banned using antibiotics as growth promoters. These new regulations may cause loses in income for some companies. Natural products seem to be good substitutes to replace synthetic antibiotics in animal production. Natural plant remedies are being developed and classes of active natural principles are being identified.

Carotenoids

Carotenoids are natural fat-soluble yellow, orange, or red pigments that are synthesised in plants, algae, fungi, yeasts and bacteria. In addition, the characteristic yellowish colours of many birds, insects, fish and crustaceans are due to the presence of carotenoids in their bodies; however, animals and humans cannot synthesise carotenoids de novo and are dependent on the dietary sources.[4,5] In nature, the total production of carotenoids has been estimated at 10^8 ton per year, moreover this mass is concentrated mainly in four carotenoids: lutein, violaxanthin and neoxanthin in green leaves and fucoxanthin, which predominates in marine algae. Over 600 of these compounds have been identified in nature; however, only about 40 are present in a typical human diet. Human plasma and tissues contain only 20 carotenoids, which are represented mainly by β-carotene, lycopene, lutein, β-cryptoxanthin and α-carotene.[6,7]

Carotenoids belong to the tetraterpenes family and these are characterised by a polyisoprenoid structure with a long conjugated chain of double bond and a near bilateral symmetry around the central double bond. According to their chemical composition, they are classified in two classes: carotenes, which contain only carbon and hydrogen atoms and xanthophylls (oxycarotenoids) that have carbon, hydrogen and at least one oxygen atom. Due to the presence of the conjugated double bonds, carotenoids can undergo *cis-trans* isomerisation.[7,8] Carotenoids usually occur in their all-*trans* configuration, which make them the thermodynamically more stable isomer. However, there is some evidence for the presence of *cis*-isomers in plants (e.g., palm oil fruit, algae), especially in chlorophyll-containing tissues but also in a number of fruit. Cis-isomers may also arise as a result of processing.[9,10]

They are present in various types of food, but the major sources of dietary carotenoids include orange and yellow fruits and vegetables as well as green leafy vegetables. Smaller amounts can be extracted from milk and foods containing dairy fat, egg yolks, sea fish and carotenoids added as colorants to foods during processing.[11] Table 2 presents some of the richest dietary sources of these compounds in a human diet.

The intake of carotenoids in the human diet is more variable than the intake of many other dietary constituents. However, the total carotenoid intake by humans is estimated to be between 6-11 mg per day (based on five major carotenoids consumed in US). The dietary intake of lycopene is about 2-5 mg per day, whereas the intakes of β-carotene and lutein are slightly lower, approximately 3 and 2-3 mg per day respectively, with the remainder from β-cryptoxanthin and α- carotene.[11,12]

Carotenoids are important components of the human diet because they have been linked to a multitude of health benefits. They play an important role in the cell communication and protection against photooxidative processes by acting as singlet molecular oxygen, as well as, peroxyl radical scavengers and can interact synergistically with other antioxidants.[14] In mammals, some of them can be metabolised to retinol and function as vitamin A precursors. Positive effects between high dietary consumption and tissue concentration of carotenoids and reduced risk of chronic diseases such as age-related macular degeneration were also observed. There is also strong evidence showing that a rich diet in carotenoids prevents cardiovascular diseases and certain cancers like lung, colon, breast and prostate cancer.[7,14,15]

Carotenoids have been shown to possess antioxidant activity and their antioxidant properties are thought to be the main mechanism by which they exert their beneficial effects. Recent studies have also shown that they may mediate their effects via other mechanisms such as gap junction communication, cell growth regulation, modulating gene expression and immune response and as modulators of Phase I and II drug metabolising enzymes.[7,16]

A considerable pool of data shows a relationship between the risk of cancer and dietary carotenoid intake. Some studies confirmed their protective role, whereas others found no correlation or even describe an adverse activity. A case-control study (2.706 cases of cancer of the oral cavity, pharynx, oesophagus, stomach, colon and rectum vs 2.879 controls) indicated that a high intake of tomatoes and tomato-based food, both of which are rich sources of lycopene, was strongly associated with reduced risk of digestive tract cancers, especially stomach, colon and rectal.[17] Another study reported that a diet supplemented with β-carotene at a dose of 50 mg per day over five years showed no effect on the occurrence of new basal- or squamous-cell carcinoma in well-nourished patients who previously had skin cancer. On the other hand, it was also shown that administration of 25 mg of β-carotene per day with or without vitamin C (1 g per day) and α-tocopherol (400 mg per day) for 5-8 years did not reduce the occurrence of colorectal adenoma in patients who had a prior history of adenomas.[11] A large-scale study—The Alpha-Tocopherol, Beta-Carotene (ATBC) study suggested that β-carotene, under certain circumstances, enhances carcinogenesis. This research involved 29.133 male smokers from Finland, whose diet was supplemented with β-carotene (20 mg per day) or α-tocopherol (50 mg per day), or both, or placebo for 5-8 years. Surprisingly, the subjects who received β-carotene alone or in a combination with α-tocopherol showed 18% increased incidence of lung cancer and 8% increased in total mortality.[16,18]

There are several reports that lycopene intake is correlated with a decreased risk of prostate cancer. An important study monitored dietary habits and the incidence rate of prostate cancer in approximately 48.000 men for four years. Researchers observed that subjects who ate ten or more servings of tomato products per week (tomatoes, tomato sauce and pizza sauce) were up to 34% less likely to develop prostate cancer. Men, whose intake was 4-7 servings per week, were 20% less likely to develop the cancer. Tomato sauce was the strongest dietary predictor of reduced prostate cancer risk (66%) and the major predictor of serum lycopene levels. On the other hand, other studies found no protective effect of serum level and dietary lycopene intakes on prostate cancer risk.[17]

Some of the epidemiological studies, which considered the role of carotenoids in the development of Cardiovascular Disease (CVD) supported the hypothesis that these compounds had preventive activity. There is evidence that the consumption of processed tomato products causes a reduction in lipoprotein sensitivity to oxidative damage.[4] A study that compared Lithuanian and Swedish populations showed that low lycopene levels were connected with an increased risk and mortality from Coronary Heart Disease (CHD). Another study also confirmed the positive role of lycopene. This substance caused a reduction of total serum cholesterol levels and thereby decreased the risk of CVD[7]. However, there are also studies that found no association between carotenoid dietary intake, their plasma concentration and the risk of CVD as well as CHD.[12]

Vitamin A is an essential nutrient for human health and is responsible for the promotion of growth, cellular differentiation, morphogenesis, embryonic development and visual function. Some carotenoids can be converted into active forms of this vitamin and, in doing so, may prevent vitamin A deficiency. They are termed provitamin A compounds. β-carotene, α-carotene and similarly β-cryptoxanthin as well as nearly 50 other carotenoids with β-ring end groups belong to this family, whereas lycopene is not a provitamin A compound. It has been estimated that carotenoids from fruits and vegetables provide more that 70% of the vitamin A intake in Third World countries, whereas in Western societies, the contribution is much lower.[4,5,11]

The macula of the eye contains only two carotenoids: lutein and zeaxanthin. These molecules are thought to protect the eye by acting as filters for damaging blue light and quenching reactive oxygen species (ROS). There is some evidence that dietary carotenoids affect the human eye and

Table 2. Concentrations of selected carotenoids in fruit, vegetables and food products (data modified from refs. 4,7,12,13)

	Carotenoid Content mg/100 g Wet Weight				
Food Source	**α-Carotene**	**β-Carotene**	**Lycopene**	**Lutein**	**Zeaxanthin**
Pepper	16.7	-	-	50.30	160.80
Green bean	7.0	-	-	49.4	-
Sweet corn	6.0	5.9	-	52.20	43.7
Banana	5.0	-	-	3.3	0.40
Apricot	3.7	2.6	-	10.1	3.10
Carrots, cooked	3.7	-	-	-	-
Carrots, raw	-	18.3	-	-	-
Apricot, dried	-	17.6	-	-	-
Mangos, canned	-	13.1	-	-	-
Sweet patato, cooked	-	9.5	-	-	-
Carrots, cooked	-	8.0	-	-	-
Pumpkin, canned	-	6.9	-	-	-
Kale, cooked	-	6.2	-	-	-
Spinach, raw	-	5.6	-	586.90	-
Spinach, cooked	-	5.2	-	12.47	-
Winter butternut squash	-	4.6	-	-	-
Swiss chard, raw	-	3.9	-	-	-
Pepper, red, raw	-	2.4	-	-	-
Pepper, red, cooked	-	2.2	-	-	-
Cantaloupe, raw	-	1.6	-	-	-
Broccoli, cooked	-	1.3	-	-	-
Tomato paste	-	1.2	-	-	-
Tomato powder, drum or spray dried	-	-	112.63-126.49	-	-
Sun-dried tomato in oil	-	-	46.50		-
Pizza sauce, canned	-	-	12.71	-	-
Ketchup	-	-	9.90-13.44	-	-
Tomato soup, condensed	-	-	7.99	-	-
Tomato sauce	-	-	6.20	-	-
Tomato paste	-	-	5.40-150.00	-	-
Guawa, fresh	-	-	5.40	-	-
Tomato juice	-	-	5.00-11.60	-	-
Tomatoes, cooked	-	-	3.70	-	-
Guawa, juice	-	-	3.34	-	-

continued on next page

Table 2. Continued

	Carotenoid Content mg/100 g Wet Weight				
Food Source	α-Carotene	β-Carotene	Lycopene	Lutein	Zeaxanthin
Grapefruit, raw pink	-	-	3.36	-	-
Watermelon, fresh	-	-	2.30-7.20	-	-
Papaya, fresh	-	-	2.00-5.30	-	-
Tomatoes, fresh	-	-	0.88-4.20	-	-
Apricot, dried	-	-	0.86	-	-
Apricot, canned	-	-	0.06	-	-
Brussel sprout	-	-	-	61.0	-
Green collard	-	-	-	16.30	-
Beet, green	-	-	-	7.70	-
Green peas, cooked	-	-	-	1.69	-
Broad bean	-	-	-	50.6	-
Broccoli	-	-	-	161.40	-
Green cabbage	-	-	-	8.00	-
Lettuce	-	-	-	11.00	-
Parsley	-	-	-	581.20	-
Pea	-	-	-	163.30	-
Watercress	-	-	-	1.07	-
Orange	-	-	-	6.40	5.00
Peach	-	-	-	7.80	4.20
Tomato	-	-	-	7.80	-

cause modification of macular pigments. For example, supplementation with lutein (30 mg per day over 140 days) resulted in increased serum level of lutein and corresponding increase in the concentration of lutein in the macula on the human eye.[5,12]

Carotenoids appear to be responsible for the potential health benefits such as reduction of the incidence of age-related diseases of the eye, like cataract and age-related macular degeneration disease (AMD), probably by their ability to quench active oxygen species. A study on men consuming high levels of lutein and zeaxanthin in a diet over eight years showed that those carotenoids lowered the risk of cataract by 19%. However, research with β-carotene and cataract by US physicians over 12 years showed no benefit in healthy men, but the excess risk of cataract in smokers was attenuated by about one quarter.[4] Another study showed that women with the highest intake of lutein and zeaxanthin had a 22% decreased risk of cataract compared to those in the lowest quintile.[12]

In a case-control study, AMD patients and matched control subjects (who had eye problems) were divided into five groups and given various nutrients from foods. Carotenoids turned out to be the nutrient class with the strongest protective effect against AMD. People with the highest carotenoid intake showed a 43% lower risk of developing AMD compared to those with the lowest administration. It was found that lutein and zeaxanthin were responsible for this beneficial effect.[12]

Phenolic Compounds

The healthful properties of plants might in part be due to their content of polyphenols. Plants produce phenolic compounds as secondary metabolites to interact with their environment. To be more precise, these compounds are responsible both for plant organoleptic qualities such as the colouring of leaves and fruits and for other physiological processes, including vasodilatatory, anti-inflammatory, anti-bacterial, anti-viral, anti-proliferative as well as attracting and repelling insects and protecting plants from herbivores.

Fresh fruits, vegetables, leaves, nuts, seeds, flowers and barks are rich sources of these compounds which possess antioxidant activity via their ability to scavenge free radicals or bind pro-oxidant metal ions by means of their OH groups. Also soya sprouts are a good source of phenolic compounds. They have been used for centuries in many dishes in the Orient, but recently have also become popular in Western countries.[19,20]

Flavonoids

Flavonoids represent the largest group of phenolic compounds and the most abundant species in the human diet. Over 6.000 different flavonoids have been identified but, only a small number of them are important from a dietary point of view.[21] According to their structure, flavonoids can be divided into six major subclasses: flavones, flavonols, flavanones, flavanols (catechins), anthocyanidins and isoflavones.[8,22] Various studies have shown that the consumption of flavonoid-rich foods may have positive effects on human health, even if their bioavailability is only partial. The absorbed portion of the amount which is ingested ranges from 0.2% to 0.9% for tea catechins to 20% for quercetin and isoflavones. Moreover, the bioavailability of certain flavonoids differs markedly, depending on the food source. For instance, the absorption of quercetin from onions has been shown to be four-fold higher than absorption from apples or tea.[23,24]

Comparing with other dietary antioxidant compounds such as tocopherols and vitamin C, the flavonoid concentration in blood plasma is very low. The concentration of flavonols, flavanols and flavanones ranges between 0.06-0.07 μM. In the case of anthocyanidins it is lower than 0.15 μM. For α-tocopherol and vitamin C the plasma concentration values are much higher and range between 30-150 or 15-40 μM, respectively.[25]

Although flavonoids are widespread in nature, their total daily intake is changeable. Their consumption mainly depends on food composition, food content as well as on dietary habits and preferences.[26] Table 3 shows the dietary flavonoids intake resulting from various studies on flavones consumption in Europe, Asia and US.

The estimated daily intake of flavonoids ranges from 3 mg in Finland, through 20 mg in Holland and USA up to 68 mg in Japan. Tea drinkers have a greater flavonoid intake estimated as 430 mg per day.[23]

Table 3. Flavones: apigenin, luteolin content in various products (data from ref. 26)

Country	Dietary Intake of Flavonoids* (mg/day)	Major Dietary Sources of Flavonoids
Danmark	26	Tea, onions, apples
Finland	3-10; 0-41	Fruit and vegetables; apples, onions
Greece	15	Fruit and vegetables
Italy	23-34; 35	Red wine; fruit and vegetables
Japan	60-68; 17	Green tea
Netrerlands	23; 33	Tea, onions, apples
United States	20	Onions, black tea

*not total flavonoid intake, this values refer mostly to three flavonols (quercetin, myricetin and kaempferol) and two flavones (apigenin and luteolin).

The most common flavonol in human diet is quercetin, which is present in various fruits and vegetables, but the highest concentration of this molecule has been found in onion (284-486 mg/kg). Onion is usually consumed in low quantities, but, on the other hand, tea and wine which contain lower amounts of quercetin, are consumed in high quantities in some countries.[21] The richest dietary sources of flavonols in a human diet are shown in Table 4.

Flavonoids possess a wide range of biological activities on humans. Most of those beneficial health effects are attributed to their antioxidant, free radical scavenging properties as well as metal

Table 4. Flavonols: quercetin, myricetin, kaempferol content in various products (data from ref. 25)

Product	Flavonols Content (mg/100 g or mg/100 ml)
Vegetables	
Capers	316
Dock leaves	120
Fennel	84,4
Hartwort leaves	38,9
Red onions	38,8
Kale	34,5
Sweet potato leaves	30,2
Chives	21,7
Hot peppers	16,0–50,0
Onions	15,4
Broccoli raw	9,37
Spinach	4,86
Celery	3,50
Broccoli cooked	2,44
Cherry tomatoes	2,30–20,3
Lettuce	1,20–9,40
Fruits	
elderberries	42,0
cranberries	18,4
currants	13,5
apples	4,42
bilberries	4,13
blueberries	3,93
apricots	2,55
grapes	2,54
cherries	1,25
plums	1,20
blackberries	1,10
Cereal	
Buckwheat	23,1
Spices and herbs	
Dill weed	55,0
Others	
Cocoa powder	20,3
Green tea	2,69
Black tea	2,07
Red wine	1,50

chelating abilities.[27] The antioxidant properties of flavonoids depend on their chemical structure and on the number and arrangement of functional groups. The carbonyl group at C-4 and the double bond between C-2 and C-3 are particularly important in this sense. Moreover, the presence of a catechol moiety controls the efficiency of the physical quenching of singlet oxygen ($^1O_2$) by flavonoids and the presence of a 3-hydroxyl largely determines the efficiency in the reaction with $^1O_2$. The antioxidant activity of flavonoids[24] usually increases with the number of hydroxyl groups.[24,28] The ability of flavonoids to inhibit low-density lipoprotein (LDL) oxidation is thought to play an important role in the prevention of cardiovascular diseases. For example, high flavonoid intake has been related to a decreased incidence of myocardial infarction and a reduced risk of coronary heart disease (CHD).[27] Several studies have shown that flavonoids might act as antiproliferative and anticarcinogenic agents. The inhibition of carcinogenesis, both in 'in vitro' and 'in vivo' experiments takes place by affecting the molecular events in the initiation, promotion and progression cellular phases.[22] Moreover, some flavonoids have been reported to have anti-inflammatory activities, mainly related to their ability to inhibit the production of inflammatory mediators such as prostaglandins, leukotrienes and nitric oxide. Flavonoids also possess antiallergic, antidiabetic, antineoplastic, antiviral, antibacterial, hepato- and gastro-protective activities.[24,29]

Phenolic Acids

Most plant phenolic acids such as gallic acid, vanillic acid, procatechuic acid and syringic acid are derived from hydroxybenzoic acid. Others, such as p-coumaric acid, caffeic acid and ferulic acid, are derived from hydroxycinnamic acids.[19] Caffeic and chlorogenic acids occur in fruits, soybeans and coffee beans. Ferulic acid is present in fruits, soybean and rice. Gallic acid occurs in big amounts in guava and *Geraniaceae* species.[30]

Various studies suggest the anticancer properties of phenolic acids. According to these studies the carcinogenesis is avoided through several mechanisms: the inhibition of carcinogen uptake, the inhibition of carcinogen activation, the deactivation or organism detoxification in presence of carcinogenic species. Phenolic acids also affect apoptosis in tumour cells, prevent the carcinogen binding to DNA and enhance the level or fidelity of DNA repair. Besides, phenolic acids have good antioxidant properties due to their chemical structure.[19]

Catechins

Catechins are a class of compounds derived from the di- or tri-hydroxyl group substitution and the meta-5,7-dihydroxy group substitution in the B and A rings of flavanols, respectively. They are responsible for the bitterness and the astringency of food. Substantial quantities of these compounds are present in different blends of tea. The (–)-epigallocatechin-3-gallate (EGCG), (–)-epigallocatechin (EGC), (–)-epicatechin-3-gallate (ECG), (–)-epicatechin (EC), (+)-gallocatechin (GC) and (+)-catechin (C) are the major catechins present in green tea. Besides their presence in green and black teas, catechins are present in large quantities in fruits such as berries, grapes, spotted knapweed, shea, cocoa, carob and grape seeds.[31] Table 5 shows some of the richest sources of catechins in the human diet.

Green tea catechins are natural chemopreventive agents recognized as potent inhibitors of the nuclear transcription factor kappa-light-chain-enhancer of activated B-cells (NF-κB) which plays an important role in regulating the immune response to infection. The incorrect regulation of NF-κB has been linked to cancer, inflammatory and autoimmune diseases, septic shock, viral infection and improper immune development. Catechins may block one or more steps in the NF-κB signalling pathway such as the activation of NF-κB signalling cascade, the translocation of NF-κB into the nucleus, the DNA binding of the dimers, or the interactions with the basal transcriptional machinery. Moreover, they suppress the enzyme Cyclooxygenase-2 (COX-2) which is responsible for the formation of prostanoids, engaged in inflammation. The epigallocatechin-3-gallate, epigallocatechin, epicatechin-3-gallate and theaflavins from black tea, also inhibited COX-dependent arachidonic acid metabolism in human colon mucosa and colon tumours. In addition, catechins inhibit DNA methylation, which is observed in a wide variety of cancers.[31] Catechin and epicatechin have also been shown to be prooxidants as they induced oxidative damage to DNA in

Table 5. Catechins: (-)-epicatechin, (-)-epigallocatechin gallate content in various products (data from ref. 25)

Product	Catechins Content (mg/100 g or mg/100 ml)
Fruits	
Blackberries	18.7
Grapes	17.6
Cherries	11.7
Apricots	11.0
Raspberries	9.23
Apples	9,0
Plums	6.19
Cranberries	4.20
Raisins	3.68
Pears	3.43
Nectarines	2.75
Peaches	2.30
Blueberries	1.11
Others	
Green tea	132
Dark chocolate	53,5
Black tea	33,0
Milk chocolate	13,4
Red wine	12,0

a human leukemia cell line by generating H_2O_2 during redox reactions. Epigallocatechin, the major form of catechins, besides antioxidant activity also exerts an immunomodulatory function and antimicrobial activity against some pathogens as well. Epigallocatechin gallate and epigallocatechin have been reported to inhibit cell-mediated oxidation of low-density lipoprotein in a concentration of 0-400 μM.[19]

Green tea extracts, gallocatechin gallate, gallocatechin, (−)-epigallocatechin-3-gallate, catechin-3 gallate and catechin at 100 μM concentration have been shown to inhibit antiapoptotic Bcl-2 proteins. Polyphenols present in green tea were also reported to induce apoptosis through accumulation of Apoptotic Protease Activating Factor 1 protein which is a precursor of apoptosis. Effects of catechins on DNA and gene expression may also contribute to their anticarcinogenic properties.[19]

Catechins are poorly absorbed from the human body and undergo a substantial biotransformation to species that include glucuronides, sulfates and methylated compounds.[30] In blood serum, the concentration of catechin ingested in the form of tea, ranges between 0.2% and 2% with maximum concentrations achieved 1.4-2.4 h after consumption. A study performed on humans has shown that the consumption of eight cups of black tea a day caused a rise of catechins in serum from 0.08 to 0.2 μM.[19]

Some epidemiological experiments have demonstrated that green tea can be effective for the chemoprevention of prostate cancer since the incidence of prostate cancer is lower in Japanese and Chinese populations[32]—the major consumers of green tea. Moreover, green tea ointment and capsules have been shown to be effective for treating cervical lesions, suggesting that green tea extracts can be a potential therapy regimen for patients with HPV (*Human Papilloma Virus*). Epigallocatechin gallate is also interesting because of its sunscreen protective activity against UVB signal transduction.[30]

Stilbenes

Stilbenes are organic compounds containing the functional group 1,2-diphenylethylene.[19] Stilbene itself exists as two possible isomers. The first is *trans*-1,2-diphenylethylene, called (E)-stilbene or *trans*-stilbene. The second is *cis*-1,2-diphenylethylene, called (Z)-stilbene or *cis*-stilbene. Many stilbene derivatives, such as the hydroxylated compounds, are present naturally in plants. The most representative of the latter is resveratrol (3,5,4-trihydroxystilbene), that exists in the form of *cis*- and *trans*- isomers. In fruits it is present in several complex and substituted forms. *Trans*-resveratrol is a natural component of grapes *Vitis vinifera* L. Especially grapes' skin and the leaf epidermis are rich sources of this compound. Resveratrol is not unique to *Vitis* species but is also present in at least 72 other plant species, distributed in 12 families and 31 genera, e.g., *Veratrum, Arachis, Morus, Vaccinium* and *Trifolium*—some which for instance, mulberries and peanuts are constituents of the human diet.[32]

Resveratrol was first found by Michio Takaoka in 1940 as a component of the roots of white hellebore (*Veratrum grandiflorum* (Maxim. ex Baker) Loes. (*Liliaceae*)) and identified by Nonomura et al later (in 1963) in the dried roots of Japanese knotweed *Polygonum cuspidatum* Sieb. et Zucc. (*Polygonaceae*). In Japanese and Chinese traditional medicine, it is used for the treatment of dermatitis, gonorrhea favus, hyperlipemia, favus, athlete's foot and arteriosclerosis, as well as in allergic and inflammatory diseases and other pathologies. In 1976, *trans*-resveratrol was detected in grapevines by Langcake and Pryce. It is synthesised by leaf tissues in response to fungal infection by *Botrytis cinerea*, or after exposure to ultraviolet light. In 1992, the presence of resveratrol was reported in red wines. A number of epidemiological studies suggested that the moderate consumption of red wine by French and other Mediterranean populations was connected with the reduced incidence of cardiovascular diseases, despite high-fat diet, little exercise and widespread smoking. Resveratrol may protect against coronary heart disease by way of its significant antioxidant activity, modulation of lipoprotein metabolism, vasodilatory and platelet antiaggregatory properties.[32] In its antioxidant activity a key role is played by the number and position of the hydroxyl groups in the molecule.[33]

Resveratrol is absorbed and transferred across the small intestine as a glucuronide, probably cleaved back to the bioactive aglycone form—resveratrol—by the β-glucuronidases found in a variety of organs and body fluids, such as macrophages, blood cells, liver, lung and serum.[19] A bioavailability study conducted on human healthy subjects, consuming 25 mg of pure *trans*-resveratrol in three different matrices (white wine with 11.5% ethanol, grape juice and vegetable juice/homogenate), showed that the concentration of free (1.7 to 1.9% of total polyphenol levels) and conjugated polyphenols in serum were similar for all three sources and reached a maximum concentration after 30 minutes from the ingestion decreasing to the base level within 4 hours. Also urinary 24-h excretion after oral consumption did not show any matrix effect and accounted for 16.5% of the dose administered.[33]

So far, little is known about pharmacological activity of the *cis*- form. The *cis* isomer has not been detected in grapes but is present in wines, being probably produced from *trans*-resveratrol by yeast isomerases during fermentation, or released from viniferins, or from resveratrol glucosides.[32] Resveratrol glycosides, present in grape juice have shown lower bioavailability than the aglycone form, the cumulative excretion of resveratrol being only 5 vs 50% of the administered dose.[33]

The anti-initiation activity of resveratrol might be related to its antioxidant and anti-mutagenic effects.[19] Both *cis*- and *trans*-isomers exhibit typical antioxidant activity, i.e., they block extra- and intracellular production of reactive oxygen species, through the inhibition of both NAD(P)H oxidase activity and nitric oxide production. In addition, resveratrol inhibits lipid peroxidation.[32]

Resveratrol and its dimer vineferin, have been reported to inhibit the activities of cytochromes involved in bioactivation or deactivation of numerous carcinogens. The ability to induce apoptosis might also contribute to anticancer effects ascribed to resveratrol. Another possible explanation for the anticancer effects of resveratrol is the regulation of production or activation of specific enzymes. Moreover, resveratrol has been shown also to have estrogenic activity and immunosuppressive and immunoenhancing properties.[19]

Curcumin

Curcumin (diferuloylmethane) is a yellow pigment of the rhizome of turmeric (*Curcuma longa* L.) widely used as a spice in food. This phenolic compound shows antioxidant, anti-inflammatory and chemopreventive activity. The anticancer potential comes from its ability to suppress proliferation of a wide variety of tumour cells, to regulate transcription factors, chemokines, cell surface adhesion molecules, growth factor receptors and kinases. The daily oral dose of 3.6 g is recommended in the prevention or treatment of cancer.[30]

Curcumin as well as various other curcuminoids mediate their therapeutic effects by regulating the transcription factor NF-κB and NF-κB regulated gene products COX-2, cyclin D1 gene, adhesion molecules, matrix metalloproteinase endopeptidases, inducible nitric oxide synthase, B-cell lymphoma 2, B-cell lymphoma-extra large and tumour necrosis factors. Moreover, they could suppress the activator protein 1 transcription factor activation process and the activation of Antiapoptotic Kinase protein which plays an important role in various mammalian cancers.[31]

Curcumin and resveratrol are also known to modulate the activity of the tumor protein 53 which is a tumour-suppressor and transcription factor. Tumor protein 53 is a critical regulator in many cellular processes including cell signal transduction, cellular response to DNA-damage, genomic stability, cell cycle control and apoptosis. The protein activates the transcription of genes, which induce the apoptotic process, inhibiting the growth of cells with damaged DNA or cancer cells.

Curcumin, resveratrol, quercetin and green tea polyphenols also target the chemokines which are directly involved in the migration of leukocytes activating the inflammatory responses and participate in the regulation of tumour growth.[31]

Finally curcumin is able to modulate the mitogen-activated protein kinases signalling pathway contributing, in this way, to the inhibition of inflammation and has a direct action on liver injury and liver detoxification enzyme system.[31,33]

Phytoestrogens

Phytoestrogens are natural estrogen-like substances found in numerous plants especially of the *Leguminosae* family, but also in some vegetables, fruits and berries. Four different families of phenolic compounds are considered as phytoestrogens and include isoflavonoids, stilbenes, lignans and coumestans. These compounds are structurally similar to estradiol and thanks to this similarity they exhibit estrogenic activity due to the presence of two hydroxyl groups that provide the correct geometry to bind estrogen receptors.[16,34]

Human cells have two types of estrogen receptors: ERα and ERβ. Phytoestrogens bind mainly to the ERβ, whereas mammalian estradiol has a higher binding affinity for ERα.[16]

The estrogenic activity of these compounds was first discovered in the 1940s, when they were seen as the cause of infertility and miscarriage problems of Australian sheep feeding on clover leaves. Clover leaves contain large amounts of the two isoflavones biochanin A and formononetin that affect the estrogen-mediated response by activating the aryl hydrocarbon receptor involved in the reproductive physiology in multiple ways (AhR).[35]

Moreover, dietary phytoestrogens could play an important role in the prevention of breast cancer and prostate cancer. They may also affect bone density, cardiovascular health, cognitive ability and diminish the symptoms of the menopause.[23,35]

Isoflavones

Isoflavones are diphenolic secondary metabolites synthesised from the products of the shikimic acid and malonyl pathways by the fusion of a phenylpropanoid with three malonyl coenzyme A residues. In nature as well as in processed food, only a small fraction appears as glucoside-free (aglycones). Most compounds exist primarily as glycosides with 1-4 sugar substitutions, which may occur also in acetylated or malonylated form. However, the glycosides are readily hydrolysed in the intestine to their aglycones, which are easily transported across intestinal epithelial cells.[36,37]

Table 6. Isoflavones content in selected vegetable and food (data from ref. 40)

	Concentration (mg/100 g)			
Food	**Daidzein**	**Genistein**	**Glycetin**	**Isoflavones Total**
Roasted soybeans	56,3	86,9	19,3	162,5
Textured vegetable protein	47,3	70,7	20,2	138,2
Green soybean	54,6	72,9	7,9	135,4
Soyflour	22,6	81,0	8,8	112,4
Tempeh	27,3	32,0	3,2	62,5
Tofu	14,6	16,2	2,9	33,7
Tofu yoghurt	5,7	9,4	1,2	16,4
Soynoodle (dry)	0,9	3,7	3,9	8,5

Isoflavones are found in a variety of vegetables and fruits. Table 6 shows some of the richest dietary sources of selected isoflavones. These compounds are found at high amounts in some plants of the subfamily *Papilionoideae* of the *Leguminosae*, which includes soybean. Their content in soybean ranges from 0.14 to 1.53 mg/g and in soy flour from 1.3 to 1.98 mg/g.[36,38] Some *Trifolium* species may contain even up to 10 mg/g of isoflavones in aerial parts.[39]

Unlike string beans or snap peas, soybeans cannot be eaten as raw plant material, because they contain trypsin inhibitors, which can disrupt digestion activities in the stomach, leading to cramping and associated discomfort. For this reason, soybean seeds are usually consumed as fermented products.[36,38]

Numerous epidemiological studies have demonstrated an association between the consumption of soybean and the reduction of risk for breast and prostate cancers, cardiovascular diseases and atherosclerosis probably connected with the availability of isoflavonoids.[36,37] Isoflavones possess numerous biological activities. They play an important role in the prevention of endometrial, breast and prostate cancer, reduce risk for cardiovascular disease and affect the health problems associated with menopause and osteoporosis. Some studies also report that isoflavones might support cognitive functions.[35,36]

In Western diets, the consumption of these compounds is very low, while it is very high in Asian diets. Asian people have a daily intake of flavonoids ranging from 20 to 100 times more than the European intake.[36,37]

Several studies demonstrate that isoflavones also exhibit hormonal activity as they stimulate the synthesis of the sex hormone binding globulin.[41]

There are correlations between high isoflavone consumption diet and the reduced incidence of breast and prostate cancer in humans. In Asian countries, these types of cancer are considerably lower, while an increased risk appears for Asian immigrants living in the US probably due to the change in their dietary habits.[35,42]

Some studies have also revealed that Japanese women in their homeland have a high number of in situ tumours but with fewer nodal metastases. Tumors presenting metastases have less nodal spread than women with breast cancer in the US or Great Britain.[41]

Several studies have covered the relation between soy and isoflavones intake and prostate cancer risk, the incidence of which is lower in Asians, Africans and native Americans with compareed to western countries. A study performed on Japanese men living in Hawaii has shown a decreased risk of prostate cancer in those consuming rice and tofu. A similar effect has been reported for a group of Seventh Day Adventist men consuming beans, lentils and peas, tomatoes, as well as dried fruits. In a similar vein, Japanese immigrating to North America at a younger age have reported an increase in the incidence of prostate cancer.[41]

There are been also some studies investigating the influence of an isoflavone supplemented diet on other types of cancer such as gastric, colon and endometrial.

Soy consumption seems significantly protective against stomach cancer, but there is also some evidence that fermented soy foods can increase the risk of this cancer. This effect has been explained by hypothesizing the presence of a high concentration of N-nitroso salts and other unidentified compounds in fermented foods. Moreover, studies assessing the relationship between soy intake and the risk of colon and rectal cancers, have generally not been supportive of a protective effect: while, on the one hand, it has been reported that a rich-soy diet effectively reduces the risk of endometrial cancer in the multiethnic population of Hawaii and in Chinese women,[43] on the other hand it has been reported that miso, a traditional food made with soy, significantly increases the risk of rectal cancer.[44,45]

Isoflavones also possess estrogenic activity, thus, they deeply influence women's emotional state and health. In particular they play an important role in maintaining bone density,[37] lipoprotein metabolism and regulate serum cholesterol level which is one of the main risk factors for the development of cardiovascular diseases.[42] Menopausal disorders such as osteoporosis, hot flushes and mood swings caused by the lowering of estrogenic hormones are usually treated with synthetic estrogens (hormone replacement therapy).

Phytoestrogen compounds used to treat women with climacteric syndrome should contain 40 to 100 mg of isoflavones consisting of variable combinations of different aglycons (i.e., genistein, daidzein, glycitein, formononetin and biochanin A): this daily dose is toxicologically safe and consistent with the dietary content of isoflavones in Asian countries.[42] Since it is commonly known that Asian women have considerably fewer menopausal symptoms than Western women, it has been hypothesised that this difference is due to their high intake of phytoestrogen, particularly isoflavones. This medical study relates to the efficacy of isoflavone-rich soy or isoflavone supplements as a possible dietary alternative to hormone replacement therapy.[44,46]

The beneficial effects of isoflavones on cholesterol level and bone density have been reported in various studies demonstrating their influence on the improvement of vascular reactivity,[37] the reduction of platelet activation and aggregation as well as on the induction of vasodilation through the stimulation of the endothelial enzyme nitric oxide synthase 3.[42] With regard to the beneficial effects in the prevention of osteoporosis, studies demonstrate that isoflavones increase the synthesis of vitamin D in a number of nonrenal cell types thus contributing to bone mineral density.[35-37] However, contrary or uncertain effects have also been reported.[37]

A few studies have also examined the effect of phytoestrogens on the improvement of cognitive function, even if there is no clear evidence to support this thesis. In this study, a group of male and female students were assigned both a high (100 mg per day) and low (0.5 mg per day) isoflavones diet. The results showed that students on the high isoflavone diet group showed significant improvement in short and long-term memory and mental flexibility.[35]

Finally, isoflavones may have a protective effect on skin health, improving skin elasticity by increasing local blood flow. In addition, the antioxidant and anti-inflammatory properties of phytoestrogens may alleviate the carcinogenic effect of various exogenous noxious agents such as ionizing radiation or chemical compounds.[42]

Lignans

Lignans constitute a class of phytoestrogens derived from phenylalanine by dimerization of substituted cinnamic alcohols to a dibenzylbutane skeleton. The cyclisation or modification of dimers results in a large variety of chemical structures in which different sub-groups can be identified such as furofuran, furan, dibenzylbutane, dibenzylbutyrolactone, aryltetralin, arylnaphthalene, dibenzocyclooctadiene and dibenzylbutyrolactol.[43,47]

Lignans are widespread in many plant families: in particular, they are synthesised and accumulated in the heartwood region of trees as a metabolic event in its formation and protection against fungi causing rotting.[43,47]

In mammals, lignans are metabolized by the gut flora producing the so-called 'mammalian lignans': enterodiol and enterolactone. These compounds differ from their precursors by the presence of some hydroxyl groups replacing methyl groups on the aromatic rings.[35,48] Enterodiol and enterolactone exert estrogenic and/or anti-estrogenic activity by regulating the unconjugated/conjugated sex hormones levels, high values of which are considered a risk factor for breast and prostate cancers.[35]

Lignans exhibit also antioxidant activity and prevent colon and thyroid cancers.[49] In the treatment of cancer an important role is played by the semisynthetic derivatives of podophyllotoxin (PTOX) which is a non-alkaloid toxin in the lignan family. This molecule is an active antiwart agent as well as the pharmacological precursor for the important anti-cancer drug etoposide.[43]

The chemical synthesis of PTOX is not a cheap process and thus it is highly desirable to develop alternative sources for this compound. One solution is the extraction from plant roots and rhizomes such as *Podophyllum peltatum* propagated under in vitro conditions, cell suspension cultures of *Linum album* or other species within the *Linum* genus.[43]

Polyunsatured Fatty Acids

Polyunsaturated fatty acids (PUFAs) are carboxylic acids of 20 and 22-carbon chains having various double bonds. The most important for human diet are Omega-3 and Omega-6 in which the first double bond is located at the positions 3 and 6, respectively. The most well-known among the Omega-6 PUFAs are the linoleic and arachidonic acids. The most well-known Omega-3's are α-linolenic (ALA), eicosapentaenoic (EPA) and docosahexaenoic (DHA) acids. The human body cannot synthesize PUFAs but can convert linoleic acid to arachidonic acid and ALA to EPA.[50] Thus, only the precursors need be provided in the diet. Good sources of omega-3 PUFAs are dark fleshed fishes such as herrings, mackerels, sardines and salmons, marine foods and fish oil supplements;[50-53] α-linolenic acid is present also in leafy green vegetables, purslane, walnuts and perilla seed, pumpkin seed, flaxseed and rapeseed oils, soybean and canola oils.[50,51,54] However, excessive consumption of fish should be avoided especially by women of childbearing-age since they contain a high amount of mercury, causing an improper foetus development.[55]

Omega-6 PUFAs, rather, are contained in most vegetable oils,[54] cereals, eggs, animal fat, whole grain breads, sunflower and corn oils.[50,52,54]

The conversion of ALA to EPA and EPA to DHA by a human organism is low, ranging from 0.2% to 15%. This applies to ALA even when is fed at a high level. However, high intakes of ALA have been reported to result in significant increases in very long chain omega-3 fatty acids in various body compartments.[51]

An optimal balance between Omega-3 and Omega-6 fatty acids is extremely important for human body, since they contribute to the maintenance of the whole hormone balance.[50] Moreover, EPA and DHA consumption has also been connected with reduction of the risk of cardiovascular diseases.[52,53] The cardioprotective effect includes anti-arhythmic, anti-thrombotic and anti-inflammatory activities, lowering blood pressure, endothelial function improvement and the slowing down of atherosclerotic plaque growth.[51]

PUFAs affect also cell membranes fluidity and the behavior of enzymes and receptors bound to membranes. Furthermore, they have antibiotic-like properties. α-linolenic acid, for example, rapidly kills *Staphylococcus aureus* and suppresses proinflammatory cytokines such as interleukins and tumour necrosis factor.[50]

An important role is played by Omega-3 also in the development and activity of the central nervous system and the improvement of cognitive development and memory-related learning.[54,56] Docosahexaenoic acid, especially, is involved in the growth and function of nervous tissue affecting both the neurogenesis and the neurotransmission. This accounts for considering DHA as an essential supplement in the diet of preschool and young infants: their organism is not able to produce DHA at the rate required by the rapid growth of their brain, thus, it must provided by appropriate dietary supplements.[53]

In Western diets, the ratio of omega-6 to omega-3 PUFAs ranges from 15-16:1 while recent dietary data suggests the healthy range of 1-4:1.[50]

The recommended daily amount of EPA and DHA for adults is at least 220 mg. Two or three servings of fatty fish per week (about 1250 mg EPA and DHA per day) are regarded as protection from psychiatric and neurological disorders. The same amount of EPA and DHA is provided in form of 3000 to 4000 mg standardised fish oils ingested per day. In the case of ALA, the adequate daily intake for adults should be roughly 2220 mg per day. Regarding flaxseed oil, doses up to 3000 mg per day are recommended to prevent neurodegenerative disorders and doses up to 6000 mg per day may be recommended to treat these conditions. If it is taken in the form of flaxseeds, the recommended dose is 1 tablespoon two to three times per day or 2 to 4 tablespoons once per day and seeds should be ground before eating and taken with lots of water. Decoction of flaxseeds can be ingested too. 100 g of raw flaxseed provides 22800 mg of ALA.[54]

Conjugated Linoleic Acids

Conjugated Linoleic Acids (CLA) are a mixture of positional and geometrical isomers of linoleic acid discovered accidentally in grilled beef.[16,57] At least 28 structures are known for these compounds mainly contained in ruminant meats (beef, lamb) and dairy products and in lower amounts in nonruminant meat.[57-59] Table 7 present some of the richest dietary sources of CLA.

The different content of CLA in foods is mainly due to the type of feeding and the quality of feedstuff chosen for animals. For instance, it has been proven that modern agricultural feeding practice limits CLA content in animal tissues and their products.[61]

In the laboratory, CLA can be produced from pure linoleic acid, sunflower oil or corn oil heated in the presence of alkali. As an alternative, the isomerisation of linoleic acid induced by the bacteria *Lactobacillus plantarum* can be used.[58,59]

Conjugated Linoleic Acids have been linked to a multitude of health benefits which range from anticarcinogenic, antitherogenic, antiadipogenic and antidiabetogenic activity to the prevention of heart diseases, improvement of the immune system, treatment of obesity and the increase of lean body mass.[57,59] Mostly, these results have been obtained from studies using mixtures of CLA isomers. Indeed, the different isomers show different activities and mechanism of action on specific tissues and organs and the major effects can be attributed to the two species *cis*-9, *trans*-11 and *trans*-10, *cis*-12 conjugated linoleic acid.[57,62,63]

The evidence of CLA effects on the balance of lean and fat body mass has been obtained by studies performed on different rodents (mice, rats) and pigs and on overweighed humans fed with various isomers of CLA in variable dosages. In the former, a lowering of fat body mass together with an increase of lean body mass has been observed in CLA-treated animals compared to nontreated ones.[62-64] In the latter, contrasting results have been obtained: some sustaining a reduction in body fat mass, during the first 6 months of treatment, others reporting no effect on fat mass, fat-free mass, percent body fat, body weight or blood lipids.[63,64]

The antiatherogenic effect has been shown, studying the influence on thromboxane production, platelet aggregation and cholesterol level in the blood plasma of rabbits and hamsters. In the first case, the results have suggested that a diet with 0,5 g of CLA per day causes a reduction of 12% in the atherosclerotic lesions of the aortic surface. In the second case, hamsters fed with 20% butterfat for 12 weeks, have shown a significant lowering in the aortic lipid deposition.[57-59]

Similar studies performed on humans with a 0.7% CLA supplementation per day, have demonstrated a decrease in the level of total cholesterol, triglycerides as well as HDL cholesterol in serum. However, opposite results have been also obtained reporting no changes in plasma lipid and lipoprotein concentrations after the consumption of 3.9 g of CLA per day.[57]

Among the beneficial effects of CLA is also the action on the immune system: researchers have observed that these compounds in addition to decreasing the production of inflammatory mediators like prostaglandins, show an inhibitor effect on proinflammatory cytokines in animal and human cell lines, decrease the regulation of hypersensitivity reactions caused by the body immune response to allergens and inhibit hepatoma, colorectal, breast, prostate and lung carcinogenesis at

Table 7. The concentration of conjugated linoleic acids (CLA) in various products (data modified from refs. 60,62,64)

Product	CLA (mg/g fat)
Dairy products	
Condensed milk	7,0
Butter fat	6,1
Homogenized milk	5,5
Cultured buttermilk	5,4
Plain yoghurt	4,8
Butter	4,7
Sour cream	4,6
Low-fat yogurt	4,4
2% milk	4,1
Ice cream	3,6
Cheeses	
Ricotta	5,6
Swiss	5,4
Edamer	5,3
Mozzarella	4,9
Cottage	4,8
Brie	4,7
Cheddar sharp	4,5
Cheddar medium	4,0
Parmesan	4,0
Meat	
Lamb	5,8
Fresh ground beef	4,3
Veal	2,7
Fresh ground turkey	2,6
Pork	0,6
Chicken	0,9
Egg yolk	0,6
Salmon	0,3
Vegetable oils	
Safflower oil	0,7
Sunflower oil	0,4
Peanut	0,2
Olive	0,0

various levels: initiation, promotion and progression. In addition, the intake of *cis*-9, *trans*-11 and *trans*-10, *cis*-12 isomers should influence the immune function decreasing the T-cell lymphocyte activation. However, no effects on immune system after influenza vaccination have been displayed in some young healthy women and no alteration in prostaglandin production and lymphocyte proliferation have been noticed in stimulated peripheral blood mononuclear cells.[57,63]

Finally, CLA might affect bone metabolism as some studies demonstrate that: (1) human intestinal cells treated with CLA are able to mobilise higher amounts of Calcium;[57] (2) supplemented-CLA chicks show a higher dry and fat-free bone ash comparing to control fed chicks;[61] (3) male Wistar rats on CLA diet, show an enhancement in Calcium absorption even if measurable effects on bone mass cannot be observed.[57]

Positive effects have also been noticed in postmenopausal women consuming higher amounts of CLA in their diet; an improvement of bone mineral density was observed.[57]

Some studies also indicate that CLA might possess an antidiabetic effect. A lowering of glucose level, plasma insulin and free fatty acids was observed in obese and diabetic rats after the supplementation of their diet with 1.5% of CLA isomers.[58,59] Increased insulin sensitivity after administering *trans*-10, *cis*-12 isomer was also noted.[63] However, caution is of the utmost since another study reports no changes in plasma glucose or insulin levels in cows fed with 10 g per day of CLA isomers.[64]

Tocopherols and Tocotrienols

Tocopherols and tocotrienols belong to the family of vitamin E which was discovered in 1922 by Evans and Bishop as an essential factor for the reproduction in rats. The family consists of eight lipid-soluble molecules composed of a chromonal ring with a side chain. In tocopherols, the ring has a 15-carbon side chain at the C-2 position; tocotrienols are structurally similar except for the presence of three *trans* double bonds in the hydrocarbon tail.[65,66]

Tocopherols are generally present in nuts (almonds) and common vegetable oils (wheat germ, sunflower). Tocotrienols are present in cereal grains (oat, barley and rye) and some vegetable oils (palm oil and rice bran oil)[66-68] (Table 8).

Both tocopherols and tocotrienols are synthesised in photosynthetic organisms including bacteria, algae, plants and fungi. Vitamin E was chemically synthesised for the first time in 1938.[67]

The overall intake of vitamin E derivatives largely depends on the dietary preferences in different countries. For instance, the daily intake in European and US countries ranges from 6.4 mg to 12.6 mg. The daily recommended dose according to different Medicine and Nutrition institutes should vary from 12 to 15 mg per day to the upper limit of 1000 mg per day. This amount has been raised to 1073 mg in some clinical trials, with no negative effects observed on the patients.[10,66,70]

Most of the scientific literature on vitamin E and its derivatives concerns the tocopherols. Vitamin E as well as tocotrienol and tocopherol homologues possess antioxidant activity, protect against cardiovascular disease, atherosclerosis and certain types of cancer. Moreover, they reveal

Table 8. Composition and amounts of vitamin E in dietary vegetable oils (data from ref. 69)

Vegetable Oil	Vitamin E Total (mg/Tb)	Tocopherols (%) α	β	γ	δ	Tocotrienols (%) α	β	γ
Wheat germ	33	45	27	10	10	1	7	0
Palm	12	21	0	32	0	15	3	29
Soybean	11	8	0	64	28	0	0	0
Cottonseed	10	53	0	47	0	0	0	0
Corn	10	20	6	74	0	0	0	0
Canola	8	38	0	60	2	0	0	0
Safflower	7	69	0	31	0	0	0	0
Sunflower	7	89	0	9	2	0	0	0
Peanut	5	45	0	50	5	0	0	0
Olive oil	1	100	0	0	0	0	0	0
Coconut	0,5	26	0	0	14	12	3	45

Kilocalorie consideration: Tablespoon oil = 126 kilocalories (1 Tablespoon oil = 14 g; fat has 9 kilocalories/g).

beneficial effects in neurodegenerative diseases such as Alzheimer's or Parkinson's. Vitamin E also shows beneficial effects as anti-tumourigenic, photoprotective and skin barrier stabilizer that accounts for its wide use in cosmetic and skin care products.[65,66,70,71]

α-tocopherol exerts its antioxidant activity by both scavenging those radicals which are responsible for the propagation of lipid peroxidation chain reaction and decreasing the assembly of active Nicotinamide Adenine Dinucleotide Phosphate-oxidase responsible for the reactive oxygen species production, which is involved in lipid peroxidation.[72]

Some epidemiological studies support also the health benefits of tocopherols and tocotrienols in cardiovascular disease prevention and cholesterol lowering. In particular, an inverse association between vitamin E intake and the risk of cardiovascular disease has been demonstrated and the inhibition of the hepatic enzyme 3-hydroxy-3-methylglutaryl-coenzyme A, responsible for cholesterol synthesis, has been observed even in presence of micromolar amounts of tocotrienol.[65,68]

With regard to tocopherols' anti-tumourigenic activity, some studies report a protective effect of α-tocopherol and γ -tocopherol against prostate cancer and of γ -tocopherol against colon cancer. In the first case, the therapeutic effect seems to pertain to the γ-tocopherol form, while no protective effects are ascribed to α-tocopherol.[73]

Limonene

Limonene is a hydrocarbon classified as a cyclic monoterpene. The molecular skeleton of monoterpenes consists of two isoprene units but oxygen-containing compounds such as alcohols, aldehydes or ketones (terpenoids) are also found. These compounds are very widespread in the essential oils of many plants and show chemoattractant or chemorepellent activity as well as providing plants with their distinctive fragrance.[20]

Limonene is particularly abundant in citrus fruits such as lemon, sweet orange, grapefruit and the mandarin *C. clementina*.[31,74-76]

Some analyses on the volatile components emitted by citrus have shown that the highest level of limonene condenses in the gynaecium (62.5%), stamens (22.9%) and petals (3.1%), while no production of limonene has been found in the pollen. Moreover, young leaves contain more limonene (65.3%) than the old ones (30.1%).[77,78] A considerable amount is present also in celery (*Apium graveolen*), cardamom, caraway, dill, peppermint and mopanes;[31,74] in the latter, limonene and α-pinene are presumably responsible for the strong turpentine odor of the pods.[79]

On the basis of its properties, limonene is applied as a flavouring agent for fruit juices, soft drinks, baked goods and dairy products as well as for cosmetics and cleaning products.[74]

However, limonene is also known to medicine for its chemotherapeutic activity and minimal toxicity in preclinical studies: the D-limonene isomer significantly increases tumour latency and reduces tumour multiplicity by regulating the signal transduction and cell growth;[74,75] moreover it improves bile flow, the immune system, the metabolism of cholesterol and can help to dissolve gallstones.[80]

Allicin and Diallyl Disulfide

Allicin belong to the group of thiosulfinate compounds, appearing in high amounts in garlic extracts. The species γ-glutamylcysteine present in raw garlic and onion is the precursor of various sulfur-containing sulfoxides which are responsible for the characteristic odor and flavor of garlic and onion.[81]

Thiosulfinates- including allicin- are volatile, unstable compounds which rapidly transform into other types of organosulfur compounds.[82] For instance, allicin is easily converted to allin by the enzyme alliinase released from vacuoles when garlic tissues are destroyed by cutting, crushing or chewing; after keeping allicin at 20°C for 20 h, it rapidly decomposes to diallyl disulfide, diallyl sulfide, diallyl trisulfide and sulfur dioxide.[81-83] Moreover, it can be easily denatured by boiling.[84]

The stable forms of allicin and of other highly reactive thiosulfinates are responsible for most of the health benefits attributed to onion and garlic. For many years garlic has been used as a spice and as a medicinal plant against headache, bites, intestinal worms, tumours and antiseptic remedy for wounds and ulcers. In China, onion and garlic tea have been recommended at length for fever, cholera and dysentery.[81]

Other properties of garlic and onion include antimicrobial, antioxidant, antimutagenic, antiasthmatic, immunomodulatory and prebiotic.[81,84,85] Garlic inhibits the growth of both gram-positive and gram-negative bacteria; onion is effective against gram-positive ones.[81] In 'in vitro' conditions, fresh garlic extract (even applied at 128 times dilution) inhibits the growth of a broad spectrum of bacteria such as *Staphylococcus aureus, Klebsiella pneumoniae, Escherichia coli, Pseudomonas, Proteus, Salmonella, Micrococcus, Bacillus subtilis, Mycobacterium* and *Clostridium*.[84]

Organo-sulfur compounds are protective agents in cardiovascular diseases, as they reduce serum cholesterol and triacylglycerol levels, show hypolipidemic, anti-hypertensive, anti-diabetic and antithrombotic activity as well as antioxidant properties preventing atherosclerosis.[81,84,85]

Diallyl sulfide and diallyl disulfide protect the stomach against *Helicobacter pylori* infection, causing dyspepsia, gastric and duodenal ulcer and probably gastric cancer.[84] Furthermore, an antiviral effect (against herpes simplex types 1 and 2 viruses, influenza A virus, influenza B virus, human cytomegalovirus, vesicular stomatitis virus, rhinovirus, human immunodeficiency virus, pneumonia virus and rotavirus[81]) has been observed, as well as insecticide and protective effects against parasites, fungi and yeasts. Protozoan parasites include: *Opalina ranarum, Opalina dimidicita, Balantidium entozoon, Entamoeba histolytica, Tripanosoma brucei, Leishmania, Leptomonas colosoma, Crithidia fasciculata, Giardia lamblia, Giardia intestinalis, Cryptosporidium baileyi, Tetratrichomonas gallinarum* and *Trichomonas vaginalis*.[81,86] Fungi and yeasts include: *Candida, Trichophyton, Torulopsis, Rhodotorula, Cryptococcus, Aspergillus* and *Trichosporon*.[81]

Anticarcinogenic benefits are mainly concerned with oesophagus, stomach, colon, bladder, prostate, liver, lungs, mammas, brain and skin sarcomas and carcinomas, as sulfur compounds influence DNA repair, acting as antimutagenic agents, stimulate T-cell proliferation and macrophages cytotoxicities on tumour cell lines and stimulate the activity of detoxifying enzymes.[81,83,85] Sometimes, the supplementation of raw-garlic extracts leads to allergic reactions such as contact dermatitis, bronchial asthma and chemical burns on the skin. These drawbacks can be overcome using different pharmaceutical products retaining garlic's health benefits. An example is given by aged garlic extracts, for which additional therapeutic effects have been found: they possess hepatoprotective, neuroprotective and antioxidative activities probably linked to the formation of new compounds during the long-term extraction. Additionally, no severe toxic side effects were observed even in cases of high ingestion dosages.[81,82]

Glucosinolates

Glucosinolates are nitrogen-sulfur containing compounds[87] occurring in high concentration in all cruciferous vegetables such as cabbages, broccoli, cauliflowers, Brussels sprouts, radishes, turnips, cress and in some relishes and oilseeds.[10] At least 120 compounds are known, sharing a common chemical structure made of a sulfonated moiety, a β-D-thioglucose group and a variable side chain derived from one amino acid. Depending on the type of amino acid precursor, glucosinolates can be classified into three chemical groups: aliphatic (those derived from Ala, Leu, Ile, Met, or Val), aromatic (those derived from Phe or Tyr) and indolic (those derived from Trp).[87,88] Upon plant damage, (e.g., by cutting, crushing, or chewing) glucosinolates are hydrolysed by the enzyme thioglucoside glucohydrolase and degraded into a variety of products. Some of them, like isothiocyanates and indolic compounds seem to be responsible for the majority of glucosinolates' beneficial effects.[16,87]

The decrease of risk of lung, stomach, colon and rectum carcinomas are some of the properties ascribed to these molecules.[16,88] Some compounds are identified as potent cancer prevention agents due both to their ability to improve the activity of detoxification enzymes, such as quinone reductase, glutathione-*S*-transferase and glucuronosyl transferases and due to their property of preventing tumour growth by blocking cell cycle and promoting apoptosis. Moreover, some compounds exhibit potential for treating gastritis and stomach cancer caused by *Helicobacter pylori* by reducing the number of precancerous lesions.[87] Despite this, no absolute certainty exists about the healthy properties of glucosinolates since an inverse relation has been found between the concentration of isothiocyanate metabolites in the urine of a group of Chinese men and the incidence of lung cancer.[88]

Capsacinoids

Capsaicinoids are nitrogen-containing plant secondary metabolites homologous with branched- or straight-chain alkylvanillylamides. They are found mainly in the *Capsicum* family, with the highest concentration present in chilli peppers. Capsacinoids are responsible for peppers' typical pungency which depends on the type of pepper and its geographical origin. This family of molecules include various compounds such as capsaicin, nordihydrocapsaicin, homocapsaicin, nonivamide and homodihydrocapsaicin.[16,89,90]

In addition to their flavouring property which makes them particularly useful as food additives, capsaicinoids reveal interesting pharmacological actions. For instance, capsaicin acts on the peripheral part of the sensory nervous system reducing the painful states caused by rheumatoid arthritis, postherpetic neuralgia, postmastectomy pain syndrome and diabetic neuropathy.[90,91] Moreover, it demonstrates a chemoprotective effect due to the modulation of the metabolism of carcinogens and/or mutagens interacting with DNA: the action mechanism shown by capsaicin is the suppression of the carcinogen binding to DNA that has been observed in the case of many polycyclic aromatic hydrocarbons.[16,90,91] However, opposite results have also been reported demonstrating that capsaicin induces gastric cancer in rats and could increase the risk factor for gastric cancer also in people consuming chilli.[16,90]

Conclusion

Living plants produce a series of secondary metabolites which appear to be important in numerous fields: toxicology, environment, pharmacology, cosmetics, medicine, food chemistry, etc. An important role played by these compounds is their beneficial effect on human health and their ability to prevent diseases.

In Western Europe, the aging of population has been and currently is a driving force for the study of new phytochemical products, able to fight age-related conditions and prevent dseases such as high blood cholesterol level, high blood pressure, arthritis, obesity and cancer. In fact, only a few plants have been the subject of detailed investigations but more than 1000 species have been claimed to offer special benefits.[92]

In 2002 the market for functional food in Europe exceeded 2 billion US$ (33 billion US$ in USA), representing less than 1% of the European food market,[93] but it is expected to reach 5% in the near future.

However, in many cases no certainty exists about the real effects of dietary phytochemicals and more epidemiological surveys and clinical studies should be performed. Moreover, in various cases contrary effects are reported. Probably the main reason for this discrepancy is that many clinically tested "products" are, in fact, multicomponent mixtures since it is frequently very difficult to isolate only one component to study its biological effect.

For this reason, standardised analytical procedures are needed and 'in vitro' and 'in vivo' models for fast screening of biological activity of minute amounts of phytochemicals need to be developed. Furthermore, the bioavailability and the concentration of different phytochemicals at the action sites needs to be better understood to offer improved methods to design functional foods: this could become particularly important when combining phytochemicals to avoid improper combinations that might cause lower absorbance or even nullify the effect of one or all the supplements.[92]

Finally, advances in biotechnology should be better exploited, even though genetically modified organisms are hardly accepted in European markets. The reason is that, indeed, biotechnology offers massive opportunities to produce the desired metabolites, modifying the genoma in such a way that their biosynthesis is improved and the processes of extraction and isolation become easier.

Acknowledgements

We are grateful to European Community for support (EU N. FOOD-CT-2005-023044). Fai-MoMa was financed by ASI for Nutraceuticals in support to astronauts in space.

References

1. Jones WHS ed. Hippocrates. Nutriment 1923; 351.
2. Funk C, The etiology of deficiency diseases. State Medicine 1912; 20:341-368.
3. Bidlack WR, Wang W. Designing functional food to enhance health. z Bidlack WR, Omaye ST, Meskin MS et al eds, Phytochemicals As Bioactive Agents. Lancaster: Technomic Publishing Co., Inc, 2000:241-270.
4. Fraser PD, Bramley PM. The biosynthesis and nutritional uses of carotenoids. Progress in Lipid Research 2004; 43:228-265.
5. Stahl W, Sies H. Bioactivity and protective effects of natural carotenoids. Biochimica et Biophysica Acta 2005; 1740:101-107.
6. Delgado-Vargas F, Jiménez AR, Paredes-López O. Natural pigments: carotenoids, anthocyanins and betalains-characteristics, biosynthesis, processing and stability. Crit Rev Food Sci Nutr 2000; 40(3):173-289.
7. Rao AV, Rao LG. Carotenoids and human health. Invited review. Pharmacological Research 2007; 55:207-216.
8. Tapiero H, Tew KD, Nguyen BG et al. Polyphenols: do they play a role in the prevention of human pathologies? Biomed Pharmacother 2002; 56:200-207.
9. Schieber A, Carle R. Occurrence of carotenoid cis-isomers in food: technological, analytical and nutritional implications. Trends Food Sci Technol 2005; 16:416-422.
10. Stahl W, van den Berg H, Arthur J et al. Bioability and metabolism. Mol Aspects Med 2002; 23:39-100.
11. Rock CL. Carotenoids: biology and treatment. Phannncol Ther 1997; 75(3):185-197.
12. Krinsky NI, Johnson EJ. Carotenoid actions and their relation to health and disease. Mol Aspects Med 2005; 26:459-516.
13. Roldán JM, Luque de Castro MD. Lycopene: the need for better methods for characterization and determination. Trends Analyt Chem 2007; 26(2):163-170.
14. Tapiero H, Townsend DM, Tew KD. The role of carotenoids in the prevention of human pathologies. Biomed Pharmacother 2004; 58:100-110.
15. Palace VP, Khaper N, Qin Q et al. Antioxidant potentials of vitamin A and carotenoids and their relevance to heart disease. Free Radic Biol Med 1999; 26(5/6):746-761.
16. Eskin NAM, Tamir S. Dictionary of Nutraceuticals and Functional Foods, Taylor and Francis Group, Boca Raton, London, New York 2006; 1-507.
17. Omoni AO, Aluko RE. The anticarcinogenic and anti-atherogenic effects of lycopene: a review. Trends Food Sci Technol 2005; 16:344-350.
18. Astorg P. Food carotenoids and cancer prevention: an overview of current research. Trends Food Sci Technol 1997; 8:406-413.
19. Nichenametla SN, Taruscio TG, Barney DL et al. A review of the effects and mechanisms of polyphenolics in cancer. Crit Rev Food Sci Nutr 2006; 46:161-183.
20. Bachioca M, Biagioti E, Ninfali P. Nutritional and technological reasons for evaluating the antioxidant capacity of vegetable products. Ital J Food Sci 2006; 2(18):209-217.
21. Erlund I. Review of the flavonoids quercetin, hesperetin and naringenin. Dietary sources, bioactivities, bioavailability and epidemiology. Nutrition Research 2004; 24:851-874.
22. Lin JK, Weng MS. Flavonoids as Nutraceuticals. In: Grotewold E. ed. The Science of Flavonoids. New York:Springer, 2006:213-238.
23. Le Marchand L. Cancer preventive effects of flavonoids—a review. Biomed Pharmacother 2002; 56:296-301.
24. Yao LH, Jiang YM, Shi J et al. Flavonoids in food and their health benefits. Plant Foods Hum Nutr 2004; 59:113-122.
25. Lotito SB, Frei B. Consumption of flavonoid-rich foods and increased plasma antioxidant capacity in humans: cause, consequence, or epiphenomenon? Free Radic Biol Med 2006; 41:1727-1746.
26. Aherne SA, O'Brien NM. Dietary flavonols: chemistry, food content and metabolism. Nutrition 2002; 18:75-81.
27. Heim KE, Tagliaferro AR, Bobilya DJ. Flavonoid antioxidants: chemistry, metabolism and structure-activity relationships. J Nutr Biochem 2002; 13:572-584.
28. Harborne JB, Williams CA. Advances in flavonoid research since 1992. Phytochemistry 2000; 55:481-504.
29. Di Carlo G, Mascolo N, Izzo AA et al. Flavonoids: old and new aspects of a class of natural therapeutic drugs. Life Sciences 1999; 65(4):337-353.

30. Russo M, Tedesco I, Iacomino G et al. Dietary phytochemicals in chemoprevention of cancer. Curr Med Chem—Immun, Endoc and Metab Agents 2005; 5:61-72.
31. Aggarwal BB, Shishodia S. Molecular targets of dietary agents for prevention and therapy of cancers. Biochem Pharmacol 2006; 71:1397-1421.
32. Orallo F. Comparative studies of the antioxidant effects of cis- and trans-resveratrol. Curr Med Chem 2006; 13:87-98.
33. Vitaglione P, Morisco F, Caporaso N et al. Dietary antioxidant compoundsand liver health. Crit Rev Food Sci Nutr 2004; 44:575-586.
34. Martin JHJ, Crotty S, Warren P et al. Does an apple a day keep the doctor away because a phytoestrogen a day keeps the virus at bay? A review of the antiviral properties of phytoestrogens. Phytochemistry 2007; 68:266-274.
35. Cornwell T, Cohick W, Raskin I. Dietary phytoestrogens and health. Phytochemistry 2004; 65:995-1016.
36. McCue P, Shetty K. Potential health benefits of soybean isoflavonoids and related phenolic antioxidants. In: Shetty K, Paliyath G, Pometto AL, Levin RE, eds. Functional Foods and Biotechnology. London:Taylor and Francis Group, 2007:133-150.
37. Brouns F. Soya isoflavones: a new and promising ingredient for the health foods sector. Food Res Int 2002; 35:187-193.
38. Birt DF, Hendrich S, Wang W. Dietary agents in cancer prevention: flavonoids and isoflavonoids. Pharmacology and Therapeutics 2001; 90:157-177.
39. Oleszek W, Stochmal A, Janda B. Concentration of isoflavones and other phenolics in the aerial parts of Trifolium species. J Agric Food Chem 2007; 55:8095-8100.
40. Knight DC, Eden JA. A review of the clinical effects of phytoestrogens. Obstet Gynecol 1996; 87(5/2):897-904.
41. Davis SR, Murkies AL, Wilcox G. Phytoestrogens in clinical practice. Integr Med 1998; 1(1):27-34.
42. Tempfer CB, Bentz EK, Leodolter S et al. Phytoestrogens in clinical practice: a review of the literature. Fertil Steril 2007; 87(6):1243-1249.
43. Fuss E. Lignans in plant cell and organ cultures: an overview. Phytochemistry Reviews 2003; 2:307-320.
44. Duncan AM, Phipps WR, Kurzer MS. Phytoestrogens. Best Pract Res Clin Endocrinol Metab 2003; 17(2):253-271.
45. Lof M, Weiderpass E. Epidemiologic evidence suggests that dietary phytoestrogen intake is associated with reduced risk of breast, endometrial and prostate cancers. Nutr Res 2006; 26:609-619.
46. Beck V, Rohr U, Jungbauer A. Phytoestrogens derived from red clover: an alternative to estrogen replacement therapy? J Steroid Biochem Mol Biol 2005; 94:499-518.
47. Suzuki S, Umezawa T. Biosynthesis of lignans and norlignans. J Wood Sci 2007; 53:273-284.
48. Meagher LP, Beecher GR. Assessment of data on the lignan content of foods. J Food Compost Anal 2000; 13:935-947.
49. Niemeyer HB, Metzler M. Differences in the antioxidant activity of plant and mammalian lignans. J Food Eng 2003; 56:255-256.
50. El-Badry AM, Graf R, Clavien PA. Omega 3—Omega 6: what is right for the liver? J Hepatol 2007; 47:718-725.
51. Balk EM, Lichtenstein AH, Chung M et al. Effects of omega-3 fatty acids on serum markers of cardiovascular disease risk: a systematic review. Atherosclerosis 2006; 189:19-30.
52. Kinney AJ. Metabolic engineering in plants for human health and nutrition. Curr Opin Biotechnol 2006; 17:130-138.
53. Sijtsma L, de Swaaf ME. Biotechnological production and applications of omega-3 polyunsatureted fatty acid docosahexaenoic acid. Appl Microbiol Biotechnol 2004; 64:146-153.
54. Mazza M, Pomponi M, Janiri L et al. Omega-3 fatty acids and antioxidants in neurological and psychiatric diseases: an overview. Prog Neuropsychopharmacol Biol Psychiatry 2007; 31:12-26.
55. Domingo JL. Omega-3 fatty acids and the benefits of fish consumption: is all that glitters gold? Environ Int 2007; 33:993-998.
56. Beltz BS, Tlusty MF, Benton JL et al. Omega-3 fatty acids upregulate adult neurogenesis. Neurosci Lett 2007; 415:154-158.
57. Bhattacharya A, Banu J, Rahman M et al. Biological effects of conjugated linoleic acids in health and disease. J Nutr Biochem 2006; 17:789-810.
58. Nagao K, Yanagita T. Conjugated fatty acids in food and their health benefits. J Biosci Bioeng 2005; 100(2):152-157.
59. Whigham LD, Cook ME, Atkinson RL. Conjugated linoleic acid: implications for human health. Pharmacol Res 2000; 42(6):503-510.
60. Sieber R, Collomb M, Aeschlimann A et al. Impact of microbial cultures on conjugated linoleic acid in dairy products—a review. Int Dairy J 2004; 14:1-15.

61. Cook ME, Pariza M. The role of conjugated linoleic acid (CLA) in health. Int Dairy J 1998; 8:459-462.
62. Rainer L, Heiss CJ. Conjugated linoleic acid: health implications and effects on body composition. J Am Diet Assoc 2004; 104(6):963-968.
63. Collomb M, Schmid A, Sieber R et al. Conjugated linoleic acids in milk fat: variation and physiological effects. Int Dairy J 2006; 16:1347-1361.
64. Evans ME, Brown JM, McIntosh MK. Isomer-specific effects of conjugated linoleic acid (CLA) on adiposity and lipid metabolism. J Nutr Biochem 2002; 13:508-516.
65. Sen CK, Khanna S, Roy S. Tocotrienols: vitamin E beyond tocopherols. Life Sci 2006; 78:2088-2098.
66. Zingg JM. Vitamin E: an overview of major research directions. Mol Aspects Med 2007; 28 (5-6):400-422.
67. Saldeen K, Saldeen T. Importance of tocopherols beyond α-tocopherol: evidence from animal and human studies. Nutr Res 2005; 25:877-889.
68. Theriault A, Chao JT, Wang Q et al. Tocotrienol: a review of its therapeutic potential. Clin Biochem 1999; 32(5):309-319.
69. Kline K, Lawson KA, Yu W et al. Vitamin E and breast cancer prevention: current status and future potential. J Mammary Gland Biol Neoplasia 2003; 8(1):91-102.
70. Tucker JM, Townsend DM. Alpha-tocopherol: roles in prevention and therapy of human disease. Biomed Pharmacother 2005; 59:380-387.
71. Reiter E, Jiang Q, Christen S. Anti-inflammatory properties of α- and γ-tocopherol. Mol Aspects Med 2007; 28(5-6):668-691.
72. Lapointe A, Couillard C, Lemieux S. Effects of dietary factors on oxidation of low-density lipoprotein particles. J Nutr Biochem 2006; 17:645-658.
73. Azzi A, Gysin R, Kempná P et al. The role of α-tocopherol in preventing disease: from epidemiology to molecular events. Mol Aspects Med 2003; 24:325-336.
74. Crowell PL. Monoterpenes in breast cancer chemoprevention. Breast Cancer Res Treat 1997; 46:191-197.
75. Vigushin DM, Poon GK, Boddy A et al. Phase I and pharmacokinetic study of D-limonene in patients with advanced cancer. Cancer Chemother Pharmacol 1998; 42:111-117.
76. Merle H, Moro M, Amparo Blázquez M et al. Taxonomical contribution of essential oils in mandarins cultivars. Biochem Syst Ecol 2004; 32:491-497.
77. Flamini G, Tebano M, Cioni PL. Volatiles emission patterns of different plant organs and pollen of citrus limon. Anal Chim Acta 2007; 589:120-124.
78. Tirillini B, Pellegrino R, Pagiotti R et al. Volatile compounds in different cultivars of Apium graveolens L. Ital J Food Sci. 2004; 4(16):477-482.
79. Ferreira D, Marais JPJ, Slade D. Phytochemistry of the mopane, colophospermum mopane. Phytochemistry 2003; 64:31-51.
80. Friedman MI, Preti G, Deems RO et al. Limonene in expired lung air of patients with liver disease. Dig Dis Sci 1994; 39(8):1672-1676.
81. Corzo-Martínez M, Corzo N, Villamiel M. Biological properties of onions and garlic. Trends Food Sci Technol 2007; 18:609-625.
82. Amagase H. Clarifying the real bioactive constituents of garlic. J Nutr 2006; 136:716S-725S.
83. Wu CC, Chung JG, Tsai SJ et al. Differential effects of allyl sulfides from garlic essential oil on cell cycle regulation in human liver tumor cells. Food Chem Toxicol 2004; 42:1937-1947.
84. Sato T, Miyata G. The Nutraceutical Benefit, part IV, Garlic Nutrition 2000; 16(9):787-788.
85. Jakubowski H. On the Health Benefits of Allium sp. Nutrition 2003; 19(2):167-168.
86. Anthony JP, Fyfe L, Smith H. Plant active components—a resource for antiparasitic agents? Trends Parasitol 2005; 21(10):299-322.
87. Halkier BA, Gershenzon J. Biology and biochemistry of glucosinolates. Annu Rev Plant Biol 2006; 57:303-333.
88. Johnson IT. Glucosinolates in the human diet. Bioavailability and implication for health. Phytochemistry Reviews 2002; 1:183-188.
89. Surh YJ, Ahn SH, Kim KCh et al. Metabolism of capsaicinoids: evidence for aliphatic hydroxylation and its pharmacological implications. Life Sci 1995; 56(16):305-311.
90. Surh YJ, Lee SS. Capsaicin in hot chili pepper: carcinogen, cocarcinogen or anticarcinogen? Fd Chem Toxic 1996; 34(3):313-316.
91. Surh YJ, Lee SS. Capsaicin, a double-edged sword: toxicity, metabolism and chemopreventive potential. Life Sci 1995; 56(22):1845-1855.
92. Andlauer W, Fürst P. Nutraceuticals: a piece of history, present status and outlook. Food Res Int 2002; 35(2-3):171-176.
93. Menrad K. Market and marketing of functional food in Europe. J Food Eng 2002; 56:181-188.

Chapter 8

Bioactive Compounds from Northern Plants

Anja Hohtola*

Abstract

Northern conditions are characterised by long days with much light and low temperatures during the growing season. It has been claimed that herbs and berries grown in the north are stronger tasting compared to those of southern origin. The compounds imparting aroma and color to berries and herbs are secondary metabolites which in plants mostly act as chemical means of defense. Recently, the production of secondary metabolites using plant cells has been the subject of expanding research. Light intensity, photoperiod and temperature have been reported to influence the biosynthesis of many secondary metabolites. Native wild aromatic and medicinal plant species of different families are being studied to meet the needs of raw material for the expanding industry of e.g., health-promoting food products known as nutraceutics. There are already a large number of known secondary compounds produced by plants, but the recent advances in modern extraction and analysis should enable many more as yet unknown compounds to be found, characterised and utilised.

Rose root (*Rhodiola rosea*) is a perennial herbaceous plant which inhabits mountain regions throughout Europe, Asia and east coastal regions of North America. The extract made from the rhizomes acts as a stimulant like the Ginseng root. Roseroot has been categorized as an adaptogen and is reported to have many pharmacological properties. The biologically active components of the extract are salitroside tyrosol and cinnamic acid glycosides (rosavin, rosarin, rosin).

Round-leaved sundew (*Drosera rotundifolia* L.) has circumboreal distribution. It inhabits nutrient-poor, moist and sunny areas such as peat bogs and wetlands. Sundew leaves are collected from the wild-type for various medicinal preparations and can be utilized in treating e.g., as an important "cough-medicine" for different respiratory diseases. The antimicrobial activity of extracts of aerial parts against various bacteria has been investigated. *Drosera* produces various secondary metabolites. The most abundant, among these compounds, are the naphthoquinones. Bilberry (*Vaccinium myrtillus*) is a characteristic field layer species in boreal forests. Bilberry and other northern *Vaccinium* species, berries and leaves, contain high amounts of phenolic compounds. Bilberries are known for its exceptionally high amounts of anthocyanins with powerful antioxidant capacity. They have been shown to possess beneficial health effects, like having a protective role in cardiovascular diseases and cancer. Many flavonoids also seem to have antiviral, antibacterial, antifungal and antiallergenic properties. The effect of ingested cranberry (*V. oxycoccus*) juice has been shown to prevent urinary tract infections in women.

*Anja Hohtola—Department of Biology, University of Oulu, Finland.
Email: anja.hohtola@oulu.fi

Bio-Farms for Nutraceuticals: Functional Food and Safety Control by Biosensors
edited by Maria Teresa Giardi, Giuseppina Rea and Bruno Berra.

Introduction

Plants produce bioactive compounds that are important in the healthcare, food, flavour and cosmetics industry. Many herbs and some common dietary supplements are good sources of antioxidants and anti-inflammatory compounds. Warm and temperate regions are exceptionally rich in herbal plants but even in the northern hemisphere several valuable plants with bioactive properties are found. The compounds produced by plants are separated into primary and secondary metabolites. Primary metabolites are necessary for the growth and basic metabolism in all plants, while secondary metabolites have a restricted distribution within the plant and in the plant kingdom. The term secondary metabolite reflects the fact that most of these compounds are not essential, but may play crucial roles in other functions. As organic compounds, secondary metabolites are regarded as a carbon resource of plants and many of them act in chemical defense of plants. In addition to chemical defense against herbivores, pathogens and other plants,[1] secondary metabolites can also act as UV-screens or give color to flowers attracting pollinators.[2] Thus, plant secondary metabolites are important for plant interactions with their environments and are, in turn, important for their survival and propagation.[3] Various factors, such as age of the plant, season, microbial attack, grazing, radiation, competition, nutritional status and many environmental[4] stresses have impact on the level of secondary metabolism. Carbon/nutrient balance (CNB) theory postulates that carbon-based secondary compounds tend to accumulate under conditions of low availability of inorganic nutrients and bright light. Similarly, if growth is more inhibited by low temperature than photosynthesis, fixed carbon may accumulate as secondary metabolites. Intensive research on secondary metabolites has been carried out dealing with the defense reactions in plants: that is, from the point of view of plants and their predators. In addition to predators, plants are exposed to natural climatic stress due to high irradiation, heat, chilling, frost, drought, flooding and nutrient imbalances. Active oxygen species have been proposed as a central component of plant adaptation to both biotic and abiotic stresses. Under stress conditions, active oxygen species can play two opposing roles: worsen damage or signaling the activation of defense responses.[5] Such a dual function was first described in pathogenesis but has now also been demonstrated during several abiotic stress responses. The control of oxidant levels is achieved by antioxidative systems in plant cells. Another study frame is the exploitation of bioactive compounds including medicinal products, nutraceutics, functional food and their metabolism in the body. As shown, many plants are especially good sources of antioxidants and anti-inflammatory compounds[6] containing carotenoids, ascorbic acid, phenolic compounds and various nitrogen compounds like e.g., alkaloids and amino acids.

How External Factors Influence the Biosynthesis of Secondary Metabolites

Light quality and duration of radiation is different in different global areas. Northern conditions are characterised by long days with much light and low temperatures during the growing season. There are a number of reports on the effect of light intensity and duration on the plant secondary metabolism. The effect of the day length, strength and composition of light and mild temperature has been tested with some vegetables, berries and herbs[7-10] and big differences have been found e.g., in flavour compounds. Long-day conditions enhanced growth and increased the total amount of monoterpenes in peppermint.[11,12] In dill (*Anethun graveolens* L.), growth and essential oil accumulation was reported to increase with increase in the light level and was the highest under full sunlight.[13]

Both the photoperiod and the spectral quality of the light have been shown to have an effect on secondary metabolite concentrations. The allantoin content in the roots of *Symphytum officinale* was highest when plants were grown on a 20-hours photoperiod.[14] The increase was about 80-fold compared to plants grown on an 8-hours photoperiod. Cultured, green *Lavendula vera* cells grown in the light were found to accumulate a high level of free biotin.[15] Increasing light intensity (106-402 μmol m^{-2} s^{-1}) resulted in a continuous increase in the level of leaf hypericins in St. John's wort.[16]

Temperature stress is known to cause many physiological, biochemical and molecular changes in plant metabolism and possibly alter the secondary metabolite production in plants. Plants growing in cold climates can maintain higher photosynthetic rates at low temperatures than plants growing in warmer surroundings and thus there is more fixed carbon also for secondary metabolites. Laine and Henttonen[17] studied the effects of temperature on total phenolic substances, but studies on specific compounds are quite few. Burbott and Loomis[7] suggest that warm nights cause depletion of respiratory substrates, resulting in oxidizing conditions, while cool nights preserve high levels of respiratory substrates and thus maintain reducing conditions. Saastamoinen[18] showed that the oil content of cereals increased at low temperature. Carrots from northern latitudes seem to have a common genetic response to cold stress.[8]

Anthocyanin belongs to the class of flavonoids which are important pigments in flowers, fruits and leaves serving as visual signals for pollinators and seed dispersers.[19] Like all flavonoids, anthocyanins have a strong antioxidant capacity and, by way of preventing the formation of free radicals, they have been shown to prevent number of diseases.[20-22] The effect of temperature on the colouration of the berry or fruit skin has been shown in many publications. In particular, the skin colouration in grape and apple has been widely studied. Anthocyanin accumulation in the grape skins was significantly higher at 20°C than at 30°C.[23]

Large seasonal variation is seen in the biosynthesis of secondary metabolites, with a trend towards peak levels in late spring and early summer and generally low levels during autumn and winter.[24] According to Bartley and Ishida[25] the tomato tissues produced lycopene when the plant was kept at 16-22°C but not at higher temperatures. It is also quite common that the content of bioactive compounds varies from year to year,[26-27] because chemical content can be influenced by a variety of environmental factors including water and nutrient availability.[9-10] However, according to Tuomi et al[28] growth always takes priority over secondary metabolite production.

Research and Exploitation

Throughout the history of mankind, many diseases have been treated with plant extracts. There is, at the present, a renewed interest in using plants as source of medicine[29] and nutraceutics.[30] Today it is known that plants are responsible for the production of at least 50,000 compounds including pharmaceuticals, aromas, pigments, cosmetics and food additives.[31] As many of these compounds are valuable for their industrial or agricultural properties they increase the commercial value of the plant crops involved.[32-33] Flavonoids and related phenolic substances are of special interest for their antioxidative properties. Native wild aromatic plant species of different families are studied nowadays to meet the needs of raw material of industries. Dietary supplement sales are booming in Western countries and consumption of botanical products is expected to grow exponentially.

Plants containing interesting compounds can be either traditional, long-known species or innovations like, for example, *Taxus*. Recent advances in modern extraction and structural analysis methods enable us to identify more, previously unknown, compounds in familiar plants.[29,34] One good example is the characterization of the functional properties of plant phenolics from extracts of northern plants having relevance to food quality or human health. Antimicrobial activity was found e.g., in *Filipendula ulmaria*, *Rubus chamaemorus* and *Rubus idaeus*. Common spices like *Rhus coriaria* and *Thymus vulgaris* show antibacterial effects and can also be regarded as nutraceutics as reported by Adwan et al.[35] Many other spices show antimicrobial activity.[36]

The biosynthetic routes and mechanisms describing the production of plant secondary metabolites are often very complex and can involve dozens of enzymes. Knowledge of plant metabolic pathways is still limited or partial and extensive work is going on in various fields. The first prerequisite for exploitation is an adequate understanding of biosynthetic routes and their regulation. With increasing knowledge exciting possibilities and technological advances can be foreseen. The availability of new molecular tools and technological advances has made the exploitation of novel bioactive molecules more attainable.

Plants growing in the wild may suffer from excessive gathering, so compensatory measures may become desirable. Domestication of a new crop is a process which takes many years, but if successful the benefits can make a significant impact in many areas. Research is important to gain knowledge about the optimal growth conditions for obtaining the best possible harvest of desirable product. Production of nutraceutics and biologically active compounds in cell cultures can be a possible alternative for slowly growing plants and assure a more uniform supply of the biologically active compounds. Recently, the production of secondary metabolites using plant cell cultures has been the subject of expanding research.[37] The basic steps in carrying out this type of research are usually: (1) choose an interesting plant, (2) isolate and identify relevant substances from the plant and (3) develop means to quantify these substances.[38]

Examples of Northern Plants Containing Bioactive Compounds

Rhodiola rosea

Rhodiola rosea L. (golden root, roseroot) is a perennial herbaceous plant which inhabits mountain regions throughout Europe, Asia and east coastal regions of North America (Fig 1). It grows also in northern countries, e.g., in Finland in the wet and sandy grounds of Enontekiö and Kittilä.[39] The rhizomes contain various medicinally active compounds belonging to several chemical groups like flavolignans, flavonoids, phenolic compounds, phenylpropanoids, coumarins, sterols, monoterpenes and lactones.[40] The main bioactive compounds, e.g., salidroside and cinnamyl alcohol glycosides, are products of phenylpropanoid metabolism, being derived from phenylalanine, which is a derivative of the shikimic-chorismic acid pathway. There are still many unknown points on the pathway, e.g., the enzymes involved in the formation of the glycosides of cinnamyl alcohol have not yet been described. However, enhancement of the yields of bioactive compounds has been achieved in tissue cultures using precursors.[41,42] Glucosylation of the precursors resulted in high yields of rosin and salidroside and low amounts of rosavin.

Figure 1. *Rhodiola rosea*.

Roseroot has been used for a long time in Chinese traditional medicine and in Eastern Europe as a stimulant, decreasing depression, enhancing work performance etc.[40,43] The plant has been categorized as an adaptogen and is reported to have many pharmacological properties, like stimulation of the nervous system, promotion of longevity, improvement of sleep. Cardioprotective characteristics have also been noticed.[44,45] The extract of the rhizomes acts as a stimulant similar to the ginseng root and has also been found to prevent tumour growth. The biologically active components of the extract are salitroside tyrosol and cinnamic acid glycosides (rosavin, rosarin and rosin). This plant, also called the "Ginseng of the North", seems to be a very potential source of bioactive compounds without substantial adverse effects.[46]

Field cultivation of *Rhodiola rosea* has been tested with relatively good results.[47,48] The concentration of pharmacologically active compounds is dependent on the age and genotype. Photoculture enhanced the content of salidroside in the rhizomes exceeding by 1.5-3 times the levels observed in the plants grown in the field stands.[49] Using field cultivation of similar plants it takes several years to obtain a satisfactory amount of the compounds of interest. Tissue culture offers an alternative method for the production of the genetically identical plants with required qualities.[50-51] Production of biologically active compounds in bioreactors can be a possible alternative for slow growing plant and assure more uniform supply of the products.[41,42] While the concentration of interesting compounds is low in undifferentiated cell cultures, stimulation of the biosynthesis of cinnamyl glycosides can be enhanced by biotransformation, in this case by adding cinnamyl alcohol. As a result, three to six-fold rosin content was achieved in the roseroot callus cultures compared to the content of wild growing plants. Also, production of new compounds is possible using elicitors and precursors.[52]

Drosera rotundifolia

Round-leaved sundew (*Drosera rotundifolia* L.) inhabits nutrient-poor, moist and sunny areas such as peat bogs and abandoned sandpits low in available nutrients (nitrogen and phosphorus and calcium) (Fig 2). *D. rotundifolia* is carnivorous: it gains nutrients, especially nitrogen and phosphorus from captured insects. Adaptation to nutrient-poor conditions allows it to be very competitive and persistent in acid wetlands. Naphthoquinones are the main secondary metabolites produced by *Drosera*.[53] These phenolic compounds are formed through the acetate-malonate and shikimic acid pathways. The naphthoquinones of sundew are derived from acetate, which is formed from L-tyrosine most likely by a homogentisate ring-cleavage pathway. The key enzyme of this ring-cleavage reaction is homogentisate oxidase.[54] Naphthoquinones are synthesized and accumulated also in plant tissue cultures of sundew, but the naphthoquinone levels under in vitro conditions were lower than upon transfer ex vitro.[55] The amount of naphthoquinones varies among different *Drosera* species[56] and in different tissues of the plants.[57,58] In some sundew species, the concentration of naphthoquinones varies during the growing season, but in *D. rotundifolia* the amount is fairly constant.[59] The concentration of 7-methyljuglone in Finnish *Drosera* populations was higher than in Italian species.[27,59] The strongest influences on the concentration were determined by the weather; warm and dry summer seems to favour 7-methyljuglone biosynthesis in the northernmost Finnish habitats.[59] There was no visible north-south trend in the concentration of 7-methyljuglone between the Finnish populations.

The effects of different levels of light and inorganic nutrients in the growth medium on the investment in carnivory, amount of secondary metabolite, growth and reproduction in *D. rotundifolia* were investigated by Thoren et al.[60] It was found that the carbon-based secondary compound, 7-methyljuglone, in the leaves did not respond to shading or nutrient addition. On the other hand, plants grown in the shade or in media with inorganic nutrients had less sticky leaves and had reduced their investment in carnivory. Favourable conditions for photosynthesis seem to increase the allocation of energy for defence compounds. Nutrient addition to the growth medium did not affect the growth or reproduction significantly. One tool to study the biosynthesis and its regulation is to use the genetic transformation technique which has also been developed for *D. rotundifolia*.[61] The sundews are endangered in many countries and they are protected, for example, in Belgium.[62]

Figure 2. *Drosera rotundifolia*.

The round-leaved sundew is presently not in danger of extinction in Finland,[39] but the small size of plants makes collection from natural stands laborious and therefore, various possibilities for cultivation have been studied extensively. Results were in general positive.[48]

The naphthoquinones of sundew have many physiological effects. Juglone, for example, which is toxic and an effective inhibitor of seed germination for many plants, has also been shown to be inhibitory to several insects and to be highly toxic to fungi as well as different fungal pathogens.[63] Sundew naphthoquinones have also important pharmacological effects. Sundew leaves are collected from the wild and various medicinal preparations and can be utilized as medicinal compounds in treating various respiratory diseases having anti-inflammatory and spasmolytic effects.[64-66] Antimicrobial activity of extracts of aerial parts against oral bacteria[67] and other microbes has also been investigated.

Vaccinium myrtillus

Bilberry (*Vaccinium myrtillus L,* european blueberry) (Fig 3) and other northern and native *Vaccinium* species, berries and leaves, contain large amounts of phenolic compounds, especially proanthocyanidins and flavonols which belong to the flavonoids.[68-70] Pathways of flavonoid biosynthesis start generally with phenylpropanoid metabolism and lead to major subgroups. Flavonoids constitute a relatively diverse family of aromatic molecules that are derived from phenylalanine and malonyl-coenzyme A.[71] The expression of five flavonoid biosynthesis genes during the development of bilberry fruit in relation to the accumulation of anthocyanins, proanthocyanidins and flavonols in wild berries has been studied. The results show a clear correlation between anthocyanin accumulation and expression of the flavonoid pathway genes during the ripening of berries.

Flavonoids have a function in plant stress reactions, such as defence against pathogen attack, wounding or against excessive UV-light. Flavonoids not only protect the plant from the harmful effects of UV irradiation but also play a crucial role in the sexual reproduction process. A special

Figure 3. *Vaccinium myrtillus.*

class of flavonoid polymers, the tannins, plays a structural role in the plant. Yet other classes of flavonoids, flavonols and anthocyanins, have been implicated in the attraction of pollinators. Certain flavonoids participate in the interaction between plants and other organisms such as symbiotic bacteria and parasites. This raises the intriguing question as to how these different compounds arose and evolved.[72] It has been found that contents of catechins, flavonols and hydroxycinnamic acids are higher in the bilberry leaves exposed to direct sunlight.[70] Taulavuori et al[73] noticed that the long-term exposure to UV-B radiation in the subarctic did not cause oxidative stress in bilberry. Organ dependent distribution of phenolic compounds was noticed in bilberry and blueberry, the anthocyanin content being especially high in bilberry pulp.[74] Blueberries, like *V. angustifolium, V. corymbosum* and, *V. ashei*, are widely cultivated and breeding work is going on in order to increase e.g., antioxidant activity in berries. Among blueberry cultivars, a remarkable genotypic and environmental variation in total phenolic content and antioxidant capacity has been found.[26] Also, a marked year-to-year variability was noticed, so the weather conditions may have a significant role in biosynthesis of phenolics.

Bilberries contain exceptionally high amounts of anthocyanins compared to other blueberries like *Vaccinium corymbosum.*[75] There are more than 250 naturally occurring anthocyanins which consist of one of six aglycons glycosylated with various sugar substitutes. Anthocyanins are known for their powerful antioxidant capacity and are believed to provide many health benefits, like having a protective role in cardiovascular diseases and cancer.[20,21,76] Reactive oxygen species play a critical role in the impairment of nitric oxide-mediated vascular functions and cardiovascular disease.[77] Anthocyanins of certain berries are exceptionally potent oxygen radical scavengers.[78] It has been shown by in vitro studies that bilberry anthocyanins cause the heart arteries to relax and may protect normal nitric oxide-mediated dilation.[79] The results indicate that colourful berries could be valuable as dietary preventative agents for general health of the heart and vascular system.

In spite of a low bioavailability, anthocyanins of *Vaccinium* species seem to have a positive effect on plasma antioxidant capacity.[80,81]

Anthocyanins and the ellagic acid of many berries including bilberry have been shown to exhibit chemopreventive activity, to reduce cancer cell proliferation and to inhibit tumour formation both in vitro and in vivo studies.[82-84] The berry phenolics have proved effective for inhibiting cancer cell—breast, colon, prostate and oral cancer—proliferation in vitro and stimulating apoptosis.[85] A number of studies have suggested that bilberry extracts improve vision in reduced light conditions, but more recent trials have failed to find such an effect.[86] Many flavonoids also have shown to have antiviral, antibacterial, antifungal and antiallergenic properties.[87,88]

Bilberry fruit and leaves have traditionally been used as a treatment for diabetes in adults. However, controlled trials are lacking or patchy for humans. Only recently extracts of blueberry (*Vaccinium angustifolium* Ait) have been shown to improve glucose transport and the insulin response of muscle and pancreas cells in vitro thus protecting these cells against glucose toxicity typical for diabetes.[89] In addition, pancreatic beta cells were increased in number by treatment with blueberry extracts. Hence, phenolic antioxidants may be promising as anti-diabetic agents.

Conclusion

What is the impact of Northern cool and long days on the production aroma substances is not yet clearly shown but it is evident that plants have remarkable phenotypic flexibility in their responses also to transiently changing environmental conditions. Warm and temperate nature is rich in herbal plants which are good sources of antioxidants and anti-inflammatory compounds, commonly used as dietary supplements. Nevertheless, even in northern climes several plants can give many valuable products.

References

1. Biere A, Marak HB, van Damme JMM. Plant chemical defense against herbivores and pathogens: generalized defense or trade-offs? Oecologia 2004; 140:430-441.
2. Waterman PG, Mole S. Analysis of phenolic plant metabolites. Blackwell Scientific Publishing, London, England 1994.
3. Harborne JB. Recent advances in chemical ecology. Natural Products Report 1996; 12:83-98.
4. Bryant JP, Chapin III FS, Klein DR. Carbon/nutrient balance of boreal plants in relation to vertebrate history. Oikos 1983; 40:357-368.
5. Dat J, Vandenabeele S, Vranová E et al. Dual action of the active oxygen species during plant stress responses. Cellular Mol Life Sci 2000; 57:779-795.
6. Velioglu YS, Mazza G, Gao L et al. Antioxidant activity and total phenolics in selected fruits, vegetables and grain products. J Agric Food Chem 1998; 46:4113-4117.
7. Burbott AJ, Loomis WD. Effects of light and temperature on the monoterpenes of peppermint. Plant Physiol 1967; 42:20-28.
8. Hårdh JE, Persson AR, Ottoson L. Quality of vegetables cultivated at different latitudes in Scandinavia. Acta Agr Scand 1977; 27:81-96.
9. Clark RJ, Menary RC. Environmental effects on peppermint (Mentha piperita L.). I. Effect of day-length, photon flux density, night temperature and day temperature on the yield and composition of peppermint oil. Aust J Plant Physiol 1980; 7:685-692.
10. Watson R, Wright CJ, McBurney T et al. Influence of harvest date and light integral on the development of strawberry flavour compounds. J Exp Bot 2002; 53:2121-2129.
11. Voirin B, Brun N, Bayet C. Effects of daylength on the monoterpene composition of leaves of Mentha x piperita. Phytochemistry 1990; 29:749-755.
12. Fahlén A, Welander M, Wennersten R. Effects of light-temperature regimes on plant growth and essential oil yield of selected aromatic plants. J Sci Food Agr 1997; 73:111-119.
13. Halva S, Craker LE, Simon JE et al. Light quality, growth and essential oil in dill (Anethum graveolens L.) J Herbs Spices Med Plants 1992; 1:59-69.
14. Castro AHF, Young MCM de Alvarenga AA, Alves JD. Influence of photoperiod on the accumulation of allantoin in comfrey plants. R Bras Fisiol Veg 2001; 13:49-54.
15. Watanabe K, Yano SI, Yamada Y. The selection of cultured plant cell lines producing high levels of biotin. Phytochemistry 1982; 21:513-516.
16. Briskin DP, Gawienowski MC. Differential effects of light and nitrogen on production of hypericins and leaf glands in Hypericum perforatum. Plant Biochem Physiol 2001; 39:1075-1081.

17. Laine K, Henttonen H. Phenolics/nitrogen ratios in the blueberry Vaccinium myrtillus in relation to temperature and microtine density in Finnish Lapland. Oikos 1987; 50:389-395.
18. Saastamoinen M. Effects of environmental factors on grain yield and quality of oats (Avena sativa L.) cultivated in Finland. Acta Agric Scand Sect B Soil Plant Sci 1998; 48:129-137.
19. Holton TA, Cornish EC. Genetics and biochemistry of anthocyanin biosynthesis. Plant Cell 1995; 7:1071-1083.
20. Bagchi D, Sen CK, Bagchi M et al. Anti-angiogenic, antioxidant and anti-carcinogenic properties of a novel anthocyanin-rich berry extract formula. Biochemistry 2004; 69:75-80.
21. Kähkönen MP, Hopia AI, Vuorela HJ et al. Antioxidant activity of plant extracts containing phenolic compounds. J Agric Food Chem 1999; 47:3954-3962.
22. Prior RL et al. Antioxidant capacity as influenced by total phenolic and anthocyanin content, maturity and variety of Vaccinium species. J Agric Food Chem 1998; 46:2686-2693.
23. Yamane Y, Jeong ST, Goto-Yamamoto N et al. Effects of temperature on anthocyanin biosynthesis in rape berry skins. Am J Enol Vitic 2006; 57:54-59.
24. Bjerke JW, Elvebakka A, Domínguez E et al. Seasonal trends in usnic acid concentrations of Arctic, alpine and Patagonian populations of the lichen Flavocetraria nivalis. Phytochemistry 2005; 66:337-44.
25. Bartley GE, Ishida BK. Developmental gene regulation during tomato ripening and in-vitro sepal morphogenesis. BMC Plant Biol 2003; 3:4.
26. Connor AM, Luby JJ, Tong CBS. Genotypic and environmental variation in antioxidant activity, total phenolic content and anthocyanin content among blueberry cultivars. J Amer Soc Hort Sci 2002; 127:89-97.
27. Kämäräinen T, Uusitalo J, Jalonen J et al. Regional and habitat differences in 7-methyljuglone content of Finnish Drosera rotundifolia. Phytochemistry 2003; 63:309-314.
28. Tuomi J, Fagerström T, Niemelä P. Carbon allocation, phenotypic plasticity and induced defences. In: Tallamy DW and Raupp MJ, eds. Phytochemical Induction by Herbivores. New York: John Wiley and Sons 1991; 85-104.
29. Gurib-Fakim A. Medicinal plants: traditions of yesterday and drugs of tomorrow. Mol Aspects Med 2006; 27:1-93.
30. Espín JC, García-Conesa MT, Tomás-Barberán FA. Nutraceuticals: Facts and fiction. Phytochemistry 2007; 68:2986-3008.
31. Pichersky E, Gang DR. Genetics and biochemistry of secondary metabolites in plants: an evolutionary perspective. Trends in Plant Sci 2000; 5:439-445.
32. Paganga G, Miller N, Rice-Evans CA. The polyphenolic content of fruit and vegetables and their antioxidant activities. What does a serving constitute? Free Radic Res 1999; 30:153-162.
33. Bingham SA, Atkinson C, Liggins J et al. Phyto-oestrogens: Where are we now? Br J Nutr 1998; 79:393-406.
34. Rauha JP, Remes S, Heinonen M et al. Antimicrobial effects of Finnish plant extracts containing flavonoids and other phenolic compounds. Int J Food Microbiol 2000; 56:3-12.
35. Adwan G, Abu-Shanab B, Adwan K et al. Antibacterial effects of nutraceutical plants growing in Palestine on Pseudomonas aeruginosa. Turk J Biol 2006; 30:239-242.
36. Arora DJ, Kaur J. Antimicrobial activity of spices. Int J Antimicrob Agents 1999; 12:257-262
37. Veerporte R, Contin A, Memelink J. Biotechnology for the production of plant secondary metabolites. Phytochemistry Rev 2002; 1:13-25.
38. Ruffoni B, Pistelli L, Bertoli A et al. Plant cell cultures: Bioreactors for industrial production, this volume.
39. Hämet-Ahti L, Suominen J, Ulvinen T et al. (eds.). Retkeilykasvio (Field Flora of Finland). Finnish Museums of Natural History. Botanical Museum. Helsinki 1998.
40. Brown RP, Gerbarg PL, Ramazanov Z. Rhodiola rosea. A phytomedical overview. HerbalGram 2002; 56:40-52.
41. György Z, Tolonen A, Pakonen M et al. Enhancing the production of cinnamyl glycosides in compact callus aggregate cultures of Rhodiola rosea by biotransformation of cinnamyl alcohol. Plant Sci 2004; 166:229-236.
42. György Z, Tolonen A, Neubauer P et al. Enhanced biotransformation capacity of Rhodiola rosea callus cultures for glycosid production. Plant Cell Tiss Org Cult 2005; 83:129-135.
43. Earnest CP, Morss GM, Wyatt F et al. Effects of a commercial herbal-based formula on exercise performance in cyclists. Med Sci Sports Exerc 2004; 36:504-509.
44. Spasov AA, Wikman GK, Mandrikov VB et al. A double-blind, placebo-controlled pilot study of the stimulating and adaptogenic effect of Rhodiola rosea SHR-5 extract on the fatigue of students caused by stress during an examination period with a repeated low-dose regimen. Phytomedicine 2000; 7:85-89.
45. Kelly GS. Rhodiola rosea: A possible plant adaptogen. Alternative Medicine Review 2001; 6:293-302.

46. Udintsev SN, Schakhov VP. Decrease of cyclophosphamide haematotoxicity by Rhodiola rosea root extract in mice with Ehrlich and Lewis transplantable tumors. Eur J Cancer 1991; 27:1182.
47. Furmanova M, Oledzka H, Michalska M et al. Rhodiola rosea L. (Roseroot): In vitro regeneration and the biological activity of roots. In: Bajaj YPS, ed. Biotechnology in agriculture and forestry. Berlin: Springer-Verlag 1995; 412-426.
48. Galambosi B, Galambosi Sz-Sz, Varga E et al. Cultivation methods, root yield and flavonoid content of roseroot (Rhodiola rosea L.) cultivated in Finland. Book of Abstracts. Cultivation, harvesting and of medicinal plants. Slovakia 1999.
49. Kovaleva NP, Tikhomirov AA, Dolgushev VA. Specific characteristics of Rhodiola rosea growth and development under the photoculture conditions. Russian J Plant Physiol 2003; 50:527-531.
50. Duskova J, Dusek J. Testing of potential growth regulators on plant cultures in vitro. Folia Pharm Univ Carol XXIII 1998; 15-20.
51. Furmanova M, Hartwich M, Alfermann AW et al. Rosavin as a product of glycosylation by Rhodiola rosea (roseroot) cell cultures. Plant Cell Tiss Org Cult 1999; 56:105-110.
52. Tolonen A, György Z, Jalonen J et al. LC/MS/MS identification of glycosides produced by biotransformation of cinnamyl alcohol in Rhodiola rosea compact callus aggregates. Biomedical Chromatography 2004; 18:550-558.
53. Hook ILI. Naphthoquinone contents of in vitro cultured plants and cell suspensions of Dionaea muscipula and Drosera species. Plant Cell Tiss Org Cult 2001; 67:281-285.
54. Durand R, Zenk MH. Enzymes of the homogentisate ring-cleavage pathway in cell suspension cultures of higher plants. FEBS Lett 1974; 39:218-220.
55. Wawrosch C, Vackar E, Grauwald B et al. Variations of naphthoquinone levels in micropropaqated drosera species in vitro, under greenhouse and outdoor growth conditions. Sci Pharm 2005; 73:251-262.
56. Bonnet M, Coumans M, Hofinger M et al. High-performance gas chromatography of 1,4-naphthoquinones from Droseraceae. Chromatographia 1984; 18:621-622.
57. Hook I, Walsh J, Kavanagh P et al. Naphthoquinone production by cultures of cape sundew (Drosera capensis). Pharmaceut Pharmacol Lett 1997; 7:93-95.
58. Repcák M, Galambosi B, Takkunen N. The production of 7-methyljuglone, quercetin and kaempferol by Drosera anglica and D. rotundifolia. Biologia 2000; 55:429-433.
59. Caniato R, Filippini R, Capelletti E. Naphthoquinone contents of cultivated Drosera species Drosera binata, D. binata var. dichotoma and D. capensis. Int J Crude Drug Res 1989; 27:129-136.
60. Thorén LM, Tuomi J, Kämäräinen T et al. Resource availability affects investment in carnivory in Drosera rotundifolia L. New Phytol 2003; 159:507-511.
61. Hirsikorpi M, Kämäräinen T, Teeri T et al. Agrobacterium tumefaciens—mediated transformation of round leaved sundew (Drosera rotundifolia L.). Plant Sci 2002; 162:537-542.
62. Leclerq J, Angenot L. A propos du Drosera peltata et de la standardisation de la teinture de Drosera. Journal de Pharmacie de Belgique 1984; 39:269-274.
63. Seigler S. Plant Secondary Metabolism. London, Boston, Dordrecht: Kluwer Academic Publishers 1998; 1-713.
64. Melzig MF, Pertz HH, Krenn L. Anti-inflammatory and spasmolytic activity of extracts from Droserae Herba. Phytomedicine 2001; 8:225-229.
65. Krenn L, Beyer G, Pertz HH et al. In vitro antispasmodic and anti-inflammatory effects of Drosera rotundifolia. Arzneimittel-Forschung 2004; 54:402-5.
66. Paper DH, Karall E, Kremser M et al. Comparison of the antiinflammatory effects of Drosera rotundifolia and Drosera madagasgariensis in the HET-CAM assay. Phytotherapy Res 2005; 19:323-326.
67. Didry N, Dubreuil L, Trotin F et al. Antimicrobial activity of aerial parts of Drosera peltata Smith on oral bacteria. J Ethnopharmagol 1998; 60:91-96.
68. Witzell J, Gref R, Näsholm T. Plant-part specific and temporal variation in phenolic compounds of boreal bilberry (Vaccinium myrtillus) plants. Biochem Syst Ecol 2003; 31:115-127.
69. Jaakola L, Määttä K, Pirttilä AM et al. Expression of genes involved in anthocyanin biosynthesis in relation to anthocyanin, proanthocyanidin and flavonol levels during bilberry fruit development. Plant Physiol 2002; 130:729-739.
70. Jaakola L, Määttä-Riihinen K, Kärenlampi S et al. Activation of flavonoid biosynthesis by solar radiation in bilberry (Vaccinium myrtillus L.) leaves. Planta 2004; 218:721-728.
71. Winkel-Shirley B. Flavonoid biosynthesis. A colorful model for genetics, Biochem. Cell Biol Biotechnol Plant Physiol 2001; 126:485-493.
72. Koes R, Quattrocchio R, Mol J. The flavonoid biosynthetic pathway in plants: function and evolution. BioEssays 1994; 16:123-132.
73. Taulavuori E, Bäckman J, Taulavuori K et al. The long-term exposure to enhanced ultraviolet-B radiation in the subarctic does not cause oxidative stress in Vaccinium myrtillus. New Phytol 1998; 140:691-697.

74. Riihinen K, Jaakola L, Kärenlampi S et al. Organ-specific distribution of phenolic compounds in bilberry (Vaccinium myrtillus) and 'northblue' blueberry (Vaccinium corymbosum x V. angustifolium). Food Chem 2008; 110:156-160.
75. Kalt W, Dufour D. Health functionality of blueberries. HortTechnol 1997; 7:216-221.
76. Bomser J, Madhavi DL, Singletary K et al. In vitro anticancer activity of fruit extracts from Vaccinium species. Planta Med 1996; 62:212-216.
77. Shah AM, Channon KM. Free radicals and redox signalling in cardiovascular disease. Heart 2004; 90:486-487.
78. Nakajima JI, Tanaka I, Seo S et al. LC/PDA/ESI-MS Profiling and Radical Scavenging Activity of Anthocyanins in Various Berries. J Biomed Biotechnol 2004; 5:241-247.
79. Bell DR, Gochenaur K. Direct vasoactive and vasoprotective properties of anthocyanin-rich extracts. J Appl Physiol 2006; 100:1164-1170.
80. Kay CD, Holub BJ. The effect of wild blueberry (Vaccinium angustifolium) consumption on postprandial serum antioxidant status in human subjects. British J Nutr 2002; 88:389-397.
81. Talavera S, Felgines C, Texier O et al. Bioavailability of a bilberry anthocyanin extract and its impact on plasma antioxidant capacity in rats. J Sci Food Agric 2006; 86:90-97.
82. Hou DX. Potential mechanisms of cancer chemoprevention by anthocyanins. Curr Mol Med 2003; 3:149-159.
83. Kang SY, Seeram NP, Nair MG et al. Tart cherry anthocyanins inhibit tumor development in ApcMin mice and reduce proliferation of human colon cancer cells. Cancer Lett 2003; 194:13-19.
84. Zafra-Stone S, Yasmin T, Bagchi M et al. Berry anthocyanins as novel antioxidants in human health and disease prevention. Mol Nutr Food Res 2007; 51:675-683.
85. Seeram NP, Adams LS, Zhang Y et al. Blackberry, black raspberry, blueberry, cranberry, red raspberry and strawberry extracts inhibit growth and stimulate apoptosis of human cancer cells in vitro. J Agric Food Chem 2006; 54:9329-9339.
86. Canter PH, Ernst E. Anthocyanosides of Vaccinium myrtillus (bilberry) for night vision—a systematic review of placebo-controlled trials. Survey of Ophthalmology 2004; 49:38-50.
87. Morazzoni P, Bombardelli E. Vaccinium myrtillus L. Fitoterapia 1996; 67:3-29.
88. Puupponen-Pimiä R, Nohynek L, Hartmann-Schmidlin S et al. Berry phenolics selectively inhibit the growth of intestinal pathogens. J Appl Microbiol 2005; 98:991-1000.
89. Martineau LC, Couture A, Spoor D et al. Anti-diabetic properties of the Canadian lowbush blueberry Vaccinium angustifolium Ait. Phytomedicine 2006; 13:612-23.

CHAPTER 9

Nutraceutical Use of Garlic Sulfur-Containing Compounds

Eleftherios Touloupakis* and Demetrios F. Ghanotakis

Abstract

Garlic is one of world's oldest medicines that have been employed not only for flavouring but also as a medical herb for its prophylactic and therapeutic actions. Most of garlic beneficial effects are due to the presence of the organosulphate molecule allicin. Allicin is a highly unstable molecule and, during processing, is rapidly transformed into a variety of organosulfur components. The enzyme alliinase, which is responsible for the conversion of alliin to allicin, is irreversibly destroyed at the acidic environment of stomach. This is the reason why most garlic supplements contain garlic powder or granules, but do not contain allicin itself. Garlic alliinase could be encapsulated and coated with materials which would protect it in the harsh conditions of the stomach. The objective of this chapter is to summarize the most important garlic health benefits and to discuss promising encapsulation/stabilization approaches.

Introduction

Garlic (*Allium sativum* L.) is a perennial, erect bulbous herb, 30-60 cm tall which is native to the Mediterranean regions of Africa and Europe. The underground portion consists of a bulb with numerous fibrous rootlets. The bulb contains numerous cloves and gives rise above ground to a number of narrow, grass-like leaves (Fig. 1). Although formally classified in the family *Liliaceae*, represented by 280 separate genera and 4000 species, recent taxonomic revisions have seen members of the genus *Allium* placed in the family *Alliaceae*. Originated from Central Asia, garlic has been used as a spice, food and folklore medicine for over 5000 years and is the most widely researched medicinal plant. Through trade, garlic spread its popularity throughout Asia and eventually to Egypt and Europe. There are various ancient medical texts from Greece, Egypt, Rome, China and India each prescribing medical applications for garlic. Excavations of ancient Greek temples have unearthed garlic and the palace of Knossos in Crete, dating to 1400-1800 BC, contained well-preserved garlic when it was excavated.[1] The *Codex Ebers*, an Egyptian medical papyrus translated in 1937, contained over 800 medical formulations, 22 of which contained garlic. The Romans also extolled the virtues of garlic. Pliny the elder, a roman naturalist, described in his *Historia Naturalis* how garlic could be used for gastrointestinal disorders, dog and snake bites, scorpion stings, asthma, madness, convulsions and tumours. Dioscorides, physician to the Roman army in first century A.D., also recommended garlic to be used as a vermifuge. Use of garlic has been recorded by Hippocrates, the father of modern medicine, as a laxative and a diuretic and also by Aristophanes and Galenal for the treatment of uterine tumours.

*Corresponding Author: Eleftherios Touloupakis—Department of Chemistry, University of Crete, PO Box 2208, 71003 Voutes-Heraklion, Greece. Email: toulou_e@chemistry.uoc.gr

Bio-Farms for Nutraceuticals: Functional Food and Safety Control by Biosensors
edited by Maria Teresa Giardi, Giuseppina Rea and Bruno Berra.

Figure 1. *Allium sativum* bulbs.

The majority of garlic supplements sold today contain either a garlic powder or granules. Garlic powder can serve as an important nutritional supplement. However if garlic powder is stored for long periods, active ingredients present in freshly ground garlic are often eliminated or otherwise rendered inactive.

Most of garlic beneficial effects are due to allicin. Allicin is produced during the crushing of garlic cloves by the enzyme alliinase. Once formed, allicin begins reacting with itself, resulting in more stable organosulfur compounds. This is the reason that no garlic supplement actually contains allicin itself.

In this mini review we summarize the most important garlic health benefits and we discuss promising encapsulation/stabilization approaches.

Garlic Chemical Compounds

Numerous chemical substances are present in fresh garlic. Some of them are: ajoene, allicin, alliin, allyl disulfides, allyl sulfides, allyl trisulfides, cycloalliin, cysteine, cysteine sulfoxides, cystine, diallyl sulfides, dimethyl sulfides, disulfides, glutathione, methionine, methyl sulfides, pseudoscordinine, scordinine, sulfanes, tetrathiol, thiosulfinates and trisulfides.[2,3] Garlic is known to contain high levels of phosphorous, calcium and iron. Vitamins like riboflavin, thiamine, nicotinic acid and vitamin-C are also present. In addition, garlic contains the minerals selenium and germanium. Garlic also contains linalool, citral, α-phellandrene, geraniol, propionic aldehyde, valeraldehyde.[4] These phytochemicals can be useful for the treatment or prevention of a number of diseases, including cancer, coronary heart disease, obesity, hypercholesterolemia, diabetes type 2, hypertension, cataract and disturbances of the gastrointestinal tract (e.g., colic pain, flatulent colic and dyspepsia).

The health benefits of garlic are derived from a wide variety of components and from the different ways it is administered. It is possible that such a variety of effects partly depends on the processing of garlic extract preparations.[5] There are many garlic product processes; those used more frequently are shown in (Table 1).[6] The oil obtained by the maceration of garlic contains 2-vinyl-[4H]-1,3-dithiin and 3-vinyl-[4H]-1,2-dithiin, allyl sulfides and ajoene. Steam-distilled garlic oil contains diallyl disulfide and diallyl trisulfide. The lipid-lowering effects of some oil-soluble sulfur compounds observed in vitro are associated with cytotoxicity.[7] Water-soluble garlic extracts are not cytotoxic.[8] Moreover, the oil-soluble organosulfur compounds of garlic, such as allicin, ajoene and their intermediate products, are not found in urine or blood even after consumption of a large amount of garlic because alliinase, the enzyme that transforms alliin in

Table 1. The most used garlic product processes

Method	Characteristics
Freeze-drying	No change in chemical composition, the resulting product is often used for culinary purposes.
Low Temperature Drying	This process involves drying sliced fresh cloves at 50°C for 3-4 days. Some allicin is formed due to the slicing process.
Distillation	Steam-distilled garlic contains mainly allyl sulfides. Allicin is probably lost or converted to the allyl sulfide degradation compounds.
Maceration in Oil	Chopped garlic is homogenized and slowly extracted in soybean or another vegetable oil. Such products contain vinyldithins, allyl sulfides and ajoene.
Hydroalcoholic Short Extraction	The stability of the constituents in fluid and dry extracts and their final chemical content could be variable.
Hydroalchoholic Long Maceration	Sliced garlic is placed in 20% ethanol and macerated for a long period, filtered and subsequently concentrated. Allicin is completely converted to allyl sulfides, which are largely all volatilized or converted to other compounds.

allicin, is inactivated by the low pH in the stomach.[9] The water-soluble compounds, besides being more stable, appear to have a higher bioavailability and appear to be able to enter the blood and reach target organs. In fact, after garlic consumption, a water-soluble compound, such as N-acetyl-S-ally-cysteine, was found in human blood and urine.[10] Raghavan et al, reported that drying the garlic and converting it to powder results in 30-35% loss of volatile flavour.[11] Garlic powder could be stabilized with respect to flavour and colour by mixing it with encapsulated garlic oil.[12,13]

Garlic Biological Activities

A variety of biological activities have been reported for garlic extracts.[14,15] The most important ones are summarized below.

Antimicrobial. Garlic is regarded as a natural antibiotic, effective against a wide number of microorganisms.[16,17]

Antibacterial. Garlic extracts have been demonstrated to inhibit bacterial growth. Sulfur compounds in garlic are known to destroy thiol groups in bacterial enzymes. Garlic preparations have been shown to exhibit antibacterial activity against *Helicobacter pylori*, *Shigella dysenteriae*, *Shigella flexneri*, *Shigella sonnei* and *Escherichia coli.*[18] Garlic has been found to be effective against clinical strains of *Staphylococcus*, *Escherichia*, *Proteus*, *Pseudomonas* and *Klebsiella* bacteria.[19]

Antiparasitic. Garlic has been found to be effective against gastrointestinal parasites of humans and animals.[20] In addition, it is effective against *Entamoeba histolytica*, *Hymenolepsis nana* and *Giardia lamblia.*[21] Garlic can be used as an alternative treatment against nematode parasites in humans.[22] Nok et al, also investigated the garlic induced death of protozoans.[23]

Antifungal. Garlic extracts have a strong antifungal effect and inhibit the formation of mycotoxins like the aflatoxin of *Aspergillus parasiticus.*[24] Davis et al, reported that garlic extract displayed significant in vitro fungicidal and fungistatic activity against 3 different isolates of *Cryptococcus neoformans.*[25]

Antiviral. Garlic extracts have an antifungal effect against human cytomegalovirus, influenza B, herpes simplex virus type 1, herpes simplex virus type 2, parainfluenza virus type 3, vaccinia virus, vesicular stomatitis virus and human rhinovirus type 2.[26] Weber et al, reported the efficacy of allicin and its various transformation products against Herpes simplex virus 1 and 2, Vesicular stomatitis virus, Vaccinia virus and Parainfluenza virus.[27]

Anti-carcinogenic. Garlic can be use as a preventive against stomach, colon and other types of cancers, due to possible immune enhancing and antitumorigenic effects of its sulfur containing compounds.[28-30] Allyl sulfur and other garlic compounds can slow or prevent the growth of tumor cells.[31,32] Aqueous extracts of garlic are reported to cause significant improvement in prostate cancer patients.[33] Some researchers have stated that garlic's potential anticancer effects may be due to its content of the trace mineral selenium.[34]

Anti-inflammatory. Garlic promotes an anti-inflammatory environment by cytokine modulation as well as bringing in modulation in human blood that leads to an overall inhibition of NF-κB activity in surrounding tissue.[35,36]

Anti-thrombotic. Several studies have indicated that whole garlic and garlic aqueous extract have an antiplatelet action.[37-39] This property is due in part to the compounds alliin and ajoene, which have fibrinolytic activity. Ajoene inhibits thromboxane synthesis through the inhibition of the cyclo-oxygenase and lipoxygenase enzymes.[40] Garlic extract can be used as a potential treatment against atherosclerosis.[41]

Cholesterol lowering. Garlic and its constituents inhibit key enzymes involved in cholesterol and fatty acid synthesis in cultured rat hepatocytes and human HepG2 cells.[7,42] Garlic treatment reduces circulating triglycerides and cholesterol.[43-45] Anim-Nyame et al, have shown that garlic will increase peripheral blood flow mediated by interleukin-6, an important property attributed to be beneficial for persons with cardiovascular diseases.[46]

Garlic and diabetes. Garlic was effective in reducing blood glucose in streptocin-induced as well as alloxan-induced diabetes mellitus in rats and mice models.[47-49] Kasuga et al, have reported that treatment with aged garlic extract is effective against adrenal hypertrophy, hyperglycemia and production of elevated levels of corticosterones in stress induced hyperglycemic mice.[50] Due to its hypoglycemic effect garlic may be useful as coadjuvant therapy in the treatment of type 2 diabetes and some of its physiological complications.[51,52]

Antioxidant. The antioxidant properties of garlic and different garlic preparations are well documented.[53-57] The high antioxidant properties of garlic presents evidence that it may either prevent or delay chronic diseases associated with aging. Among garlic-derived products, aged garlic extract is the preparation with higher antioxidant activity. Aged garlic extract is obtained by storage at room temperature of sliced and soaked, in a water/ethanol mixture, garlic for longer than 10 months.[58] The potent antioxidant activity is due to the extraction procedure which increases stable and highly bioavailable water-soluble organo-sulphur compounds content, such as SAC and S-allylmercaptocysteine.[59] This garlic preparation has other compounds with antioxidant effect, such as stable lipid-soluble allyl sulphides, as diallyl polisulphides, tetrahydro-b-carboline derivatives, which are formed during the natural aging process, flavonoids; saponins; and essential micronutrients (selenium, Se) and macronutrients, as lectins, whose antiperoxide effect has been demonstrated in the liver, kidney and heart of rats.[10,60] Another recently identified antioxidant compounds of aged garlic extract are N-fructosyl glutamate, N-fructosyl arginine.[61-62] Garlic extract can be used as a prophylactic treatment against the common cold and in the treatment of sickle-cell anemia,[63,64] to treat hypertension and reduce the risk of other cardiovascular diseases[62,65,66] and against heavy metal poisoning.[67] Garlic compounds may also have neuroprotective effects.[68,69] Garlic possesses a mild diuretic action[70,71] and can be used as a renoprotective agent. It has been also found that the aged garlic extract can be a potential protective regimen for arsenic mediated toxicity.[72]

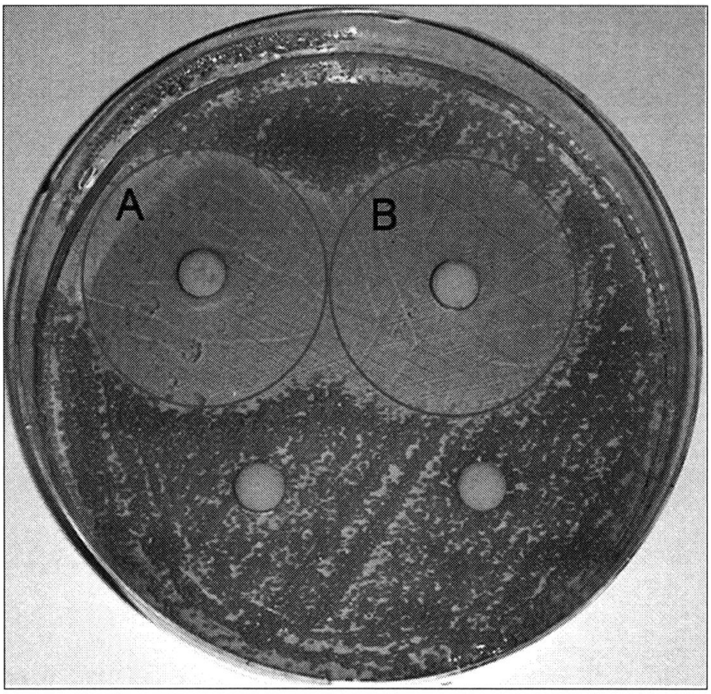

Figure 2. Effect of garlic crude extract on *Saccharomyces cerevisiae* culture. Inhibition zone of A) 50 μl of garlic extract and B) 10 μg of miconazole.

In our laboratory we have demonstrated the inhibitory action of garlic extract against *Saccharomyces cerevisiae*. The antibiogram method was used for measuring the diameter of the garlic extract zone of inhibition (Fig. 2).

Alliin and Allicin

Alliin ((+)-S-allyl-L-cysteine sulfoxide) and allicin (S-(+)-2-propenyl-L-cysteine sulfoxide) are two important sulfur-containing compounds found in garlic (Fig. 3). Both alliin and allicin are known as "organosulfur" compounds. Many medicinal organosulfur compounds are found in the *Alliums*, but garlic contains the highest concentration of them. Alliin is further classified as a "cysteine sulfoxide". Garlic contains other cysteine sulfoxides, but alliin is the most abundant of these, as well as the most abundant sulfur compound in garlic overall. The amounts of alliin present in different strains of garlic were studied by numerous investigators. Considerable variations have been reported, ranging from 2.8 to 7.7 mg g^{-1}.

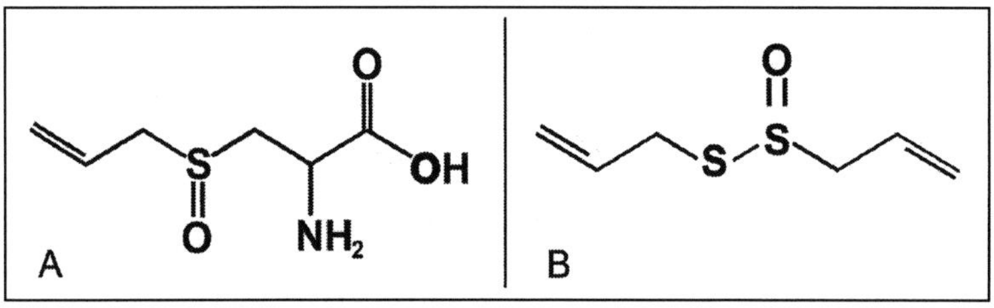

Figure 3. Molecular formulas of: A) alliin and B) allicin.

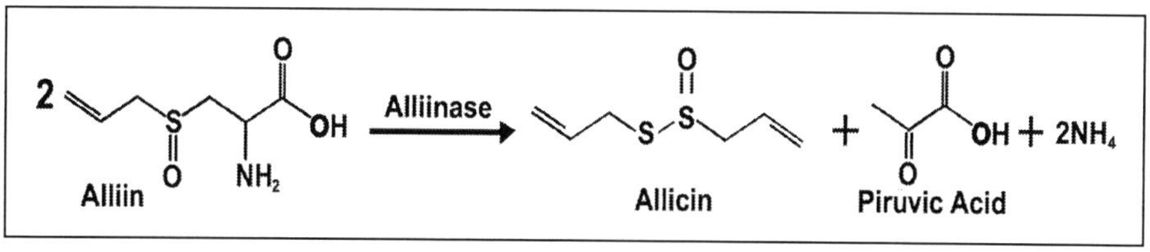

Figure 4. Enzymatic reaction of garlic alliinase.

Alliin demonstrates a variety of important activities. It reduces serum glucose and serum enzyme levels.[73] Alliin at levels of 200 mg kg^{-1} body weight can significantly reduce the blood glucose level in diabetic mice, rats and rabbits.[50] Alliin has an inhibitory effect upon key enzymes involved in cholesterol biosynthesis; it also possesses an antioxidant activity.[59,74] Kim and Kyung demonstrated that alliin has an antifungal activity.[75] Alliin was found to be the stable precursor. When garlic is cut, chopped or crushed, the clove's membrane disrupts and alliin is transformed enzymatically into allicin by allinase (Fig. 4).[76] The transformation of alliin into the biologically active allicin molecule upon crushing of a garlic clove is very rapid and is complete in seconds. Garlic cloves are odour-free until crushed or processed when garlic supplements are manufactured and cross-section studies have indicated that the substrate alliin and the enzyme alliinase are located in different compartments. This organization suggests that it is designed as a potential defence mechanism against microbial pathogens in the soil. Invasion of the cloves by fungi and other soil pathogens causes the interaction between alliin and alliinase that rapidly produces allicin, which in turn inactivates the invader. The produced reactive allicin molecules have a very short half-life, as they react with many of the surrounding proteins, including alliinase. Allicin is often thought of as the "active ingredient" in garlic and has been the subject of numerous scientific studies, but many of its transformation products have medicinal properties as well.

Allicin, which is classified as a "thiosulfinate", is responsible for the typical odor of garlic. Thiosulfinates contain two sulfur atoms bonded together, one of which is also bonded to an oxygen atom. Allicin accounts for about 70-80% of total thiosulfinates in garlic. The total allicin yield has been determined as 2.5 mg g^{-1} of fresh crushed garlic or about 5-20 mg per clove. Allicin is unstable and converts readily into mono-, di- and trisulfides and other compounds such as ajoene.[9] Further transformation of organosulfur compounds can occur after interaction with free sulfhydryl groups, including those present in cysteine, glutathione or proteins.

Allicin possesses a variety of important activities. It has a high antioxidant activity and has been reported to induce caspase mediated apoptosis in cancer cells.[77-79] The antibacterial action of garlic is mainly due to allicin.[16] Yamada and Azuma demonstrated that pure allicin is effective in vitro against species of *Candida*, *Cryptococcus*, *Trichophyton*, *Epidermophyton* and *Microsporum* at low concentration.[80] Allicin may be an effective therapeutic candidate to control the pain, promote ulcer healing and prevent the recurrence of recurrent aphthous ulcer.[81] Garlic antiatherosclerotic properties are mainly attributed to allicin.[82] Many of allicin transformation products have medicinal properties as well.[83]

Garlic Alliinase

Alliinase (Aliin lyase) (EC 4.4.1.4) is a homodimeric glycoprotein found in many plants of the genus *Allium* such as garlic (*Allium sativum*), onion (*Allium cepa*) and Chinese chives (*Allium tuberosum*). The three-dimensional structure of the garlic alliinase has been obtained at 1.5 Å resolution.[84] The active enzyme is a pyridoxal-5′-phosphate-dependent protein and belongs to the class I family of pyridoxal-5′-phosphate-dependent enzymes. In garlic, each alliinase subunit consists of 448 aminoacids accounting for a molecular weight of 51,500 Da, four putative N-glycosylation sites and one pyridoxal 5-phosphate (PLP) molecule as a cofactor.[85] Alliinase belongs to the family of mannose-rich glycoproteins with an estimated carbohydrate content of about 5.5-6.0% and can form a stable complex with garlic lectins.[86] The lectin-alliinase complex offers a high stability to the enzyme and its enzymatic activity. Alliinase catalyses the production of allicin from alliin (Fig. 4). The reaction catalyzed by alliinase is categorized as α,β-elimination reaction and involves an aminoacryl intermediate bound to PLP. Alliinase converts the substrate alliin to allylsulfenic acid and aminoacrylate. Two molecules of allylsulfenic acid condense spontaneously to form the sulfur containing volatile, allicin. The aminoacrylate is lysed from the PLP and spontaneously decompose to pyruvate and ammonia. The K_m of alliinase is 1.1 mM, its pH optimum 6.5 and its isoelectric point is 6.35. Alliinase is present in large amounts in garlic cloves: at least 10% of the total protein content (10 mg g^{-1} fresh weight). With its C-S lyase activity, alliinase is able to cleave the Cb-Sc bond of sulfoxide derivatives of the amino acid cysteine thus giving rise to all the garlic sulfur compounds.

Garlic Supplements

Garlic supplements vary widely in their chemical composition, depending on the age of the garlic and how it is processed. Significant amounts of alliin and alliinase can survive the drying and processing necessary to manufacture garlic supplements. Freeze-drying, for example, preserves higher levels of these phytochemicals compared to other types of drying, but it is an expensive process. When dried garlic powder is added to water, alliin and alliinase quickly react and allicin is produced. Most of garlic supplements found in the market do not contain allicin. Some products are standardized to contain a certain amount of alliin and to have an uncertain allicin yield (amount of allicin generated inside the human body). No allicin was detected in human blood after oral intake of commercially available products standardized for allicin potential or allicin yield.[10,87] Even if the supplements contain plenty of alliin and active alliinase, they must be enteric-coated to protect alliinase from the low pH in the stomach, to reach small intestine, where it can act on the alliin and convert it to allicin.[87] Enteric-coated garlic products resist human stomach conditions in order to prolong the shelf life and protect alliinase activity and deliver allicin directly into the intestinal tract. Enteric coatings are cellulose or poly(acrylic acid) esters that are acid-resistant but readily dissolve at neutral pH. Garlic extract can be also encapsulated.[88,89] In a similar manner alliinase and/or alliin can be encapsulated. Encapsulation enables alliinase and/or alliin to maintain their viability for extended periods of time protecting, for example, from ions or free radicals. Capsule structure is divided into the core and the coating material. The coating material is the external layer or layers that cover the core material. Coating substances are basically film forming materials that can be selected from a wide variety of natural or synthetic polymers, depending on the material to be coated and the characteristics desired to the final capsules.[90] An effective coating material should have good rheological properties at high concentration and easy manipulation during the process of encapsulation. It should be selected so that it produces a stable emulsion with the active ingredient and does not react or degrade the active material during process and storage. The interior content of the capsules can be released under different mechanisms, like diffusion, dissolution, fracturation and biodegradation.

Two of the most promising encapsulation materials are layered double hydroxides and alginate gels. These materials are useful because they are biocompatible inorganic matrices that are cheap and easy to work with.[91-94]

Layered Double Hydroxides

Layered double hydroxides (LDH), also known as hydrotalcite-like compounds, belong to a class of lamellar compounds that enables the incorporation of several species by guest-host chemistry.[95] LDH structure can be described by considering the brucite-like structure, $Mg(OH)_2$, in which the Mg(II) cations are in the center of edge-sharing octahedra, with hydroxyl groups in their vertices, resulting in a planar structure (Fig. 5). The chemical composition of this class of intercalation compounds can be expressed by the general formula $[M^{2+}_{1-x}M^{3+}_x(OH)_2]^{x+}A^{m-}_{x/m}\,nH_2O$: where M^{2+} and M^{3+} are respectively the divalent and trivalent cations in the brucite-like layers and A^{m-} is the charge balancing interlayer anion. The specific charge of the layer is directly related to the exchange ratio [x in the general formula, $x = M3/(M^{2+} + M^{3+})$]. Thus, a wide variety of LDHs can be obtained by varying the M^{2+} and M^{3+} cations species and their ratio, as well as the intercalated anion A^{m-}, which can be exchanged by an ion-exchange process.[95] Due to their properties, several new LDHs

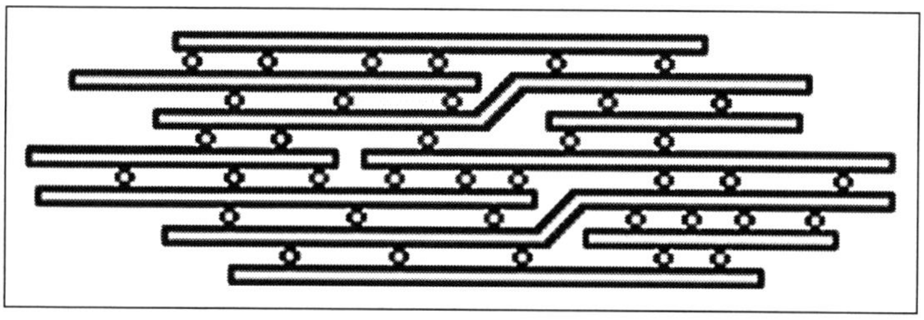

Figure 5. Schematic planar structure of layered double hydroxides.

compounds have been prepared with potential technological applications, such as heterogeneous catalysis, host structures for photoactivation and photocatalysis and ion exchange.[96,97] Inorganic layered double hydroxides can be also used as nonviral vectors.[98] Enteric-coated layered double hydroxides can be used as a controlled release drug delivery system.[93]

Alginates

Alginate is the term usually used for the salts of alginic acid. Alginate is present in the cell walls of brown algae as the calcium, magnesium and sodium salts of alginic acid. Alginates constitute a family of unbranched binary copolymers of 1-4-linked β-D-mannuronic acid (M) and a-L-guluronic acid (G). The monomers are arranged in a pattern of blocks along the chain, with homopolymeric regions (termed M and G blocks) (Fig. 6). The way in which these M and G units are arranged in the chain and the overall ratio, M/G, of the two units in a chain can vary from one species of seaweed to another.

The uses of alginates are based on three main properties. The first is their ability, when dissolved in water, to thicken the resulting solution. The second is their ability to form gels; gels form when a calcium salt is added to a solution of sodium alginate in water. The gel forms by chemical reaction, the calcium displaces the sodium from the alginate, holds the long alginate molecules together and a gel is the result. No heat is required and the gels do not melt when heated. The third property of alginates is the ability to form films of sodium or calcium alginate and fibres of calcium alginates. Alginates are used in a wide variety of foodstuff such as pet food chunks, onion rings, stuffed olives, low fat spreads, sauces and pie fillings. Alginates are used as food additives to improve, modify and stabilize the texture of foods. The fact that alginates belongs to EU-approved additives makes it an ideal support material. Alginates are widely used to immobilize cells or enzymes.[99,100]

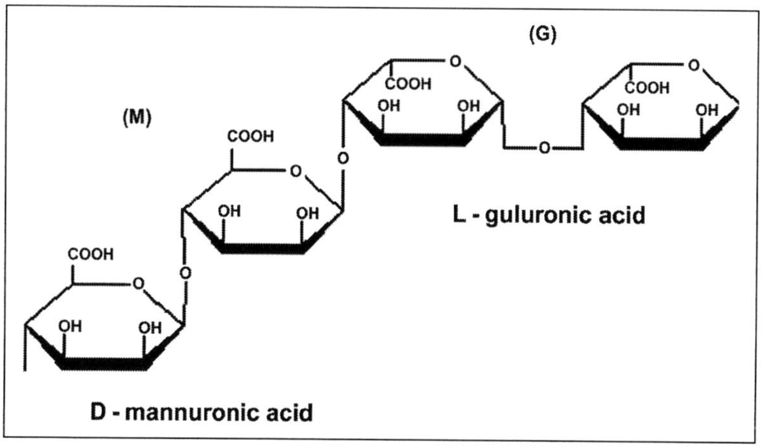

Figure 6. Chemical structures of alginic acid.

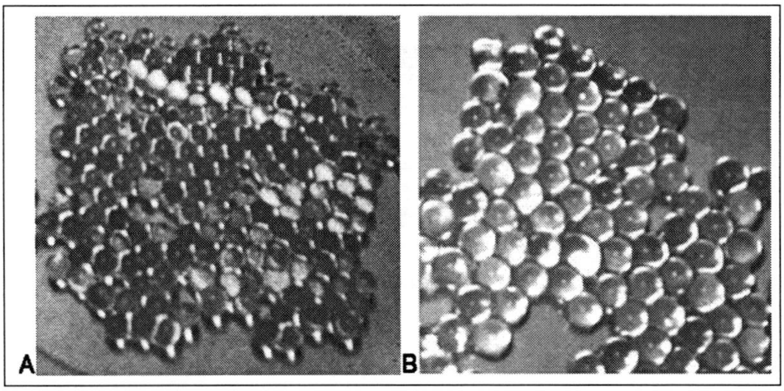

Figure 7. Alginate bead with (A) and without (B) entrapped alliinase. The yellow color is due to the presence of the alliinase cofactor PLP. A color version of this image is available at www.landesbioscience.com/curie.

In a series of immobilization experiments we managed to stabilize purified alliinase from garlic by entrapping it in calcium alginate beads (3 mm diameter, 1.5 mg protein per 1g beads) (Fig. 7). The entrapped enzyme was highly active and efficiently converted alliin to allicin (manuscript in preparation).

Conclusion

Garlic contains a lot of volatile organosulfur compounds with a vast variety of biological functions. Most of garlic beneficial effects are due to the presence of alliin and allicin. Allicin is a very unstable molecule and this is the reason most of garlic supplements in the market do not contain this important molecule.

Alliinase and alliin are present in high quantities in garlic and are easily isolated in pure and active forms. The garlic enzyme alliinase can be stabilized in its active form into biocompatible materials, like layered double hydroxides or alginate beads. The immobilized enzyme along with purified alliin can be used in garlic nutraceutical products. An enteric-coating of the immobilized molecules, would allow them to bypass the acidic environment of the stomach and thus result to the production of allicin in the human intestine.

References

1. Moyers S. Garlic in health, history and world cuisine. St. Petersburg, FL. Suncoast Press, 1996;1-36.
2. Rose P, Whiteman M, Moore PK et al. Bioactive S-alk(en)yl cysteine sulfoxide metabolites in the genus Allium: the chemistry of potential therapeutic agents. Nat Prod Rep 2005; 22:351-368.
3. Lanzotti V. The analysis of onion and garlic. J Chromatogr A 2006; 1112;3-22.
4. Milner JA. Garlic: its anticarcinogenic and antitumorigenic properties. Nutr Rev 1996; 54:S82-S86.
5. Gorinstein S, Jastrzebski Z, Leontowicz H et al. Comparative control of the bioactivity of some frequently consumed vegetables subjected to different processing conditions. Food Control 2009; 20:407-413.
6. Staba EJ, Lash L, Staba JE. A Commentary on the effects of garlic extraction and formulation on product composition. J Nutr 2001; 131:1118S-1119S.
7. Liu L, Yeh Y. Water-soluble organosulfur compounds of garlic inhibit fatty acid biosynthesis in cultured rat hepatocytes. Lipids 2001; 36(4):395-400.
8. De Martino A, Filomeni G, Aquilano K et al. Effects of water garlic extracts on cell cycle and viability of HepG2 hepatoma cells. J Nutrit Biochem 2006; 17:742-749.
9. Lawson LD, Hughes BG. Characterization of the formation of allicin and other thiosulfinates from garlic. Planta Med 1992; 58:345-350.
10. Amagase H, Petesch BL, Matsuura H. Intake of garlic and its bioactive components. J Nutr 2001; 131:955S-926S.
11. Kaymak-Ertekin F, Gedik A. Kinetic modelling of quality deterioration in onions during drying and storage. J Food Eng 2005; 68(4):443-453.

12. Ramalakshmi K, Prabhakar Rao PG, Raghavan B. Effect of storage on the quality of stabilized garlic powder. Indian Spices 1999; 34(4):5-7.
13. Yu LI, Shi-ying XU. Preparation of garlic powder with high allicin content. Agric Sci China 2007; 6(7):890-898.
14. Tattelman EMD. Health effects of garlic. Am Fam Phys 2005; 72:103-106.
15. Agarwal KC. Therapeutic actions of garlic constituents. Med Res Rev 1996; 16:111-124.
16. Aukris S. Antimicrobial properties of Allicin from garlic. Microbes infect 1999; 11:125-129.
17. Davis SR, Perrie R, Apitz-Castro R. The in vitro susceptibility of Scedosporium prolificans to ajoene, allitridium and a raw extract of garlic (Allium sativum). J Antimicrob Chemother 2003; 51(3):593-597.
18. Chowdhury AKZ, Ahsan M, Islam SN et al. Efficacy of aqueous extract of garlic and allicin in experimental shigellosis in rabbits. Indian J Medic Res 1991; 93:33-36.
19. Sivam GP. Recent protection against Helicobacter pylori and other bacterial infections by garlic. J Nutr 2001; 131:1106S-1108S.
20. Zenner L, Callait MP, Granier C et al. In vitro effect of essential oils from Cinnamomum aromaticum, Citrus limon and Allium sativum on two intestinal flagellates of poultry, Tetratrichomonas gallinarum and Histomonas meleagridis. Parasite 2003; 10(2):153-157.
21. Soffar SA, Mokhtar GM. Evaluation of the antiparasitic effect of aqueous garlic (Allium sativum) extract in hymenolepiasis nana and giardiasis. J Egypt Soc Parasitol 1991; 21:497-502.
22. Ayaz E, Turel I, Gul A et al. Evaluation of the anthelmentic activity of garlic (Allium sativum) in mice naturally infected with Aspiculuris tetraptera. Recent Pat Anti-Infect Drug Discov 2008; 3(2):149-152.
23. Nok AJ, Williams S, Onyenekwe PC. Allium sativum-induced death of African trypanosomes. Parasitol Res 1996; 82:634-637.
24. Lawson LD. The composition and chemistry of garlic cloves and processed garlic. In: Koch HP, Lawson LD, eds. Garlic: The Science And Therapeutic Application Of Allium Sativum L. Baltimore: Williams and Wilkins, 1996;37-108.
25. Davis LE, Shen J, Royer RE. In vitro synergism of concentrated Allium sativum extract and amphotericinB against Cryptococcus neoformans. Planta Med 1994; 60:546-549.
26. Harris JC, Cottrell SL, Plummer S et al. Antimicrobial properties of Allium sativum (garlic). Appl Microbiol Biotechnol 2001; 57:282-286.
27. Weber ND, Andersen DO, North JA et al. In vitro virucidal effects of Allium sativum (Garlic) Extract and Compounds. Planta Med 1992; 58:417-423.
28. Thomson M, Ali M. Garlic [Allium sativum]: a review of its potential use as an anti-cancer agent. Curr Cancer Drug Targets 2003; 3(1):67-81.
29. Li H, Li HQ, Wang Y et al. An intervention study to prevent gastric cancer by microselenium and large dose of allitridum. Chin Med J 2004; 117(8):1155-1160.
30. Milner JA. Preclinical perspectives on garlic and cancer. J Nutr 2006; 136:827S-831S.
31. Lea MA, Ayyala US. Differentiating and growth inhibitory effects of diallyl disulfide on cancer cells. Int J Oncol 1997; 11:181-185.
32. Sigounas G, Hooker J, Anagnostou A et al. S-Allylmercaptocysteine inhibits cell proliferation and reduces the viability of erythroleukemia, breast and prostate cancer cell lines. Nutr Cancer 1997; 27:186-191.
33. Durak I, Yilmaz E, Devrim E et al. Consumption of aqueous garlic extract leads to significant improvement in patients with benign prostate hyperplasia and prostate cancer. Nutrit Res 2003; 23:199-204.
34. Dong Y, Lisk D, Block E et al. Characterization of the biological activity of gammaglutamyl-S e-methylselenocysteine: a novel, naturally occurring anticancer agent from garlic. Cancer Res 2001; 61(7):2923-2928.
35. Hodge G, Hodge S, Han P. Allium sativum (garlic) suppresses leukocyte inflammatory cytokine production in vitro: Potential therapeutic use in the treatment of inflammatory bowel disease. Cytometry 2002; 48:209-215.
36. Keiss HP, Dirsch VM, Hartung T et al. Garlic (Allium sativum L.) modulates cytokine expression in lipopolysaccharide activated human blood thereby inhibiting NF-κB activity. J Nutr 2003; 133:2171-2175.
37. Allison GL, Lowe GM, Rahman K. Aged garlic extract and its constituents inhibit platelet aggregation through multiple mechanisms. J Nutr 2006; 136:782S-788S.
38. Ohaeri OC, Adoga GI. Anticoagulant modulation of blood cells and platelet reactivity by garlic oil in experimental diabetes mellitus. Biosci Rep 2006; 26(1):1-6.
39. Chan K, Yin M, Chao W. Effect of diallyl trisulfide-rich garlic oil on blood coagulation and plasma activity of anticoagulation factors in rats. Food Chem Toxicol 2007; 45:502-507.
40. Srivastava KC, Tyagi OD. Effects of a garlic-derived principle (ajoene) on aggregation and arachidonic acid metabolism in human blood platelets. Prostaglandins Leukot Essent Fatty Acids 1993; 49:587-595.
41. Koscielny J, Klussendorf D, Latza R et al. The antiatherosclerotic effect of Allium sativum. Atherosclerosis, 1999; 144(1):237-249.

42. Yeh YY, Liu L. Cholesterol-lowering effects of garlic extracts and organosulfur compounds: human and animal studies. J Nutr 2001; 131:989S-993S.
43. Chetty KN, Calahan L, Harris KC et al. Garlic attenuates hypercholesterolemic risk factors in olive oil fed rats and high cholesterol fed rats. Pathophysiology 2003; 9(3):127-132.
44. Kannar D, Wattanapenpaiboon N, Savige GS et al. Hypocholesterolemic effect of an enteric-coated garlic supplement. J Am Coll Nutr 2001; 20(3):225-231.
45. Qidwai W, Qureshi R, Hasan SN et al. Effect of dietary garlic (Allium sativum) on the blood pressure in humans—a pilot study. J Pak Med Assoc 2000; 50(6):204-207.
46. Anim-Nyame N, Sooranna SR, Johnson MR et al. Garlic supplementation increases peripheral blood flow; a role for interleukin-6? J Nutr Biochem 2004; 15:30-36.
47. Ohaeri OC. Effect of garlic oil on the levels of various enzymes in the serum and tissue of streptozotocin diabetic rats. Biosci Rep 2001; 21:19-24.
48. Bhagyalakshmi N, Thimmaraju R, Venkatachalam L et al. Nutraceutical applications of garlic and the intervention of biotechnology. Crit Rev Food Sci Nutr 2005; 45(7):607-621.
49. Eidi A, Eidi M, Esmaeili E. Antidiabetic effect of garlic (Allium sativum L.) in normal and streptozotocin-induced diabetic rats. Phytomedicine 2005; 13:624-629.
50. Kasuga S, Ushijima M, Morihara N et al. Effect of aged garlic extract (AGE) on hyper glycemia induced immobilization stress in mice. Nippon yakurigaku Zassh 1999; 114:192-197.
51. Ryan EA, Pick Mea, Marceau C. Use of alternative medicines in diabetes mellitus. Diabet Med 2001; 18(3):242-245.
52. Baluchnejadmojarad T, Roghani M. Garlic extract attenuates time-dependent changes in the reactivity of isolated aorta in streptozotocin-diabetic rats. Life Sci 2003; 73(18):2281-2289.
53. Borek C. Antioxidant health effects of aged garlic extract. J Nutr 2001; 131:1010S-1015S.
54. Banerjee SK, Mukherjee PK, Maulik SK. Garlic as an antioxidant: the good, the bad and the ugly. Phytother Res 2003; 17(2):97-106.
55. Benkeblia N. Free-radical scavenging capacity and antioxidant properties of some selected onions (A. cepa L.) and garlic (A. sativum L.) extracts. Braz Arch Biol Technol 2005; 48(5):753-759.
56. Bozin B, Mimica-Dukic N, Samojlik I et al. Phenolics as antioxidants in garlic (A. sativum L., Alliaceae). Food Chem 2008; 111:925-929.
57. Queiroz YS, Ishimoto EY, Bastos DHM et al. Garlic (Allium sativum L.) and ready-to-eat garlic products: In vitro antioxidant activity. Anal Meth 2009; 115(1):371-374.
58. Amagase H. Clarifying the real bioactive constituents of garlic. J Nutr 2006; 136:716S-725S.
59. Chung LY. The antioxidant properties of garlic compounds: Allyl cysteine, alliin, allicin and allyl disulfide. J Med Food 2006; 9:205-213.
60. Ichikawa M, Yoshida J, Ide N et al. Tetrahydro-b-carboline derivatives in aged garlic extract show antioxidant properties. J Nutrit 2006; 136:726S-731S.
61. Ryu K, Ide N, Matsuura H et al. N Alpha-(1-deoxy-D-fructos-1-yl)-L-arginine, an antioxidant compound identified in aged garlic extract. J Nutr 2001; 131:972S-976S.
62. Corzo-Martınez M, Corzo N, Villamiel M. Biological properties of onions and garlic. Trends Food Sci Tech 2007; 18:609-625.
63. Josling P. Preventing the common cold with a garlic supplement: a double-blind, placebo controlled survey. Adv Ther 2001; 18(4):189-193.
64. Takasu J, Uykimpang R, Sunga M et al. Aged garlic extract therapy for sickle cell anemia patients. BMC Blood Disord 2002; 2(1):3.
65. Gardner CD, Messina M, Lawson LD et al. Soy, garlic and ginkgo biloba: their potential role in cardiovascular disease prevention and treatment. Curr Atheroscler Rep 2003; 5(6):468-475.
66 Rahman K, Lowe GM. Garlic and cardiovascular disease: A critical review. J Nutr 2006; 136:736S-740S.
67. Schulz V, Hansel R, Tyler V et al. Rational phytotherapy: A physician's guide, 5th ed. Berlin: Springer-Verlag, 2004;128-138.
68. Ito Y, Ito M, Takagi N et al. Neurotoxicity induced by amyloid beta-peptide and ibotenic acid in organotypic hippocampal cultures: protection by S-allyl-L-cysteine, a garlic compound. Brain Res 2003; 985(1):98-107.
69. Peng Q, Buzzard AR, Lau BH. Neuroprotective effect of garlic compounds in amyloidbeta peptide-induced apoptosis in vitro. Med Sci Monit 2002; 8:328-337.
70. Pedraza-Chaverri J, Yam-Canul P, Chirino YI et al. Protective effects of garlic powder against potassium dichromate-induced oxidative stress and nephrotoxicity. Food Chem Toxicol 2008; 46:619-627.
71. Kabasakal L, Sehirli O, Cetinel S et al. Protective effect of aqueous garlic extract against renal ischemia/reperfusion injury in rats. J Med Food 2005; 8:319-326.
72. Chowdhury R, Dutta A, Chaudhuri SR et al. In vitro and in vivo reduction of sodium arsenite induced toxicity by aqueous garlic extract. Food Chem Toxicol 2008; 46:740-751.

73. Nasim SA, Dhir B, Samar F et al. Sulphur treatment alters the therapeutic potency of alliin obtained from garlic leaf extract. Food Chem Toxicol 2009; 47:888-892.
74. Salman H, Bergman M, Bessler H et al. Effect of a garlic derivative (alliin) on peripheral blood cell immune responses. Int J Immunopharmacol 1999; 21(9):589-597.
75. Kim JW, Kyung KH. Antiyeast activity of heated garlic in the absence of alliinase enzyme action. J Food Sci 2006; 68(5):1766-1770.
76. Kuettner EB, Hilgenfeld R, Weiss MS. The active principle of garlic at atomic resolution. J Biol Chem 2002; 277(48):46402-46407.
77 Oommen S, Anto RJ, Srinivas G et al. Allicin (from garlic) induces caspase- mediated apoptosis in cancer cells. European Pharmacol 2004; 485:97-103.
78. Miron T, Wilchek M, Sharp A et al. Allicin inhibits cell growth and induces apoptosis through the mitochondrial pathway in HL60 and U937 cells. J Nutrit Biochem 2008; 19:524-535.
79. Okada Y, Tanaka K, Sato E et al. Kinetic and mechanistic studies of allicin as an antioxidant. Org Biomol Chem 2006; 4:4113-4117.
80. Yamada Y, Azuma K. Evaluation of the in vitro antifungal activity of allicin. Antimicrob. Agents Chemother 1997; 11:743-749.
81. Jiang XW, Hu J, Mian FI. A new therapeutic candidate for oral aphthous ulcer: Allicin. Med Hypoth 2008; 71:897-899.
82. Gonen A, Harats D, Rabinkov A et al. The antiatherogenic effect of allicin: possible mode of action. Pathobiology 2005; 72:325-334.
83. Li M, Ciu JR, Ye Y et al. Antitumor activity of Z-ajoene, a natural compound purified from garlic: antimitotic and microtubule-interaction properties. Carcinogenesis 2002; 23(4):573-579.
84. Shimon LJW, Rabinkov A, Miron T et al. Alliin lyase (alliinase) from garlic (Allium sativum): crystallization and preliminary X-ray characterization. Acta Cryst 2002; 58:1335-1337.
85. Kuettner EB, Hilgenfeld R, Weiss MS. Purification, characterization and crystallization of alliinase from garlic. Arch Biochem Biophys 2002; 402(2):192-200.
86. Smeets K, Van Damme M, Van Leuven F et al. Isolation and characterization of lectins and lectin-alliinase complexes from bulbs of garlic (Allium sativum) and ramsons (Allium ursinum). Glycoconjugate J 1997; 14:331-343.
87. Lawson LD, Wang JZ. Low allicin release from garlic supplements: a major problem due to the sensitivities of alliinase activity. J Agric Food Chem 2001; 49:2592-2599.
88. Li Y, Xu SY, Sun DW. Preparation of garlic powder with high allicin content by using combined microwave-vacuum and vacuum drying as well as microencapsulation. J Food Engin 2007; 83(1):76-83.
89. Ayala-Zavala JF, Soto-Valdez H, González-Leon A et al. Microencapsulation of cinnamon leaf (Cinnamomum zeylanicum) and garlic (Allium sativum) oils in β-cyclodextrin. J Incl Phenom Macrocycl Chem 2008; 60(3-4):359-368.
90. Gibbs BF, Kermasha S, Alli I et al. Encapsulation in the food industry: a review. Int J Food Sci Nutrit 1999; 50:213-224.
91. Becker TA, Kipke DR, Brandon T. Calcium alginate gel: a biocompatible and mechanically stable polymer for endovascular embolization. J Biomed Mater Res 2001; 54(1):76-86.
92. Orive G, Ponce S, Hernandez RM et al. Biocompatibility of microcapsules for cell immobilization elaborated with different type of alginates. Biomaterials 2002; 23(18):3825-3831.
93. Li B, He J, Evans DG et al. Enteric-coated layered double hydroxides as a controlled release drug delivery system. Int J Pharm 2004; 287(1-2):89-95.
94. Choy JH, Park M, Oh JM. Bio-nanohybrids based on layered double hydroxide. Curr Nanosci 2006; 2(3):275-281.
95. Crepaldi EL, Valim JB. Layered double hydroxides: Structure, synthesis, properties and applications. Quim Nova 1998; 21:300-311.
96. Trikeriotis M, Ghanotakis DF. Intercalation of hydrophilic and hydrophobic antibiotics in layered double hydroxides. Int J Pharm 2007; 332(1-2):176-184.
97. Kantonis G, Trikeriotis M, Ghanotakis DF. Biocompatible protoporphyrin IX-containing nanohybrids with potential applications in photodynamic therapy. J Photochem Photobiol A Chem 2007; 185(1):62-66.
98. Choy JH, Kwak SY, Jeong YJ et al. Inorganic layered double hydroxides as nonviral vectors. Angew Chem Int Ed 2000; 39:4041-4045.
99. Sankalia MG, Mashru RC, Sankalia JM et al. Papain entrapment in alginate beads for stability improvement and site-specific delivery: Physicochemical characterization and factorial optimization using neural network modeling. AAPS Pharm Sci Tech 2005; 6(2):E209-E222.
100. Orive G, Hernandez RM, Gascon AR et al. Encapsulation of cells in alginate gels. In Guisan JM, ed. Methods in biotechnology. Immobilization of enzymes and cells, 2nd ed. Totowa: Humana Press, 2006; 22:345-355.

Chapter 10

Genetic Engineering to Enhance Crop-Based Phytonutrients (Nutraceuticals) to Alleviate Diet-Related Diseases

Autar K. Mattoo,* Vijaya Shukla, Tahira Fatima, Avtar K. Handa and Surender K. Yachha

Abstract

Nutrition studies have provided unambiguous evidence that a number of human health maladies including chronic coronary artery, hypertension, diabetes, osteoporosis, cancer and age- and lifestyle-related diseases are associated with the diet. Several favorable and a few deleterious natural dietary ingredients have been identified that predispose human populations to various genetic and epigenetic based disorders. Media dissemination of this information has greatly raised public awareness of the beneficial effects due to increased consumption of fruit, vegetables and whole grain cereals—foods rich in phytonutrients, protein and fiber. However, the presence of intrinsically low levels of the beneficial phytonutrients in the available genotypes of crop plants is not always at par with the recommended daily allowance (RDA) for different phytonutrients (nutraceuticals). Molecular engineering of crop plants has offered a number of tools to markedly enhance intracellular concentrations of some of the beneficial nutrients, levels that, in some cases, are closer to the RDA threshold. This review brings together literature on various strategies utilized for bioengineering both major and minor crops to increase the levels of desirable phytonutrients while also decreasing the concentrations of deleterious metabolites. Some of these include increases in: protein level in potato; lysine in corn and rice; methionine in alfalfa; carotenoids (β-carotene, phytoene, lycopene, zeaxanthin and lutein) in rice, potato, canola, tomato; choline in tomato; folates in rice, corn, tomato and lettuce; vitamin C in corn and lettuce; polyphenolics such as flavonol, isoflavone, resveratrol, chlorogenic acid and other flavonoids in tomato; anthocyanin levels in tomato and potato; α-tocopherol in soybean, oil seed, lettuce and potato; iron and zinc in transgenic rice. Also, molecular engineering has succeeded in considerably reducing the levels of the offending protein glutelin in rice, offering proof of concept and a new beginning for the development of super-low glutelin cereals for celiac disease patients.

*Corresponding Author: Autar K. Mattoo—Sustainable Agricultural Systems Lab, The Henry A. Wallace Beltsville Agricultural Research Center, United States Department of Agriculture, Agricultural Research Service, Beltsville, Maryland 20705-2350, USA. Email: autar.mattoo@ars.usda.gov

Bio-Farms for Nutraceuticals: Functional Food and Safety Control by Biosensors edited by Maria Teresa Giardi, Giuseppina Rea and Bruno Berra.

Introduction

Nutrition and human health have become synonymous due to the realization that certain nutrients when consumed in specific quantities may contribute to optimal health.[1] Poor nutrition is particularly relevant in regard to the diet-related diseases.[2] Targeting specific dietary treatment as therapeutic has important clinical and economic relevance, particularly in high prevalent diseases including coronary artery disease, hypertension, diabetes, osteoporosis and possibly cancer in adults and celiac disease (CD), cow's milk protein allergy (CMPA), galactosemia and glycogen storage disorders in children.[3] Prevention and therapy of age- and lifestyle-related diseases by individualized tailoring of optimal diets or nutrients seem conceivable but will need scientific validation through human nutrition trials and evaluation of dosage with minimal side effects.

Majority (97%) of the genes associated with human disease have been classified under monogenic diseases because each is a result of a dysfunctional gene.[4] Although human health disorders are basically genetic, there is a definite interplay of disorders/disease with either insults contributed by consumption of certain, commonly used foods, or preventive strategies involving dietary intervention, through nutritionally-enhanced food. In several cases, the offending as well as favorable dietary agents over and above genetic predisposition have been identified in human populations.[5-7] For instance: a gluten-free diet reverses the adverse health effects in CD; restricting intake of galactose and lactose helps in the control of galactosemia; and milk-free diet remedies CMPA. Diseases classified as polygenic in nature such as epithelial cancers, diabetes and heart disease seem to be ameliorated by the intake of dietary antioxidants, vitamins and other phytonutrients.[8,9] Dietary phytonutrients constitute variable antioxidants that prevent accumulation of damaging oxidants (reactive oxygen species, ROS) potentially preventing oxidation of cellular and sub-cellular macromolecules. Thus, phytonutrients, particularly those with antioxidant properties, have become a household name and are in demand as desirable nutraceuticals worldwide (see Chapter by Adams and Adams).

Analysis of about 200 dietary studies concluded that fruit and vegetable consumption is significantly protective against various types of cancers.[10] Notably, persons who consumed lower quantities of fruits and vegetables were at twice the risk of experiencing cancer than those that consumed higher proportion.[10] Nonetheless, a wide variation in response to food components as modifiers of cancer risk and tumor behavior has also been observed in other preclinical and clinical studies. These inconsistencies can be due to a number of factors, including: (i) the kind and level (based on the serving size) of antioxidant constituents of food; (ii) specificity and temporal relationships among dietary constituents as modifiers of molecular events; (iii) multi-factorial nature of signals that determine phenotypic expression; (iv) the nature of antioxidant response elements (AREs) present in the human genome which are polymorphic.[11] Clearly, there is a need for a better understanding of the relationship between dietary intervention and disease progression to develop fundamental understanding of disease-related process(es) impacted by one or more dietary constituent.[12]

There is a major gap in our understanding of the exact pathophysiology and a paucity of scientific tools to target prevention, enzyme replacement for specific deficiency, or gene therapy. However, definite progress has been made where it is possible to effectively manage several human disorders by dietary therapy and potentially prevent other serious diseases by consumption of nutritionally-enriched vegetables and fruits. The concept of developing nutritionally functional food requires: (1) the understanding of the mechanisms of prevention and protection; (2) the identification of the biologically active molecules and (3) the demonstrated efficacy of these molecules with human subjects. Accurate information is also highly desired to have in place about the required daily allowance (RDA) for effective phytonutrients, to name a few, vitamin C, carotenoids such as lycopene, β-carotene and lutein, phenolics, flavonoids, minerals, etc., in order to design biotechnological strategies for manipulating their contents in vegetables and fruits. Nutritional therapies are prone to several limitations currently existing either in achieving an appropriate control—prevention of ongoing organ damage, best survival and normalcy—or in acceptability, convenience and affordability.

Databases are growing with information on the content of phytonutrients in edible vegetables and fruits as well as with transgenic technology developments for enhancing the nutritional content of vegetable crops via engineering of specific metabolic pathways. Metabolic pathway engineering approaches have demonstrated the power of genetic manipulation in enhancing the content of nutrients beneficial for human health in transgenic crops. In the same vein, suppression of key genes to inhibit production of allergenic proteins or toxins in crops is highly sought.

Diet and Human Diseases

Diseases such as CD, galactosemia and phenylketonuria occur because of intolerance in a certain population to specific food components (Table 1).[2] Preventive measures include dietary restrictions and eliminating foods containing the disagreeable compound. CD is caused in genetically susceptible individuals (HLA predisposing genotypes, DQ2 and DQ8) who are sensitive to gluten, the storage protein present in wheat, barley and rye. CD is a common lifelong disorder in European-origin countries, affecting approximately 1% of the general population and prevalent also in North Africa, Middle East, India and Saharawi people.[6] Continued consumption of gluten-rich diets in many developing countries will result in increasing the frequency of this disease in the future. CD manifests as chronic diarrhea, anemia, growth failure, vitamin (fat and water soluble) deficiencies and other systemic disorders.[7] If left untreated, higher incidence of malignancy is seen. A definitive treatment for CD is life-long elimination of gluten from the diet. Since wheat, barley or rye form part of common diets, the administration of gluten-free diet is expensive and difficult, inconvenient, socially problematic and with high chances of noncompliance.[13] There is, thus, a need for better, convenient and long lasting treatment options.

Galactosemia is an autosomal recessive disorder of galactose metabolism, caused by a deficiency of the enzyme galactose-1-phosphate uridyltransferase (GALT; EC 2.7.7.12).[14] Galactose accumulation leads to toxic metabolites that cause damage to patient's liver and central nervous system. It is in the neonatal period that its clinical manifestation occurs, via ingesting breast milk and top milk causing release and accumulation of galactose from lactose by the intestinal lactase. The afflicted children suffer from jaundice, hepatosplenomegaly, liver dysfunction, hypoglycemia, renal tubular dysfunction, muscle hypotonia, cataract and *Escherichia coli* sepsis (Tables 1 and 2). If left untreated, galactosemic patients rapidly develop liver cirrhosis and early death. GALT activity in red blood cells together with determination of urinary sugars and sugar alcohols (showing elevated concentrations of galactose and

Table 1. Diet-related monogenic diseases in humans

Disease	Cause	Therapy
Celiac disease (CD)	In genetically susceptible individuals upon eating gluten present in barley, rye and wheat	Gluten-free diet
Cow's milk protein allergy (CMPA)	Allergic disorder to milk protein, β-lactoglobulin	Milk-free diet supplemented with calcium, riboflavin and vitamin D
Galactosemia	Deficiency of galactose-1-phosphate uridyl transferase	Galactose and lactose-free diet
Glycogen storage disorder (GSD)	Dysfunction of enzymes in glycogen metabolism	Nasogastric feeding or augmented diet with corn starch
Phenylketonuria	Defective phenylalanine metabolism, defective or deficient phenylalanine hydroxylase	Phenylalanine-free diets supplanted with tyrosine

Modified from reference 2.

Table 2. Major types of glycogen storage disorders showing defect and organ involved

Type	Defect	Predominantly Affected Organ
I: *Types*: Ia,Ib,Ic,Id	Glucose-6-phosphatase system deficiency	Liver
II: (4 onset forms) Pompe disease	Defective lysosomal α-glucosidase	Muscle ± cardiac muscle
III: (6 forms)	Amylo-1,6-glucosidase deficiency	Only liver Type IIIa: Both liver + muscle
IV: (Many forms) Andersen disease/Amylopectinosis	Deficiency of the glycogen-branching enzyme	Cirrhosis of liver: Both liver + muscle
V: (2 forms) McArdle disease	Muscle phosphorylase deficiency	Muscle
VI	Liver phosphorylase deficiency	Liver
VII: (2 subtypes)	Phosphofructokinase deficiency	Muscle
IX	Tissue-specific phosphorylase kinase deficiency	Liver
0: (2 subtypes)	Glycogen synthase deficiency	Liver
XI	Defect in glucose transporter 2 (GLUT-2)	Liver

galactitol) are diagnostic tools to identify classical galactosemia patients. The only therapy for such patients is that galactose must be removed from the diet as soon as the diagnosis is made and patients put on lactose-free diet. Even when the diet restriction is followed, complications such as retarded mental development, verbal dyspraxia, motor abnormalities and hypergonadotrophic hypogonadism develop in the long term. In the current recommendations, the restriction on galactose-containing fruit and vegetables has not been specified.

CMPA is a common disorder due to the milk protein, β-lactoglobulin. This disorder makes infants suffer from gastrointestinal symptoms with or without skin and respiratory symptoms.[5,15] Disease symptoms are chronic diarrhea, vomiting, anemia, failure to thrive and sometimes bleeding per rectum.[15] A prevalence of 13% among children under 2 years of age with malabsorption has been reported from developing countries.[16] CMPA patients are put on milk-free diet under the supervision of a dietician, who provides advice on milk-free recipes together with a list of alternate products for adequate calcium, riboflavin and vitamin D intake to avoid development of deficiency. Thus, restriction of milk to mothers breastfeeding such children becomes at times necessary, which is cumbersome. CMPA children are fed soy formulas, extensive hydrolysates (casein, whey or mixed) and amino acid based formulas.[15,17] There is a need for alternative and better dietary therapy. Advances in food biotechnology can pave a way to develop β-lactoglobulin-free bovine/cow's milk or devise biological processing to eliminate allergenic protein fraction of milk.

Dysfunctional enzymes involved in the synthesis and/or degradation of glycogen lead to glycogen storage disorder (GSD), a disease characterized by abnormal inherited glycogen metabolism in the liver, muscle and brain. Based on each defective metabolic step in carbohydrate pathways gives rise to different types (I-XI) of GSD[18,19] (Tables 1 and 3). Detailed dietary management for various types of GSD disorder is summarized in Table 3. There is a need for genetically modified food item to suit the requirement of the GSD children and development of cornstarch that has a longer life at ambient temperatures.

Phenylketonuria is another monogenic diet-related disease, a genetic disorder caused by the deficiency of or defective phenylalanine-metabolizing enzyme, phenylalanine hydroxylase.[20,21]

Table 3. Dietary management and its importance in various types of glycogen storage disorders

Type	Management	Targeted Dietary Importance
I	Restriction of dietary fructose and galactose Challenge to maintain normal blood glucose levels: • Infusions—intravenous/nasogastric and • corn starch	++
II	Enzyme replacement therapy	–
III	Hypoglycemia: corn starch supplement	+
IV	Treatment of liver disease	–
V	Supportive	High protein diet
VI	Hypoglycemia: high carbohydrate diet + frequent feeding	+
VII	Supportive	High protein diet
IX	Hypoglycemia: high carbohydrate diet + frequent feeding	+
0	Night time corn starch to maintain normal blood glucose levels	+
XI	Restrict galactose and fructose intake. Corn starch for improved growth	+

Aromatic amino acids including phenylalanine are generated from phosphoenolpyruvate, which offers a branch point for the synthesis of pyruvate and oxaloacetate and that of the shikimate pathway.[22] Phenylalanine accumulation in the blood drastically increases the risk of neurological damage. Therefore, higher dietary levels of phenylalanine are not desirable for patients suffering from phenylketonuria.[2,23] Treatment of this disorder calls for diets absent in phenylalanine and supplemented with tyrosine.[21]

In addition to the above described, diet-induced monogenic diseases, there is more prevalence of polygenic diseases such as cancer, diabetes and heart disease, which are often linked to dysfunctional biological networks.[24] These diseases are not easily treatable. However, dietary intervention has a good potential in preventing their onset. A case in point is the recent demonstration of a positive effect of antioxidant supplementation in relieving pain and reducing oxidative distress in patients with chronic pancreatic disease, pancreatitis.[25] The prevention strategy will be furthered by the knowledge developed about the elucidation of how a single nutrient (nutraceutical) functions in a biological system and the nature of interactions between a mixture of nutrients in a diet and prevention of a disease.[1,26]

Phytonutrients and Antiproliferative Activity

Cell proliferation is a characteristic manifestation of any malignancy and, therefore, anti-proliferative activity of individual antioxidants and diets is a means to discover potential anti-cancer agents. Because phytonutrients in fruit and vegetables seem to have synergistic effects on the antioxidant activities and may reduce the risk of chronic diseases, there is considerable interest to test their performance in mitigating cell proliferation.[27,28] Understandably, research has also intensified on analyzing and enhancing nutritional, antioxidant metabolites in food crops. It is well known that vegetables, fruits, nuts and various herbs are dietary sources of one or more of the following nutraceuticals: vitamins C, B and E, β-carotene, lycopene, polyphenols, flavonoids, folates, isothyocyanates, glucosinolates and minerals. Nonetheless, antioxidant capacity of fruits is a function of genotype/cultivar, growth condition:[29-32] the content of lycopene, ascorbic acid (vitamin

C) and phenolics in different tomato genotypes is variable[33] while the phenolic (flavonoid and capsaicinoid) content of hot pepper is affected by the color and maturation stage, the antioxidant activity at red stage being greater than the green maturity stage.[34] Likewise, antioxidant capacity due to stilbenes (resveratrol, pterostilbene and piceatannol) in various selections and cultivars of berries from USA and Canada indicated highest resveratrol content in grapes and berries.[35] These findings highlight the importance of taking into consideration the fact that antioxidant capacity of fruits significantly varies between genotypes, species and developmental stage, which can affect their nutritive value. Also of relevance is the finding that pomegranate juice shows higher bioactivity than individual polyphenols extracted from the fruit.[36]

Resveratrol was found to have growth inhibitory and antiproliferative activity in vitro on human promyelocytic leukemia (HL-60) cells.[37] Among fruit extracts tested on in vitro proliferation of HepG2 human liver-cancer cells, cranberry had highest inhibitory activity followed by lemon, apple, strawberry, red grape, banana, grapefruit and peach.[38] Also, it has been noted that the antiproliferative effects of fruit extracts and their antioxidant capacity are not always tightly correlated. Total antioxidant capacity (due to total phenolics and flavonoids) in fruit extracts of eight different strawberry cultivars did not correlate with their antiproliferative activity.[39] Similar results were previously obtained with raspberries.[40] These studies point out that multiple and synergistic interactions among chemicals promote antiproliferative activity of a fruit compared to an isolated antioxidant.[36,41,42]

Lycopene-rich foods lower biomarkers of oxidative stress and carcinogenesis in healthy and type II diabetic patients and prostate cancer patients, respectively. Dietary fats are recommended together with lycopene to enhance the beneficial aspects of lycopene.[9] Rats fed 0.7 mg/d lycopene for 4 days showed a significant reduction in the androgen status of the animal.[43] Induced oxidative stress in COS-7 cells was antagonized by blue berry extracts that lowered activation of cyclic AMP responsive element binding protein (pCREB) and possibly pPKCgamma kinase.[44]

Polyphenols[45,46] and dietary flavonols[47] were found to bind human lower density lipoproteins (LDL) whose cellular oxidation is considered a key step in atherogenicity.[48] Intake of grape juice,[49,50]tea, or red wine,[51] or an antioxidant nutrient like vitamin E (tocopherol)[52] enriches LDL levels and protects them against oxidation.

Genetic Engineering to Improve Nutrient (Nutraceutical) Content in Produce

Staple foods such as rice, corn, wheat and cassava are relatively poor sources of health-promoting nutrients. In the developing world this contributes to health problems because the traditional diets are deficient in one or more essential nutrients, causing malnutrition and disease. Thus, a simple solution entails food biofortification with externally supplied vitamins and antioxidants, or via genetic manipulation of nutraceuticals in staple food. Off-the-shelf supplements are commonly consumed because it is perceived that they improve whole body health. However, the bioavailability of nutrients in a diet determines how much of a good nutrient's potential is realized.[53] The refined tools of genomics, proteomics and metabolomics are now in vogue to enable in-depth studies on the relationships between diet, genetics and metabolism. A biotechnological approach is to genetically engineer the required threshold of nutrients in the edible portions of a grain crop or a fruit/vegetable crop. Thus, development of 'golden' rice with added pro-vitamin A, β-carotene and protein is an innovation that might help conquer malnutrition in the developing world.[54,55] A recent advance has been the development of transgenic multivitamin corn, a study that introduced four cDNAs encoding enzymes in the biosynthetic pathways of vitamins β-carotene, ascorbate and folate.[56] The transgenic corn endosperm accumulated up to 59.32 µg/gDW of β-carotene, 22.78 µg/gDW of lycopene, 106.94 µg/gDW of vitamin C (ascorbate) and 35.76 µg/gDW of zeaxanthin. We summarize below examples of the application of genetic engineering and biotechnology approaches in enhancing the levels of various phytonutrients (nutraceuticals) in cereal, vegetable and fruit crops. Table 4 lists cereal crops and Table 5 lists vegetable and fruit crop—however, not all those listed have been elaborated upon. The results clearly indicate the power of genetic engineering in achieving the desired levels of nutraceuticals in crop plants.

Table 4. Genetically modified enhancement in nutrients of major crops

Target Nutrient/Crop	Gene Introduced/ Modified	Promoter Used	Transgenic vs. WT	Reference
Amino Acids/ Rice	*GluB* (RNA silencing)	CaMV35S (constitutive)	Low glutelin content than Lgc-1 mutant	57
	CTA-(altered) tRNA^{lys}	Ubi	41.42% vs. 0.81% lysine content of prolamines	58
Amino Acids/ Corn	*LKR/SDH*/Corn (RNAi suppression)	B32 (endosperm specific)	Free lysine: ~650 ppm vs. 30 ppm	59
	SB401/Potato	P19Z (seed specific)	Lysine: 54.8% vs. 16.1%; total protein content: 39% vs. 11.6%	117
	Synthetic porcine α-lactalbumin	γ-zein (endosperm specific)	Lysine: 47% vs. 29%	118
	LKR1/SDH/Corn (dsRNA) + *CordapA/ Corynbacterium glutamicum*	Pe35S	Lysine: 4000 ppm vs. 100 ppm	60
Carotenoid/ Rice	PSY/daffodil	CaMV35S or Gt1 (endosperm specific)	Phytoene: 0.74 μg g^{-1}DW (with pGt1); 0.32 μg g^{-1}DW (with CaMV35S); nd in WT	63
	Psy/daffodil + *Crt1/ Erwinia uredevora*	Gt1 and CaMV35S	Higher β-carotene and lutein and zeaxanthin	53
	Psy/daffodil + *Crt1/E. uredevora* (construct 1); *Lcy*/daffodil (construct 2)	Gt1 and CaMV35S (construct 1); Gt1 (construct 2)	From variable carotenoid to only β-carotene accumulation: 1.6 μg · g^{-1} in heterozygous lines	53
	psy/corn + crtl/daffodil	Glu1 (endosperm specific)	Total carotenoids: 37 μg g^{-1}DW; β-carotene: 84 μg · g^{-1} DW (nd in WT)	54
Carotenoid/ Corn	*Psy1*/corn + *crt1/E. uredevora*	LMW glutelin/ wheat and D-hordein/barley (endosperm specific)	β-carotene: 4.79 vs. 0.09 μg g^{-1}DW; Lycopene: 22.78 vs. 0 μg g^{-1}DW	55
	crtB + *crt1/E. uredevora*	γ-zein	Provitamin A: 7 vs. 0.39 μg g^{-1}DW; total carotenoids: 33 vs. 1 μg g^{-1}DW	66

continued on next page

Table 4. Continued

Target Nutrient/ Crop	Gene Introduced/ Modified	Promoter Used	Transgenic vs. WT	Reference
Folate/Rice	*GTPCHI* and *ACDS*	A single T-DNA expression cassette	Folate: 38.3 vs. 0.38 nmol g^{-1}FW; 5methyl-tetrahydrofolate, 89% of total folate	89
	GTPCHI/Arabidopsis	glb-1 (endosperm specific)	Pterin: 0.23 vs. 0.92 nmol g^{-1}FW	89
Folate/Corn	*GCH1/E. coli*	D-hordein	1.94 vs. 0.94 μg g^{-1}DW	55
Ascorbate/Corn	*DHAR*/wheat	Ub-(ubiquitin) or Sh2 (shrunken 2)	Ascorbate: ~80 vs. 42 nmol g^{-1} FW; 20-106 fold increase in DHAR activity	93
	DHAR/rice	D-hordein	106 vs. 17.53 μg g^{-1}DW	55

nd: not detected; DW: dry weight; FW: fresh weight.

Proteins and Amino Acids

Major Crops

The seed storage protein, glutelin, which is a cause for CD, commonly occurs in cereals. Its reduction is therefore a promising avenue for CD patients. In this context, Low Glutelin Content-1 (*Lgc1*) is a dominant mutation that reduces glutelin content in rice grain while increasing prolamine and other seed storage proteins.[57] RNA interference (RNAi) approach was used to silence glutelin *GluB* gene, producing low glutelin protein phenotypes of rice (Table 4).[58] Among the transformants characterized, *GluB4-*Δ*GluB5* and Δ*GluB5* were found to accumulate glutelin content lesser than the LGC-1 mutant. These authors suggested that these transgenic lines are a good resource to produce "super low-protein rice" and may be more beneficial than LGC-1 for kidney disease patients.

The content of essential amino acids such as lysine is low in staple crops such as rice. To increase the lysine/amino acid content of rice seeds, $tRNA^{lys}$ with altered anti-codons (chimeric $tRNA^{lys}$) was expressed under ubiquitin promoter in rice callus.[59] Such a strategy resulted in significant increase in the content of lysine without affecting the quantity or the quality of the storage proteins. A different approach to increase lysine content was to inhibit its catabolism. Thus, an inverted repeat (IR) sequence corresponding to part of the saccharophine dehydrogenase (SDH) region of corn (*IR-SDH*) was introduced to enable endosperm-specific suppression of the corn bifunctional lysine degradation enzyme, lysine-ketoglutarate reductase and SDH (LKR-SDH), in transgenic corn kernels by RNAi.[60] This approach led to a 20-fold increase in free lysine content in homozygous transgenic corn kernels over the null control. These studies were complemented by a study that employed a single transgene cassette for simultaneous expression of the deregulated lysine degradation enzyme, *CordapA* and suppression of LKR/SDH, in transgenic corn.[61] This approach of combining LKR/SDH silencing and *CordapA* expression led to over 40-fold (4000 ppm in transgenic versus 100 ppm in nontransgenic control) accumulation of free lysine in corn grain.

Minor Crops

Potato tops the list of vegetable crops consumed in the world. Apart from being a rich source of starch and vitamin C, it lacks protein and other nutrients. In an effort to produce a high quality protein-rich potato, Amaranthus seed albumin gene, *AmA1*, was engineered and expressed

Table 5. Genetically modified enhancement in nutrients of vegetables and fruits

Target Nutrient/Crop	Gene Introduced/Modified	Promoter Used	Transgenic vs. WT	Reference
Amino Acids/Potato	*AmA1/A. hypochondriacus*	CaMV35S (constitutive) and GBSS (tuber specific)	Total protein: ~16.5 vs. 11.5 mg g^{-1}FW; 2.5-4 fold increase in essential amino acids: lysine, methionine, cysteine and tyrosine	61
Amino Acids/Alfalfa	*AtCGS*/Arabidopsis	Rubisco SSU	Methionine: 4846 vs. 2198 nmol g^{-1}FW; cysteine: 1873 vs. 782 nmol g^{-1}FW	62
Amino Acids-Choline/Tomato	*ySAMDC, Spe2*/Yeast	E8 (fruit specific)	Increase in glutamate, asparagine and choline	22
Carotenoid/Brassica	ε-*CYC/B. napus* (RNAi)	P35S (constitutive)	β-carotene: 90.76 vs. 0.49 μg g^{-1}FW; lutein: 76.22 vs. 3.30 μg g^{-1}FW	119
Carotenoid/Flaxseed	*crtB/P. ananatis*	CaMV35S and FAE1 (seed specific)	156.3 vs. 8.4 μg g^{-1}DW	71
Carotenoid/Canola	*crtB*/bacteria	Napin (seed specific)	1000 vs. 33 μg g^{-1}FW	69
β-Carotene/Canola	*crtI* + *crtb*/bacteria	Seed specific	857 vs. 5 μg g^{-1}FW	70
	DET1 (RNAi suppression)	CaMV35S	7 vs. 0.5 μg g^{-1}FW	120
Lycopene/Canola	*crtI + crtb/bacteria*	Seed specific	44 vs. 33 μg g^{-1}FW	72
β-Carotene/Tomato	*DET1*/Tomato (RNAi)	P119 (fruit specific)	20.5 vs. 2.4 μg g^{-1}FW	79
	βLcy/Arabidopsis	Pds (fruit specific)	50.8 vs. 8.0 μg g^{-1}FW	73
	βLcy/Arabidopsis + *βLcy*/Pepper	Pds	63 vs. 5 μg g^{-1}FW	74
	Cry-2/Tomato	CaMV35S	101 vs. 78 μg g^{-1}DW	121
	CrtB/E. uredovora	PG (fruit specific)	1700 vs. 33 μg g^{-1}DW	77

Continued on next page

Table 5. Continued

Target Nutrient/Crop	Gene Introduced/Modified	Promoter Used	Transgenic vs. WT	Reference
	β-cyclase/E. herbicola	Atpl (chloroplast targeted)	286 vs. 69 μg g^{-1}DW	122
	Crt1/E. uredovora	CaMV35S	520 vs. 271 μg g^{-1}DW	76
Lutein/Tomato	*Cry-2*/Tomato	CaMV35S	36 vs. 23 μg g^{-1}DW	122
	β*Lcy*/Arabidopsis (antisense)	Pds	97 vs. 54 μg g^{-1}FW	73
	Cry-2/Tomato (gene silencing)	CaMV35S	1,353 vs. 775 μg g^{-1}DW	122
Lutein/Tomato	*DET1*/Tomato	2A11, P119, TFM7 (fruit specific)	2-fold	79
	CUL4/Tomato (RNAi) *HP1*/ Tomato (RNAi)	CaMV35S TFM7	300 vs. 150 μg g^{-1}FW	82
Lycopene/Tomato	*ySAMDC, Spe2*/Yeast	E8	2 to 3-fold: 150-175 vs. 25-50 μg g^{-1}FW	83
β-Cryptoxanthin, zeaxanthin/ Tomato	β*Lcy*/Arabidopsis + β*Chy*/ Pepper	Pds	24 μg g^{-1}FW (nd in WT)	74
β-Cryptoxanthin, zeaxanthin/ Tomato	β*Lcy*/Arabidopsis + β*Chy*/ pepper	Pds	24 μg g^{-1}FW	74
Carotenoids/Potato	*Or*/cauliflower	GBSS (tuber specific)	31 vs. 5.5 μg gDW^{-1}	85,86
	CrtB/E. uredovora	Patatin (tuber specific)	35 vs. 5.6 μg g^{-1}FW	72
	CrtI,CrtB and *CrtY*	Pat1 (tuber specific)	Total carotenoids: 114 vs. 5.8 μg g^{-1}DW; β-carotene: 47 vs. 0.013 μg g^{-1}DW;	123
	LCY-e/potato (antisense)	Patatin B33 (tuber specific)	Total carotenoids: 12272 vs. 4672 ng g^{-1}DW; β-carotene: 43 vs. 3 ng g^{-1}DW; zeaxanthin: 990 vs. 262 ng g^{-1}DW	124

Continued on next page

Table 5. Continued

Target Nutrient/Crop	Gene Introduced/Modified	Promoter Used	Transgenic vs. WT	Reference
Phytoene/Tomato	*crtB/E. uredovora*	PG	47 vs. 28 μg g^{-1}DW	77
	Psy-1/Tomato	CaMV35S	240.2 vs. 102 μg g^{-1}FW	125
Phytoene/Potato	*dxs/E. coli*	Patatin (tuber specific)	3.0 vs. 0.4 μg g^{-1}FW	126
Ketocarotenoid/Potato	*crtO/Cyanobacterium* (cotrans-formation of *zep* transformants)	CaMV35S	Ketocarotenoids accumulate up to 10-12% of total carotenoids	127
Ketocarotenoid/Carrot	*CrtO/H. pluvialis*	Ubi/CaMV35S/RolD	70% of total carotenoid converted to keto-carotenoids	87
Folate/Tomato	*GCH1*/mammalian	E8	Total folate: 4.5 vs. 2.3 nmol g^{-1}FW	90
	GCH1/Mammalian + *ADCS*/ Arabidopsis	E8	25 vs. 1 nmol g^{-1}FW	91
Folate/Lettuce	*Gch1*/avian	CaMV35S	1.885 vs. 0.34 μg g^{-1}FW	92
Iron/Lettuce	*pfe*/Soybean	CaMV35S	398 vs. 227 μg g^{-1}DW	128
Ascorbate/Lettuce	Gul oxidase/rat	CaMV35S	4 to 7-fold	94
Tocopherols/Lettuce	γ *TMT*/Arabidopsis	CaMV35S	Over 2-fold in vitamin E; complete conversion of γ-tocopherol to α-tocopherol	107
Tocopherols/Soybean	*AtVTE3* + *AtVTE4*/Arabidopsis	Napin (seed specific)	α-tocopherol: 91.4 vs. 10.5%	106
	γ *TMT*/Arabidopsis	CaMV35S	4 fold increase in α-tocopherol	108
	γ *TMT/P. frutescens*	Vicilin (seed specific)	α-tocopherol: 450.1 vs. 43.9 pmol mg^{-1}FW; β-tocopherol: 55.7 vs. 4.3 pmol mg^{-1}FW; 4.8-fold higher vitamin E activity	110

Continued on next page

Table 5. Continued

Target Nutrient/Crop	Gene Introduced/Modified	Promoter Used	Transgenic vs. WT	Reference
Tocopherols/Oilseed	γ *TMT*/Arabidopsis	CaMV35S	α-tocopherol: 62 vs. 9.8%; γ-tocopherol: 35.84 vs. 86.82%	109
Tocopherols/Tomato	*hmgr-1*/Arabidopsis	CaMV35S	Phytosterol: 1800 vs. 1182 μg g^{-1}DW; campesterol: 161.7 vs. 96 μg g^{-1}DW; cycloartenol: 234 vs. 162 μg g^{-1}DW	78
Flavonoids/Tomato	*HQT*/Tomato	CaMV35S	Chlorogenic acid: 1.85 μg g^{-1}FW	96
	DET1/Tomato (RNAi suppression)	TFM7, 2A11 and P119	1.9 to 3.5-fold	79
	Cry-2/Tomato	CaMV35S	Total: 5.31 vs. 1.85 μg mg^{-1}DW	121
	chi/Petunia	CaMV35S	Quercetin 16.52 vs. 0.25 mg g^{-1}DW; kaempferol: 2.05 vs. 0.03 mg g^{-1}DW	95
Flavonoids/Potato	*DFR*/*P. hybrida*	CaMV35S	Chlorogenic acid: 2.5 vs.1.5 mg g^{-1}DW; pelargonidin: 0.4 vs.0.1 mg g^{-1}DW; petunidin: 3.0 vs. 0.5 mg g^{-1}DW	103
Flavonol/Tomato	*Chi*/Petunia	CaMV35S	1.9 vs. 0.09 mg g^{-1}DW	95
Flavonol/Tomato	*MYB12*/Arabidopsis	CaMV35S	Quercetin: 20 vs. 0.30 mg g^{-1} DW; kaempferol: 28 vs. 0.05 mg g^{-1} DW; chlorogenic acid: 1.17 vs. 0.04 mg g^{-1} DW	129
Anthocyanin/Tomato	*Del*/Snapdragon + *Ros1*/Snapdragon	E8	2.83 mg g^{-1}FW (nd in WT)	104
Isoflavone/Tomato	*IFS*/Soybean	CaMV35S	Genistin: 0.45 $nmol^{-1}$gFW; quercetin glucoside: >200 nmol g^{-1}FW	98

Continued on next page

Table 5. Continued

Target Nutrient/Crop	Gene Introduced/Modified	Promoter Used	Transgenic vs. WT	Reference
Stilbenes/Tomato	*STS*/V. vinifera *CHS*/Petunia + *CHR*/alfalfa *CHI*/petunia + *FNS-II*/gerbera	Pd35S	Stilbenes: 0.01mg g^{-1}FW (whole fruit); 0.037 mg g^{-1}FW (peel); deoxychalcones: 0.26 mg g^{-1}FW (peel) luteolin aglycon: 0.34 mg g^{-1}FW; luteolin 7-glucoside: 0.15 mg g^{-1}FW (peel); rutein: 0.90 vs. 56 mg g^{-1}FW (peel)	97
Resveratrol/Tomato	*StSy/V. vinifera*	CaMV35S	4-53 μg g^{-1}FW (nd in WT)	99
Phenolics and Stilbenes/Tomato	*StSy/V. vinifera*	CaMV35S	Caffeic acid: 0.058 mg g^{-1} FW; trans-piceid: 126.58 mg kg^{-1} FW; trans-resveratrol: 48.48 mg kg^{-1} FW	100
Piceid/Apple	*Vst1/V. vinifera*	Vst1(gene specific)	20-80 μg g^{-1}FW	101
Phenylpropanoids/Strawberry	*CHS*/Strawberry (antisense)	CaMV35S	100 vs. 1%	102
Tocopherols/Potato	*crtB*—TP/*E. uredovora*	Patatin (tuber specific)	α-tocopherol: 7.5 μg g^{-1} FW	72
	AtHPPD and *AtHPT*/Arabidopsis	CaMV 35S	α-tocopherol: (with *AtHPDD*) 338 vs. 432 ng g^{-1}FW; (with *AtHPT*) 578 vs. 284 ng g^{-1}FW	111
Fatty acyl composition/Canola	*CrtB*/bacteria	Napin	18:1 (oleic acid) component	69
Linolenic acid/Sweet potato	*NtFAD3*/Tobacco	E12Ω	22.1 vs. 10%	130

nd: not detected; DW: dry weight; FW: fresh weight.

constitutively as well as in a seed-specific manner in potato (Table 5).[62] The transgenic potato transformants accumulated protein and had increased levels of essential amino acids. The constitutive (16.12 mg/g tuber) as well as tuber-specific (16.58 mg/g tuber FW) expression of *AmA1* increased the total protein content of tuber to the same extent.

Methionine level was increased in alfalfa by introducing Arabidopsis cystathionine γ-synthase (*AtCGS*), the enzyme that controls the synthesis of the first metabolite in the methionine pathway, under the control of rubisco small subunit promoter.[63] This approach enhanced the methionine and cysteine contents in transgenics and the authors remarked that the 2-fold higher protein-bound methionine (3.65 ± 0.11 g methionine/100 g of protein) in transgenic alfalfa line L-1 amounts to an adequate supply of methionine in diet.

Genetically engineered accumulation of polyamines, spermidine and spermine, in tomato fruit resulted in the higher content of amino acids including asparagine, glutamate and glutamine, and choline.[23]

Carotenoids

Major Crops

Carotenoids have received considerable attention in recent years because of their antioxidant property as well as their potential in preventing malnutrition and disease. For instance, β-carotene is a pro-vitamin A and essential for vision. Lycopene is a potent antioxidant that has the potential to prevent epithelial cancers in addition to providing benefits to cardiovascular and immunity-based health. Rice is devoid of β-carotene. Therefore, genetic engineering was used to produce β-carotene in rice endosperm. Daffodil (*Narcissus pseudonarcissus*) *Psy* gene, which encodes phytoene synthase and produces the first carotenoid phytoene, a key precursor of β-carotene, was expressed in rice under a CaMV 35S (constitutive) or Gt1 (endosperm-specific) promoter. The transgenic rice accumulated 0.74 μg/g dry weight phytoene in rice seed while other carotenoids were not detected (Table 4).[64] This study catalyzed efforts to engineer the carotenoid biosynthetic pathway in rice endosperm using daffodil *Psy* gene and bacterial (*Erwinia uredevora*) phytoene desaturase *Crt1* gene.[54,65] Transgenic rice seeds were yellow in color due to the accumulation of carotenoids mainly β-carotene and, to some extent, lutein and zeaxanthin. Cotransformation with the vector carrying *psy* and *crt1* genes and that carrying daffodil lycopene β-cyclase gene led to segregating transgenic lines with an accumulation of carotenoid up to 1.6 μg/g.

Several studies have shown that *psy* gene product catalyzes a limiting and regulatory step in carotenoid biosynthesis. Several sources of *Psy* cDNA were used to transform rice in combination with the bacterial *crt1* gene expressed under the endosperm-specific promoter. The maximum carotenoid accumulation was achieved with maize *psy*, increasing the total carotenoid level to 37 μg/g, some 23-fold higher than a previous report on the Golden Rice.[55] Production of Golden Rice 2 is a landmark achievement and defuses the criticism that genetically modified rice is not good enough to fulfill the RDA of provitamin A in a normal daily diet.[65,66]

Bacterial (*Erwinia uredevora*) *crtB* and *crt1* genes were also expressed in white corn under the control of endosperm-specific duplicated zein promoter, which resulted in a color change in the kernels and a level of β-carotene that ranged from 7-10 μg/g dry seed.[67]

Minor Crops

Vegetables and fruits such as tomato and carrots are good sources of carotenoids, however the levels are below the RDA.[68,69] Transformation of canola seed with bacterial phytoene synthase (*crtB*) gene using seed-specific napin promoter resulted in orange colored seed with a 50-fold higher (1000-1500 μg/gFW) total carotenoid level than the wild type (33 μg/gFW) (Table 5).[70] Canola seeds coexpressing *crtI* and *crtB* gene accumulated lycopene (29 μg/gFW) and β-carotene (857 μg/gFW).[71] In a seed-specific manner, bacterial *crtB* gene expression in flaxseeds resulted in 7.8 to18.6-fold increase in the carotenoid levels.[72]

crtB gene was also used in a tuber-specific manner to increase carotenoids, violaxanthin, lutein and β-carotene, in transgenic potato.[73] A number of studies have focused on increasing the levels of β-carotene in tomato (Table 5). Namely, fruit-specific[74,75] or constitutive[76] expression of lycopene β-cyclase and constitutive expression of phytoene desaturase[77] led to high levels of β-carotene in tomato fruit. The bacterial phytoene synthase fused to the tomato polygalacturonase promoter was introduced into tomato and the resulting transformants accumulated significantly higher levels of carotenoids, phytoene, lycopene, β-carotene and lutein.[78]

Transformation of tomato has also involved beneficial genes fused to regulatable promoters[79] or in using suppression RNAi to downregulate photomorphogenesis regulatory protein gene DE-ETIOLATED1 (*DET1*), a negative regulator of light mediated responses,[80] which showed the promise of biotechnology to obtain carotenoid levels high enough to meet the RDA dose. UV-DAMAGED DNA BINDING PROTEIN-1/high pigment-1 (*DDB1/HP1*) gene is a homologous gene of *DET1* that forms a complex with *Cul4* (Cullin 4, a component of Cul4 based E3 ligases, facilitates transfer of ubiquitin to the substrate for subsequent degradation by proteasome).[81,82] RNAi suppression of *CUL4* constitutively results in increased carotenoid accumulation.[83]

Downstream from the β-carotene pathway are important photoprotective compounds whose levels have also been successfully manipulated via genetic intervention. Using the fruit-specific Pds promoter to drive the expression of β-*Lcy* and β-*Chy* genes from Arabidopsis and pepper, respectively, about 100-fold increase in β-cryptoxanthin and zeaxanthin was achieved.[75]

Expression of yeast SAM dcarboxylase (*ySAMdc*) fused to a ripening-specific promoter E8 in tomato resulted in the accumulation of higher polyamines spermidine and spermine. The increase in the therapeutic polyamines in the transgenic lines signaled higher accumulation of other nutraceuticals such as lycopene, choline and various other health promoting phytonutrients.[2,84,85]

The cauliflower *Or* gene, *Orange* gene mutation responsible for β-carotene accumulation in cauliflower, when introduced in a tuber-specific manner into potato led to a 6-fold higher level of total carotenoids in transgenic versus the controls.[86,87] Ketocarotenoids are rare in plants but are strong antioxidants. They are chemically synthesized and used as dietary supplements and pigments in aquaculture and nutraceutical industry. Ketocarotenoid biosynthesis pathway was engineered in carrot by introducing an algal *Haematococcus pluvialis* β-carotene ketolase gene under the control of constitutive, ubiquitin or rolB promoter.[88] This resulted in 70% conversion of total β-carotene to ketocarotenoid that accumulated to a level of 2,400 μg/g root dry weight. These transgenic carrots have good potential for biopharming ketocarotenoid production for functional food, nutraceutical and aquaculture industries.

Folates

Major Crops

Plants are the main source of folates for human health. Many diseases are associated with or as a consequence of folate deficiency including neural tube defects, coronary and cardiovascular disease, Alzheimer's and a whole range of cancers and leukemia.[89]

Genetic enhancement of folates in plants is therefore seen as a major contribution to alleviate diseases associated with folate deficiency. Transgenic rice plants were generated with a construct consisting of Arabidopsis GTP cyclohydrolase I (GTPCH I) and aminodeoxychorismate synthase (ADCS) on a single T-DNA cassette under the control of rice endosperm-specific globulin (glb-1) and glutelin B1 (gluB1) promoters, respectively, to increase the folate content of rice seeds. The resulting transgenic seeds accumulated folates ranging from 6.0-38.3 nmol/g, respectively 15-100 times higher than the wild type. Interestingly, 89% of total folates in the transgenic rice was in the form of 5-methyltetrahydrate, which makes biofortified folates superior to folic acid-fortified rice.[90]

In the corn study, expression of *E. coli folE* gene encoding GTP cyclohydrolase (GCH1) under the control of the barley D-hordein promoter produced a 2-fold increase in folate content.[56]

Minor Crops

Folate enhancement in tomato was tested via engineering the GTP cyclohydrolase I (*gchI*) in pteridine, p-aminobenzoate (PABA) pathway.[91] Their strategy employed synthetic gene on the basis of feedback-insensitive mammalian *gchI* and the fruit-specific E8 promoter to drive the expression. The transgenics accumulated twice as much folate as the nontransgenic control. The same laboratory also employed Arabidopsis aminodeoxychorismate synthase *ADCS* gene,[92] the first step in PABA biosynthesis to transform tomato, to increase further the folate levels. Upon crossing the gchI and ADCS lines to pyramid the two genes, the progeny obtained was found to accumulate 25-fold more folate than the controls. Such a remarkable enhancement in tomato folate level is at par with the RDA for adults.

Synthetic codon-optimized *gchI* based on native (*Gallus gallus*) chicken gene was introduced into lettuce to enhance folate levels. The transgenic lettuce lines thus generated had significantly increased folate content, which ranged from 2.1 to 8.5-fold higher than the nontransgenic lines.[93] The high-folate enriched lettuce would be sufficient to supply 26% of the RDA for folate, provided its bioavailability and bioactivity are optimal and equivalent.

Vitamin C (Ascorbate)

Major Crops

Dehydroascorbate reductase (DHAR) regenerates ascorbic acid (AsA) from its oxidized form and has been used to increase the level of AsA. Upon 100-fold increase in the expression of wheat *dhar* cDNA under the control of ubiquitin (ub) or Shrunken (Sh2) promoter in corn, 2 to 4-fold increase in AsA was quantified in kernels.[94] Expression of rice *dhar* (under the control of barley D-hordein promoter) in corn resulted in 6-fold increase in AsA (ascorbate).[56]

Minor Crops

Ascorbic acid levels were engineered in lettuce by constitutive expression of the rat AsA biosynthetic pathway gene, GulL oxidase; transgenic lettuce accumulated 4 to 7-fold higher AsA than the control plants.[95]

Polyphenolics

Major/Minor Crops

Polyphenols are also potent antioxidants and their potential as preventive for disease incidence in humans has catalyzed a number of studies to overaccumulate specific phenolics in crop plants. Most studies have employed key genes to transform vegetables and fruits. Constitutive over-expression of petunia chalcone isomerase *chi-a* gene, involved in flavonol biosynthesis, resulted in a 78-fold increase in flavonol content of tomato peel.[95] Chlorogenic acid and other flavonoids were also increased several-fold via constitutive expression of hydroxycinnamoyl transferase *HQT* gene.[97]

Plants are poor in flavones. However, genetic engineering has been able to produce novel polyphenolics/flavonoids in high concentrations in transgenic tomato fruit transformed with various, heterologous genes of flavonoid biosynthesis pathway.[98] Soybean isoflavone synthase (*IFS*) gene, whose protein product is a catalyst for the formation of genistein, was constitutively expressed in tomato. This resulted in the accumulation of isoflavone in leaves and fruit peels of transgenic tomato.[99] Similarly, introduction of grape stilbene synthase (*StSY*) gene in tomato generated resveratrol, which is not naturally present in tomato.[100] Transgenic fruit produced up to 53 μg/gFW of resveratrol which was accompanied with additional increases in the levels of ascorbate and glutathione. Trans-resveratrol, trans-piceid and cinnamic acid derivatives of flavonols and flavonones were identified by HPLC in the resveratrol-producing tomato transgenics.[100] When the same *StSy* gene was put under the control of its own wound, pathogen and UV inducible promoters and introduced into apple, the resultant transgenic apple synthesized phytoalexin resveratrol glucoside, piceid.[102]

In a different approach, constitutively expressed anti-sense strawberry chalcone synthase gene (*CHS*) led to the depletion of anthocyanin, flavonol and proanthocyanidines while phenylpropanoid pathway was upregulated.[103] Constitutive expression in potato of petunia dihydroflavonol reductase gene (*DFR*), which encodes for a key enzyme in flavonoid biosynthesis, significantly enhanced anthocyanin and phenolic levels.[104]

Delila (*Del* encodes a basic helix-loop-helix transcription factor) and *Rosea1* (*Ros 1*-MYB-related transcription factor) genes from the snapdragon were fused to the ripening-specific E8 promoter and introduced into tomato.[105] Del and Ros1 are known to activate a broad spectrum of phenylpropanoid/flavonoid pathway genes in tomato. This approach resulted in high induction of anthocyanin accumulation in transgenic tomatoes.

Tocopherols

Major/Minor Crops

Tocopherols show anticarcinogenic properties and have the potential as adjuvant chemotherapeutics.[106] Several studies on enhancing the content of α-tocopherol, one of the most bioactive forms of tocopherol, have been reported. Over-expression of the *At-VTE4* (γ-*TMT*), which catalyzes the final step in the biosynthesis of α-tocopherol, in combination with *At VTE3* (2-methyl-6-phytylbenzoquinol methyltransferase) in soybean seed led to a 8-fold increase in α-tocopherol.[107] Subsequently, Arabidopsis γ-*TMT* was constitutively expressed in various crops, namely, soybean, oil seed and lettuce and in each case a higher level of α-tocopherol was obtained compared to the nontransgenic counterparts.[108-110] Similarly, *Perilla frutescens* γ-*TMT* introduced seed-specifically in soybean also produced higher levels of α-tocopherol and β-tocopherol.[111] In potato, tocopherol biosynthesis was manipulated by constitutively expressing *Arabidopsis* p-hydroxyphenylpyruvate dioxygenase (*At-HPPD*) or homogentisate phytyltransferase (*At-HPT*) genes, which led to respective accumulation of tuber α-tocopherol at 1030 ng/gFW and 579 ng/gFW compared to 281.5 ng/gFW in the controls.[112]

Iron

Major Crops

Rice endosperm is poor in iron content (7-24 mg/kg).[113] In an effort to improve the iron content of rice, soybean ferritin *pfe* gene was expressed under the control of seed storage protein promoter, glutelin GluB-1.[114] The iron content of transgenic rice seeds was three times (38.1 μg/gDW) as much as the wild type (14.3 μg/gDW) plants.

Phaseolus vulgaris ferritin *pfe* gene expressed under the control of Gt1 promoter also increased iron accumulation, varying from 11.53-22.07 μg/g rice seed, which is twice as much as in the wild type.[115] Following milling, wild type rice seeds contain low amounts of iron, 0.2-2.8 mg/100 g.[116] Expression of the soybean ferritin gene under the control of an endosperm-specific GluB-1 promoter led to higher amounts of iron and zinc in transgenic rice, each g of unpolished grain contained 71 μg of iron and 55.5 μg of zinc.[117]

Conclusion

Phytonutrients are now recognized as important determinants of human health. This has catalyzed investigations into broader aspects of plant-based nutraceuticals. These include: elucidation of biochemical pathways and identifying the rate-limiting steps; engineering metabolic pathways to direct the intermediary metabolism flux towards a particular nutrient (nutraceutical); testing efficacy of either an isolated and purified nutraceutical or a crop engineered with an enhanced nutrient level in animal and human models; testing crops silenced for health-detrimental factors including allergens; comparing bioavailability of an individual nutrient (nutraceutical) fed either as a food supplement or in the form of a fortified food. A common goal of such studies is to enable dietary intervention in human health to combat monogenic or polygenic diseases.

The molecular genetics and modern biotechnology approaches in conjunction with deciphering the metabolome of a crop plant are powerful tools that will help in specific redesigning of metabolism in food crops to accumulate desired, or close-to-the-desired, levels of a particular phytonutrient. Such studies will also help in driving higher production levels of nutraceuticals in heterologous systems such as bacteria, algae and yeast. Crops can also be redesigned to be free of disease-causing food constituents. This should allow a neat treatment for patients suffering from monogenic, diet-related diseases, to name a few, CD, galactosemia, CMPA, glycogen storage disorder and phenylketonuria.

Progress in merging agriculture with preventive medicine will depend on intense collaborative research between physicians, nutritionists and plant biologists so that planning strategies are rationally designed and the developments in genetic technology successfully applied for better crop engineering. Engineering metabolism of plants can have a major impact of global proportions, in redesigning pathways to provide healthier and functional foods, improve the quality of raw materials and plant production factories to produce novel pharmaceuticals.

References

1. Mutch DM, Wahli W, Williamson G. Nutrigenomics and nutrigenetics: the emerging faces of nutrition. The FASEB J 2005; 19:1602-1616.
2. Mattoo AK, Yachha SK, Fatima T. Genetic manipulation of vegetable crops to alleviate diet-related diseases. In: F.T. Barberan, F. and M.I. Gil, eds. Improving the health-promoting properties of fruit and vegetable products, Woodhead Publ. Ltd., Cambridge, 2008; 327-345.
3. Simopoulos AP. Omega-6/omega-3 essential fatty acid ratio and chronic diseases. Food Rev Int 2004; 20:77-90.
4. Jimenez-Sanchez G, Childs B, Valle D. Human disease genes. Nature 2001; 409:853-855.
5. Schrander JJ, van den Bogart JP, Forget PP et al. Cow's milk protein intolerance in infants under 1 year of age: a prospective epidemiological study. Eur J Pediatr 1993; 152:640-644.
6. Catassi C. Celiac disease in Europe. In: Catassi C, Fasano A, Corazza GR, eds. The Global Village of Coeliac Disease, Italian Coeliac Society, AIC press, 2005; 71-85.
7. Yachha SK. Celiac disease: India on the global map. J Gastroenterol Hepato 2005; 221:1511-1513.
8. Fjeld CR, Lawson RH. Food, phytonutrients and health. Nutr Revs 1999; 57, ILSI, Washington, DC.
9. Basu A, Imrhan V. Tomatoes versus lycopene in oxidative stress and carcinogenesis: conclusions from clinical trials. Eur J Clin Nutr 2007; 61:295-303.
10. Block G, Patterson B, Subar A. Fruit, vegetables and cancer prevention: a review of the epidemiological evidence. Nutr Cancer 1992; 18:1-29.
11. Wang X, Tomso DJ, Chorley BN et al. Identification of polymorphic antioxidant response elements in the human genome. Human Mol Genet 2007; 16:1188-1200.
12. Milner JA. Nutrition and cancer: essential elements for a roadmap. Cancer Lett 2008; 269:189-198.
13. Yachha SK, Poddar U. Celiac disease in Asia. In: Catassi C, Fasano A, Corazza GR, eds. The Global Village of Coeliac Disease, Italian Coeliac Society, AIC press, 2005; 101-108.
14. Bosch AM. Classical galactosemia revisited. J Inherit Metab Dis 2006; 29:516-25.
15. Kemp AS, Hill DJ, Allen KJ et al. Guidelines for the use of infant formulas to treat cows milk protein allergy: an Australian consensus panel opinion. Med J Aust 2008; 188:109-112.
16. Yachha SK, Misra S, Malik A et al. Spectrum of malabsorption syndrome in north Indian children. Indian J Gastroenterol 1993; 12:120-125.
17. Pedrosa M, Pascual CY, Larco JI et al. Palatability of hydrolysates and other substitution formulas for cow's milk-allergic children: A comparative study of taste, smell and texture evaluated by healthy volunteers. J Investig Allergol Clin Immunol 2006; 16:351-356.
18. Chen Y-T. Glycogen storage diseases and other inherited disorders of carbohydrate metabolism. In: Braunwald E, Fauci A, Kasper DL et al, eds. Harrison's Principles of Internal Medicine, New Delhi: McGraw-Hill, 2001; 2:2281-2288.
19. Shin YS. Glycogen storage disease: clinical, biochemical and molecular heterogeneity. Semin Pediatr Neurol 2006; 13:115-120.
20. Levy HL. Phenylketonuria: Old disease, new approach to treatment. Proc Natl Acad Sci USA 1999; 96:1811-1813.
21. Gillies PJ. Nutrigenomics: the rubicon of molecular nutrition. J Am Diet Assoc 2003; 103:S50-S55.
22. Heldt H-W. Plant Biochemistry, Elsevier New York: Academic Press, 2005; 299-302.
23. Mattoo AK, Sobolev AP, Neelam A et al. NMR spectroscopy-based metabolite profiling of transgenic tomato fruit engineered to accumulate spermidine and spermine reveals enhanced anabolic and nitrogen-carbon interactions. Plant Physiol 2006; 142:1759-1770.
24. Segal E, Friedman N, Koller D et al. A module map showing conditional activity of expression modules in cancer. Nat Genet 2004; 36:1090-1098.
25. Bhardwaj P, Garg PK, Maulik SK et al. A randomized controlled trial of antioxidant supplementation for pain relief in patients with chronic pancreatitis. Gastroenterol 2009; 136:149-159.

26. Arab L. Individualized nutritional recommendations: do we have the measurements needed to assess risk and make dietary recommendations? Proc Nutr Soc 2004; 63:167-172.
27. Liu YS, Roof S, Ye ZB et al. Manipulation of light signal transduction as a means of modifying fruit nutritional quality in tomato. Proc Natl Acad Sci USA 2004; 101:9897-9902.
28. de Kok TM, Van Breda SG, Manson MM. Mechanism of combined action of different hemopreventive dietary compounds. Eur J Nutr 2008; (47):51-59.
29. Wang SY, Lin HS. Antioxidant activity in fruits and leaves of blackberry, raspberry and strawberry varies with cultivar and developmental stages. J Agric Food Chem 2000; 48:140-146.
30. Wolfe K, Wu X, Liu RH. Antioxidant activity of apple peels. J Agric Food Chem 2003; 51:609-614.
31. Scalzo J, Politi A, Pellegrini N et al. Plant genotype affects total antioxidant capacity and phenolic contents in fruit. Nutrition 2005; 21:207-213.
32. Tsao R, Yang R, Xie S et al. Which polyphenolic compounds contribute to the total antioxidant activities of apple? J Agric Food Chem 2005; 53:4989-4995.
33. George B, Kaur C, Khurdiya DS et al. Antioxidants in tomato (Lycopersicum esculentum) as a function of genotype. Food Chem 2004; 84:45-51.
34. Materska M, Perucka I. Antioxidant activity of the main phenolic compounds isolated from hot pepper fruit (Capsicum annum L.). J Agric Food Chem 2005; 53:1750-1756.
35. Rimando AM, Kalt, W, Magee JB, Dewey J et al. Resveratrol, pterostilbene and piceatannol in Vaccinium berries. J Agric Food Chem 2004; 52:4713-4719.
36. Seeram N, Adams L, Henning S et al. In vitro antiproliferative, apoptotic and antioxidant activities of punicalagin, ellagic acid and a total pomegranate tannin extract are enhanced in combination with other polyphenols as found in pomegranate juice. J Nutri Biochem 2005; 16:360-367.
37. Surh YJ, Hurh YJ, Kang JY et al. Resveratrol, an antioxidant present in red wine, induces apoptosis in human promyelocytic leukemia (HL-60) cells. Cancer Lett 1999; 140:1-10.
38. Sun J, Chu YF, Wu X et al. Antioxidant and antiproliferative activities of common fruits. J Agric Food Chem 2002; 50:7449-7454.
39. Meyers KJ, Watkins CB, Pritts MP et al. Antioxidant and antiproliferative activities of strawberries. J Agric Food Chem 2003; 51:6887-6892.
40. Liu M, Li XQ, Weber C et al. Antioxidant and antiproliferative activities of raspberries. J Agric Food Chem 2002; 50:2926-2930.
41. Liu YS, Roof S, Ye ZB et al. Manipulation of light signal transduction as a means of modifying fruit nutritional quality in tomato. Proc Natl Acad Sci USA 2004; 101:9897-9902.
42. Campbell JK, Stroud CK, Nakamura MT et al. Serum testosterone is reduced following short-term phytofluene, lycopene or tomato powder consumption in F334 rats. J Nutr 2006; 136:2813-2819.
43. Joseph JA, Fisher DR, Bielinski D. Blueberry extract alters oxidative stress-mediated signaling in COS-7 cells transfected with selectively vulnerable muscarinic receptor subtypes. J Alzheimer Dis 2006; 9:35-42.
44. Visioli F, Bellomo G, Montedoro GF et al. Low density lipoprotein oxidation is inhibited in vitro by olive oil constituents. Atherosclerosis 1995; 117:25-32.
45. Covas MI, Fitó M, Lamuela-Raventós RM et al. Virgin oil phenolic compounds: binding to human low density lipoprotein (LDL) and effect on LDL oxidation. Intl J Clin Pharmacol Res 2000; 20:49-54.
46. Lamuela-Raventós RM, Covas MI, Fitó M et al. Detection of dietary antioxidant phenolic compounds in human LDL. Clin Chem 1999; 45:1870-1872.
47. Steinberg D, Parthasarathy S, Carew STE et al. Beyond cholesterol. Modifications of low-density lipoprotein that increase its atherogenicity. New England J Med 1989; 320:915-924.
48. Stein JH, Keevil JG, Wiebe DA et al. Purple grape juice improves endothelial function and reduces the susceptibility of LDL cholesterol to oxidation in patients with coronary artery disease. Circulation 1999; 100:1050-1055.
49. Vinson JA, Yang J, Proch J et al. Grape juice, but not orange juice has in vitro, ex vivo and in vivo antioxidant properties. J Med Food 2000; 3:167-171.
50. Fuhrman B, Lavy A, Aviram M. Consumption of red wine with meals reduces the susceptibility of human plasma and low-density lipoprotein to lipid peroxidation. Am J Clin Nutr 1995; 61:549-554.
51. Jilal I, Fuller CJ, Huet BA. The effect of alpha tocopherol supplementation on LDL oxidation. A dose response study. Areteriosclerosis Thrombosis Vas Biol 1995; 15:190-198.
52. Rao AV, Agarwal S. Bioavailability and in vivo antioxidant properties of lycopene from tomato products and their possible role in the prevention of cancer. Nutr Cancer 1998; 31:199-203.
53. Ye X, Al-Babili S, Kloti A et al. Engineering the provitamin A (β-carotene) biosynthetic pathway into (carotenoid-free) rice endosperm. Science 2000; 287:303-305.
54. Paine JA, Shipton CA, Chaggar S et al. Improving the nutritional value of golden rice through increased pro-vitamin A content. Nat Biotech 2005; 23:482-487.

55. Naqvi S, Zhu C, Farre G et al. Transgenic multivitamin corn through biofortification of endosperm with three vitamins representing three distinct metabolic pathways. Proc Natl Acad Sci USA 2009; 106:7762-7767.
56. Iida S, Amano E, Nishio T. A rice (Oryza sativa L.) mutant having a low content of glutelin and a high content of prolamine. Theor Appl Genet 1993; 87:374-378.
57. Kusaba M, Miyahara K, Iida S et al. Low glutelin content1: A dominant mutation that suppresses the glutelin multigene family via RNA silencing in rice. Plant Cell 2003; 15:1455-1467.
58. Wu XR, Chen ZH, Folk WR. Enrichment of cereal protein lysine content by altered $tRNA^{lys}$ coding during protein synthesis. Plant Biotechnol J 2003; 1:187-194.
59. Houmard NM, Mainville JL, Bonin CB et al. High-lysine corn generated by endosperm specific suppression of lysine catabolism using RNAi. Plant Biotechnol J 2007; 5:605-614.
60. Frizzi A, Huang S, Gilbertson LA et al. Modifying lysine biosynthesis and catabolism in corn with a single bifunctional expression/silencing transgene cassette. Plant Biotechnol J 2008; 6:13-21.
61. Chakraborty S, Chakraborty N, Datta A. Increased nutritive value of transgenic potato by expressing a nonallergenic seed albumin gene from Amaranthus hypochondriacus. Proc Natl Acad Sci USA 2000; 97:3724-2729.
62. Avraham T, Badani H, Galili S et al. Enhanced levels of methionine and cysteine in transgenic alfalfa (Medicago sativa L.) plants over-expressing the Arabidopsis cystathionine γ-synthase gene. Plant Biotechnology J 2005; 3:71-79.
63. Burkhardt PK, Beyer P, Wünn J et al. Transgenic rice (Oryza sativa) endosperm expressing daffodil (Narcissus pseudonarcissus) phytoene synthase accumulates phytoene, a key intermediate of provitamin A biosynthesis. Plant J 1997; 11:1071-1078.
64. Beyer P, Al-Babili S, Ye X, Lucca P et al. Golden rice: Introducing the β-carotene biosynthesis pathway into rice endosperm by genetic engineering to defeat vitamin A deficiency. J Nutr 2002; 132:506S-510S.
65. Dawe D, Robertson R, Unnevehr L. Golden rice: what role could it play in alleviation of vitamin A deficiency? Food Policy 2002; 27:541-560.
66. Aluru M, Xu Y, Guo R et al. Generation of transgenic maize with enhanced provitamin A content. J Exp Bot 2008; 59:3551-3562.
67. Edwards RS, Reuter FH. Pigment changes during the maturation of tomato fruit. Food Technol Australia 1967; 19:352-357.
68. Johjima T, Matsuzoe N. Relationship between color value (a/b) and colored carotene content in fruit of various tomato cultivars and breeding line. Acta Hort 1995; 412:152-159.
69. Shewmaker CK, Sheehy JA, Daley M et al. Seed-specific-expression of phytoene synthase: increase in carotenoids and other metabolic effects. Plant J 1999; 20:410-412.
70. Ravanello MP, Ke D, Alvarez J et al. Coordinated expression of multiple bacterial carotenoid genes in canola leading to altered carotenoid production. Metabolic Engg 2003; 5:255-263.
71. Fujisawa M, Watanabe M, Choi SK et al. Enrichment of carotenoids in flaxseeds by metabolic engineering with introduction of bacterial phytoene synthase gene crtB. J Biosci Bioeng 2008; 105:636-641.
72. Ducreux LJ, Morris WL, Hedley PE et al. Metabolic engineering of high carotenoid potato tubers containing enhanced levels of β-carotene and lutein. J Exp Bot 2005; 56:81-89.
73. Rosati C, Aquilani R, Dharmapuri S et al. Metabolic engineering of beta-carotene and lycopene content in tomato fruit. Plant J 2000; 24:413-419.
74. Dharamapuri S, Rosati C, Pallara P et al. Metabolic engineering of xanthophyll content in tomato fruits. FEBS Lett 2002; 519:30-34.
75. D'Ambrosio C, Giorio G, Marino I et al. Virtually complete conversion of lycopene into β-carotene in fruits of tomato plants transformed with the tomato lycopene β-cyclase (tcy-b) cDNA. Plant Sci 2004; 166:207-214.
76. Römer S, Fraser PD, Kiano JW et al. Elevation of the provitamin A content of transgenic tomato plants. Nat Biotechnol 2000; 18:666-669.
77. Fraser PD, Romer S, Shipton CA et al. Evaluation of transgenic tomato plants expressing an additional phytoene synthase in a fruit-specific manner. Proc Natl Acad Sci USA 2002; 99:1092-1097.
78. Enfissi EMA, Fraser PD, Lois LM et al. Metabolic engineering of the mevalonate and nonmevalonate isopentenyl diphosphate-forming pathways for the production of health-promoting isoprenoid in tomato. Plant Biotechnol J 2005; 3:17-27.
79. Davuluri GR, Van Tuinen A, Fraser PD et al. Fruit-specific RNAi-mediated suppression of DET1 enhances carotenoid and flavonoid content in tomatoes. Nat Biotechnol 2005; 23:890-895.
80. Bernhardt A, Lechner E, Hano P et al. CUL4 associates with DDB1 and DET1 and its downregulation affects diverse aspects of development in Arabidopsis thaliana. Plant J 2006; 47:591-603.
81. Chen H, Shen Y, Tan X et al. Arabidopsis CULLIN4 forms an E3 ubiquitin ligase with RBX1 and the CDD complex in mediating light control of development. Plant Cell 2006; 18:1991-2004.

82. Wang S, Liu J, Feng Y et al. Altered plastid levels and potential for improved fruit nutrient content by downregulation of the tomato DDB1-interacting protein CUL4. Plant J 2008; 55:89-103.
83. Mehta RA, Cassol T, Li N et al. Engineered polyamine accumulation in tomato enhances phytonutrient content, juice quality and vine life. Nat Biotechnol 2002; 20:613-618.
84. Mattoo AK, Chung SH, Goyal RK et al. Overaccumulation of higher polyamines in ripening transgenic tomato revives metabolic memory, upregulates anabolism related genes and positively impacts nutritional quality. JOAC International 2008; 90:456-464.
85. Lu S, Van Eck J, Zhou X et al. The cauliflower Or gene encodes a DnaJ cysteine-rich domain-containing protein that mediates high levels of β-carotene accumulation. Plant Cell 2006; 18:3594-3605.
86. Lopez AB, Van Eck J, Conlin BJ et al. Effect of the cauliflower OR transgene on carotenoid accumulation and chromoplast formation in transgenic potato tubers. J Exp Bot 2008; 59:213-223.
87. Jayraj J, Devlin R, Punja Z. Metabolic engineering of novel ketocarotenoid production in carrot plants. Transgenic Res 2008; 17:489-501.
88. Storozhenko S, Ravanel S, Zhang G-F et al. Folate enhancement in staple crops by metabolic engineering. TRENDS Food Sci Technol 2005; 16:271-281.
89. Storozhenko S, De Brouwer V, Volckarrt M et al. Folate fortification of rice by metabolic engineering. Nat Biotechnol 2007; 25:1277-1279.
90. Díaz de la Garza R, Quinlivan EP, Klaus SM et al. Folate biofortification in tomatoes by engineering the pteridine branch of folate synthesis. Proc Natl Acad Sci USA 2004; 101:13720-13725.
91. Díaz de la Garza RI, Gregory JF III, Hanson AD. Folate biofortification of tomato fruit. Proc Natl Acad Sci USA 2007; 104:4218-4222.
92. Nunes ACS, Kalkmann DC, Aragão FJL. Folate biofortification of lettuce by expression of a codon optimized chicken GTP cyclohydrolase I gene. Transgenic Res 2009; 18:661-667. DOI 10.1007/s11248-009-9256-1.
93. Chen Z, Young TE, Ling J et al. Increasing vitamin C content of plants through enhanced ascorbate recycling. Proc Natl Acad Sci USA 2003; 100:3525-3530.
94. Jain AK, Nessler CL. Metabolic engineering of an alternative pathway for ascorbic acid biosynthesis in plants. Mol Breeding 2000; 6:73-78.
95. Muir SR, Collins GJ, Robinson S et al. Overexpression of petunia chalcone isomerase in tomato results in fruit containing increased levels of flavonols. Nat Biotechnol 2001; 19:470-474.
96. Niggeweg R, Michael AJ, Martin C. Engineering plants with increased levels of the antioxidant chlorogenic acid. Nat Biotechnol 2004; 22:746-754.
97. Schijlen E, Ric de Vos CH, Jonker H et al. Pathway engineering for healthy phytochemicals leading to the production of novel flavonoids in tomato fruit. Plant Biotechnol J 2006; 4:433-444.
98. Shih CH, Chen Y, Wang M et al. Accumulation of isoflavone genistin in transgenic tomato plants overexpressing a soybean isoflavone synthase gene. J Agric Food Chem 2008; 56:5666-5661.
99. Giovinazzo G, D'Amico L, Paradiso A et al. Antioxidant metabolite profiles in tomato fruit constitutively expressing the grapevine stilbene synthase gene. Plant Biotechnol J 2005; 3:57-69.
100. Nicoletti I, De Rossi A, Giovinazzo G et al. Identification and quantification of stilbenes in fruits of transgenic tomato plants (Lycopersicon esculentum Mill.) by reversed phase HPLC with photodiode array and mass spectrometry detection. J Agric Food Chem 2007; 55:3304-3311.
101. Rühman S, Treutter D, Fritsche S et al. Piceid (resveratrol glucoside) synthesis in stilbene synthase transgenic apple fruit. J Agric Food Chem 2006; 54:4633-4640.
102. Lunkenbein S, Coiner H, Ric de Vos CH et al. Molecular characterization of a stable antisense chalcone synthase phenotype in strawberry (Fragaria x ananassa). J Agric Food Chem 2006; 54:2145-2153.
103. Lukaszewicz M, Matysiak-Kata I, Skala J et al. Antioxidant capacity manipulation in transgenic potato tuber by changes in phenolic content. J Agric Food Chem 2004; 52:1526-1533.
104. Butelli E, Titta L, Giorgio M et al. Enrichment of tomato fruit with health promoting anthocyanins by expression of select transcription factors. Nat Biotechnol 2008; 26:1301-1308.
105. Constantinou C, Papas A, Constantinou AI. Vitamin E and cancer: An insight into the anticancer activities of vitamin E isomers and analogs. Intl J Cancer 2008; 123:739-752.
106. Van Eenennaam AL, Lincoln K, Durrett TP et al. Engineering vitamin E content: from Arabidopsis mutant to soy oil. Plant Cell 2003; 15:3007-3019.
107. Cho EA, Chong AL, Kim YS et al. Expression of γ-tocopherol methyltransferase transgene improves tocopherol composition in lettuce (Lactuca sativa L.). Mol Cells 2005; 19:16-22.
108. Kim YJ, Seo HY, Park TI et al. Enhanced biosynthesis of α-tocopherol in transgenic soybean by introducing γ-TMT gene. J Plant Biotechnol 2005; 7:1-7.
109. Yusuf MA, Sarin N-B. Antioxidant value addition in human diets: genetic transformation of Brassica juncea with γ-TMT gene for increased α-tocopherol content. Transgenic Res 2007; 16:109-113.
110. Tavva VS, Kim YH, Kagan IA et al. Increased α-tocopherol content in soybean seed overexpressing the Perilla frutescens γ-tocopherol methyltransferase gene. Plant Cell Rep 2007; 26:61-70.

111. Crowell EF, McGrath JM, Douches DS. Accumulation of vitamin E in potato (Solanum tuberosum) tubers. Transgenic Res 2008; 17:205-217.
112. Graham RD, Senandhira D, Beebe S et al. Breeding for micronutrient density in edible portions of staple food crops: conventional approaches. Field Crop Res 1999; 60:57-80.
113. Goto F, Yoshihara T, Shigemoto N et al. Iron fortification of rice seed by the soybean ferritin gene. Nat Biotechnol 1999; 17:282-286.
114. Lucca P, Hurrel R, Potrykus I. Genetic engineering approaches to improve the bioavailability and the level of iron in rice grains. Theor Appl Gnet 2001; 102:392-397.
115. Lucca P, Hurrel R, Potrykus I. Fighting iron deficiency anemia with iron-rich rice. J Am College Nutr 2002; 21:184S-190S.
116. Vasconcelos M, Datta K, Oliva N et al. Enhanced iron and zinc accumulation in transgenic rice with the ferritin gene. Plant Sci 2003; 164:371-378.
117. Yu J, Peng P, Zhang X et al. Seed-specific expression of a lysine rich protein sb401 gene significantly increases both lysine and total protein content in maize seeds. Mol Breeding 2004; 14:1-7.
118. Bicar EH, Woodman-Clikeman W, Sangtong V et al. Transgenic maize endosperm containing a milk protein has improved amino acid balance. Transgenic Res 2008; 17:59-71.
119. Yu B, Lydiate DJ, Young LW et al. Enhancing the carotenoid content of Brassica napus seeds by down-regulating lycopene epsilon cyclase. Transgenic Res 2008; 17:573-585.
120. Wei S, Li X, Gruber MY et al. RNAi mediated suppression of DET1 alters the levels of carotenoids and sinapate esters in seeds of Brassic napus. J Agric Food Chem 2009; 57:5326-5333doi:10.1021/jf803983w.
121. Giliberto L, Perrotta G, Pallara P et al. Manipulation of the blue light photoreceptor cryptochrome 2 in tomato affects vegetative development, flowering time and fruit antioxidant content. Plant Physiol 2005; 137:199-208.
122. Wurbs D, Ruf S, Bock R. Contained metabolic engineering in tomatoes by expression of carotenoid biosynthesis genes from the plastid genome. Plant J 2007; 49:276-288.
123. Diretto G, Al-Babili S, Tavazza R et al. Metabolic engineering of potato carotenoid content through tuber-specific overexpression of a bacterial mini-pathway. PLoS ONE 2007; 2:e350. doi:10.1371/journal.pone.0000350.
124. Diretto G, Tavazza R, Welsch R et al. Metabolic engineering of potato tuber carotenoids through tuber specific silencing of lycopene epsilon cyclase. BMC Plant Biol 2006; 6:13.doi: 10.1186/1471-229-6-13
125. Fraser PD, Enfissi EMA, Halket JM et al. Manipulation of phytoene levels in tomato fruits: effects on isoprenoids, plastids and intermediary metabolism. Plant Cell 2007; 19:3194-3211.
126. Morris WL, Ducreux LJM, Hedden P et al. Overexpression of a bacterial 1-deoxy-D-xylulose-5-phosphate synthase gene in potato tubers perturbs the isoprenoid metabolic network: implications for the control of the tuber life cycle. J Exp Bot 2006; 57:3007-3018.
127. Gerjets T, Sandmann G. Ketocarotenoid formation in transgenic potato. J Exp Bot 2006; 57:3639-3645.
128. Goto F, Yoshihara T, Saiki H. Iron accumulation and enhanced growth in transgenic lettuce plants expressing the iron-binding protein ferritin. Theor Appl Genet 2000; 100:658-664.
129. Luo J, Butelli E, Hill L et al. AtMBY12 regulates caffeoyl quinic acid and flavonol synthesis in tomato: expression in fruit results in very high levels of both types of polyphenol. Plant J 2008; 56:316-326.
130. Wakita Y, Otani M, Hamada T et al. A tobacco microsomal ω-3 fatty acid desaturase gene increases the linolenic acid content in transgenic sweet potato (Ipomoea batatas). Plant Cell Rep 2001; 20:244-249.

Chapter 11

Perspective for the Use of Genetic Transformants in Order to Enhance the Synthesis of the Desired Metabolites:

Engineering Chloroplasts of Microalgae for the Production of Bioactive Compounds

Udo Johanningmeier* and Dirk Fischer

Abstract

Eukaryotic microalgae have recently gained particular interest as bioreactors because they provide attractive alternatives to bacterial, yeast, plant and other cell-based systems currently in use. Over the last years there has been considerable progress in genetic engineering technologies for algae. Biotechnology companies start to apply these techniques to alter metabolic pathways and express valuable compounds in different cell compartments. In particular, the eukaryotic unicellular alga *Chlamydomonas reinhardtii* appears to be a most promising cell factory since high amounts of foreign proteins have been expressed in its chloroplast compartment. For this alga the complete nuclear, plastidal and mitochondrial genome sequences have been determined and databases are available for any searching or cloning requirements. Apart from being easily transformable, stable transgenic strains and production volumes in full containment can be obtained within a relatively short time. Furthermore, *C. reinhardtii* is a green alga which belongs to the category of organisms generally recognized as safe (GRAS status). Thus, enhancing food with edible algae like *Chlamydomonas* engineered to (over)produce functional ingredients has the potential to become an important factor in food and feed technologies.

Introduction

Genetic engineering of plants and algae by introducing and controlling foreign genes has been developed to an extent that not only allows pest protection or herbicide tolerance but also provides ample opportunities to improve yields and nutritional contents and to exploit these organisms as bioreactors for the production of high-value compounds. Although initially most efforts have been concentrated on the manipulation of higher plant systems, photosynthetic microalgae have recently gained special attention because a variety of molecular tools including new transformation methods and complete genome sequences are now available for constructing recombinant strains.[1] As compared to higher plants, the use of microalgae as green cell factories assures a significantly faster generation of stable transgenic lines and is often coupled with high yields, lower

*Corresponding Author: Udo Johanningmeier—Institut für Pflanzenphysiologie, Martin-Luther Universität Halle-Wittenberg, Weinbergweg 10, D-06120 Halle (Saale), Germany.
Email: johanningmeier@pflanzenphys.uni-halle.de

Bio-Farms for Nutraceuticals: Functional Food and Safety Control by Biosensors
edited by Maria Teresa Giardi, Giuseppina Rea and Bruno Berra.

costs and rapid, controllable growth in simple media. Microalgae are not hosts to major pathogens and recombinant strains can be grown in full containment, excluding the risk of contaminating natural populations. Moreover, methods exists which allow to construct transgenic algae without employing antibiotic resistance genes[2-4] thus minimizing any consumer-based concerns regarding the transfer of marker genes from a food product into the cells of the body.

Microalgae have long been used as food or food additives. They comprise a diverse group of prokaryotic and eukaryotic organisms and represent an almost untouched source of foodstuffs, industrial chemicals, therapeutic compounds and even renewable energy in the form of hydrogen gas. For human nutrition, edible microalgae like *Arthrospira* species ("Spirulina") and *Chlorella* are marketed as tablets, capsules and liquid or added to e.g., noodles, breads and candies to improve their nutritive and health values. Other major commercial strains used as food ingredients are *Dunaliella* and *Aphanizomenon flos-aquae*, whose extracts exhibit health promoting effects.[5] The acceptance of new microalga strains by the European Commission for Food Safety has recently been demonstrated with the marine microalga *Odontella aurita* certified in 2002 as novel food by the French company INNOVALG.

While the exploitation of products derived from natural microalgae has a long history, the generation of transgenic microalgae for biotechnological applications has just started to become an attractive system for expressing foreign proteins or other high-value compounds with e.g., antioxidant, colorant, provitamin or therapeutic properties. However, no microalgae are available on the market as genetically modified organisms so far, partly because of public acceptance issues, but also because progress in genetic engineering of microalgae initially was slow. This sluggish development was largely due to the time needed for the development of new transformation techniques, search for suitable promoters or selectable marker genes and often adjustments of reading frames to an unusual codon usage.

Microalgae as Transgenic Bioreactors

An obvious prerequisite for genetic engineering of plants and algae is the ability to transform these organisms with the gene(s) of interest. In the early 1980s there was no method available for generating transformed plants. However, this situation changed significantly when *Agrobacterium*-mediated and direct, i.e., biolistic or PEG-mediated DNA-transfer techniques were established.[6] While *Agrobacterium*-mediated transformation is now the standard method for nuclear transformation, the biolistic approach using accelerated particles (particle gun) turned out to be the preferential procedure for organell transformation. Sophisticated protocols developed along with these techniques now allow transforming all three different plant genomes residing in the nucleus as well as in plastids and mitochondria. The pioneer organism for which all the above mentioned problems were solved first was the unicellular green alga *Chlamydomonas reinhardtii*. This alga has a long history as a powerful model system for diverse areas of plant and also animal research.[7-9] The importance of this small organism like "a plant" has now been very much increased by the development of an extensive molecular toolkit and a draft of the complete genome sequence (http://genome.jgi-psf.org/Chlre3/Chlre3.home.html).[10] Such comprehensive resources are not yet available for other microalgae.

Genetic Engineering of Plant and Microalgal Chloroplasts

The organellar genomes, specifically those of plastids, are particularly attractive for genetic engineering purposes. It is generally agreed that plastids are the result of an endosymbiotic event between a eukaryotic host cell and an ancestor of the cyanobacteria[11] and thus have prokaryotic features. Plastids come in various forms and functions, the most common of which are the photosynthetically active chloroplasts supplying the organism with energy and carbohydrates. Transcription and translation rates in chloroplasts are generally high in order to produce large amounts of the enzyme ribulose bisphosphate carboxylase and to allow a rapid turnover of electron transfer components like the D1 subunit of photosystem II; this core photosynthetic protein gets damaged by reactive oxygen species and must be resynthesized constantly for ensuring uninterrupted electron flow.[12] Furthermore, chloroplasts are also able to synthesize various plant compounds like amino acids, fatty acids, lipids, plant hormones, nucleotides, vitamins and secondary metabolites. Thus, chloroplasts offer ample opportunities to use this organelle for the expression of foreign proteins and for engineering metabolic pathways.

Not only do chloroplasts naturally produce high amounts of protein, but they are also unipa-rentally inherited and integrate properly flanked foreign genes via homologous recombination into their plastome. As opposed to plant nuclear genomes, these properties provide distinct advantages of chloroplast transgenes with respect to biosafety and epigenetic stability.[13] Genetic engineering of chloroplasts has made particular progress with genes conferring agronomically valuable traits like e.g., resistance to herbicides,[14,15] to fungal and bacterial deseases[16] or to insects.[17,18] Chloroplasts have also been employed to overproduce biopharmaceuticals like somatotropin or human serum albumin, resulting in an overaccumulation of up to 7% and 11% of total soluble protein in tobacco plants.[19,20] An exceptionally high expression level of up to 45% of the total soluble protein has been obtained by expressing the *Bacillus thuringiensis cry* operon in tobacco plastids, resulting in the formation of protein crystals inside the chloroplast.[21] Furthermore, the engineering of new and also complex metabolic pathways has been demonstrated recently.[22-25]

Chloroplast genetic engineering is currently most advanced in higher plants, particularly in tobacco.[13] Also chloroplasts of edible plants like tomato and some other food crops have been stably transformed.[26-32] While higher plants offer several advantages over expression systems in other organisms, there are distinct drawbacks like e.g., the length of time necessary for their generation or concerns about containment of transgenic plants in the environment even in transplastomic lines.[33-35] Here the use of microalgae like *C. reinhardtii* provides interesting alternatives: from vector construction to lab-scale culture volumes of the transgenic strain it takes about 6 weeks (Fig. 1) and cells can be easily contained in photobioreactors.

C. reinhardtii was the first organism for which stable chloroplast transformation was reported.[36] Boynton and coworkers used a new technique employing cell bombardment with DNA-coated tungsten particles[37] and succeeded in the restoration of photosynthetic growth of an *atpB* deletion mutant of *C. reinhardtii*. While early selection methods used cloned chloroplast genes to rescue photosynthetic mutants or chloroplast gene constructs that confer resistance to herbicides[38] or antibiotics,[39] later on transformant selection was mainly based on the expression of bacterial markers like the *aadA* gene conferring spectinomycin and streptomycin resistance[40] or the *aphA-6* marker conferring kanamycin or amikacin resistance.[41] During preparation of this manuscript there were only two reports available describing stable chloroplast transformation in eukaryotic algae other than *Chlamydomonas*: the unicellular red algae *Porhyridium spp.* was transformed using a mutant

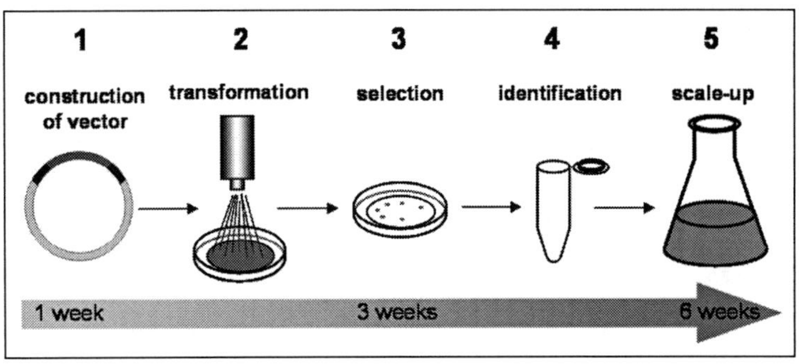

Figure 1. Timeline for the production of foreign proteins in *C. reinhardtii* chloroplasts. 1) Ligation of transgene into a vector containing chloroplast DNA sequences for homologous recombination. 2) Particle gun transformation of vector DNA into cells immobilized on filters. 3) Growth of transformants on selective media. 4) Screening for transgene insertion and protein expression. 5) Cultivation of transformants in larger volumes of liquid media.

form of the gene encoding acetohydroxyacid synthase as a dominant selectable marker[42] and the unicellular protist *Euglena gracilis* using an *aadA* cassette, which persisted in the chloroplast as episomal element without integration into the chloroplast genome.[43]

Expression of Recombinant Proteins in *C. reinhardtii* Chloroplasts

Recent progress in engineering *C. reinhardtii* chloroplasts is the result of 20 years research on developing transformation techniques, improving codon usage and finding efficient promotors and untranslated regions (UTRs) for boosting foreign gene expression. Initially, stable recombinant mRNA accumulation without protein accumulation could be detected in chloroplasts.[44,45] Expression of the *aadA* gene represented the first example of stable, but low foreign protein accumulation as judged only from the appearance of enzymatic activity.[40] Later on the *uidA* and *rluc* genes coding for β-glucuronidase and Renilla luciferase were successfully expressed in *C. reinhardtii* chloroplasts, resulting in correctly folded proteins with the desired reporter properties.[46,47] Although foreign protein expression in these studies was low, the expected products could be detected by Western blot analysis. So far all expressed genes were used as derived from their natural sources without adjusting their codon usage for an optimized expression in *C. reinhardtii* chloroplasts. The importance of codon optimization for enhanced protein production was demonstrated in 2002.[48] In this study it was shown that codon optimization of the *gfp* gene results in a 80-fold increase in green fluorescent protein (GFP) accumulation as compared with the a non-optimized version. Increased protein production could be also observed for other codon-optimized genes, underlining the importance of codon adjustment for high level protein production in plastids.[49,50]

Apart from optimal codon adjustment other factors are known to be crucial for gene expression. The 5′- and 3′-UTR′s are important for mRNA stability[51,52] and transcriptional efficiency is regulated by both chloroplast gene promoters and internal sequences of the 5′-UTR.[53] Containing bacterial like −10 and −35 elements most promoters found in *C. reinhardtii* chloroplast genes resemble bacterial sigma-70-type promoters.[54] More than the transcriptional the translational machinery is limiting chloroplast gene expression.[55] In this context the particular relevance of the 5′-UTR could be demonstrated, whereas the 3′-UTR plays a comparatively smaller role in foreign protein production.[56]

Today several transgenes have been successfully expressed in *C. reinhardtii* chloroplasts (Table 1). Although most of them are intended to be used as reporters and thus serve basic research purposes, now several proteins for pharmaceutical applications appear in the list. A latest and remarkable addition to the record is the expression of a bioactive mammalian protein, whose level is estimated to be above 5% of total cellular protein.[57] In the near future it appears possible that such transgenic Chlamydomonas strains could be ingested orally without the need for extensive purification of the bioactive compound. This perspective is supported by experiments, in which Chlamydomonas cells expressing an epitope of a pathogenic bacterium infecting salmonids were fed to trout. An immune response was observed when transgenic cells were added to the fish food (Patent application US020030022359).[58]

Application to Food Technology

In light of the recent progress in chloroplast genetic engineering described above it seems promising to develop Chlamydomonas-based expression systems to obtain products enriched with proteins or peptides of specific function that could be used e.g., as nutraceutical additives. This may provide a cost-effective means, especially when the whole organism, or a partially processed form of it, can be used as food ingredient without involving costly and complex isolation and purification procedures. As an example, physiologically active peptides derived from plant and animal proteins represent potential health enhancing components for food applications.[70-72] In vitro or in vivo hydrolysis of proteins from these sources and subsequent analyses of peptide fragments for bioactivity show that certain peptide fragments exert a multitude of health effects like antioxidant or antithrombotic activities, cholesterol-lowering abilities or antimicrobial properties, to mention only a few. Bioactive peptides

Table 1. Foreign proteins expressed in* C. reinhardtii *chloroplasts

Year	Proteins	Comments	Refs
1991	Aminoglycoside adenine transferase	Reporter activity, spectinomycin and streptomycin resistance	40
1999	β-Glucuronidase	Reporter activity, conversion of substrates to colored products	46
1999	Renilla luciferase	Reporter activity, luminescence activity	47
2000	Aminoglycoside phosphotransferase	Reporter activity, kanamycin and amikacin resistance	41
2002	Green fluorescent protein	Reporter protein, fluorescent	48
2003	HSV8-lsc	Pharmaceutical activity, first mammalian protein expressed	49
2003	Geranylgeranyl Pyrophosphate Synthase	Prenyltransferase, key enzyme in plant terpenoid biosynthesis	59
2003	Cholera toxin B subunit fused to foot and mouth disease VP1	Pharmaceutical activity, vaccine	60
2004	Bacterial luciferase	Real-time reporter activity, bioluminescence activity	50
2005	HSV8-scFv	Pharmaceutical activity, classic single-chain antibody	61
2005	Allophycocyanin	Fluorescent protein	62
2006	Human metallothionine-2	Pharmaceutical activity, UV protection	63
2006	Firefly luciferase	Real-time reporter activity, bioluminescence activity	64
2006	Human tumor necrosis factor-related apoptosis-inducing ligand (TRAIL)	Pharmaceutical activity	65
2007	Lac repressor	Repressor activity, control of transgene expression	66
2007	Classical swine fever virus (CSFV) structural protein E2	Vaccine, immune response activity in animals	67
2008	Human glutamic acid decarboxylase 65 (hGAD65)	Pharmaceutical activity, first full-length autoantigenic protein expressed	68
2008	metallothionein-like gene from *Festuca rubra*	Metal binding protein, heavy metal tolerance induction	69

are already commercially available in functional foods or as food ingredients as e.g., sold by DMV ("C12 Peption" with a Casein-derived dodecapeptide FFVAPFPEVFGK) or Ingredia ("ProDiet F200" with peptide YLGYLEQLLR). A detailed compilation of bioactive peptides in commercial products is given in a recent review by Hartmann and Meisel.[72]

The practicability of expressing bioactive peptides in chloroplasts has already been demonstrated in a study employing the antimicrobial peptide MSI-99.[73] This 22-amino-acid-long peptide is an analog of magainin, which has been isolated from skin secretions of the African clawed frog *Xenopus laevis*. Its activity is thought to arise from the ability to form pores in membranes with negatively charged phospholipid headgroups. This effect is not observed with membranes of plants and animals, which have no net charge in their outer membrane leaflets.[74] Magainin is effective against diverse bacteria, fungi and protozoa and has anticancer activity.[75] Transgenic tobacco plants expressing a magainin analog in their chloroplasts were indeed protected against different plant pathogens, indicating that in plastids a sufficiently stable peptide can be produced and retains its function, although it is located in a foreign environment. If microalgae like *Chlamydomonas reinhardtii* would express the preferred bioactive peptides in adequate quantities, they could be extremely useful as additives in food industries.

Conclusion

Microalgae have long been used as nutritional supplement or food and feed sources. Recent progress in genetic engineering technologies make eukaryotic microalgae efficient bioreactors for the production of various bioactive compounds. The green alga Chlamydomonas reinhardtii serves as a pioneering model organism because its complete genome sequence is known, transformation procedures are established and extensive molecular toolkits are available. Specifically the chloroplast compartment as a subcellular bioreactor provides distinct advantages with respect to biosafety as well as epigenetic stability and offers new opportunities for the expression of foreign proteins and for engineering metabolic pathways.

Acknowledgments

Dedicated to Prof. Dr. Drs. h.c. mult. Achim Trebst on the occasion of his 80th birthday.

This work was supported by the NUTRA-SNACKS project funded within the EU Framework Programme 6.

References

1. Walker TL, Purton S, Becker DK et al. Microalgae as bioreactors. Plant Cell Rep 2005; 24:629-41.
2. Debuchy R, Purton S, Rochaix JD. The argininosuccinate lyase gene of Chlamydomonas reinhardtii: an important tool for nuclear transformation and for correlating the genetic and molecular maps of the ARG7 locus. EMBO J 1989; 8:2803-2809.
3. Kindle KL, Schnell RA, Fernandez E et al. Stable nuclear transformation of Chlamydomonas using the Chlamydomonas gene for nitrate reductase. J Cell Biol 1989; 109:2589-2601.
4. Ferris PJ. Localization of the nic-7, ac-29 and thi-10 genes within the mating-type locus of Chlamydomonas reinhardtii. Genetics 1995; 141:543-549.
5. Spolaore P, Joannis-Cassan C, Duran E et al. Commercial applications of microalgae. J Biosci Bioeng 2006; 101:87-96.
6. Newell CA. Plant transformation technology: Developments and applications. Mol Biotechnol 2000; 16:53-65.
7. Rochaix JD. Chlamydomonas reinhardtii as the photosynthetic yeast. Annu Rev Genet 1995; 29:209-30.
8. Grossman AR, Harris EE, Hauser C et al. Chlamydomonas reinhardtii at the crossroads of genomics. Eukaryot Cell 2003; 2:1137-50.
9. Harris EH. Chlamydomonas as a Model Organism. Rev Plant Physiol Plant Mol Biol 2001; 52:363-406.
10. Merchant SS, Prochnik SE, Vallon O et al. The Chlamydomonas genome reveals the evolution of key animal and plant functions. Science 2007; 318:245-50.
11. Raven JA, Allen JF. Genomics and chloroplast evolution: what did cyanobacteria do for plants? Genome Biol 2003; 4:209-213.
12. Mattoo A, Giardi MT, Raskind A et al. Dynamic metabolism of photosystem II reaction center proteins and pigments. A review. Physiol Plant 1999; 107:454-461.
13. Maliga P. Plastid transformation in higher plants. Annu Rev Plant Biol 2004; 55:289-313.
14. Daniell H et al. Containment of herbicide resistance through genetic engineering of the chloroplast genome. Nat Biotechnol 1998; 16:345-3.
15. Lutz KA et al. Expression of bar in the plastid genome confers herbicide resistance. Plant Physiol 2001; 125:1585-1590.
16. DeGray G et al. Expression of an antimicrobial peptide via the chloroplast genome to control phytopathogenic bacteria and fungi. Plant Physiol 2001; 127:852-862.
17. Kota M et al. Overexpression of the Bacillus thuringiensis (Bt) Cry2Aa2 protein in chloroplasts confers resistance to plants against susceptible and Bt-resistant insects. Proc Natl Acad Sci USA 1999; 96:1840-1845.

18. McBride KE et al. Amplification of a chimeric Bacillus gene in chloroplasts leads to an extraordinary level of an insecticidal protein in tobacco. Bio/Technology 1995; 13:362-365.
19. Daniell H, Khan MS, Allison L. Milestones in chloroplast genetic engineering: an environmentally friendly era in biotechnology. TRENDS in Plant Science 2002; 7:84-91.
20. Staub JM et al. High-yield production of a human therapeutic protein in tobacco chloroplasts. Nat Biotechnol 2000; 18:333-338.
21. De Cosa B, Moar W, Lee S-B et al. Overexpression of the Bt cry2Aa2 operon in chloroplasts leads to formation of insecticidal crystals. Nat Biotechnol 2001; 19:71-74.
22. Nakashita H, Arai Y, Shikanai T et al. Introduction of bacterial metabolism into higher plants by polycistronic transgene expression. Biosci Biotechnol Biochem 2001; 65:1688-1691.
23. Lössl A, Eibl C, Harloff HJ et al. Polyester synthesis in transplastomic tobacco (Nicotiana tabacum L.): significant contents of polyhydroxybutyrate are associated with growth reduction. Plant Cell Rep 2003; 21:891-899.
24. Arai Y, Shikanai T, Doi Y et al. Production of polyhydroxybutyrate by polycistronic expression of bacterial genes in tobacco plastid. Plant Cell Physiol 2004; 45:1176-84.
25. Wurbs D, Ruf S, Bock R. Contained metabolic engineering in tomatoes by expression of carotenoid biosynthesis genes from the plastid genome. Plant J 2007; 49:276-288.
26. Sidorov VA, Kasten D, Pang S-Z et al. Stable chloroplast transformation in potato: use of green fluorescent protein as a plastid marker. Plant J 1999; 19:209-216.
27. Ruf S, Hermann M, Berger IJ et al. Stable genetic transformation of tomato plastids and expression of a foreign protein in fruit. Nat Biotechnol 2001; 19:870-875.
28. Dufourmantel N, Pelissier B, Garcon F et al. Generation of fertile transplastomic soybean. Plant Mol Biol 2004; 55:479-489.
29. Kumar S, Dhingra A, Daniell H. Plastid-expressed betaine aldehyde dehydrogenase gene in carrot cultured cells, roots and leaves confers enhanced salt tolerance. Plant Physiol 2004; 136:2843-2854.
30. Lelivelt CLC, McCabe MS, Newell CA et al. Stable plastid transformation in lettuce (Lactuca sativa L.). Plant Mol Biol 2005; 58:763-774.
31. Kanamoto H, Yamashita A, Asao H et al. Efficient and stable transformation of Lactuca sativa L. cv. Cisco (lettuce) plastids. Transgenic Res 2006; 15:205-217.
32. Nugent GD, Coyne S, Nguyen TT et al. Nuclear and plastid transformation of Brassica oleracea var. botrytis (cauliflower) using PEG-mediated uptake into protolasts. Plant Sci 2006; 170:135-142.
33. Ellstrand NC. When transgenes wander, should we worry? Plant Physiol 2001; 125:1543-5.
34. Ellstrand NC. Current knowledge of gene flow in plants: implications for transgene flow. Philos Trans R Soc Lond B Biol Sci 2003; 358:1163-70.
35. Ruf S, Karcher D, Bock R. Determining the transgene containment level provided by chloroplast transformation. Proc Natl Acad Sci USA 2007; 104:6998-7002.
36. Boynton JE, Gillham NW, Harris EH et al. Chloroplast transformation in Chlamydomonas with high velocity microprojectiles. Science 1988; 240:1534-1538.
37. Klein TM, Wolf ED, Wu R et al. High-velocity microprojectiles for delivering nucleic acids into living cells. Nature 1987; 327:70-73.
38. Przibilla E, Heiss S, Johanningmeier U et al. Site-specific mutagenesis of the D1 subunit of Photosystem II in wildtype Chlamydomonas. Plant Cell 1991; 3:169-174.
39. Newman SM, Gillham NW, Harris EH et al. Targeted disruption of chloroplast genes in Chlamydomonas reinhardtii. Mol Gen Genet. 1991; 230:65-74.
40. Goldschmidt-Clermont M. Transgenic expression of aminoglycoside adenine transferase in the chloroplast: a selectable marker of site-directed transformation of Chlamydomonas. Nucleic Acids Res 1991; 19:4083-9.
41. Bateman JM, Purton S. Tools for chloroplast transformation in Chlamydomonas: expression vectors and a new dominant selectable marker. Mol Gen Genet 2000; 263:404-10.
42. Lapidot M, Raveh D, Sivan A et al. Stable chloroplast transformation of the unicellular red alga Porphyridium species. Plant Physiol 2002; 129:7-12.
43. Doetsch NA, Favreau MR, Kuscuoglu N et al. Chloroplast transformation in Euglena gracilis: splicing of a group III twintron transcribed from a transgenic psbK operon. Curr Genet 2001; 39:49-60.
44. Blowers AD, Bogorad L, Shark KB et al. Studies on Chlamydomonas chloroplast transformation: foreign DNA can be stably maintained in the chromosome. Plant Cell 1989; 1:123-132.
45. Blowers AD, Ellmore GS, Klein U et al. Transcriptional analysis of endogenous and foreign genes in chloroplast transformants of Chlamydomonas. Plant Cell 1990; 2:1059-1070.
46. Ishikura K, Takaoka Y, Kato K et al. Expression of a foreign gene in Chlamydomonas reinhardtii chloroplast. J Biosci Bioeng 1999; 87:307-314.
47. Minko I, Holloway SP, Nikaido S et al. Renilla luciferase as a vital reporter for chloroplast gene expression in Chlamydomonas. Mol Gen Genet 1999; 262:421-425.

48. Franklin S, Ngo B, Efuet E et al. Development of a GFP reporter gene for Chlamydomonas reinhardtii chloroplast. Plant J 2002; 30:733-744.
49. Mayfield SP, Franklin SE, Lerner RA. Expression and assembly of fully active antibody in algae. Proc Natl Acad Sci USA 2003; 100:438-442.
50. Mayfield SP, Schultz J. Development of a luciferase reporter gene, luxCt, for Chlamydomonas reinhardtii chloroplast. Plant J 2004; 37:449-458.
51. Salvador ML, Suay L, Anthonisen IL et al. Changes in the 5'-untranslated region of the rbcL gene accelerate transcript degradation more than 50-fold in the chloroplast of Chlamydomonas reinhardtii. Curr Genet 2004; 45:176-182.
52. Klein U, Salvador ML, Bogorad L. Activity of the Chlamydomonas chloroplast rbcL gene promoter is enhanced by a remote sequence element. Proc Natl Acad Sci USA 1994; 91:10819-10823.
53. Suay L, Salvador ML, Abesha E et al. Specific roles of 5' RNA secondary structures in stabilizing transcripts in chloroplasts. Nucleic Acids Res 2005; 33:4754-4761.
54. Klein U, De Camp JD, Bogorad L. Two types of chloroplast gene promoters in Chlamydomonas reinhardtii. Proc Natl Acad Sci USA 1992; 89:3453-3457.
55. Eberhard S, Drapier D, Wollman FA. Searching limiting steps in the expression of chloroplast encoded proteins: relations between gene copy number, transcription, transcript abundance and translation rate in the chloroplast of Chlamydomonas reinhardtii. Plant J 2002; 31:149-160.
56. Barnes D, Franklin S, Schultz J et al. Contribution of 5'- and 3'-untranslated regions of plastid mRNAs to the expression of Chlamydomonas reinhardtii chloroplast genes. Mol Genet Genomics 2005; 274:625-636.
57. Mayfield SP, Manuell AL, Chen S et al. Chlamydomonas reinhardtii chloroplasts as protein factories. Curr Opin Biotechnol 2007; 18:126-33.
58. Griesbeck C, Kobl I, Heitzer M. Chlamydomonas reinhardtii: a protein expression system for pharmaceutical and biotechnological proteins. Mol Biotechnol 2006; 34:213-23.
59. Fukusaki EI, Nishikawa T, Kato K et al. Introduction of the Archaebacterial Geranylgeranyl Pyrophosphate Synthase Gene into Chlamydomonas reinhardtii chloroplast. J Biosci Bioeng 2003; 95:283-287.
60. Sun M, Qian K, Su N et al. Foot and mouth disease virus VP1 protein fused with cholera toxin B subunit expressed in Chlamydomonas reinhardtii chloroplast. Biotechnol Lett 2003; 25:1087-1092.
61. Mayfield SP, Franklin SE. Expression of human antibodies in eukaryotic micro-algae. Vaccine 2005; 23:1828-1832.
62. Su ZL, Qian KX, Tan CP et al. Recombination and heterologous expression of allophycocyanin gene in the chloroplast of Chlamydomonas reinhardtii. Acta Biochim Biophys Sin (Shanghai) 2005; 37:709-712.
63. Zhang YK, Shen GF, Ru BG. Survival of human metallothioneine-2 transplastomic Chlamydomonas reinhardtii to ultraviolet B exposure. Acta Biochim Biophys Sin (Shanghai) 2006; 38:187-193.
64. Matsuo T, Onai K, Okamoto K et al. Real-time monitoring of chloroplast gene expression by a luciferase reporter: evidence for nuclear regulation of chloroplast circadian period. Mol Cell Biol 2006; 26:863-870.
65. Yang Z, Li Y, Chen F et al. Expression of human soluble TRAIL in Chlamydomonas reinhardtii chloroplast. Chin Sci Bull 2006; 51:1703-1709.
66. Kato K, Marui T, Kasai S et al. Artificial control of transgene expression in Chlamydomonas reinhardtii chloroplast using the lac regulation system from Escherichia coli. J Biosci Bioeng 2007; 104:207-213.
67. He DM, Qian KX, Shen GF et al. Recombination and expression of classical swine fever virus (CSFV) structural protein E2 gene in Chlamydomonas reinhardtii chloroplasts. Colloids Surf B Biointerfaces 2007; 55:26-30.
68. Wang X, Brandsma M, Tremblay R et al. A novel expression platform for the production of diabetes-associated autoantigen human glutamic acid decarboxylase (hGAD65). BMC Biotechnol 2008; 8:87-89.
69. Han S, Hu Z, Lei A. Expression and function analysis of the metallothionein-like (MT-like) gene from Festuca rubra in Chlamydomonas reinhardtii chloroplast. Sci China Ser C-Life Sci 2008; 51:1076-1081.
70. Korhonen H, Pihlanto A. Food-derived bioactive peptides-opportunities for designing future foods. Curr Pharm 2003; 9:1297-1308.
71. Korhonen H, Pihlanto A. Bioactive peptides: production and functionality. Int Dairy J 2006; 16:945-960.
72. Hartmann R, Meisel H. Food-derived peptides with biological activity: from research to food applications. Curr Opin Biotechnol 2007; 18:163-169.
73. DeGray G, Rajasekaran K, Smith F et al. Expression of an antimicrobial peptide via the chloroplast genome to control phytopathogenic bacteria and fungi. Plant Physiol 2001; 127:852-62.
74. Zasloff M. Antimicrobial peptides of multicellular organisms. Nature 2002; 415:389-95.
75. Jacob L, Zasloff M. Potential therapeutic applications of magainins and other antimicrobial agents of animal origin: antimicrobial Peptides. Ciba Found Symp 1994; 186:197-223.

Chapter 12

Biological Elicitors of Plant Secondary Metabolites:

Mode of Action and Use in the Production of Nutraceutics

Simone Ferrari*

Abstract

Many secondary metabolites of interest for human health and nutrition are produced by plants when they are under attack of microbial pathogens or insects. Treatment with elicitors derived from phytopathogens can be an effective strategy to increase the yield of specific metabolites obtained from plant cell cultures. Understanding how plant cells perceive microbial elicitors and how this perception leads to the accumulation of secondary metabolites, may help us improve the production of nutraceutics in terms of quantity and of quality of the compounds. The knowledge gathered in the past decades on elicitor perception and transduction is now being combined to high-throughput methodologies, such as transcriptomics and metabolomics, to engineer plant cells that produce compounds of interest at industrial scale.

Introduction

Plants are able to accumulate an astonishing number of small compounds, often called secondary metabolites,[1] that are of great interest for their potential use as pharmaceutical drugs, nutrients and food additives. Many secondary metabolites have important ecological roles, deterring herbivores and protecting against microbial pathogens; thus it is evident that specialized metabolites indeed have major functions and the term "natural products" is therefore preferable to "secondary metabolites".[2] Besides direct extraction and chemical synthesis, plant cell cultures have been developed as a promising alternative for producing metabolites that are difficult to obtain by traditional means and cell and tissue cultures today are evaluated for the commercial production of natural products. Unlike the production of plants, that is seasonally limited, cell cultures can be used year-round independently of seasons, geographic location and political situations. Cells can be induced in a coordinated manner to produce metabolites of interest, that can be directly isolated from the culture medium or from the cells. In spite of several decades of efforts, however, production of plant secondary metabolites by plant cell culture technology has still many restrictions. To date, only the production of shikonin by *Lithospermum erythrorthizon* cell cultures and of taxol by *Taxus* cell cultures is successfully industrialized for commercial application.[3] One of the major obstacles is the low yield of secondary metabolites in plant cell cultures. Since the major role of these metabolites is to protect plants from attacks by insect, herbivores and pathogens, the ability of specific plant- or pest-derived molecules to induce the accumulation of high levels of

*Simone Ferrari—Dipartimento di Biologia Vegetale, Università degli Studi di Roma "La Sapienza" Piazzale Aldo Moro, 5 00185 Rome, Italy. Email: simone.ferrari@uniroma1.it

Bio-Farms for Nutraceuticals: Functional Food and Safety Control by Biosensors edited by Maria Teresa Giardi, Giuseppina Rea and Bruno Berra.

specific metabolites has been exploited to increase their production, both in whole plants and in cell cultures. Basic and applied plant biotechnological research aimed at identifying the most effective chemical inducers of metabolites of interest and the molecular and biochemical mechanisms underlying this induction, is currently unvealing several aspects of the fundamental biology of the interaction between plants and pests and is also suggesting new strategies to exploit this knowledge in order to obtain such compounds on industrial scale.

A large number of low molecular weight compounds produced by plants have evolved as a consequence of the the everlasting arms race with their natural enemies, namely microbial pathogens, herbivores and other competing plants. Some of these compounds accumulate in specific plant tissues regardless the presence of a biotic or abiotic agent of disease. These compounds, known as phytoanticipins, include, for instance, saponins, a class of triterpenoid glycoalkaloids with antimicrobial activity that accumulate in different plant tissues (i.e., α-tomatine in tomato leaves, avenacine A-1 in oat roots).[4] Other examples of phytoanticipins include flavonoids, tannins, glucosinolates and cyanogenic glycosides. The role of these molecules is to inhibit or deter potential pathogens or herbivores that should try to attack the plant. Despite their effectiveness against at least some pests, the constitutive accumulation of defence compounds has a significant energetic cost and normally they are found, in healthy plants, only in those parts that are more susceptible to attack (e.g., the epidermis) or that are more vulnerable (young leaf primordia, flowers). However, when a plant is challenged by an invading microorganism or it is wounded, it undergoes a metabolic reprogramming, leading to the accumulation, at the site of attack, of high levels of specific antimicrobial, antifeeding or generally toxic compounds. When these compounds are induced by microbial pathogens, they are usually referred to as phytoalexins and include a wide array of chemical species, such as simple phenylpropanoid derivatives, flavonoid- and isoflavonoid-derived compounds, sesquiterpenes and polyketides (for an exhaustive review on phytoalexins, see ref. 5). The phytoalexin hypothesis was first formulated by Müller and Borger,[6] who demonstrated that the infection of potato tuber tissue with an incompatible race of *Phytophthora infestans* induces resistance to a subsequent challenge by virulent pathogens. This observation led to the hypothesis that the tuber tissue, in response to the incompatible interaction, produced nonspecific substances (phytoalexins) that inhibited the growth of the pathogen and also protected the tissue against later infection by other pathogens. Over 15 years later, Müller demonstrated that bean pod tissues infected with incompatible pathogens produce fungistatic substances.[7] The study of phytoalexins has been extensive over the past 70 years. Much of this work has provided new insights into the regulation of gene activation, phytochemical diversity and the chemistry and biochemistry of secondary metabolites. In 1981 a very general definition of phytoalexins was presented, simply stating that phytoalexins are antimicrobial compounds produced after infection or elicitation by abiotic agents.[8] Beside phytoalexins, other secondary metabolites induced upon biotic stress are important for, but not limited to, defence against insects (e.g., toxic alkaloids). Furthermore, a plant undergoing an attack usually displays the accumulation of high levels of reactive oxygen species (ROS), either as a result of the pest activity, or as a defensive mechanism to restrict pathogen growth or insect feeding.[9] As a consequence, the production of antioxidant molecules, aimed to restrict the damage caused by ROS, is often observed in plant tissues challenged with a microbial pathogen or an insect. These antioxidant molecules are also of technological interest, since they may be used as drugs or as food additives.

The most actively pursued strategies to increase the production of target natural products in plant cell cultures are the application of chemical elicitors[10] and the study of signal transduction pathways and transcription factors required for the expression of genes involved in the biosynthesis of specific compounds, in order to genetically engineer the cell culture.[11] Generally speaking, activation of the biosynthesis of plant natural products requires that an extracellular or intracellular signal is perceived by a receptor on the surface of the plasma membrane. Elicitor perception initiates a signal transduction cascade that leads to the activation or to the de novo biosynthesis of transcription factors, which in turn directly regulate the expression of biosynthetic genes involved in secondary metabolism. Much effort has been put into cloning biosynthetic genes, identifying transcription factors, revealing the signal transduction steps underlying elicitor activation of plant secondary metabolism and also into the manipulation of regulatory and biosynthetic genes to engineer plant cells and enhance the production of target secondary metabolites.[12] These efforts will hopefully improve the production of natural products of interest and eventually be applied on an industrial scale. However, due to lack of detailed information about the biosynthetic genes and their regulation, such as the signal transduction pathways and transcription factors involved, success in metabolic engineering of plant secondary metabolites is so far very limited. Studying upstream signalling cascades that regulate biosynthetic genes is therefore a necessary step towards the improvement of plant cell biotechnology. It is expected that a better understanding of the signal transduction pathways linking plant cell stimulation and biosynthesis of natural compounds may help develop new strategies to alter the production of target compounds by either activation or suppression of certain metabolic pathways. In this review we will summarize the current knowledge on the molecules that trigger the activation of defence responses in plants, the signal transduction that links the perception of these molecules to the expression of genes involved in secondary metabolism and on new technologies available to dissect the mechanism of action of elicitors and improve the accumulation of small molecules in plants.

Biological Elicitors of Defence Responses in Plants

Elicitors are chemicals that trigger the activation of plant defence responses, including phytoalexin accumulation. These chemicals can be abiotic agents, such as metal ions and inorganic compounds, or can be derived from other organisms, such as molecules derived from fungi, bacteria, viruses or herbivores, as well as plant-derived chemicals that are released at the attack site or accumulate systemically upon pathogen or herbivore attack. In this review we will focus only on those molecules derived directly or indirectly from pathogens and pests, e.g., fungal or plant cell wall components, bacterial, virus or herbivore constituents, collectively called microbe-associated molecular patterns, MAMPs, or pathogen-associated molecular patterns, PAMPs.[13,14]

Fungal culture filtrates, yeast crude extract (YE) and killed bacteria can elicit potent defence responses in plant cells.[15] However, these elicitors are complex mixtures of chemicals, only a part of which is effective in inducing the accumulation of specific compounds. For instance, YE contains several components that can elicit plant defence responses, including chitin, N-acetylglucosamine oligomers, β-glucan, glycopeptides and ergosterol.[16] However, the only detectable YE component inducing terpenoid indole alkaloids (TIA) biosynthetic gene expression in *Catharanthus roseus* is a water-soluble, low molecular weight component, which is probably a small peptide.[17] For this reason, great effort has been put into the identification of the specific components that are responsible for the biological activity of the above mentioned elicitors. Several recent reviews describe the structure and action of microbial-derived elicitors in plant-pathogen interactions.[14,15,18] So far, different classes of compounds derived from microbial, insect or plant cells have shown eliciting activity (Table 1). The first class includes poly- or oligosaccharides such as chitin, chitosan (obtained after de-acetylation of chitin) and their fragments (chitooligosaccharides), derived from fungal and insect cell walls; laminarin and other β-glucans, such as those found in the cell wall of algae and oomycetes and their fragments; xyloglucans and oligogalacturonides (OGs), which are released from plant cell wall structural carbohydrates (hemicellulose and pectin, respectively) following enzymatic degradation. Another major class of elicitors is represented by proteins (e.g., cryptogein and polygalacturonases from fungi; harpin and flagellin from bacteria) or peptides derived from those proteins. Finally, several lipidic compounds, including syringolides, Nod factors and lipopolysaccharides, have elicitor activity. Other elicitors, specifically involved in the recognition of insects, include fatty acid-amino acid conjugates from *Manduca* spp, volicitin from *Spodoptera exigua*, a β-glucosidase from *Pieris brassicae* oral secretions and bruchins from cowpea weevil oviposition fluid (reviewed in ref. 19). Plant ATPase fragments can also mediate plant perception of insect attacks.[20] All these molecules can elicit direct defence responses, such as the accumulation of antifeeding compounds,[21] or indirect defence responses, such as compounds that attract predators and parasitoids.[22] Endogenous elicitors of defence responses can be released

Table 1. Examples of biological elicitors of plant defence responses

Elicitor	Description	Reference
Oligogalacturonides (OGs)	Pectic fragments from plant cell wall	135
Chitosan	Deacetylated chitin fragments from fungal cell wall	136
β-glucan	Component of the mycelia cell walls of *Phytophthora megasperma* and other oomycetes	137
Lipopolysaccharides (LPS)	Gram-negative bacteria	73
Elicitins	Proteins from oomycetes	138
Pep-13	Oligopeptide of 13 amino acids within a 42-kDa transglutaminase secreted by *Phytophthora sojae*	139
Flg22	22-amino acid N-terminal fragment of bacterial flagellin	140
Xylanase	Xylan-degrading enzyme from *Trichoderma* spp.	141
BcPG1	Endopolygalacturonase from *Botrytis cinerea*	142
EF-Tu	Bacterial elongation factor	143

in wounded plants as a consequence of herbivore feeding. The best studied example is tomato systemin, an 18-amino acid peptide,[23] which activates signalling pathways leading to defence responses effective against insects.[24-26] It must be noted that the accumulation of metabolites of interest can vary to a large extent in different species treated with different elicitors. For instance, YE has been successfully employed to induce production of secondary metabolites in *Nicotiana* and *Escholtzia* cell cultures, but was not effective on vanilla (*Vanilla planifolia*) cultures, whereas chitosan successfully induces phenylalanine ammonia lyase (PAL) activity and phenolics production in this species.[27] It is therefore important to characterize the response to different elicitors each time that we want to produce a new metabolite or we are working with a new plant species.

Studying elicitor signal transduction while focusing on only one secondary metabolite usually gives very limited information about metabolic fluxes. Investigation of the signal transduction involved in the production of multiple secondary metabolites, for example, from two branches of a common biosynthetic pathway, can lead to more significant findings. It is often observed that elicitors induce rearrangements of metabolic fluxes between a constitutively expressed pathway and an elicitor-inducible pathway ("metabolic channeling"). This differential regulation of the biosynthesis of branch compounds reflects an important feature of elicitor-mediated induction of plant secondary metabolism: elicitors selectively induce the accumulation of phytoalexin-like compounds that act as antimicrobial or repellent agents to kill microbes or repel insects. One cost of the biosynthesis of phytoalexins is the decreased accumulation of other primary or secondary compounds. In etiolated sorghum (*Sorghum bicolor*), light induces the accumulation of cyanidin 3-dimalonyl glucoside by increasing the expression of anthocyanin biosynthesis genes, whereas fungal inoculation suppresses the expression of these genes and anthocyanin accumulation and stimulates accumulation of a phytoalexin, 3-deoxyanthocyanidin and the activation of PAL and chalcone synthase (CHS) genes.[28] Similar phenomena occur in other plants and other biosynthetic pathways. Clarifying the signalling pathways and the regulatory mechanisms involved in these switching processes will greatly help us to selectively produce certain desired natural products.

Perception and Transduction of General Elicitors

Several studies demonstrated that specific recognition between microbial elicitors and the corresponding plant receptors triggers specific signalling events leading to the activation of defence responses. The first characterized elicitor receptor in plants is the Arabidopsis FLS2, which is responsible for the perception of bacterial flagellin, a structural protein of the bacterial flagellum.[29] FLS2 is required for the activation of defence responses upon treatment of Arabidopsis plants and cell cultures with either flagellin or flg22, a 22-amino acid peptide conserved in this protein and specifically recognized by plant cells.[30] In the last few years, more plant receptors for general elicitors have been identified. Zipfel and colleagues isolated EFR1, the Arabidopsis receptor responsible for the recognition of the bacterial elongation factor EF-Tu.[31] CEBiP, a putative receptor for chitin fragments, was isolated from rice cells.[32] Few potential receptors for molecules released during insect attack have been also identified and cloned, as in the case of tomato systemin receptor, whose architecture is similar to that of FLS2 and EFR1.[33] The previously mentioned receptors are membrane-bound proteins. There are however exceptions: for instance, the bacterial glycolipid elicitor syringolide is perceived by a soluble protein.[34] Despite these progresses, the proteins responsible for the perception of several general elicitors, as in the case of OGs, remain elusive and future research should be directed to fill this gap in our knowledge.

Perception of an elicitor by its receptor(s) usually is followed by events of phosphorylation and dephosphorylation of plasma membrane and cytosolic proteins, ion fluxes, plasma membrane depolarization, extracellular alkalinization and cytoplasmic acidification.[35] These events occur within the first minutes to a few hours after elicitor perception and mediate the subsequent activation of downstream responses. Activation of NADPH oxidases and apoplastic peroxidases leads to a massive but transient accumulation of ROS ("oxidative burst"), which, together to the production of defence-specific signalling molecules, such as ethylene (ET), salicylic acid (SA) and jasmonate (JA), can contribute to induce "late" defence responses, including secondary metabolite production.[36,37] A simplified scheme illustrating the possible events linking elicitor perception to the activation or expression of enzymes involved in secondary metabolism is shown in Figure 1. It must be noted that the elements of a transduction pathway can be differentially activated in response to different elicitors. Furthermore, an extensive cross-talk among the components of the different signalling pathways occurs, leading to different target responses upon treatment with different biotic and abiotic stimuli. One specific elicitor transduction pathway can use a combination or a partial combination of such events which can differ in kinetics and intensity depending on the stimulus.

Among the earliest events induced by elicitors is the activation of G-proteins. Experiments using various activators and inhibitors suggest that G-proteins are involved in elicitor-induced biosynthesis of plant secondary metabolites, including benzophenanthridine alkaloids in bloodroot,[38] isoflavonoids in soybean[39] and β-thujaplicin in *Cupressus lusitanica*.[40] Rapid ion fluxes, namely K^+/H^+ exchange, Cl^- effluxes and increases of $(Ca^{2+})_{cyt}$, are also important elements of defence-related signalling pathways.[41,42] In particular, the role of Ca^{2+} has been extensively studied, since this ion is a key second messenger in all organisms.[41] Elicitor-induced Ca^{2+} fluxes are important for the accumulation of plant secondary metabolites and Ca^{2+} channel blockers can inhibit the induction of phytoalexins by elicitors in some plant species, such as rice[43] or *C. lusitanica*.[40] The increase of cytoplasmic Ca^{2+} levels may lead to the activation of subsequent defence responses either directly or through Ca^{2+} sensors, such as calmodulins and Ca^{2+} -dependent protein kinases (CDPKs). In tobacco, calcium/calmodulin antagonists inhibit sesquiterpene phytoalexin accumulation mediated by elicitors, as well as the induction of sesquiterpene cyclase enzyme activity, a key regulatory enzyme for sesquiterpene biosynthesis.[44] Elicitor-induced phosphoinositide breakdown by phospholipase A (PLA) and C (PLC) is another early event activated by different elicitors.[45-48] PLC can be activated by $(Ca_2^+)_{cyt}$ spiking and converts phosphatidyl-4,5-bisphosphate (PIP_2) into inositol-3-phosphate (IP_3) and diacyl glycerol (DAG). IP_3 signalling has been involved in elicitor-induced accumulation of, among others, 6-methoxymellein in carrot cells,[49] medicarpin in alfalfa cell culture[50] and thujaplicin in Mexican cypress.[51]

Reversible phosphorylation of target proteins by protein kinases (PKs) and phosphatases is a universal mechanism regulating cell functions in all eukaryotes. Phospho-proteomics experiments indicate a rapid phosphorylation of several proteins after elicitor treatment.[52-54] Protein phosphorylation can directly activate secondary metabolism biosynthetic enzymes by posttranslational modification, or partecipate to signalling cascades resulting in increased expression of the corresponding genes. Direct phosphorylation of enzymes involved in secondary metabolism has been already reported. For instance, a PK activated during elicitation of

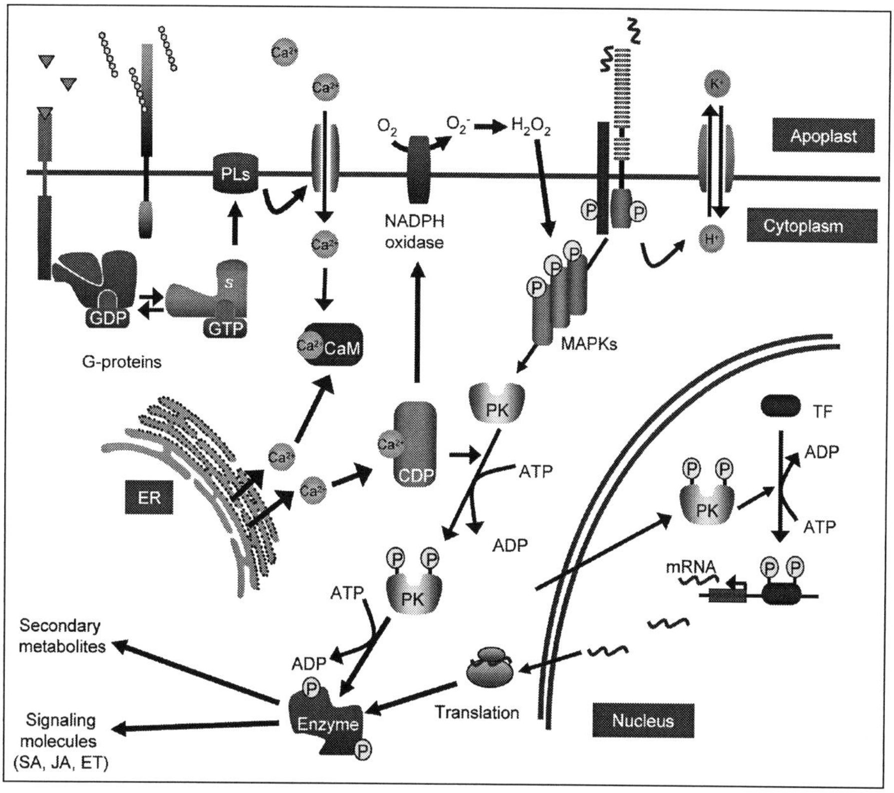

Figure 1. Early signalling events activated by elicitors. The cartoon depicts a simplified model of the early signal transduction events linking elicitor perception to the expression of genes encoding enzymes involved in the biosynthesis of secondary metabolites and secondary signals. Different elicitors are perceived by distinct membrane receptors, though they may activate the same signalling pathways. One of the first events associated to elicitor perception is the activation of trimeric G-proteins, that in turn activate phospholipases, which can release secondary messengers that activate downstream responses. Among these, one of the most important is the transient increase of intracellular calcium Ca^{2+} concentration, mediated by the activation of channels on the plasma membrane and on the membranes of internal stores, such as the endoplasmatic reticulum (ER). Ca^{2+} can activate directly or indirectly, through binding to calmodulin (CaM) or Ca^{2+} -dependent protein kinases (CDP) target proteins, such as NADPH oxidases, which are responsible for the oxidative burst, or downstream protein kinases (PK). Elicitor perception also activates plasma membrane channels, such as K^+/H^+ antiport channels, as a consequence of plasma membrane depolarization; ion fluxes are responsible for a transient cytoplasm acidification, that can act as a signal for the production of secondary metabolites. Elicitors also activate Mitogen-Activated PK (MAPK) cascades that phosphorylate transcription factors (TF) regulating the expression of genes encoding enzymes involved in the biosynthesis of specific secondary metabolites or of secondary signals, such as salicylic acid (SA), jasmonic acid (JA) and ethylene (ET).

French been cell cultures with a fungal elicitor can phosphorylate PAL and phosphorylation of a recombinant PAL from poplar appears to regulate its activity.[55] However, several targets of PKs activated after elicitation probably represent elements of signalling pathways upstream of the expression of biosynthetic activities. In particular, plant mitogen-activated protein kinase (MAPK) cascades mediate the activation of common responses after perception of pathogen and pathogen-derived elicitors.[56,57] The most extensively characterized MAPKs involved in plant defence are the tobacco SA-induced protein kinase (SIPK) and wounding-induced protein kinase (WIPK),[58-60] AtMPK6 and AtMPK3 in Arabidopsis[56,61,62] and the salt stress-induced MAPK (SIMK) and stress-activated MAP kinase (SAMK) in alfalfa.[63,64] Upon receptor-mediated activation, MAPKs are translocated to the nucleus where they might interact with transcription factors that induce expression of defense genes.[65] Asai and colleagues have identified a complete Arabidopsis MAPK cascade (MEKK1, MKK4/MKK5 and MPK3/MPK6) and the transcription factors that function downstream of the flagellin receptor FLS2.[56] Increasing evidence suggests that plant MAPK cascades mediate the expression of genes involved in secondary metabolism during elicitation. For instance, a tobacco MAPK cascade controls the expression of genes encoding 3-hydroxy-methylglutaryl-CoA reductase and PAL[66] and MAPK activity is required for saponin accumulation in ginseng cells treated with chitosan[67]. It must be noted that significant positive and negative cross-talk between MAPKs exists. For instance, the Arabidopsis MPK4 is required for JA-responsive gene expression, but it negatively regulates SA-dependent gene expression.[57] Moreover, MAPK activation is not a universal requirement for elicitor-dependent responses: for instance, in grapevine cells, PAL and stilbene synthase are activated by a fungal polygalacturonase independently of MAPK cascades.[68]

ROS, such as superoxide anion (O_2^-) and hydrogen peroxide (H_2O_2), are toxic intermediates resulting from reduction of molecular O_2 that can act as signals that regulate defence responses and phytoalexin accumulation.[69] Accumulation of high levels of ROS in plant tissues is observed during oxidative burst triggered either by pathogen attack, or by elicitor treatment.[36,46,70-76] O_2^- -generating NADPH oxidases are considered the main enzymatic sources of ROS in the oxidative burst.[77-80] In Arabidopsis, several genes encoding proteins with high similarity to the mammal NADPH oxidase gp91^{phox} subunit have been characterized. Among them, *AtrbohD* is required for the production of ROS during infection with different bacterial and fungal pathogens.[80,81] O_2^- has been implicated in the accumulation of some secondary metabolites, like furanocoumarin in parsley cell cultures,[82] isoflavonoid glyceollin in soybean,[83,84] p-coumaroyloctopamine in potato,[85] tcibulin 1 and 2 in *Allium cepa*,[86] beta-thujaplicin in *C. lusitanica*[40] and momilactones in rice.[87] However, in some instances the oxidative burst is not required for elicitor-mediated activation of secondary metabolism. In *C. reoseus*, ROS generation is neither necessary nor sufficient for the induction of genes involved in the induction of TIA biosynthesis by YE.[75] We have recently observed that a null mutation in the Arabidopsis *AtrbohD* gene completely abolishes the oxidative burst induced by OGs and flg22, but does not affect expression of genes involved in the biosynthesis of camalexin and indole glucosinolates.[88] It is therefore unlikely that, in this system, ROS have an important role in regulating secondary metabolism activated by elicitors.

How ROS mediate elicitor-induced production of secondary metabolites is still not clear. Lipid peroxidation may be initiated either by ROS or by lipoxygenases. Enzymatic oxidation of alpha-linolenate can result in the biosynthesis of cyclic oxylipins of the JA type, while free radical-catalyzed oxidation of alpha-linolenate may yield other classes of cyclic oxylipins termed phytoprostanes, which can also activate defence responses.[89] ROS and JA are key signal components in the stimulation of taxol production in *Taxus chinensis* cells.[90] On the other hand, despite fungal elicitors are able to induce paclitaxel accumulation in *T. brevifolia* cells, the oxidative burst caused by these elicitors leads to cell death and low yield. It is therefore important to analyze both the positive and negative impact of ROS production on elicitor-mediate production of secondary metabolites in plant cell cultures.

Activation of defence responses in plants is largely mediated by secondary signals, whose levels increase locally and systemically upon biotic stress. The most studied of these signalling molecules are SA, ET and JA. Many elicitors can activate the biosynthesis of one or more of these signals in plant cells. The role of JA in regulating secondary metabolism has been extensively investigated. Treatments of plant tissues with JA activate biosynthetic pathways and elicitor-mediated activation of secondary metabolism requires an intact JA-dependent signalling pathway, at least in some plant species. For instance, jasmonic acid and its methylated derivative, methyl jasmonate

(MeJA), regulate the production of indole glucosinolates in Arabidopsis,[91] TIA in *C. roseus*,[17] beta-thujaplicin in *C. lusitanica*[92] and resveratrol in grapevine.[93] SA accumulation is also observed in plant cells treated with different elicitors[94-96] and treatment of Arabidopsis leaves with different PAMPs can activate the systemic biosynthesis of SA, suggesting that their perception is important for the establishment of systemic acquired resistance.[97] SA or its derivatives can induce the accumulation of secondary metabolites, as in the case of taxuyunnanine C in *T. chinensis* cell cultures treated with trifluoroethyl salicylate,[98] or scopolamine and hyoscyamine in *Brugmansia candida* hairy roots treated with SA.[99] Therefore, at least in some cases, SA and its analogues can be used to increase the yield of specific compounds in plant cell cultures. Similarly, ET is able to induce production of some secondary compounds in several plant species.[92,100,101]

Combination of elicitors and secondary signalling chemicals often appears to be very effective in inducing the accumulation of specific metabolites,[102] supporting the idea of a synergistic interaction between general elicitors and secondary signals in the activation of secondary metabolism during pathogen infection. Different reports suggest however that general elicitors can induce defence responses independently of SA, ET and JA. Zhang and colleagues identified genes induced in Arabidopsis in response to chitosan independently of SA, ET and JA.[103] Flg22 pretreatments increase Arabidopsis resistance to bacterial infection also in mutants impaired in SA, ET and JA signalling.[104] More recently, we have shown that pathogen infection and OGs induce Arabidopsis defence responses, including the expression of *PAD3*, encoding a cytochrome P450 required for the biosynthesis of the phytoalexin camalexin,[105] independently of SA-, JA- and ET-mediated signalling, because mutants impaired in these pathways were still protected by OGs against the fungal pathogen *Botrytis cinerea* and this protection requires *PAD3*.[106] Several reports indicate that production of some secondary metabolites after treatment with fungal elicitors is also independent of SA, JA and ET, as in the case of camalexin production in response to fungal infection of Arabidopsis plants.[107]

Emerging Techniques to Improve Secondary Metabolites Production, Based on Elicitor Signalling Pathways

Dissection of signalling networks activated by elicitors will lead to discoveries of more secondary metabolite biosynthetic genes and regulatory factors and will facilitate engineering of the production of target compounds. Taking advantage of advanced metabolic engineering techniques, researchers can construct new metabolic pathways in one species by transferring heterologous genes from other species, for the production of novel compounds. Given the complicated metabolic networks of plant secondary metabolism, the cross-talk among signalling pathways, the interactions of complex primary and secondary metabolic fluxes and various levels of regulatory mechanisms of protein and gene activity, it is difficult to obtain a complete view using traditional techniques. With the development of high through-put approaches, including functional genomics, proteomics and metabolomics, many new tools can now be applied to the study of plant secondary metabolism.

Full genome sequencing has led us to understand that the complexity of plant secondary metabolism is reflected by a similar level of complexity at the gene level. Genomic and biochemical evidence indicates that individual plant genomes are likely to have the hitherto unappreciated potential to produce diverse repertoires of metabolites. This is evident from the analysis of the Arabidopsis genome, which was the first one to be sequenced.[108] Approximately 25% of known *Arabidopsis thaliana* genes are implicated in natural product synthesis. The list of Arabidopsis genes that are involved in specific natural product pathways is growing rapidly, but is by no means complete. Furthermore, over 170 natural products from seven major classes of compounds are reported in Arabidopsis.[109] Many more genomes, in particular from plants of pharmaceutical and nutraceutical interest, are still uncharacterized. From a production perspective, it is important to make sense of the large amount of data that we are currently obtaining from experiments conducted using large-scale, high through-put approaches and find ways to apply this knowledge to improve the production of specific compounds from plant cell cultures. For this reason, identification of genes involved in the biosynthesis of secondary metabolites in response to elicitors and the characterization of their regulation, can open new perspectives in the field. Transcript-profiling analysis of plants treated with pathogens or elicitors can allow us to identify genes involved in secondary metabolism or defence against oxidative stress and therefore of interest for human health. In particular, microarray analysis allows the identification of genes coregulated under specific stimuli through the simultaneous measurement of the expression of thousands of genes in response to different treatments or in different samples. Several papers published in the past few years have focused on the transcript profile of Arabidopsis plants challenged with pathogens or elicitors.[103,110-117] These studies revealed the up- or down-regulation of hundreds of genes upon different treatments. Statistical analysis then can be used to identify genes that are coregulated and whose products are possibly involved in the same transduction cascade or in the same biosynthetic pathway. For instance, a meta-analysis of publicly available sets of Arabidopsis microarray data obtained in a wide range of different stress and developmental conditions has revealed a clear and robust co-expression of genes encoding enzymes of specific metabolic pathways, in particular those involved in the biosynthesis of indoles, phenylpropanoids and flavonoids.[118] The combination of these microarray data with sequence analysis allows to draw very precise hypotheses on the function of otherwise uncharacterized genes. In some cases, functional analysis of selected genes identified by microarray analysis has confirmed their role in defence against pathogens[118,119] and possibly in secondary metabolism regulation. Even when microarrays are not available for the species of interest, other techniques can be used to identify differentially expressed genes, such as Expressed Sequence Tags (ESTs) and cDNA-amplified fragment length polymorphisms (cAFLPs).[120] Once potentially interesting genes are identified, augmenting or repressing their expression in plant cells through stable or transient genetic transformation can confirm their role in the production of natural products and also provide a way to increase their accumulation. For instance, constitutive expression of codeinone reductase in opium poppy resulted in the accumulation of higher levels of morphine.[121] This strategy can be also applied to transfer whole biosynthetic pathways into microbial organisms that can be cultured more easily and at lower cost than plant cells. For instance, two heterologous pathways for the biosynthesis of plant-derived terpenoid natural products were recently reconstructed in *Escherichia coli* by the expression of native plant cytochrome P450s.[122]

Beside the manipulation of single biosynthetic enzymes, it is possible to alter metabolic fluxes to a larger extent increasing or decreasing the expression of endogenous regulatory proteins that normally fine-tune the accumulation of secondary metabolites in response to elicitors. The most widely applied approach to this goal is the overexpression of transcription factors involved in the regulation of specific compounds (for review, see ref. 11). For instance, *C. roseus* transcription factors induced by JA treatment and able to bind to the promoter of the strictosidine synthase gene have been identified[123-126] and overexpression of one of them clearly increased the production of terpene indole alkaloids in transgenic cells.[125]

Profiling the whole or a group of secondary metabolites from plants under various treatments can also help understand metabolic fluxes and the underlying regulatory mechanisms and, in particular, in many pathways with only a limited number of characterized genes or enzymes, or in species where the available genomic information is limited. Monitoring a whole set of metabolites after treatment with different elicitors is an especially powerful way to learn the effects of the activation of defence responses and the related transduction pathways leading to the accumulation of natural products. Various high resolution spectrometry techniques and in particular gas or liquid chromatography coupled to mass spectrometry (GC-MS and LC-MS), are now available to separate and identify a large number of low molecular weight compounds in a plant sample.[127,128] A great deal of data has been collected from metabolic profile analyses of Arabidopsis plants challenged with different biotic stimuli. Walker and colleagues profiled the metabolites secreted by Arabidopsis roots after treatment with MeJA, SA and fungal elicitors.[129] Changes in the levels of indolic and phenylpropanoid secondary metabolites in wild-type and mutant Arabidopsis roots infected with *Pythium sylvaticum* have also been reported,[130] indicating that most of the indolics increase strongly upon infection, whereas the phenylpropanoids decrease. Metabolic profiles of plants of interest for the production of natural products are also available. Analysis of elicitor-induced *Taxus* cell suspension cultures, for instance, revealed that MeJA induces the accumulation of taxols oxygenated at C position 13, but not 14.[131] Recently, GC-MS-based metabolite profiling was also used to analyse the response of *M. truncatula* cell cultures to elicitation with MeJA, YE or UV light.[132]

Combining the approaches of transcription profiling, proteomics and metabolomics offers a powerful tool in studying all aspects of plant secondary metabolism as a whole. A study by Hirai and colleagues describes the analysis of metabolomics and transcriptomics data of *Arabidopsis* plants grown under sulfur deficiency.[133] This study indicates that groups of metabolites/genes regulated by the same mechanism cluster together. The same research group recently combined transcript and metabolite profiling approaches to identify two novel Arabidopsis MYB factors that regulate glucosinolate production in response to MeJA treatment.[134] This work is a good example of how metabolomics and transcriptomics can be employed to identify new genes that can be used to engineer plants with increased production of secondary metabolites.

Conclusion

In conclusion, more studies on the signal transduction pathways that link elicitor perception to the biosynthesis of specific secondary metabolites and high through-put analysis of gene transcripts, proteins and metabolites in elicited plants will be essential for better understanding the regulation of plant secondary metabolism and for engineering new plants with improved production of metabolites of interest.

Acknowledgements

This work was supported by the Italian Ministry of University and Research (grant PRIN2006) and by the Italian National Research Council.

References

1. Wink M. Evolution of secondary metabolites from an ecological and molecular phylogenetic perspective. Phytochemistry 2003; 64:3-19.
2. Morrissey JP, Osbourn AE. Fungal resistance to plant antibiotics as a mechanism of pathogenesis. Microbiol Mol Biol Rev 1999; 63:708-724.
3. Zhao J, Davis LC, Verpoorte R. Elicitor signal transduction leading to production of plant secondary metabolites. Biotechnology Advances 2005; 23:283-333.
4. VanEtten HD, Mansfield JW, Bailey JA et al. Two Classes of Plant Antibiotics: Phytoalexins versus "Phytoanticipins". Plant Cell 1994; 6:1191-1192.
5. Hammerschmidt R, Dann EK. The role of phytoalexins in plant protection. Novartis Found Symp 1999; 223:175-87.
6. Müller KO, Borger H. Experimentelle untersuchungen über die phytophthorainfestans-resistenz der kartoffel. Arb Biol Reichsanst Land Forstwirtsch 1940; 23:189-231.
7. Müller KO. Studies on phytoalexins: I. The formation and the immunological significance of phytoalexin produced by Phaseolus vulgaris in response to infections with Sclerotinia fructicola and Phytophthora infestans. Aust J Biol Sci 1958; 11:275-300.
8. Paxton J. Phytoalexins—a working redefinition. Phytopathol Z 1981; 101:106-109.
9. Lamb C, Dixon RA. The oxidative burst in plant disease resistance. Annu Rev Plant Physiol Plant Mol Biol 1997; 48:251-275.
10. Zhong JJ. Biochemical engineering of the production of plant-specific secondary metabolites by cell suspension cultures. Adv Biochem Eng Biotechnol 2001; 72:1-26.
11. Memelink J, Kijne JW, van der HR et al. Genetic modification of plant secondary metabolite pathways using transcriptional regulators. Adv Biochem Eng Biotechnol 2001; 72:103-125.
12. Sato F, Hashimoto T, Hachiya A et al. Metabolic engineering of plant alkaloid biosynthesis. Proc Natl Acad Sci USA 2001; 98:367-372.
13. He P, Shan L, Sheen J. Elicitation and suppression of microbe-associated molecular pattern-triggered immunity in plant-microbe interactions. Cell Microbiol 2007; 9:1385-1396.
14. Parker JE. Plant recognition of microbial patterns. Trends Plant Sci 2003; 8:245-247.
15. Nurnberger T, Brunner F, Kemmerling B et al. Innate immunity in plants and animals: striking similarities and obvious differences. Immunol Rev 2004; 198:249-266.
16. Granado J, Felix G, Boller T. Perception of Fungal Sterols in Plants (Subnanomolar Concentrations of Ergosterol Elicit Extracellular Alkalinization in Tomato Cells). Plant Physiol 1995; 107:485-490.

17. Menke FL, Parchmann S, Mueller MJ et al. Involvement of the octadecanoid pathway and protein phosphorylation in fungal elicitor-induced expression of terpenoid indole alkaloid biosynthetic genes in catharanthus roseus. Plant Physiol 1999; 119:1289-1296.
18. Nurnberger T, Brunner F. Innate immunity in plants and animals: emerging parallels between the recognition of general elicitors and pathogen-associated molecular patterns. Curr Opin Plant Biol 2002; 5:318-324.
19. Kessler A, Baldwin IT. Plant responses to insect herbivory: the emerging molecular analysis. Annu Rev Plant Biol 2002; 53:299-328.
20. Schmelz EA, Carroll MJ, Leclere S et al. Fragments of ATP synthase mediate plant perception of insect attack. Proc Natl Acad Sci USA 2006; 103:8894-8899.
21. Sirvent TM, Krasnoff SB, Gibson DM. Induction of hypericins and hyperforins in Hypericum perforatum in response to damage by herbivores. J Chem Ecol 2003; 29:2667-2681.
22. Turlings TC, Loughrin JH, McCall PJ et al. How caterpillar-damaged plants protect themselves by attracting parasitic wasps. Proc Natl Acad Sci USA 1995; 92:4169-4174.
23. McGurl B, Pearce G, OrozcoCardenas M et al. Structure, expression and antisense inhibition of the systemin precursor gene. Science 1992; 255:1570-1573.
24. Doares SH, Syrovets T, Weiler EW et al. Oligogalacturonides and chitosan activate plant defensive genes through the octadecanoid pathway. Proc Natl Acad Sci USA 1995; 92:4095-8.
25. McGurl B, Pearce G, Ryan CA. Polypeptide signalling for plant defence genes. Biochem Soc Symp 1994; 60:149-154.
26. Constabel CP, Bergey DR, Ryan CA. Systemin activates synthesis of wound-inducible tomato leaf polyphenol oxidase via the octadecanoid defense signaling pathway. Proc Natl Acad Sci USA 1995; 92:407-411.
27. Funk C, Brodelius P. Phenylpropanoid metabolism in suspension cultures of vanilla planifolia andr: IV. Induction of vanillic acid formation. Plant Physiol 1992;99(1):256-262.
28. Lo SC, Nicholson RL. Reduction of light-induced anthocyanin accumulation in inoculated sorghum mesocotyls. Implications for a compensatory role in the defense response. Plant Physiol 1998; 116:979-989.
29. Gomez-Gomez L, Boller T. FLS2: an LRR receptor-like kinase involved in the perception of the bacterial elicitor flagellin in Arabidopsis. Mol Cell 2000; 5:1003-1011.
30. Chinchilla D, Bauer Z, Regenass M et al. The Arabidopsis receptor kinase FLS2 binds flg22 and determines the specificity of flagellin perception. Plant Cell 2006; 18:465-476.
31. Zipfel C, Kunze G, Chinchilla D et al. Perception of the bacterial PAMP EF-Tu by the receptor EFR restricts Agrobacterium-mediated transformation. Cell 2006; 125:749-760.
32. Kaku H, Nishizawa Y, Ishii-Minami N et al. Plant cells recognize chitin fragments for defense signaling through a plasma membrane receptor. Proc Natl Acad Sci USA 2006; 103:11086-11091.
33. Scheer JM Ryan CA Jr. The systemin receptor SR160 from Lycopersicon peruvianum is a member of the LRR receptor kinase family. Proc Natl Acad Sci USA 2002; 99:9585-9590.
34. Ji C, Okinaka Y, Takeuchi Y et al. Specific Binding of the Syringolide Elicitors to a Soluble Protein Fraction from Soybean Leaves. Plant Cell 1997; 9:1425-1433.
35. Garcia-Brugger A, Lamotte O, Vandelle E et al. Early signaling events induced by elicitors of plant defenses. Mol Plant Microbe Interact 2006; 19:711-724.
36. Bolwell GP, Bindschedler LV, Blee KA et al. The apoplastic oxidative burst in response to biotic stress in plants: a three-component system. J Exp Bot 2002; 53:1367-1376.
37. Doke N, Miura Y, Sanchez LM et al. The oxidative burst protects plants against pathogen attack: mechanism and role as an emergency signal for plant bio-defence—a review. Gene 1996; 179:45-51.
38. Mahady GB, Liu C, Beecher CW. Involvement of protein kinase and G proteins in the signal transduction of benzophenanthridine alkaloid biosynthesis. Phytochemistry 1998; 48:93-102.
39. Rajasekhar VK, Lamb C, Dixon RA. Early events in the signal pathway for the oxidative burst in soybean cells exposed to avirulent pseudomonas syringae pv glycinea. Plant Physiol 1999; 120:1137-1146.
40. Zhao J, Sakai K. Multiple signalling pathways mediate fungal elicitor-induced beta-thujaplicin biosynthesis in Cupressus lusitanica cell cultures. J Exp Bot 2003; 54:647-656.
41. Trewavas AJ, Malho R. Ca2+ signalling in plant cells: the big network! Curr Opin Plant Biol 1998; 1:428-433.
42. White PJ, Broadley MR. Calcium in plants. Ann Bot (Lond) 2003; 92:487-511.
43. Umemura K, Ogawa N, Koga J et al. Elicitor activity of cerebroside, a sphingolipid elicitor, in cell suspension cultures of rice. Plant Cell Physiol 2002; 43:778-784.
44. Vogeli U, Vogeli-Lange R, Chappell J. Inhibition of phytoalexin biosynthesis in elicitor-treated tobacco cell-suspension cultures by calcium/calmodulin antagonists. Plant Physiol 1992; 100:1369-1376.
45. Chandra S, Heinstein PF, Low PS. Activation of phospholipase A by plant defense elicitors. Plant Physiol 1996; 110:979-986.

46. Kasparovsky T, Blein JP, Mikes V. Ergosterol elicits oxidative burst in tobacco cells via phospholipase A2 and protein kinase C signal pathway. Plant Physiol Biochem 2004; 42:429-435.
47. Legendre L, Yueh YG, Crain R et al. Phospholipase C activation during elicitation of the oxidative burst in cultured plant cells. J Biol Chem 1993; 268:24559-24563.
48. van der Luit AH, Piatti T, van Doorn A et al. Elicitation of suspension-cultured tomato cells triggers the formation of phosphatidic acid and diacylglycerol pyrophosphate. Plant Physiol 2000; 123:1507-1516.
49. Kurosaki F, Tsurusawa Y, Nishi A. Breakdown of phosphatidylinositol during the elicitation of phytoalexin production in cultured carrot cells. Plant Physiol 1987; 85:601-604.
50. Walton TJ, Cooke CJ, Newton RP et al. Evidence that generation of inositol 1,4,5-trisphosphate and hydrolysis of phosphatidylinositol 4,5-bisphosphate are rapid responses following addition of fungal elicitor which induces phytoalexin synthesis in lucerne (Medicago sativa) suspension culture cells. Cell Signal 1993; 5:345-356.
51. Zhao J, Guo Y, Kosaihira A et al. Rapid accumulation and metabolism of polyphosphoinositol and its possible role in phytoalexin biosynthesis in yeast elicitor-treated Cupressus lusitanica cell cultures. Planta 2004; 219:121-131.
52. Benschop JJ, Mohammed S, O'flaherty M et al. Quantitative phospho-proteomics of early elicitor signalling in Arabidopsis. Mol Cell Proteomics 2007;6(7):1198-214.
53. Nuhse TS, Boller T, Peck SC. A plasma membrane syntaxin is phosphorylated in response to the bacterial elicitor flagellin. J Biol Chem 2003; 278:45248-45254.
54. Peck SC, Nuhse TS, Hess D et al. Directed proteomics identifies a plant-specific protein rapidly phosphorylated in response to bacterial and fungal elicitors. Plant Cell 2001; 13:1467-1475.
55. Allwood EG, Davies DR, Gerrish C et al. Phosphorylation of phenylalanine ammonia-lyase: evidence for a novel protein kinase and identification of the phosphorylated residue. FEBS Lett 1999; 457:47-52.
56. Asai T, Tena G, Plotnikova J et al. MAP kinase signalling cascade in Arabidopsis innate immunity. Nature 2002; 415:977-983.
57. Petersen M, Brodersen P, Naested H et al. Arabidopsis map kinase 4 negatively regulates systemic acquired resistance. Cell 2000; 103:1111-20.
58. Droillard MJ, Thibivilliers S, Cazale AC et al. Protein kinases induced by osmotic stresses and elicitor molecules in tobacco cell suspensions: two crossroad MAP kinases and one osmoregulation-specific protein kinase. FEBS Lett 2000; 474:217-22.
59. Kumar D, Klessig DF. Differential induction of tobacco MAP kinases by the defense signals nitric oxide, salicylic acid, ethylene and jasmonic acid. Mol Plant Microbe Interact 2000; 13:347-51.
60. Romeis T, Piedras P, Zhang S et al. Rapid Avr9- and Cf-9 -dependent activation of MAP kinases in tobacco cell cultures and leaves: convergence of resistance gene, elicitor, wound and salicylate responses. Plant Cell 1999; 11:273-87.
61. Kovtun Y, Chiu WL, Tena G et al. Functional analysis of oxidative stress-activated mitogen-activated protein kinase cascade in plants. Proc Natl Acad Sci USA 2000; 97:2940-5.
62. Nuhse TS, Peck SC, Hirt H et al. Microbial elicitors induce activation and dual phosphorylation of the Arabidopsis thaliana MAPK 6. J Biol Chem 2000; 275:7521-6.
63. Cardinale, F, Jonak C, Ligterink W et al. Differential activation of four specific MAPK pathways by distinct elicitors. J Biol Chem 2000; 275:36734-36740.
64. Cardinale F, Meskiene I, Ouaked F et al. Convergence and divergence of stress-induced mitogen-activated protein kinase signaling pathways at the level of two distinct mitogen-activated protein kinase kinases. Plant Cell 2002; 14:703-711.
65. Ligterink W, Kroj T, zur NU et al. Receptor-mediated activation of a MAP kinase in pathogen defense of plants. Science 1997; 276:2054-2057.
66. Yang KY, Liu Y, Zhang S. Activation of a mitogen-activated protein kinase pathway is involved in disease resistance in tobacco. Proc Natl Acad Sci USA 2001; 98:741-746.
67. Hu X, Neill SJ, Fang J et al. Mitogen-activated protein kinases mediate the oxidative burst and saponin synthesis induced by chitosan in cell cultures of Panax ginseng. Sci China C Life Sci 2004; 47:303-312.
68. Vandelle E, Poinssot B, Wendehenne D et al. Integrated signaling network involving calcium, nitric oxide and active oxygen species but not mitogen-activated protein kinases in BcPG1-elicited grapevine defenses. Mol Plant Microbe Interact 2006; 19:429-440.
69. Levine A, Tenhaken R, Dixon R et al. H2O2 from the oxidative burst orchestrates the plant hypersensitive disease resistance response. Cell 1994; 79:583-593.
70. Apostol I, Heinstein PF, Low PS. Rapid stimulation of an oxidative burst during elicitation of cultured plant cells. Plant Physiol 1989; 90:109-116.
71. Aziz A, Poinssot B, Daire X et al. Laminarin elicits defense responses in grapevine and induces protection against Botrytis cinerea and Plasmopara viticola. Mol Plant Microbe Interact 2003; 16:1118-1128.

72. Legendre L, Rueter S, Heinstein PF et al. Characterization of the oligogalacturonide-induced oxidative burst in cultured soybean (glycine max) cells. Plant Physiol 1993; 102:233-240.
73. Meyer A, Puhler A, Niehaus K. The lipopolysaccharides of the phytopathogen Xanthomonas campestris pv. campestris induce an oxidative burst reaction in cell cultures of Nicotiana tabacum. Planta 2001; 213:214-222.
74. Mithofer A, Fliegmann J, Daxberger A et al. Induction of H(2)O(2) synthesis by beta-glucan elicitors in soybean is independent of cytosolic calcium transients. FEBS Lett 2001; 508:191-195.
75. Pauw B, van Duijn B, Kijne JW et al. Activation of the oxidative burst by yeast elicitor in Catharanthus roseus cells occurs independently of the activation of genes involved in alkaloid biosynthesis. Plant Mol Biol 2004; 55:797-805.
76. Xu X, Hu X, Neill SJ et al. Fungal elicitor induces singlet oxygen generation, ethylene release and saponin synthesis in cultured cells of Panax ginseng C. A. Meyer. Plant Cell Physiol 2005; 46:947-954.
77. Desikan R, Hancock JT, Coffey MJ et al. Generation of active oxygen in elicited cells of Arabidopsis thaliana is mediated by a NADPH oxidase-like enzyme. FEBS Lett 1996; 382:213-217.
78. Groom QJ, Torres MA, Fordham-Skelton AP et al. rbohA, a rice homologue of the mammalian gp91phox respiratory burst oxidase gene. Plant J 1996; 10:515-522.
79. Keller T, Damude HG, Werner D et al. A plant homolog of the neutrophil NADPH oxidase gp91phox subunit gene encodes a plasma membrane protein with Ca2+ binding motifs. Plant Cell 1998; 10:255-266.
80. Torres MA, Dangl JL, Jones JD. Arabidopsis gp91phox homologues AtrbohD and AtrbohF are required for accumulation of reactive oxygen intermediates in the plant defense response. Proc Natl Acad Sci USA 2002; 99:517-522.
81. Torres MA, Jones JD, Dangl JL. Pathogen-induced NADPH oxidase-derived reactive oxygen intermediates suppress spread of cell death in Arabidopsis thaliana. Nat Genet 2005; 37:1130-1134.
82. Jabs T, Tschope M, Colling C et al. Elicitor-stimulated ion fluxes and O2- from the oxidative burst are essential components in triggering defense gene activation and phytoalexin synthesis in parsley. Proc Natl Acad Sci USA 1997; 94:4800-4805.
83. Degousee N, Triantaphylides C, Montillet JL. Involvement of oxidative processes in the signaling mechanisms leading to the activation of glyceollin synthesis in soybean (glycine max). Plant Physiol 1994; 104:945-952.
84. Guo ZJ, Lamb C, Dixon RA. Potentiation of the oxidative burst and isoflavonoid phytoalexin accumulation by serine protease inhibitors. Plant Physiol 1998; 118:1487-1494.
85. Matsuda F, Miyagawa H, Ueno T. Involvement of reactive oxygen species in the induction of (S)-N-p-coumaroyloctopamine accumulation by beta-1,3-glucooligosaccharide elicitors in potato tuber tissues. Z Naturforsch (C) 2001; 56:228-234.
86. Kravchuk Z, Perkovs'ka HI, Dmytriiev OP. (Role of active forms of oxygen in the induction of phytoalexin synthesis in Allium cepa cells). Tsitol Genet 2003; 37:30-35.
87. Yamaguchi T, Tanabe S, Minami E et al. Activation of phospholipase D induced by hydrogen peroxide in suspension-cultured rice cells. Plant Cell Physiol 2004; 45:1261-1270.
88. Galletti R, Denoux C, Gambetta S et al. The AtrbohD-mediated oxidative burst elicited by oligogalacturonides in Arabidopsis is dispensable for the activation of defense responses effective against Botrytis cinerea. Plant Physiol 2008; 148:1695-1706.
89. Thoma I, Loeffler C, Sinha AK et al. Cyclopentenone isoprostanes induced by reactive oxygen species trigger defense gene activation and phytoalexin accumulation in plants. Plant J 2003; 34:363-375.
90. Wu J, Ge X. Oxidative burst, jasmonic acid biosynthesis and taxol production induced by low-energy ultrasound in Taxus chinensis cell suspension cultures. Biotechnol Bioeng 2004; 85:714-721.
91. Brader G, Tas E, Palva ET. Jasmonate-dependent induction of indole glucosinolates in Arabidopsis by culture filtrates of the nonspecific pathogen Erwinia carotovora. Plant Physiol 2001; 126:849-860.
92. Zhao J, Zheng SH, Fujita K et al. Jasmonate and ethylene signalling and their interaction are integral parts of the elicitor signalling pathway leading to beta-thujaplicin biosynthesis in Cupressus lusitanica cell cultures. J Exp Bot 2004; 55:1003-1012.
93. Tassoni A, Fornale S, Franceschetti M et al. Jasmonates and Na-orthovanadate promote resveratrol production in Vitis vinifera cv. Barbera cell cultures. New Phytol 2005; 166:895-905.
94. Clarke A, Mur LA, Darby RM et al. Harpin modulates the accumulation of salicylic acid by Arabidopsis cells via apoplastic alkalization. J Exp Bot 2005; 56:3129-3136.
95. Dorey S, Kopp M, Geoffroy P et al. Hydrogen peroxide from the oxidative burst is neither necessary nor sufficient for hypersensitive cell death induction, phenylalanine ammonia lyase stimulation, salicylic acid accumulation, or scopoletin consumption in cultured tobacco cells treated with elicitin. Plant Physiol 1999; 121:163-172.
96. Klarzynski O, Plesse B, Joubert JM et al. Linear beta-1,3 glucans are elicitors of defense responses in tobacco. Plant Physiol 2000; 124:1027-38.

97. Mishina TE, Zeier J. Pathogen-associated molecular pattern recognition rather than development of tissue necrosis contributes to bacterial induction of systemic acquired resistance in Arabidopsis. Plant J 2007; 50:500-513.
98. Qian ZG, Zhao ZJ, Xu Y et al. Novel chemically synthesized salicylate derivative as an effective elicitor for inducing the biosynthesis of plant secondary metabolites. Biotechnol Prog 2006; 22:331-333.
99. Pitta-Alvarez SI, Spollansky TC, Giulietti AM. The influence of different biotic and abiotic elicitors on the production and profile of tropane alkaloids in hairy root cultures of Brugmansia candida. Enzyme Microb Technol 2000; 26:252-258.
100. Alves LM, Kalan EB, Heisler EG. An in vitro control mechanism for potato stress metabolite biosynthesis. Plant Physiol 1981; 68:1465-1467.
101. Nakazato Y, Tamogami S, Kawai H et al. Methionine-induced phytoalexin production in rice leaves. Biosci Biotechnol Biochem 2000; 64:577-583.
102. Khosroushahi AY, Valizadeh M, Ghasempour A et al. Improved Taxol production by combination of inducing factors in suspension cell culture of Taxus baccata. Cell Biol Int 2006; 30:262-269.
103. Zhang B, Ramonell K, Somerville S et al. Characterization of early, chitin-induced gene expression in Arabidopsis. Mol Plant Microbe Interact 2002; 15:963-970.
104. Zipfel C, Robatzek S, Navarro L et al. Bacterial disease resistance in Arabidopsis through flagellin perception. Nature 2004; 428:764-767.
105. Schuhegger R, Nafisi M, Mansourova M et al. CYP71B15 (PAD3) catalyzes the final step in camalexin biosynthesis. Plant Physiol 2006; 141:1248-1254.
106. Ferrari S, Galletti R, Denoux C et al. Resistance to botrytis cinerea induced in Arabidopsis by elicitors is independent of salicylic acid, ethylene or jasmonate signaling but requires PAD3. Plant Physiol 2007; 144:367-379.
107. Thomma BP, Nelissen I, Eggermont K et al. Deficiency in phytoalexin production causes enhanced susceptibility of Arabidopsis thaliana to the fungus Alternaria brassicicola. Plant J 1999; 19:163-71.
108. The Arabidopsis Genome Initiative. Analysis of the genome sequence of the flowering plant Arabidopsis thaliana. Nature 2000; 408:796-815.
109. D'Auria JC, Gershenzon J. The secondary metabolism of Arabidopsis thaliana: growing like a weed. Curr Opin Plant Biol 2005; 8, 308-316.
110. Eulgem T. Regulation of the Arabidopsis defense transcriptome. Trends Plant Sci 2005; 10:71-78.
111. Ferrari S, Galletti R, Denoux C et al. Resistance to botrytis cinerea induced in Arabidopsis by elicitors is independent of salicylic acid, ethylene or jasmonate signaling but requires PAD3. Plant Physiol 2007; 144:367-379.
112. Huang X, von Rad U, Durner J. Nitric oxide induces transcriptional activation of the nitric oxide-tolerant alternative oxidase in Arabidopsis suspension cells. Planta 2002; 215:914-923.
113. Marathe R, Guan Z, Anandalakshmi R et al. Study of Arabidopsis thaliana resistome in response to cucumber mosaic virus infection using whole genome microarray. Plant Mol Biol 2004; 55:501-520.
114. Narusaka Y, Narusaka M, Seki M et al. The cDNA microarray analysis using an Arabidopsis pad3 mutant reveals the expression profiles and classification of genes induced by Alternaria brassicicola attack. Plant Cell Physiol 2003; 44:377-387.
115. Navarro L, Zipfel C, Rowland O et al. The transcriptional innate immune response to flg22. Interplay and overlap with Avr gene-dependent defense responses and bacterial pathogenesis. Plant Physiol 2004; 135:1113-1128.
116. Schenk PM, Kazan K, Wilson I et al. Coordinated plant defense responses in Arabidopsis revealed by microarray analysis. Proc Natl Acad Sci USA 2000; 97:11655-11660.
117. Suzuki H, Reddy MS, Naoumkina M et al. Methyl jasmonate and yeast elicitor induce differential transcriptional and metabolic reprogramming in cell suspension cultures of the model legume Medicago truncatula. Planta 2005; 220:696-707.
118. Gachon CM, Langlois-Meurinne M, Henry Y et al. Transcriptional coregulation of secondary metabolism enzymes in Arabidopsis: functional and evolutionary implications. Plant Mol Biol 2005; 58:229-245.
119. Ramonell K, Berrocal-Lobo M, Koh S et al. Loss-of-function mutations in chitin responsive genes show increased susceptibility to the powdery mildew pathogen erysiphe cichoracearum. Plant Physiol 2005; 138:1027-1036.
120. Goossens A, Hakkinen ST, Laakso I et al. A functional genomics approach toward the understanding of secondary metabolism in plant cells. Proc Natl Acad Sci USA 2003; 100:8595-8600.
121. Larkin PJ, Miller JA, Allen RS et al. Increasing morphinan alkaloid production by over-expressing codeinone reductase in transgenic Papaver somniferum. Plant Biotechnol J 2007; 5:26-37.
122. Chang MC, Eachus RA, Trieu W et al. Engineering Escherichia coli for production of functionalized terpenoids using plant P450s. Nat Chem Biol 2007; 3:274-277.

123. Chatel G, Montiel G, Pre M et al. CrMYC1, a Catharanthus roseus elicitor- and jasmonate-responsive bHLH transcription factor that binds the G-box element of the strictosidine synthase gene promoter. J Exp Bot 2003; 54:2587-2588.
124. Menke FL, Champion A, Kijne JW et al. A novel jasmonate- and elicitor-responsive element in the periwinkle secondary metabolite biosynthetic gene Str interacts with a jasmonate- and elicitor-inducible AP2-domain transcription factor ORCA2. EMBO J 1999; 18:4455-4463.
125. van der Fits L, Memelink J. ORCA3, a jasmonate-responsive transcriptional regulator of plant primary and secondary metabolism. Science 2000; 289:295-297.
126. van der Fits L, Memelink J. The jasmonate-inducible AP2/ERF-domain transcription factor ORCA3 activates gene expression via interaction with a jasmonate-responsive promoter element. Plant J 2001; 25:43-53.
127. Bino RJ, Hall RD, Fiehn O et al. Potential of metabolomics as a functional genomics tool. Trends Plant Sci 2004; 9:418-425.
128. Hall R, Beale M, Fiehn O et al. Plant metabolomics: the missing link in functional genomics strategies. Plant Cell 2002; 14:1437-1440.
129. Walker TS, Bais HP, Halligan KM et al. Metabolic profiling of root exudates of Arabidopsis thaliana. J Agric Food Chem 2003; 51:2548-2554.
130. Bednarek P, Schneider B, Svatos A et al. Structural complexity, differential response to infection and tissue specificity of indolic and phenylpropanoid secondary metabolism in Arabidopsis roots. Plant Physiol 2005; 138:1058-1070.
131. Ketchum RE, Rithner CD, Qiu D et al. Taxus metabolomics: methyl jasmonate preferentially induces production of taxoids oxygenated at C-13 in Taxus x media cell cultures. Phytochemistry 2003; 62:901-909.
132. Broeckling CD, Huhman DV, Farag MA et al. Metabolic profiling of Medicago truncatula cell cultures reveals the effects of biotic and abiotic elicitors on metabolism. J Exp Bot 2005; 56:323-336.
133. Hirai MY, Fujiwara T, Awazuhara M et al. Global expression profiling of sulfur-starved Arabidopsis by DNA macroarray reveals the role of O-acetyl-l-serine as a general regulator of gene expression in response to sulfur nutrition. Plant J 2003; 33:651-663.
134. Hirai MY, Sugiyama K, Sawada Y et al. Omics-based identification of Arabidopsis Myb transcription factors regulating aliphatic glucosinolate biosynthesis. Proc Natl Acad Sci USA 2007; 104:6478-6483.
135. Hahn MG, Darvill AG, Albersheim P. Host-Pathogen Interactions : XIX. The endogenous elicitor, a fragment of a plant cell wall polysaccharide that elicits phytoalexin accumulation in soybeans. Plant Physiol 1981; 68:1161-1169.
136. Hadwiger LA, Beckman JM. Chitosan as a component of pea-fusarium solani interactions. Plant Physiol 1980; 66:205-211.
137. Keen NT, Yoshikawa M, Wang MC. Phytoalexin elicitor activity of carbohydrates from phytophthora megasperma f.Sp. Glycinea and other sources. Plant Physiol 1983; 71:466-471.
138. Ricci P, Bonnet P, Huet JC et al. Structure and activity of proteins from pathogenic fungi phytophthora eliciting necrosis and acquired resistance in tobacco. Eur J Biochem 1989; 183:555-563.
139. Brunner F, Rosahl S, Lee J et al. Pep-13:a plant defense-inducing pathogen-associated pattern from Phytophthora transglutaminases. EMBO J 2002; 21:6681-6688.
140. Gomez-Gomez L, Felix G, Boller T. A single locus determines sensitivity to bacterial flagellin in Arabidopsis thaliana. Plant J 1999; 18:277-284.
141. Lotan T, Fluhr R. Xylanase, a novel elicitor of pathogenesis-related proteins in tobacco, uses a non-ethylene pathway for induction. Plant Physiol 1990; 93:811-817.
142. Poinssot B, Vandelle E, Bentejac M et al. The endopolygalacturonase 1 from Botrytis cinerea activates grapevine defense reactions unrelated to its enzymatic activity. Mol Plant Microbe Interact 2003; 16:553-564.
143. Kunze G, Zipfel C, Robatzek S et al. The N terminus of bacterial elongation factor Tu elicits innate immunity in Arabidopsis plants. Plant Cell 2004; 16:3496-3507.

Chapter 13

Hairy Root Cultures for Secondary Metabolites Production

Laura Pistelli* Annalisa Giovannini, Barbara Ruffoni, Alessandra Bertoli and Luisa Pistelli

Abstract

Hairy roots (HRs) are differentiated cultures of transformed roots generated by the infection of wounded higher plants with *Agrobacterium rhizogenes*. This pathogen causes the HR disease leading to the neoplastic growth of roots that are characterized by high growth rate in hormone free media and genetic stability. HRs produce the same phytochemicals pattern of the corresponding wild type organ. High stability and productivity features allow the exploitation of HRs as valuable biotechnological tool for the production of plant secondary metabolites. In addition, several elicitation methods can be used to further enhance their accumulation in both small and large scale production. However, in the latter case, cultivation in bioreactors should be still optimized. HRs can be also utilised as biological farm for the production of recombinant proteins, hence holding additional potential for industrial use. HR technology has been strongly improved by increased knowledge of molecular mechanisms underlying their development. The present review summarizes updated aspects of the hairy root induction, genetics and metabolite production.

Introduction

Agrobacterium rhizogenes is a Gram negative soil-borne bacterium of the family *Rhizobiaceae*, that causes the hairy roots disease by infecting wounded higher plants. The pathogen transfers a DNA segment (T-DNA region bounded by 25 bp direct oligonucleotide repeats) from its large root-inducing (Ri) plasmid into the genome of the infected plant. This T-DNA carries a set of genes that encode enzymes for the phytohormone auxin control and cytokinin biosynthesis (*iaaM, iaaH, ipt*) and encoding genes for opines (unusual aminoacids). The new hormonal balance induces the formation of proliferating roots, called hairy roots (HR), which emerge at the wounded site.[1] The hairy root phenotype is characterized by fast hormone-independent growth, lack of geotropism, lateral branching and genetic stability. Owing to their stable and high productivity, hairy root cultures have been investigated for several decades because of the potential to produce valuable metabolites that are present in wild type roots.[2] Long-term aseptic hairy root cultures have been established from more than 200 species of higher plants, due to their ability to synthesize a wide diversity of secondary metabolites and to adjust their metabolic activities in response to biotic and abiotic stress.[3] Recent progress allows a better understanding of the molecular mechanisms involved in the transfer of T-DNA into plants.[4] In this regard *A. rhizogenes* behaves similarly to *Agrobacterium tumefaciens. A. tumefaciens* genetically transforms

*Corresponding Author: Laura Pistelli—Department of Plant Crop Biology via Mariscoglio 34, I-56124 Pisa, Italy. Email: pistelli@agr.unipi.it

Bio-Farms for Nutraceuticals: Functional Food and Safety Control by Biosensors
edited by Maria Teresa Giardi, Giuseppina Rea and Bruno Berra.

its host by transferring a well-defined DNA segment from its tumor-inducing (Ti) plasmid to the host-cell genome. In nature, the transferred DNA (T-DNA) carries a set of oncogenes and opine-catabolism genes, whose expression in plant cells leads to neoplastic growth of the transformed tissue and opines production. These compounds are amino acid derivatives used almost exclusively by the bacteria as a nitrogen source. The molecular machinery needed for T-DNA production and transport into the host cell comprises proteins that are encoded by a set of bacterial chromosomal (*chv*) and Ti plasmid virulence (*vir)* genes. After activation by signals secreted by the host plant (acetosyringone, a phenolic molecule that is a potent inducer of *Agrobacterium*) a subset of *Agrobacterium* virulence (*vir*) genes is expressed, leading to the excision of a single-stranded copy of the T-DNA, from the T-DNA region on the Ti plasmid, by the VirD1/D2 protein complex. VirD2 is a strand-specific endonuclease which cuts the T-DNA border repeats, attaches covalently to one end and initiates packaging of one strand into a linear complex. The mobile single-stranded T-DNA (T-strand), covalently linked at its 50 end to VirD2 (VirD2-DNA complex), is then exported to the host cell by a T4SS, encoded by the virB operon and the virD4 gene. Independently of the VirD2-T-strand complex (immature T-complex), several Vir proteins, VirE2, VirE3, VirF and VirD5, are also exported by the same VirB and VirD4 channel into the host cell cytoplasm. Once inside the host cell, the VirD2-T-strand conjugate is coated with VirE2 molecules and produces a mature T-complex that is then imported into the host cell nucleus, with the assistance of several host and bacterial factors, uncoated by targeted proteolysis and integrated into the plant genome, by so far undetermined mechanism.

The T-DNA transfer is not sequence specific. This feature enables the replacement of the native T-DNA genes with any gene(s) of interest and its delivering in host cells.[5] There can be different patterns of T-DNA integration depending on the origin of a particular Ti plasmid. The Ti plasmids that specify the synthesis of the opine, octopine, T-DNA tends to be integrated into two separate sections derived from the leftmost (TL-DNA) and rightmost (TR-DNA) segments of the T-DNA. Octopine TL-DNA is long 12 kb, encodes seven genes and is alone responsible for the tumor phenotype. On the other hand the T-DNA from *A. tumefaciens* strains which encodes synthesis of the opine, nopaline, tends to integrate as a single tandem repeat. Nopaline T-DNA is 23 kb and encodes thirteen genes. The T-DNA of the *A. rhizogenes* A4 agropine strain consists of two not contiguous stretches of DNA, the T_L-DNA and the T_R-DNA. The T_R-DNA contains two genes homologous to the *iaaM* and *iaaH* genes of Ti plasmids.[6] Four loci involved in the hairy root formation have been identified through insertional mutagenesis in the TL-DNA. They do not show any homology with the T-DNA of Ti plasmids and were called root loci (rol) *A, B, C* and *D*.[7]

The neoplastic roots produced by *A. rhizogenes* infection are characterized by high growth rate and genetic stability (Fig.1). In some species HR can be regenerated into plants. Transgenic plants have been obtained after *A. rhizogenes* mediated transformation in 89 different taxa, representing 79 species from 55 genera and 21 families.[8,9] These plants display a characteristic phenotype, called HR syndrome, which includes reduced apical dominance in both stems and roots, shortened internodes, high growth rate of roots in culture, wrinkled leaves with increased width to length ratio, plagiotropic roots, with altered geotropism, altered flower morphology, late flowering, reduced fertility and reduced pollen and seed production.[10,11] Transformation of individual *rol* genes into plants has provided information on the function and phenotype induced by these genes, both individually and in combination.[11]

How *Agrobacterium rhizogenes* Genes Can Affect Plant Tissues Development?

A. rhizogenes causes the development of HR in higher plants via the transfer of one or two fragments of T-DNA from a root-inducing (Ri) plasmid to the host plant genome.[7,12,13]

Among the 18 ORFs localized in the Ri T-DNA,[14] four coincide with genetic loci (*rolA, B, C, D*) identified by transposon mutagenesis from the effects of their inactivation on the virulence properties of *A. rhizogenes*.[7] These genes correspond to ORFs 10, 11, 12 and 15 of TL-DNA.[14] Studies of transgenic plants that carry various combinations of the TL-DNA genes have shown

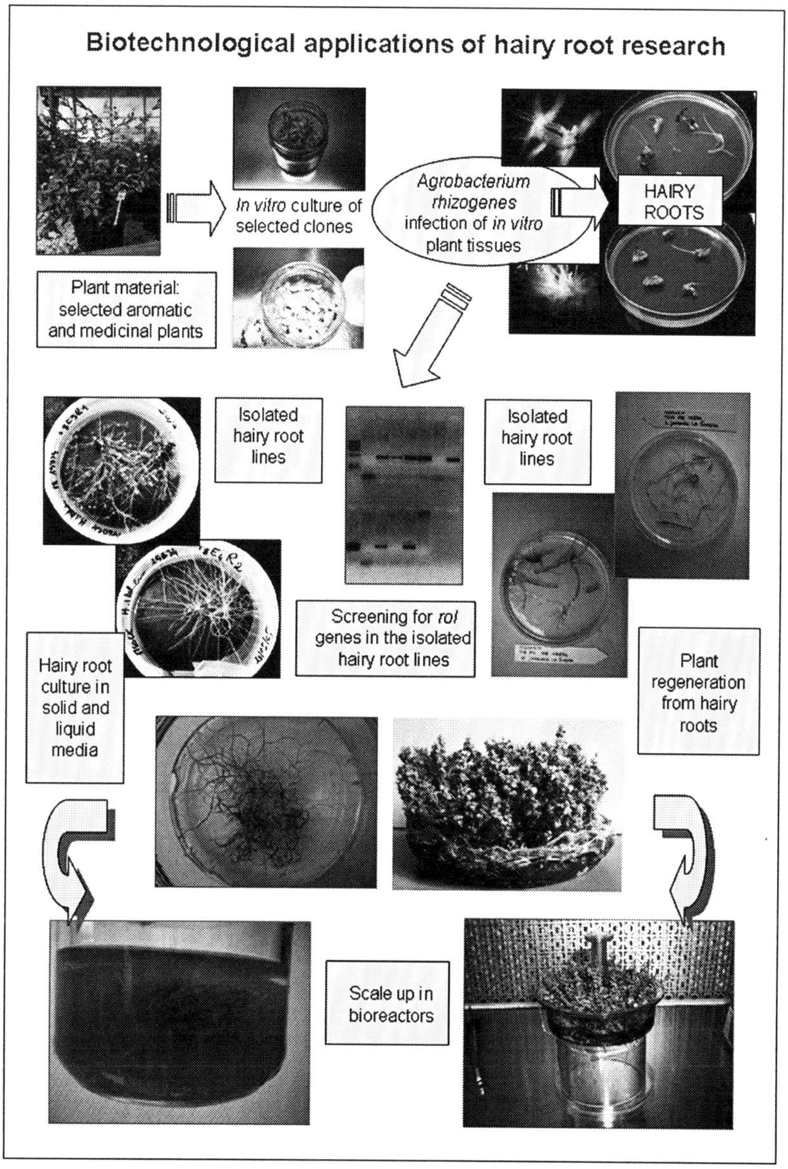

Figure 1. Biotechnological applications of hairy root research: *Agrobacterium rhizogenes* hairy root induction in selected officinal plants, isolation of hairy root lines, genetics, plant regeneration and scaling-up in bioreactors for secondary metabolite production.

that two other ORFs, ORF13 and ORF14, also play significant roles in the induction of roots on carrot disks and tobacco leaf segments.[15] Aoki and Syono[16] also showed that these genes could independently promote the induction of adventitious roots by the *rolB* gene in tobacco leaf disks.

Different *rol* gene products have either different targets, that may be involved in the control of plant, or have a qualitative various effect on the same target.[17] When individually inserted in plants, the *rol* oncogenes affect plant growth and development, each in its characteristic and distinctive way.

Transgenic plants containing each gene alone do not show all the phenotypical abnormalities of HR syndrome, meanwhile this full phenotype might be the result of the synergistic action of these genes.[18,19] Although the biochemical functions of *rol* genes remain poorly understood, they are useful tools for improving ornamental flowers, as their expression in transgenic plants yields many beneficial traits.[20]

Among the *rol* genes, *rolC* has been the most widely studied because its effects are the most advantageous in terms of improving ornamental and horticultural traits. *RolC* results in reduced apical dominance, altered leaf morphology and reduced seed production.[18] In addition to the dwarfness and the increase in lateral shoots that lead to a bushy phenotype, *rolC*-plants display more, smaller flowers and advanced flowering; surprisingly, these plants may have better rooting capacity and they show almost no undesirable traits.

Plants expressing *rolB* show flower heterostyly and abundant adventitious rooting[21,11] while inactivation of *rolB* causes the abolition of virulence.[7] Transgenic plants containing *rolA* show wrinkled leaves and reduced internode distance.[18,22] More recently, the effects of *rolD* have been shown to consist primarily in a strong acceleration and stimulation of flowering both in tobacco plants and cultured tissues[23] and in *Arabidopsis. RolD,* the least studied among the *rol* genes, offers promising applications due to its capability to promote flowering.

The formation of HR from infected cells is more likely regulated by *rolA, rolB* and *rolC*.[11,18,21] Among these, the *rolB* locus is thought to be the most important for HR induction.[15,16] A single gene *rolB* induced roots, while *rolC* and ORF13 and ORF14 of TL-DNA of Ri plasmid independently promoted the root induction by the *rolB* gene.[16] The effects of these genes on the rolB-mediated rooting were in the order ORF13>rolC<ORF14.[16]

Although *rol* genes have been the object of several studies for a long time, the function of their products is still an unsettled question.[24,25] We focused the attention on some new achievement of possible mechanism of action of the *rol* genes.

RolC Mechanism of Action

The *rolC* gene effects are well documented and have been extensively tested in many plant species. They produce essentially similar phenotypes in all the analyzed species. The *rolC* affects the plant size and architecture, these include decreased height, internode length, male fertility, apical dominance and an increased number of flowers.[26] Other morphological effects are the unique changes in leaf size, color and shape. Leaf size includes leaf length to leaf width ratio and such changes can affect the overall appearance and texture of a plant increasing its ornamental value. The modification of color includes the introduction of curly or wrinkled leaves. The reduction in height due to *rolC* effect have been observed at different degrees of dwarfness among independent transformants carrying the same *rolC* gene construct, even within the same species.[27] These differences depend from several factors as the site of integration, copy number, mutation, somaclonal variation and changes in expression level.[27,28] Most of these factors, such as site of integration, are beyond the control of the researchers, but some extent localization and expression levels could be managed using specific promoters.[26] A high number of independent regenerates improves the probability to obtain interesting genotypes. The obtained results in a wide range of genera elucidated the effects of the *rol*C gene on ornamental traits.[26] In general, rising the length of the promoter, increase also the quantity of *rolC* mRNA expression and the degree of phenotypic changes.

The effects on plant morphology of the *RolC* gene may be due to cytokinin-beta-glucosidase activity that increases cytokinin levels.[29,30] Moreover gel-permeation experiments revealed that transformation of ginseng cells by the *rolC* gene could significantly affect activity of some carbohydrates isoforms. In fact, significantly increased levels of activity of p- and a-D-galactosidase and 1,3-p-D-glucanase were detected in *rolC* gene transformed cells compared to the control nontransformed cells.[31] Such peculiarities of the *rolC* gene action stimulated investigations aimed to understand how the gene expression could affect secondary metabolism.

The *rolC* stimulatory effect on secondary metabolism was demonstrated by investigations with different groups of secondary metabolites, such as tropane alkaloids, pyridine alkaloids, indole alkaloids, ginsenosides.[32-35] The strong stimulatory effect of the *rolC* gene on synthesis of shikimate-derived anthraquinone phytoalexins in the *Rubia cordifolia* transgenic callus cultures has been shown.[36,37] The assessment of signal transduction pathways showed that the *rolC*-gene-mediated signal did not interfere with general plant defense pathways, leading to synthesis of phytoalexin-type secondary metabolites. In particular, the Ca^{2+}-dependent NADPH oxidase pathway as well as salicylic-acid-mediated and octadecanoid pathways have not been affected in *rolC*-transformed cells of *R. cordifolia*.[31,37]

Recently, an inhibitory effect of the *A. rhizogenes rolC* gene has been demonstrated on rabdosiin and rosmarinic acid production.[38] A possible interpretation of the controversial activity of the *rolC* gene on secondary metabolism is that the *rolC* mediated signal interferes with different regulatory backgrounds existing in plants. However, it is unknown which chains of regulatory pathways the gene affects. Proteins phosphatases represent a possible target of the *rolC* gene.[37]

RolB Mechanism of Action

The function of the *rolB* gene gives a clue to the mechanism of adventitious root formation in plants. The adventitious roots induced by the *rolB* gene produce abundant lateral roots under tissue culture conditions.[15] These phenomena indicate that the rolB protein has a crucial effect on the formation of both lateral and adventitious roots. Therefore, elucidation of the function of the rolB protein may provide evidences for understanding root formation in plants.

The difference in growth capacity among HR lines is caused by variation in the level of *rolB* gene expression.[39,40] A proper level of *rolB* expression for active growth of HR appears necessary because either a high or low level correlates with impaired growth of HR. Overexpression of the *rolB* gene under the control of the CAMV35S RNA promoter (P35S) suppresses adventitious root induction[11] and apparently induces cell death (necrosis) both in callus and leaves of young plants.[18] Tobacco cells transformed with the *rolB* gene show an alteration of auxin-induced hyperpolarization of the plasma membrane.[41]

RolB protein was suggested to be a beta-glucosidase, in fact, in in vitro experiments it was able to increase the levels of free indoleacetic acid (IAA, the natural endogenous auxin) by releasing it from its inactive glucose conjugates.[29] However, in plant tissues IAA-glucosides were shown not to be substrates for RolB.[42] The function of RolB protein as a tyrosine phosphatase or as an auxin-binding protein has been proposed.[43,44] Despite correlations between *rolB* function and regulation or perception of the auxin neither direct evidence for its function nor a convincing explanation between these functions and adventitious rooting has been provided. Furthermore, although auxin perception is altered in *rolB*-transformed cells, no plant factor that interacts with rolB protein has been identified.[41] The results strongly imply that *rolB* is not involved in the regulation of hormone metabolism. The auxin effects observed in *rolB*-transformed plants could be due to an altered perception of the hormone stimuli.

The nuclear localization of *rolB*, reported recently by Moriuchi,[25] raises the possibility that *rolB* might function as a transcriptional coactivator/mediator. The authors isolated cDNAs encoding six14-3-3 homologs as candidates for plant factors interacting with the 1724rolB protein and demonstrated that Nt14-3-3 omegaII interacts with the 1724rolB protein in plant cells.[25] There are some highly conserved regions and some inserted or deleted regions in the sequences of the proteins of the *rolB* family. Although these and similar motifs were absent in the 1724rolB protein, it bound specifically the Nt14-3-3-omegaII protein, indicating that the mode may be unrelated to the phosphoserine/threonine motif. Although the mechanism of adventitious root induction caused by the 1724rolB protein is still unknown, it must be correlated with the interaction between the 1724rolB protein and Nt14-3-3-omega-II and with the nuclear localization of the 1724rolB protein. The protein of *rolB* family may alter developmental plasticity in higher plants by association with plant proteins.[25] The development of additional *rolB*-transformed plant model systems is important to exploit this new biotechnological tool for genetic engineering of plant secondary metabolism.

In *Rubia cordifolia*, anthraquinone (AQ) production is greatly increased in *rolC* and *rolB*-transformed cultures compared with the nontransformed culture. Methyl jasmonate and salicylic acid strongly increased AQ accumulation in the transgenic and nontransgenic calluses in the same way, so that it might be supposed a potential significance of these proteins in plant secondary metabolism.[36]

Recently the *rolB* gene was introduced in *Vitis amurensis* cells and the *rolB*-transformed calli are able of producing up to 3.15% dry wt of resveratrol.[45] This ability is tightly correlated with the abundance of *rolB* mRNA transcripts. Tyrosine (Tyr) phosphatase inhibitors abolished the *rolB*-gene-mediated stimulatory effect, thus documenting for the first time the involvement of tyrosine phosphorylation in plant secondary metabolism.

Interestingly, the signaling pathway by which the *rolB* gene activates plant defense reactions does not depend on oxidative signals. In fact the stimulatory effect of the gene on stilbene production was abolished when the *rolB*-calli were cultivated in the presence of either PAO (phenylarsine oxide) and Na-orthovanadate inhibitors of Tyr phosphatases.[45] These results indicate that Tyr phosphorylation indeed is involved in the stimulatory function of the *rolB* gene. Thus, a component of the signaling network that controls stilbene biosynthesis in *Vitis amurensis* cells is negatively regulated by Tyr phosphorylation. The *rolB* signal seems sufficient to overcome this negative signal, thus switching the regulatory balance to pathway activation. Recent studies suggest that protein Tyr phosphorylation performs critical functions in plants, regulating activity of MAP kinases, transcription factors and ROS (reactive oxygen species) signaling.[46] Kiselev suggests an additional role of Tyr (de) phosphorylation in plant secondary metabolism.[45] This investigation is the most prominent demonstration of the potential of *rolB*-gene transformation in cases where the success in secondary metabolite production cannot be achieved by commonly used methods.

RolD Mechanism of Action

The plant oncogene *rolD* stimulates the reproductive phase transition in plants. In tomato the insertion of *rolD* has a pleiotropic effect, affecting traits of economic interest such as plant productivity as well as characters generally related to the defense response to pathogens. *RolD* was suggested to exert its effect on increased flowering through changes in the concentration of plant hormones in transformed plants.[23] In transgenic tobacco plants *rolD* induces a striking earliness in the flowering process and an increase in the number of flowers that have been related to an accumulation of proline or to a depletion of ornithine.[23]

Recent biochemical assays have shown that *rolD* encodes an ornithine cyclodeaminase (OCD) able to catalyze the NAD^+-dependent conversion of ornithine to proline. The high proline concentration in tomato flowers has led some authors to argue for a proline-mediated role of *rolD* in flowering.[47] All the plants harboring the *rolD* gene were shown to be more tolerant to the toxin in ion leakage experiments than control. The proline produced by the OCD enzymatic activity of *RolD* may prompt an explanation for the role of this protein in HR growth. In fact a substantial increase in proline concentration in the growing region of maize primary roots at low water potential has been reported. This finding suggests a role of proline biosynthesis in sustaining root growth under these conditions.[48]

An increase in the amount of proline could affect the rate of biosynthesis of hydroxyproline rich glycoproteins (HRGPs, extensins and arabinogalactan proteins). These proteins are structural constituents of the plant cell wall and are thought to play a key role in the regulation of cell division, cell wall self assembly and cell extension.[49] Alternatively, stimulation of root growth by *rolD* expression could be related to the depletion of ornithine, a polyamine precursor and to the consequent possible alterations of the polyamine pool. Polyamines involvement has been also supported by the evidence that overexpression of arginine decarboxylase, an enzyme involved in polyamine biosynthesis, resulted in an increased level of putrescine and impaired root growth in tobacco plants.[47]

RolA and ORF13 Mechanism of Action

The RolA protein, suggested to act as a transcription factor,[50] has been proposed to be implicated in the metabolism of gibberellins, as a reduction in their content has been found in *rol*A-transgenic tobacco plants.[51] This could explain the dwarfing of these plants, since a similar phenotype was obtained by applying inhibitors of gibberellin synthesis,[51] although the wild phenotype was not completely restored with the application of gibberellins.[52] The *rol*A gene was also reported to be responsible for changes in polyamine metabolism by inhibiting their conjugation.[53,54] Moreover, measurement of their transmembrane potential difference in response to auxins showed *rol*A-tobacco protoplasts to be more sensitive to auxins.[41]

The sequence of ORF13 is highly conserved in the agropine-, mannopine-, cucumopine- and mikimopine-type Ri plasmids and two plant counterparts were found in tobacco (*Nicotiana tabacum* [NtORF13]) and (*Nicotiana glauca* [NgORF13]).[55,56] Previously, it has been proposed that ORF13 encodes a growth regulator that shares properties with citokinins, suggesting an interaction with hormone signaling pathways.[57] However, divergent results have been also obtained, where a hormone-related biochemical function for the *orf13* gene product was excluded.[19]

ORF13 expression leads to the formation of spikes (protrusions between minor veins) on leaves and petals of tobacco.[50] Similar structures are formed on leaves, when KNOX (KNOTTED1-like homeobox) genes are overexpressed.[58-60]

Recently it has been demonstrated that ORF13 has a retinoblastoma RB-binding motif and binds to RB in vitro, although several, but not all, developmental changes in the phenotypes were due to a functional RB-binding motif in ORF13.[61] Furthermore the ORF13 induced ectopic expression of KNOX (KNOTTED1-like homeobox) class transcription factors, as well as of several genes involved in cell cycle regulation in tomato (*Lycopersicon esculentum*). Increased cell divisions in the vegetative shoot apical meristems and accelerated formation of leaf primordia were observed in plants expressing ORF13.[61] Simultaneously hormone homeostasis was only altered in explants of leaves, whereas in the root no effects were observed. It has been proposed that ORF13 confers meristematic competence to cells infected by *A. rhizogenes* by inducing the expression of KNOX genes and promotes the transition of infected cells from the G1 to the S phase by binding to RB.[61]

A. rhizogenes Transformed Medicinal Plants as Farm for Aromatic and Nutraceutical Metabolite Production

Sage (Salvia officinalis L.)

The genus *Salvia* (*Labiatae*) includes more than 900 species spread out all over the world. In ethnobotany, some of them are known for their medicinal potential. They are currently commercially utilized but most of them are totally unstudied. Only few reports of in vitro culture on the genus *Salvia* were found in literature. In 1990, Olszowska and Furmanowa[62] reported the micropropagation of *S. officinalis* while a micropropagation protocol was reported in 1993 by Hosoki[63] and Tahara for *S. leucantha*, in 1997 by Molina[64] for *S. canariensis* and finally in 2004 by Skaia and Wysokinska[65] for *S. nemorosa*. The possibility to manipulate the in vitro production of rosmarinic acid (RA) in cell cultures of *S. officinalis* and *S. fruticosa* was demonstrated by Kintzios.[66] Moreover RA and the lithospermic acid B have been already produced from transformed cells of *S. miltiorrhiza*.[67] On the other hand in 2003 Savona[68] reported the identification of a new secoisopimarane diterpenoid in *S. cinnabarina* and in 2004 Bisio[69] described other new terpenoids in *S. wagneriana* and it seems that these compounds have an interesting biological activity (Fig. 2 and 3).

The general *A. rhizogenes* transformation protocol has been used in the frame of a international project on sage for the development of HR in several species; HR were obtained from cocultivated stems and leaves from *S. wagneriana*, *S. cinnabarina* and *S. jamensis* "La Siesta". Several HR lines, obtained from a single transformation event, were selected for each species and PCR amplification of *A. rhizogenes rolC* gene was detected in each transgenic line. *S. cinnabarina* HR line C1 and *S. wagneriana* 15834 HR line D3 liquid cultures were established and several elicitors were added to the liquid culture medium. The secondary metabolites extracted from *S. wagneriana*

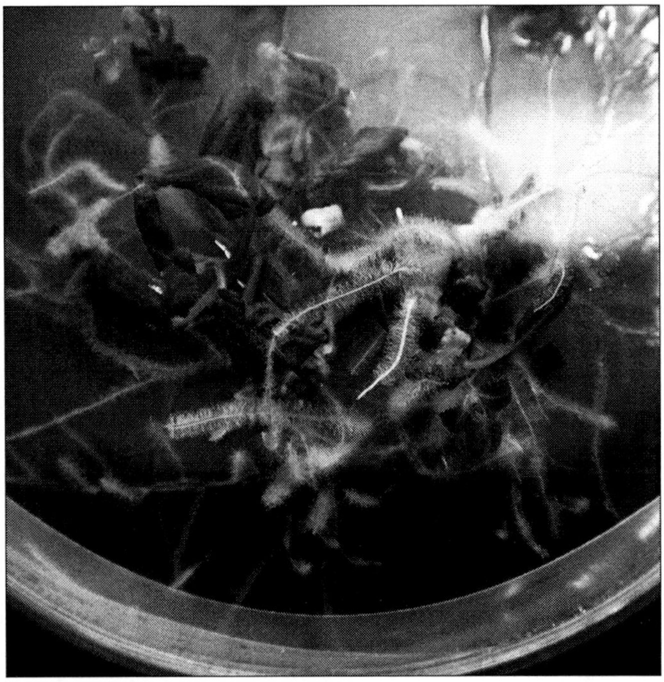

Figure 2. Development of putative hairy roots 40 days after cocultivation of *S. cinnabarina* leaf petioles with *A. rhizogenes* w.t. strain 15834.

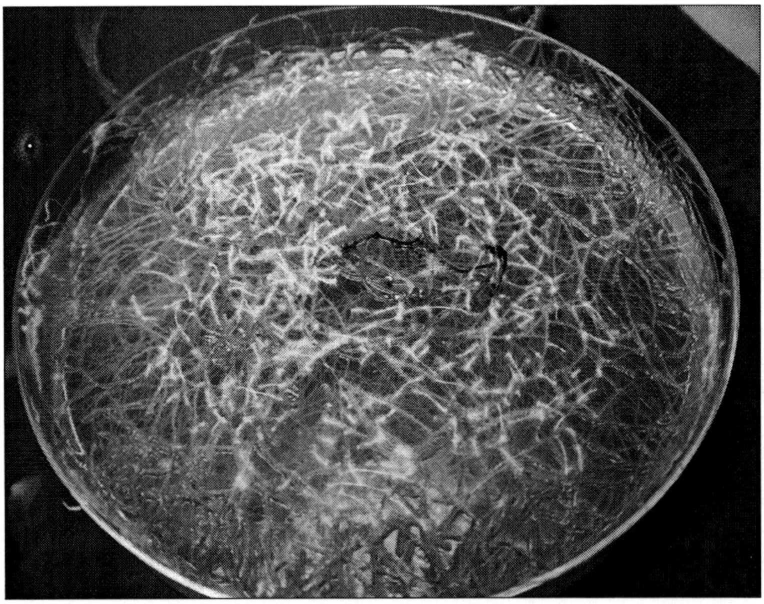

Figure 3. *S. cinnabarina* hairy root line selected after the PCR test for presence of the *A. rhizogenes rolC* gene.

Table 1. ***List of* Salvia *species transformed by* A. rhizogenes *for the production of bioactive compounds and secondary metabolites detected in cell and hairy root cultures***

Plant Name	Secondary Metabolites	Culture Type	References
Salvia broussonetii	New diterpens	Hairy root cultures	121
Salvia sclarea	Ortonaphtoquinone diterpens and novel compounds	Cell and hairy root cultures, ATCC 15834	122
Salvia cinnabarina	*p*-sitosterol, ursolic acid	Hairy root cultures	68
Salvia involucrata	Apigenin, total flavonoids	Hairy root expressing (CaMV) 35S CHI gene	123
Salvia miltiorrhiza	Cryptotanshinone, tanshinone I, tanshinone IIA and tanshinone IIB, rosmarinic acid and lithospermic acid B	Hairy root cultures, ATCC 15834	124
Salvia miltiorrhiza	Diterpenoid tanshinones	Hairy root cultures, ATCC 15834 Ag(+)	125
Salvia miltiorrhiza	Diterpenoid tanshinones	Hairy root cultures Yeast elicitor + hydrophobic polymeric resin (X-5)	126
Salvia miltiorrhiza	Diterpenoid tanshinones	Hairy root cultures Yeast elicitor + GABA (amino acid beta-aminobutyric acid)	127
Salvia wagneriana	Rosmarinic acid	Hairy root culture + elicitors	Ruffoni, personal communication

adult plants were: flavonoids, phenolic acids, triterpens, diterpens, quercetin, apigenin, luteolin, apigenin 7-glucoside, caffeic acid, chlorogenic acid, ursolic acid, betulinic acid, lupeol, *p*-sitosterol and new unknown diterpens. The results of the screening by direct infusion and LC-DAD-ESI-MS of *S. wagneriana* HR culture extracts revealed no correspondence with reference flavonoids. However, these extracts showed the presence of diterpenes especially in the samples elicitated with jasmonic acid and conditioned with mannitol; the presence of ursolic acid, *p*-sitosterol, caffeic acid and other flavonoids different from the available reference material in the samples elicitated with casein hydrolysate or jasmonic acid and conditioned with mannitol (Ruffoni, personal communication). Cell and HR cultures have been established in potentially useful *Salvia* species. Efforts are currently on the way to manage extraction and characterization of sage secondary metabolites in particular RA from in vitro cultures and to monitor the effect of elicitation on culture growth and on the production of the bioactive compounds.

In Table 1 are summarized *Salvia* species already utilized for transformation with *A. rhizogenes* and the produced secondary metabolites.

Basil (Ocimum basilicum L.)

Ocimum basilicum L. (sweet basil) is a member of the family *Lamiaceae.* It is used as a kitchen herb and as an ornamental plant in house gardens. Basil plants are valued for their pharmaceutical properties; for example, the aromatic oils produced in their leaves are used as antioxidants. Basil cultivars vary in their morphological characteristics indicating a probable different content in flavonoid fraction and other secondary metabolites as observed by other authors.[70] Phenolic acids exist constitutively in species of the *Lamiaceae* and *Boraginaceae* families.[71] Among the various secondary metabolites found in most *Lamiaceae* species, rosmarinic acid (-*o*-caffeoyl-3-4-dihydroxyphenyllactic acid) is one of the most abundant caffeic acid esters. Because of its pharmaceutical application as an antioxidant, most of the studies performed on RA have focused on understanding its production and regulation, while the biological significance of RA in plants has been virtually ignored. It has been shown to have tannin-like properties,[72] which has led to speculation that it may serve as a constitutively accumulated and inducible defense compound against pathogens and also as a feeding deterrent for herbivores.

Recently it has been observed in in vivo plants that the maximum RA accumulation occurs normally in roots, compared to leaves and shoots.[72] Meanwhile suspension cultures of sweet basil obtained from the leaves accumulated RA up to 10 mg/g^{-1} dry weight, a value up to 11 times higher than in callus cultures or in leaves of donor plants.[73] In vitro cultivation of basil can be considered as an alternative strategy to recover essential oil from the air inside the environment where the in vitro plantlets are grown.[74]

HR cultures have been developed for determining the best producing line of phenolic compound like: RA, lithospermic A acid (LA) and lithospermic B acid (LAB).[75] HR growth was fast until the third week of culture, although the higher RA and LA contents were observed later, after 6th-8th week from *inoculum*, reaching 3.5 fold higher concentration (relative to dry weight) compared to the amount detected in control plants. The LAB content was not affected by HR culture.[75] Moreover, in that study a lot of attention has been given to the choice of *A. rhizogenes* strains and the culture medium. Some authors reported that particularly in Woody Plant Medium (WPM), the biomass of the roots increased without an obvious lag time reaching the maximal biomass production at the third week of culture.[75] Although in all tested media HR produced substantial amounts of RA, particularly high levels (over 14% of dry wt) were observed in MS (clone J-l: 14.1%, at 8th week) and B5 (clone A-2: 14.0%, at 6th week) media. These levels were almost 3.5-fold higher than that (3.98% of dry wt, in leaf portion) of the intact plant and were also almost identical to those obtained in suspension cultures of *C. blumei*[76] and *S. officinalis*,[77] cultured under highly optimized conditions for RA production. In all examined cultures the maximum yield of RA was 73.5 mg/flask ~ produced by J-1 in MS medium at 5th week. LA was also produced in all clones in the three tested media, whereas LAB was certainly produced in some clones when cultured in B5 (A-2, J-1 and J-3) and WP (A-I and A-2 and J-l) media. However, in MS medium, no LAB was detected.[75] Untransformed normal roots of *O. basilicum* harbored the maximum titers of RA (0.98% g fresh weight basis) compared to leaves and shoots. HR cultures of *O. basilicum* transformed with *A. rhizogenes* (ATCC-15834) showed three-fold increases in growth and RA production compared to the untransformed normal roots.

To maximize the production of secondary metabolites, HR cultures of *O. basilicum* were challenged with salicylic acid (SA), jasmonic acid (JA), chitosan and fungal cell wall elicitors (CWE) and their effects on growth and RA accumulation were further examined. Using fungal cell wall elicitors (CWE) from *Phytophthora cinnamoni* the production of RA was enhanced 2.67 fold compared with the untreated control.[72] Meanwhile the use of SA, JA and chitosan reply in the negative way on the growth of HR cultures with subsequent blackening of the tissue in conjunction with the reduced titers of intracellular RA at the end of the time course (data not shown). This result implies that SA, JA and chitosan do not channel RA elicitation.[72]

The maximum RA accumulation occurs normally in roots, so it implies a vital biological/biochemical role of RA in the underground zone. The biological activity of RA was observed against a wide range of soil borne microorganisms.

Due to the known antimicrobial properties of phenolic acids, it has been hypothesized that RA and phenolic acids may play the role of phytoanticipins in plants.[78] Induced RA production and exudation was limited only to the roots that were challenged by in situ fungal invasion. Biologically, the findings suggest that in nature RA is a constitutive antimicrobial compound released into the surrounding rhizosphere upon microbe invasion.

A new knowledge given by Bais[72] was that RA accumulated in roots shows a growth-associated pattern and is secreted by *O. basilicum* roots as part of the root exudates upon pathogen attack. RA was found to be highly inhibitory against an array of rhizosphere microorganisms, including a potential human pathogen. The antibiotic activity was demonstrated by changes in cell morphologies of affected organisms.

These results reveal a novel defense mechanism by which roots secrete antimicrobial caffeic acid esters such as RA upon pathogen attack.

Mint (Mentha spp.)

The mint genus belong to the *Labiatae* family and include more than 25 species, without the numerous varieties obtained by spontaneous hybridization.[79,80] Mint plants are crops of considerable commercial value and are widely cultivated for their essential oils. This valuable product is mainly composed of monoterpenes and is largely used for the production of food, cosmetics and pharmaceuticals. Because the quality of oil depends on the composition of the monoterpene, it is of great interest to obtain a strain producing a better quality oil. Technological advances in genetic engineering have resulted in the potential for utilizing biotechnology to "improve" mint. The transformations of mint plants using the soil bacterium *Agrobacterium tumefaciens* have been reported.[81-89]

Few years ago it was first reported the infection with *A rhizogenes* of *M. piperita* and the regeneration of HR and shoots through the transgenic calli.[90] The first HR obtained from in vitro internodes shoots became visible one or two weeks after inoculation. The frequency of HR induction obtained by direct *Agrobacterium* infection varied according to the plant organ used: the direct infection of leaves induced more roots compared to those obtained from stem. Shoots generated roots on hormone free B5 medium within 12 weeks, then the obtained plantlets were transferred to sterile vermiculite immersed in water. The regenerated plants showed the HR syndrome. To detect the integrated DNA fragments, the polymerase chain reaction (PCR) was performed with the genomic DNA of the regenerated plants. An 1.6 kb fragment, which included the region from the *rolA* to *rolB* gene, was detected in all regenerated plants.[90]

Other mint species as *Mentha pulegium* clonal line MPH-4, a species containing high levels of endogenous phenolics was also inoculated with *A. rhizogenes*.[91] The inoculated explants exhibited higher levels of total plant phenolic components and guaiacol peroxidase activity. Enhancement of phenolic production by *A. rhizogenes* was more apparent when explants were also treated with polymeric dye R-478. Polymeric dyes are environmental pollutants widely used in the textile industry. Polymeric dyes have a structure analogous to common polycyclic aromatic hydrocarbons (PAHs). Plants that show an inherent tolerance to these dyes are currently being investigated for use in phytoremediation. The response of MPH-4 to *A. rhizogenes* and polydye R-478 are useful to determine the use of this clonal line in phytoremediation.[92]

Echinacea (Echinacea spp.)

Echinacea spp. are native of North America and Europe and are widely utilized for the immunostimolant medicinal usage. The genus *Echinacea* belongs to the *Asteraceae* (or *Compositae*) family and has nine species. The common used species are *E. angustifolia* DC (roots), *E. pallida* Nutt (roots) and *E. purpurea* (L.) Moench (roots and aerial parts).[93] *Echinacea purpurea* (L.) Moench is an important commercial specie.[94]

Several caffeic acid derivatives (CADs) such as cichoric acid, caftaric acid, chlorogenic acid and caffeic acid have been identified in *Echinacea* species. They are believed to have immunostimulatory activity.[95,96] Previous phytochemical studies showed the presence of echinacoside, caffeic acid derivatives and polysaccharides in cell suspension cultures of *E. angustifolia* derived from seedling tissues, whereas no information is available on cell culture from selected adult plants.[97]

Plants of the *Echinacea* genus are characterized by their critical germination, which represents the first important problem in promoting both in vivo and in vitro propagation. Several treatments were applied in order to overcome seed dormancy.[98-101] Because of the long growth cycle and the limitation in the environment, it is rather difficult for pharmacy to transplant *E. purpurea* and breed it. For this reason, a system of obtaining plant metabolites for pharmacy has been developed. The method of transformed HR cultures of *E. purpurea* was established by infecting different types of explants with three strains of *A. rhizogenes*.[102] The best explants utilized for the induction of HR were leaves.

Higher content of polysaccharides were detected in transformed HR than in nontransformed roots. However a low production of phenolic compounds was observed in HR, triggered by a short time of culture.[102] It is well known that *E. purpurea* presents a long balsamic period. Accumulation of secondary metabolites could be improved by physical and chemical ways in HR culture. Chemical techniques can be referred to the change of nutritional ingredients, adding hormones or precursors feeding. Physical techniques involve changes of the culture condition, such as illumination and temperature.

Recently, *E. purpurea* HRs have been produced successfully in an axenic liquid culture system and the roots were capable of producing significant amounts of CADs, especially cichoric acid and caftaric acid. Interestingly it has been found that amount of caffeic acid in HR cultures is similar to that obtained in field grown plants.[103]

In plants many phenolic compounds are produced through phenylpropanoid pathway, which is initiated by phenylalanine ammonia lyase—PAL.[104] The four identified CADs and anthocyanins in *E. purpurea* HR cultures were also biosynthesized through this pathway.[105] Therefore, there was a need to investigate the effect of light on PAL activity, which was related to the accumulation of CADs and anthocyanins in *E. purpurea* HR cultures. Very recently it has been demonstrated a definite positive effect of light on root growth, cell viability and PAL activity in relation to the biosynthesis of CADs.[106] The photoregulation of CADs biosynthesis in *E. purpurea* HR may offer additional advantages of quantitative and qualitative improvements of these medicinally important metabolites, but has not yet been explored.

E. purpurea HR exposed to light began to turn purple in color after 5 days of culture. The accumulation of purple color in light-grown HR is confined only to the outer cell layers of the cortex. In fact, by histological examination of dark-grown HR cultures it has been revealed the absence of purple color accumulation.[106] In general, the purple color is related to anthocyanin accumulation and has been shown to exist mainly in the flower of wild *E. purpurea* plants.[107] Anthocyanins were observed only in the light-grown HR cultures. Formation of anthocyanins, which can protect plant tissues from light-induced oxidative stress, was also reported in vegetative tissues and cell cultures of other plants as a response to light stress.

The *E. purpurea* HR provided another unique system to address the challenges of understanding the photoregulation of anthocyanins and other secondary metabolites.

Ginseng (Panax ginseng C.A. Meyer)

Panax ginseng C.A. Meyer belongs to the *Araliaceae* family and because of its Chinese origin is a well-known and valuable plant in oriental medicine. Crude ginseng root extracts have tonic, stimulatory and adaptogenic properties, mainly due to the presence of numerous saponins and sapogenins.[108] As wild grown plant ginseng is now rare, current ginseng supply depends almost exclusively on field cultivation which is time-consuming and labor-intensive. Agricultural ginseng roots production from seeds planting to mature roots requires 5-7 years, during which the plant growth is highly susceptible to numerous environmental factors.[109] As a result, ginseng cell, tissue and organ culture has been exploited as a biotechnological alternative, more efficient and monitored, for ginseng and its active constituents production.[110]

The induction and establishment of *P. ginseng* rhizomes HR after *A. rhizogenes* infection has been successfully performed.[111-113] These transformed roots exhibited a more rapidly growth and higher levels of ginsenosides than the normal cultured roots obtained by hormonal control. Recent investigations have shown that *inoculum* size as well as its age strongly influenced the growth of *Panax ginseng* HR cultures.[114] The first group of cited authors found that the optimal growth rate of *P. ginseng* HR was obtained when a 0.7% (w/v) was used and that 10 days was the optimal subculture cycle.[33,114]

Few attention have been given to the addition of biotic or abiotic elicitors into the HR culture media, whereas they have been extensively studied in case of cell cultures.[115] The effect of any elicitor (biotic or abiotic) is dependent on a number of factors that may interact, including the elicitor specificity and concentration and the duration of treatment and the growth stage of the culture as well.

The results obtained by Palazon[33,34] indicated that the *P. ginseng* root lines produced the highest levels of ginsenosides at the third day of culture, when they were cultured with methyl jasmonate and when the cultures were in the advanced progressive deceleration growth phase.

To improve the productivity of useful metabolites in *P. ginseng* HR cultures, further experiments were carried out using several elicitors: salicylic acid (SA), acetylsalicylic acid (ASA), yeast elicitor and bacterial elicitor. In case of SA elicitation, total ginseng saponin content increased slightly at lower elicitor dosages (0.1 to 0.5 mM). The use of ASA resulted in the inhibition of biomass growth and total ginseng saponin content increased at every elicitor dosage (0.1 to 1.0 mM) by about 1.1 times.[116] With yeast elicitor addition, HR growth evaluated in dry weight was inhibited about 0.8-fold time compared to the control, but total ginseng saponin content increased about 1.17 times compared to the control. The bacterial elicitor showed a slight inhibition of biomass growth, but total ginseng saponin content increased about 1.23 times. Thus, in general, elicitor treatments were found to inhibit the growth of the HR, although simultaneously enhancing ginseng saponin biosynthesis. Additional results were obtained by adding several abiotic elicitors. The elicitors used were: tannic acid, selenium (Se), nichel (Ni), NaCl, ascorbic acid, salicylic acid, H_2O_2, $CuSO_4$, $CaCl_2$ and combination of SA + $CaCl_2$ or SA + H_2O_2.[117] Tannic acid profoundly inhibited the HR growth during growth period. The addition of selenium at *inoculum* time did not significantly affect ginseng saponin biosynthesis, but delay in addition of selenium (after 21 d of culture), affected positively the ginseng saponin content and productivity. Also addition of 20 microM $NiSO_4$ increased ginseng saponin content and productivity.[117] Sodium chloride treatment inhibited HR growth, except the concentration of 0.3% (w/v). It was observed that all concentrations of sodium chloride added into the media increased the amount of synthesized ginseng saponins. These results suggest that processing time for the generation of ginseng saponin in a HR culture can be reduced *via* the application of an elicitor.

To stimulate growth and saponin accumulation, new compounds (oligosaccharides) have been recently used: a heptasaccharide (HS) and an octasaccharide (OS), isolated from *Paris polyphylla* var. *yunnanensis*.[118] Oligosaccharides from plant and microbial sources represent a class of the most widely recognized elicitors, acting as inducers of plant defense responses and playing regulatory roles in plant growth and development.[119,120] Separated addition of HS and OS to HR cultures at 10 days post-inoculation, increased the root biomass dry weight by more than 70% from 13 g L^{-1} to ~20 g L^{-1} and the total saponin content of roots by more than 1-fold from 1.6% to ~3.5% (w/w).[118]

The obtained results suggest that the two oligosaccharides may have plant growth regulatory activity in plant tissue cultures.

Conclusion

Plants represent a very rich source of different bioactive molecules, so that consumption of herbal medicines and medicinal plants is widespread and increasing. The medicinal plants grow spontaneously, so the main source of raw material is represented by natural and wild fields, with the consequent effects of habitat destruction and loss of genetic diversity. The use of controlled environments overcomes cultivation difficulties and thus facilitating the manipulation of plants to produce bioactive compounds and toxins.

In recent years significant progress have been made in the use of tissue culture and genetic transformation to modify pathways for the biosynthesis of target metabolites. Hairy roots are unique in their genetic and biosynthetic stability and their fast growth offers an additional advantage to use as a continuous source for the production of valuable secondary metabolites. The *Agrobacterium rhizogenes* oncogenes effects have been only recently clarified. Briefly, the RolA protein was suggested to act as a transcription factor and is implicated in the metabolism of gibberellins; the *rolC* stimulatory effect on secondary metabolism was demonstrated by investigations with different groups of secondary metabolites, such as tropane alkaloids, pyridine alkaloids, indole alkaloids, ginsenosides. The rolB protein is a beta glucosidase which has a crucial effect on the formation of both lateral and adventitious roots while recently the *rolB* gene, introduced in *Vitis amurensis* cells induced calli able to increase resveratrol production. *RolD* stimulates the reproductive phase transition in plants and it is more related to flowering process.

With these evidences, the updated results on the molecular mechanism of hairy root development, that are summarized in this report, permitted to suggest a new use of the *A. rhizogenes* genes potentiality. Instead of transformation with the bacterium wild strains, as done up to now also for the species considered in this review, it is possible to use strains with selected pool of *Rol* genes with more driven results in order to optimize metabolite production.

Acknowledgement

This work was supported by the NUTRA-SNACKS project funded within the EU Framework Programme 6.

References

1. Gaudin V, Vrain T, Jouanin L. Bacterial genes modifying hormonal balances in plants. Plant Physiol Biochem 1994; 32(1):11-29.
2. Giri A, Narasu ML. Transgenic HRs: recent trends and applications. Biotech Adv 2000; 18:1-22.
3. Guillon S, Tremouillaux-Guiller J, Pati PK et al. Hairy root research: recent scenario and exciting prospects. Commen CurrOpin Plant Biol 2006; 9:341-46.
4. Tzfira T, Citovsky V. Agrobacterium-mediated genetic transformation of plants: biology and biotechnology. Curr Opin Biotechn 2006; 17:147-154.
5. Tzfira T, Li J, Lacroix B et al. Agrobacterium T-DNA integration: molecules and models. TRENDS Gen 2004; 20:8.
6. Klee HJ, Romano CP. The roles of phytohormones in development as studied in transgenic plants. Crit Rev Plant Sci 1994; 13(4):311-24.
7. White FF, Taylor BH, Huffmman GA et al. Molecular and genetic analysis of the transferred DNA regions of the root-inducing plasmid of Agrobacterium rhizogenes. J Bacteriol 1985; 164:33-44.
8. Christey MC. Use of Ri-mediated transformation for production of transgenic plants. In vitro Cell Dev Biol Plant 2001; 37:687-700.
9. Tepfer D. Transformation of several species of higher plants by Agrobacterium rhizogenes: Sexual transmission of the transformed genotype and phenotype. Cell 1984; 37:959-67.
10. Christey MC. Transgenic crop plants using Agrobacterium rhizogenes mediated transformation. In: Doran PM, ed. Hairy roots: culture and applications. Amsterdam: 7 Harwood Academic Publishers, 1997; 99-111.
11. Spena A, Schmulling T, Koncz C et al. Independent and synergistic activity of rolA, B and C loci in stimulating abnormal growth in plants. EMBO J 1987; 6:3891-3899.
12. Huffman GA, White FF, Gordon MP et al. Hairy root inducing plasmid: physical map and homology to tumor-inducing plasmids. J Bacteriol 1984; 157:269-76.
13. Jouanin L. Restriction map of an agropine-type Ri plasmid and its homologies with Ti plasmids. Plasmids 1984; 12:91-102.
14. Slightom JL, Durand-Tardif M, Jouanin L et al. Nucleotide sequence analysis of TL-DNA of Agrobacterium rhizogenes agropine type plasmid. J Biol Chem 1986; 261:108-21.
15. Capone I, Spano L, Cardarelli M. Upstream noncoding region which confers polar expression to Ri plasmid root inducing gene rolB. Plant Mol Biol 1989; 13:43-52.
16. Aoki S, Syono K. Synergistic function of rolB, rolC, ORF13 and ORF14 of TL-DNA of Agrobacterium rhizogenes in hairy root induction in Nicotiana tabacum. Plant Cell Physiol 1999; 40:252-256.
17. Aoki S. Resurrection of an ancestral gene: functional and evolutionary analyses of the Ng rol genes transferred from Agrobacterium to Nicotiana. J Plant Res 2004; 117:329-37.
18. Schmulling T, Schell J, Spena A. Single genes from Agrobacterium rhizogenes influence plant development. EMBO J 1988; 7:2621-29.
19. Lemcke K, Schmulling T. Gain of function assays identify nonrol genes from Agrobacterium rhizogenes TL-DNA that alter plant morphogenesis or hormone sensitivity. Plant J 1998; 15:423-33.
20. Casanova E, Trillasa MI, Moysseta LR et al. Influence of rol genes in floriculture. Biotech Adv 2005; 23:3-39.
21. Cardarelli M, Mariotti D, Pomponi M. Agrobacterium rhizogenes T-DNA genes capable of inducing hairy root phenotype. Mol Gen Genet 1987; 209:475-80.

22. Sinkar VP, Pythoud F, White FF et al. RolA locus of the Ri plasmid directs developmental abnormalities in transgenic tobacco plants. Genes Dev 1988; 2:688-97.
23. Mauro ML, Trovato M, De Paolis A et al. The plant oncogene rolD stimulates flowering in transgenic tobacco plants. Dev Biol 1996; 180:693-00.
24. Nilsson O, Olsson O. Getting to the root: The role of the Agrobacterium rhizogenes rol genes in the formation of hairy roots. Physiol Plant 1997; 100:463-73.
25. Moriuchi H, Okamoto C, Nishihama R et al. Nuclear localization and interaction of rolB with plant 14-3-3 proteins correlates with induction of adventitious roots by the oncogene rolB. Plant J 2004; 38:260-75.
26. Smith AG, John KE, Gardner N. Dwarfing ornamentally crops with the rolC gene. In: Texixeira da Siva JA, ed. Floriculture, Ornamental and Plant Biotechnology, 2006; 2:54-59.
27. Kiyokawa S, Kikuchi Y, Kamada H et al. Genetic transformation of Begonia tuberhybrida by Ri rol gene. Plant Cell Rep 1996; 15:606-09.
28. Giovannini A, Secchioni N, Rabaglio M et al. Characterization of ornamental Datura plants transformed by Agrobacterium rhizogenes. In vitro cellular and developmental Biology Plant 1997; 33:101-106.
29. Estruch JJ, Schell J, Spena A. The protein encoded by rolB plant oncogene hydrolyses indole glucosides. EMBO J 1991; 10:3125-8.
30. Casanova E, Zuker A, Trillas MI et al. The rolC gene in carnation exhibits cytokinin- and auxin-like activities. Sci Hort 2003; 97:321-31.
31. Bulgakov VP, Kusaykin M, Tchernoded GK et al. Carbohydrase activities of the rolC-gene transformed and nontransformed ginseng cultures. Fitoterapia 2002; 73:638-43.
32. Bonhomme V, Matta D L, Fliniaux MA. Effects of the rolC gene on hairy root: induction development and tropane alkaloid production by Atropa belladonna. J Nat Prod 2000; 63:1249-52.
33. Palazon J, Cusido RM, Bonfill M et al. Elicitation of different Panax ginseng transformed root phenotypes for an improved ginsenoside production. Plant Physiol Biochem 2003; 41:1019-25.
34. Palazon J, Mallol A, Eibl R et al. Growth and ginsenoside production in hairy root cultures of Panax ginseng using a novel bioreactor. Planta Med 2003; 69:344-349.
35. Bulgakov VP, Khodakovskaya MV, Labetskaya NV et al. The impact of plant rolC oncogene on ginsenoside production by ginseng hairy root cultures. Phytochem 1998; 49:1929.
36. Bulgakov VP, Tchernoded GK, Mischenko NP et al. Effect of salicylic acid, methyl jasmonate, ethephon and cantharidin on anthraquinone production by Rubia cordifolia callus cultures transformed with the rolB and rolC genes. J Biotech 2002b; 97:213-21.
37. Bulgakov VP, Tchernoded GK, Mischenko NP et al. Effects of $Ca+^2$ channel blockers and protein kinase/ phosphatase inhibitors on growth and anthraquinone production in Rubia cordifolia callus cultures transformed by the rolB and rolC genes. Planta 2003; 217:349-55.
38. Bulgakov VP, Veselova MV, Tchernoded GK. Inhibitory effect of the Agrobacterium rhizogenes rolC gene on rabdosiin and RA production in Eritrichium sericeum and Lithospermum erythrorhizon transformed cell cultures. Planta 2005; 221:471-78.
39. Tanaka N, Yamakawa M, Yamashita I. Characterization of transcription of genes involved in hairy root induction on pRi1724 core-T-DNA in two Ajuga reptans hairy root lines. Plant Sci 1988; 137:95-105.
40. Tanaka N, Fujikawa Y, Aly MAM et al. Proliferation and rol gene expression in hairy root lines of Egyptian clover (Trifolium alexandrinum L.). Plant Cell Tiss Org Cult 2001; 66:175-82.
41. Mauriel C, Barbier-Brygoo H, Spena A et al. Single rol genes from the Agrobacterium rhizogenes TLDNA alter some of the cellular responses to auxin in Nicotiana tabacum. Plant Physiol 1991, 97:212-16.
42. Nilsson O, Crozier A, Schmuelling T et al. Indole-3-acetic acid homeostasis in transgenic tobacco plants expressing the Agrobacterium rhizogenes rolB gene. Plant J 1993; 3:681-89.
43. Filippini F, Rossi V, Marin O et al. A plant oncogene as a phosphatase. Nature 1996; 379:499-00.
44. Filippini F, Lo Schiavo F, Terzi M et al. The plant oncogene rolB alters binding of auxin to plant cell membranes. Plant Cell Physiol 1994; 35:767-71.
45. Kiselev KV, Dubrovina AS, Veselova MV et al. The rolB gene-induced overproduction of resveratrol in Vitis amurensis transformed cells. J Biotech 2007; 123:681-92.
46. Laloi C, Apel K, Danon A. Reactive oxygen signalling: the latest news. Curr Opin Plant Biol 2004; 7:323-328.
47. Trovato M, Maras B, Linhares F et al. The plant oncogene rolD encodes a functional ornithine cyclodeaminase. PNAS 2001; 698(23):13449-53.
48. Verslues PE, Sharp RE. Proline accumulation in maize (Zea mays L.) primary roots at low water potentials. II. Metabolic source of increased proline deposition in the elongation zone. Plant Physiol 1999; 119:1349-60.
49. Varner JE, Lin L-S. Plant cell wall architecture. Cell 1989; 56:231-39.
50. Meyer A, Tempe' J, Costantino P. Hairy root; a molecular overview. Functional analysis of Agrobacterium rhizogenes T-DNA genes. In: Stacey G, Keen NT, eds. Plant Microbe Interactions. St. Paul: APS Press, 2000:93-139.

51. Dehio C, Grossmann K, Schell J et al. Phenotype and hormonal status of transgenic tobacco plants overexpressing the rolA gene of Agrobacterium rhizogenes T-DNA. Plant Mol Biol 1993; 23:1199-10.
52. Schmulling T, Fladung M, Grossmann K et al. Hormonal content and sensitivity of transgenic tobacco and potato plants expressing single rol genes of Agrobacterium rhizogenes T-DNA. Plant J 1993; 3:371-82.
53. Sun L-Y, Monneuse M-O, Martin-Tanguy J et al. Changes in flowering and the accumulation of polyamines and hydroxycinnamic acid-polyamine conjugates in tobacco plants transformed by the rolA locus from the Ri TL-DNA of Agrobacterium rhizogenes. Plant Sci 1991; 80:145-56.
54. Martin-Tanguy J, Sun L-Y, Burtin D. Attenuation of the phenotype caused by the root-inducing, left-hand, transferred DNA and its rolA gene. Plant Physiol 1996; 111:259-67.
55. Aoki S, Kawaoka A, Sekine M et al. Sequence of the cellular T-DNA in the untransformed genome of Nicotiana glauca that is homologous to ORFs 13 and 14 of the Ri plasmid and analysis of its expression in genetic tumours of N. glauca x N. langsdorffii. Mol Gen Genet 1994; 243:706-10.
56. Fruendt C, Meyer AD, Ichikawa T et al. A tobacco homologue of the Ri-plasmid orf13 gene causes cell proliferation in carrot root discs. Mol Gen Genet 1998; 259:559-68.
57. Hansen G, Vaubert D, Heron JN et al. Phenotypic effects of overexpression of Agrobacterium rhizogenes T-DNA ORF13 in transgenic tobacco plants are mediated by diffusible factors. Plant J 1993; 4:581-85.
58. Sinha N, Williams R, Hake S. Overexpression of the maize homeobox gene, KNOTTED-1, causes a switch from determinate to indeterminate cell fates. Genes Dev 1993; 7:787-95.
59. Chuck G, Lincoln C, Hake. KNAT1 induces lobed leaves with ectopic meristems when overexpressed in Arabidopsis. Plant Cell 1996; 8:1277-89.
60. Sentoku N, Sato Y, Matsuoka M. Overexpression of rice OSH genes induces ectopic shoots on leaf sheaths of transgenic rice plants. Dev Biol 2000; 220:358-364.
61. Stieger PA, Meyer AD, Kathmann P et al. The orf13 T-DNA Gene of Agrobacterium rhizogenes Confers Meristematic Competence to Differentiated Cells. Plant Physiol 2004; 135:1798-08.
62. Olszowska O, Furmanowa M. Micropropagation of S. officinalis by shoot buds. Planta Med 1990; 56(6):637.
63. Hosoki T, Tahara Y. In vitro propagation of S. leucantha. Cav Hort Sci 1993; 28(3):226.
64. Molina M, Luis A, Luis JG. In vitro mass propagation of S. canariensis by axillary shoots. Acta Soc Bot Poloniae 1997; 66(3):351-54.
65. Skaia E, Wysokinska H. In vitro regeneraton of S. nemorosa L. from shoot tips and leaf explants. In vitro Cell and Develop. Biol Plant 2004; 40(6):596-02.
66. Kintzios S, Nikolaou A, Skoula M. Somatic embryogenesis and rosmarinic acid accumulation in S. officinalis and S. fruticosa callus cultures. Plant Cell Rep 1999; 18(6):462-466.
67. Chen H, Chen F, Zhang YL et al. Production of lithospermic acid B and rosmarinic acid in hairy root cultures of S. miltiorrhiza. J Ind Microb Biotech 1999; 22(3):133-38.
68. Savona M, Mascarello C, Bisio A et al. S. cinnabarina Martens et Galeotti: optimisation of the extraction of a new compound, tissue culture and hairy root transformation. Agr Medit 2003; 133:28-35.
69. Bisio A, De Tommasi N, Romussi G. Diterpenoids from S. wagneriana. Planta Medica 2004; 70:452-57.
70. Macchia M, Pagano A, Ceccarini L et al. Agronomic and phytochemical characteristics in some genotypes of Ocimum basilicum L. Acta Hort 2006; 723:143-49.
71. Szabo E, Thelen A, Petersen M. Fungal elicitor preparations and methyl jasmonate enhance rosmarinic acid accumulation in suspension cultures of Coleus blumei. Plant Cell Rep 1999; 18:484-89.
72. Bais HP, Walker TS, Schweizer HPB et al. Root specific elicitation and antimicrobial activity of rosmarinic acid in hairy root cultures of Ocimum basilicum. Plant Physiol Biochem 2002; 40:983-95.
73. Kintzios S. Makri O, Panagiotopoulos E et al. In vitro rosmarinic acid accumulation in sweet basil (Ocimum basilicum). Biotechn Lett 2003; 25(5):405-08.
74. Zeldin EL, Haas TB, McCown BH et al. Air recovery of essential oils from plants grown in vitro: a new production strategy. Hort Sci 1988; 23:759-62.
75. Tada H, Murakam Y, Omoto T et al. Rosmarinic acid and related phenolics in hairy root cultures of Ocimum basilicum. Phytochem 1996; 42(2):431-434.
76. Petersen M, Hausler E, Meinhard J et al. The biosynthesis of rosmarinic acid in suspension cultures of Coleus blumei. Plant Cell Tiss Organ Cult 1994; 38:171-79.
77. Hippolyte I. In vitro rosmarinic acid production. In: Kintzios S, ed. Medicinal and Aromatic Plants-Industrial Approaches: The Genus Salvia. Amsterdam: Harwood Publishers, 2000; 233-42.
78. Dixon RA. Natural products and plant disease resistance. Nature 2001; 411:843-47.
79. Kokkini S. Chemical Races Within the Genus Mentha L. In: Linskens HF, Jackson JF, eds. Essential Oils and Waxes. Heidelberg: SpringerVerlag, 1991; 12:63-7.
80. Banthorpe DV. Mentha Species (Mints): In vitro Culture and Production of lower Terpenoids and Pigments. In: Bajaj YPS, ed. Biotechnology in Agricultural and Forestry, Medicinal and Aromatic Plants IX: Heidelberg: Springer-Verlag, 1996; 37:202-25.

81. Spencer A, Harnill JD, Rhodes MJ. Production of terpenes by differentiated shoot cultures of Mentha citrata transformed with Agrobacterium tumefaciens T37. Plant Cell Rep 1990; 8:601-604.
82. Spencer A, Hami JD, Rhodes MJC. In vitro biosynthesis of monoterpenes by Agrobacterium transfomed shoot cultures of two Mentha species. Phytochem 1993; 32:911-19.
83. Diemer F, Jullien F, Faure O et al. High efficiency transformation of peppermint (Mentha piperita L.) with Agrobacterium tumefaciens. Plant Sci 1998; 136:101-08.
84. Niu X, Lin K, Hasegawa PM et al. Transgenic peppermint (Mentha x piperita L.) plants obtained by cocultivation with a Agrobacterium tumefaciens. Plant Cell Rep 1998; 17:165-71.
85. Diemer F, Caissard JC, Moja S et al. Agrobacterium tumefaciens mediated transformation of Mentha spicata and Mentha arvensis. Plant Cell Tissue Organ Cult 1999; 57:75-78.
86. Krasnyanski S, May RA, Loskutov A et al. Transformation of the limonene synthase gene into peppermint (Mentha piperita L.) and preliminary studies on the essential oil profiles of single transgenic piants. Teor Appl Genet 1999; 99:676-82.
87. Niu X, Li X, Veronese P et al. PM. Factors affecting Agrobacterium tumefaciens mediated transformation of peppermint. Plant Cell Rep 2000; 19:304-10.
88. Diemer F, Caissard JC, Sandrine M et al. Altered monoterpene composition in transgenic mint following the introduction of 4S-limonene synthase. Plant Physiol Biochem 2001; 39:603-14.
89. Mahmoud SS, Croteau RB. Metabolic engineering of essential oil yield and composition in mint by altering expression of deoxyxylulose phosphate reductoisomerase and menthofuran synthase. Proc Natl Acad Sci USA2001; 98:8915-20.
90. Inoue F, Sugiura H, Tabuchi A et al. Plant regeneration of peppermint, Mentha piperita, from the hairy roots generated from microsegment infected with Agrobacterium rhizogenes. Plant Biotech 2003; 20:169-72.
91. Strycharz S, Shetty K. Effect of Agrobacterium rhizogenes on phenolic content of Mentha pulegium elite clonal line for phytoremediation applications. Process Biochem 2002a; 38:287-93.
92. Strycharz S, Shetty K. Peroxidase activity and phenolic content in elite clonal lines of Mentha pulegium in response to polymeric dye R-478 and Agrobacterium rhizogenes. Process Biochem 2002; 37:805-12.
93. Pellati F, Benvenuti S, Magro L et al. Analysis of phenolic compounds and radical scavenging activity of Echinacea spp. J Pharm Biomed Anal 2004; 35:289-01.
94. Li TSC. Echinacea: Cultivation and medicinal value. Hort Techn 1998; 8:122-29.
95. Hu C, Kitts DD. Studies on the antioxidant activity of Echinacea root extract. J Agr Food Chem 2000; 48(5):1466-72.
96. Bergeron C, Gafner S, Batcha LL et al. Stabilization of caffeic acid derivatives in Echinacea purpurea L. glycerin extract. J Agric Food Chem 2002; 50:3967-70.
97. Smith MAL, Kobayashi H, Gawienowski M et al. An in vitro approach to investigate medicinal chemical synthesis by three herbal plants. Plant Cell Tissue Organ Cult 2002; 70:105-11.
98. Baskin CC, Baskin JM, Hoffman GR. Seed dormancy in the prairie forb Echinacea angustifolia var. angustifolia (Asteraceae): After ripening pattern during cold stratification. Int J Plant Sci 1992; 153:239-43.
99. Macchia M, Angelini LG, Ceccarini L. Methods to overcome seed dormancy in Echinacea angustifolia D.C. Sci Hort 2001; 89:317-24.
100. Feghahati SMJ, Reese RN. Ethylene-, Light- and Prechill-enhanced Germination of Echinacea angustifolia Seeds. J Amer Soc Hort Sci 1994; 119:853-58.
101. Sari AO, Morales MR, Simon JE. Ethephon can overcome seed dormancy and improve seed germination in Purple coneflower species Echinacea angustifolia and Echinacea pallida. Hort Techn 2001; 11:202-05.
102. Wang B, Zhang G, Zhu L et al. Genetic transformation of Echinacea purpurea with Agrobacterium rhizogenes and bioactive ingredient analysis in transformed cultures Colloids and Surfaces B: Biointerfaces 2006; 53:101-04.
103. Liu CZ, Abbasi BH, Gao M et al. Caffeic acid derivatives production by hairy root cultures of Echinacea purpurea. J Agric Food Chem 2006; 54:8456-60.
104. Reyes LF, Cisneros-Zevallos L. Wounding stress increases the phenolic content and antioxidant capacity of purple-flesh potatoes (Solanum tuberosum). J Agric Food Chem 2003; 51:5296-00.
105. Shirley BW. Flavonoid biosynthesis: a control model for genetics, biochemistry, cell biology and biotechnology. Plant Physiol 2001; 126:485-493.
106. Abbasi BH, Tian CL, Murch SJ et al. Light-enhanced caffeic acid derivatives biosynthesis in hairy root cultures of Echinacea purpurea. Plant Cell Rep 2007; 8:1367-1372.
107. Cheminat A, Zawatzky R, Becker H et al. Caffeoyl conjugates from Echinacea species: structure and biological activity. Phytochem 1988; 27:2787-94.
108. Tang W, Eisenbrand G. Panax ginseng. In: Mayer CA, ed. Chinese Drugs of Plant Origin. Berlin Springer-Verlag 1992:710-37.
109. Proctor JTA. Ginseng: old crop, new directions. In: Janick J, ed. Progress in New Crops. Arlington: ASHS Press, 1996:565-577.

110. Wu J, Zhong J. Production of ginseng and its bioactive components in cell culture: current technological and applied aspects. J Biotechnol 1999; 68:88-98.
111. Inomata S, Yokoyama M, Gozu Y et al. Growth pattern and ginsenoside production of Agrobacterium-transformed Panax ginseng roots. Plant Cell Rep 1993; 12:681-86.
112. Mallol A, Cusido RM, Palazon J et al. Ginsenoside production in different phenotypes of Panax ginseng transformed roots. Phytochem 2001; 57:365-71.
113. Vazquez-Flota F, Moreno-Valenzuela O, Miranda-Ham ML et al. Catharanthine and ajmalicine synthesis in Catharanthus roseus hairy root cultures. Plant Cell Tissue Organ Cult 1994; 38:273-79.
114. Jeong GT, Park D-H, Ryu H-W et al. Effects of inoculum conditions on growth of hairy roots of Panax ginseng C.A. Meyer. Appl Biochem Biotechnol 2004; 113-116:1193-03.
115. Yu KW, Gao W, Han EJ et al. Jasmonic acid improves ginsenoside accumulation in adventitious root culture of Panax ginseng C.A. Meyer. Biochem Eng 2002; 11:211-215.
116. Jeong GT, Park DH, Ryu HW et al. Production of antioxidant compounds by culture of Panax ginseng C.A. Meyer hairy roots: I. Enhanced production of secondary metabolite in hairy root cultures by elicitation. Appl Biochem Biotechnol 2005; 121-124:1147-57.
117. Jeong GT, Park DH. Enhanced secondary metabolite biosynthesis by elicitation in transformed plant root system: effect of abiotic elicitors. Appl Biochem Biotechnol 2006; 129-132:436-46.
118. Zhou E, Cao X, Zhang R et al. Stimulation of saponin production in Panax ginseng hairy roots by two oligosaccharides from Paris polyphylla var. yunnanensis. Biotechnol Lett 2007; 29:631-34.
119. John M, Rohrig H, Schmidt J et al. Cell signaling by oligosaccharides. Trends Plant Sci 1997; 2:111-15.
120. Vorwerk S, Somerville S, Somerville C. The role of plant cell wall polysaccharide composition in disease resistance. Trends Plant Sci 2004; 9:203-09.
121. Fraga BM, Diaz CE, Guadano A et al. Diterpenes from Salvia broussonetii Transformed Roots and their insecticidal activity. J Agric Food Chem 2005; 53:5200-6.
122. De Felice A, Malafronte A, De Tommasi N et al. Metabolic profiling of cells and transformed hairy roots of Salvia sclarea. 2004; XLVIII SIGA Annual Congress.
123. Li F-X, Jin Z-P, Zhao D-X et al. Overexpression of the Saussurea medusa chalcone isomerase gene in S. involucrata hairy root cultures enhances their biosynthesis of apigenin. Plant Physiol Biochem 2006; 41:1019-25.
124. Chen H, Chen F, Chiu CK et al. The effect of yeast elicitor on the growth and secondary metabolism of hairy root cultures of Salvia miltiorrhiza. Enzyme Microb Technol 2001; 22:133-138.
125. Zhang C, Qiong Y, Wai-keung C et al. Enhancement of Tanshinone Production in Salvia miltiorrhiza Hairy Root Culture by Ag+ Elicitation and Nutrient Feeding. Planta Med 2004 70(2):147-151.
126. Yan Q, Hu Z, Xiang TR et al. Efficient production and recovery of diterpenoid tanshinones in Salvia miltiorrhiza hairy root cultures with in situ adsorption, elicitation and semi-continuous operation. J Biotechn 2005; 119:416-424.
127. Ge X, Wu J. Induction and potentiation of diterpenoid tanshinone accumulation in Salvia miltiorrhiza hairy roots by p-aminobutyric acid. Appl Microbiol Biotechnol 2005; 68:183-88.

Chapter 14

Plant Tissue Culture—An Opportunity for the Production of Nutraceuticals

Mariella Lucchesini* and Anna Mensuali-Sodi

Abstract

This chapter provides a short discussion about the opportunity to cultivate in vitro plant tissue of species which synthesize secondary metabolites of nutraceutical interest. The introduction of species of particular interest in cultivation and domestication, can be an alternative to the harvest of wild species. In vitro culture techniques are a useful tool to improve production and marketing nutraceutical species which allows to make a rapid clonal propagation of plants selected for their active principles. The techniques of tissue culture are described in detail. In particular, it is underlined the necessity to clone selected plants and produce true-type plants when standardized plant products are the main goal. This can be reached by conventional micropropagation protocols culturing plants in vitro through the five culture phases. Another approach consists in applying unconventional systems in the last phase of in vitro culture which permit to develop autotrophy of the explants. Autotrophic growth improves the quality of the multiplied shoots and facilitates the acclimatization of the plantlets.

Introduction

Domestication and cultivation of plant species disclose the opportunity to solve problems by using biotechnologies to produce bioactive compounds.[1] The general interest for the traditional plant cultivation comes from the necessity to guarantee a constant supply of natural chemical compounds to be also employed as a precursor of organic synthesis; in fact, neither the collection of wild plants can satisfy the continuous demands from the industry, nor, the spontaneous plants can always supply homogenous product, depending on multiple factors. Moreover, the amount of plants growing wild more and more reduces, as a consequence of the indiscriminate collection, therefore many species are disappearing fast and those existing are often unusable due to strongly polluted areas. In such situation the introduction of species of particular interest in cultivation and domestication, can be very profitable and the technological control on their propagation can guarantee a constant biomass availability.

Why to Cultivate In Vitro Plants for the Production of Secondary Metabolites?

In vitro culture techniques are a useful tool to improve production and marketing of plant species which allows to make a rapid clonal propagation of plants selected for their active principles. Potentially any plant organism represents one source of cells for in vitro cultivation. Cells isolated

*Corresponding Author: Mariella Lucchesini—Università degli Studi di Pisa, Pisa, Italy.
Email: mlucchesini@agr.unipi.it

Bio-Farms for Nutraceuticals: Functional Food and Safety Control by Biosensors
edited by Maria Teresa Giardi, Giuseppina Rea and Bruno Berra.

from plant organs can be induced to grow and to multiply if placed on a determined culture medium maintained under aseptic conditions.

The in vitro cultivations allow supplying plant material continuously; the synthesis of target compounds such as secondary metabolites depends on adaptations to fluctuating temperatures and light conditions, stress, pathogen infections or herbivore attacks. The use of growth chambers with controlled environmental parameters decreases the problem of the alternation of bioactive compound production, which is due to the variability of the climatic conditions caused by the seasons and by the geographic area specificity. Moreover, it is possible to keep under control the presence of toxic residues typical of the conventional agriculture, or other biotic contaminants from the biomass because the in vitro cultivation takes place under conditions of sterility. The possibility to clone plants makes it feasible to manipulate the phenotypic variation in the concentration of bioactive compounds present at harvest. The aim is to increase extract potential, reduce toxin levels and increase uniformity and consistency of extracts. Moreover, the in vitro cultivation technique avoids the problems related with the collection, the transport and the storage of the plant products since the extraction can be carried out at regular intervals all over the year. The protection of the natural biological balance is guaranteed because the in vitro techniques maintain the biodiversity of germplasm avoiding the environmental impact due to intensive domestication of wild species. The employment of these techniques allows not negligible advantage, providing bioactive molecules otherwise insufficient in nature. Thanks to these techniques it is possible to carry out fast selections of cell lines in laboratory to improve the productivity instead of the plant selection in vivo, which is determined by the time of growth and development which is typical for the crop species.[2]

Many researchers all over the world are working on the development of in vitro cultivations in order to obtain uniform alimentary and agronomic compounds in a fast way and with yields higher than ones obtained in the natural processes.[3] Moreover, the in vitro cultivations can protect endangered plants indiscriminately collected contributing, therefore, to avoid their extinction. The future of the plant biotechnology is, therefore, a lot promising as far as production of active principles, propagation and selection of plants of nutraceutical interest. The realization of protocols for the in vitro plant cell cultures is based on the employment of particular strategies in order to maintain the cells in relatively low growth status corresponding to more elevated accumulation of secondary metabolites: under such conditions, in fact, the plant cells practically decelerate the primary metabolic pathways useful for growth and development, while the secondary metabolism becomes extremely active. Therefore, in order to obtain the biosynthesis of secondary metabolites it is useful and necessary to establish the conditions[4] which allow the cells to stay in a phase of their vital cycle in which they don't divide, but do not die. The growth rate is estimated experimentally as fresh or dry weight increase. There are two main systems of plant cell cultivation: the close and the open systems. The first ones are more commonly applied in the research laboratories and consist in obtaining cell growth by adding a fixed amount of cells to a culture medium; the biomass increases until one of the environmental factor becomes limiting. At this stage the metabolites can be recovered from the cell mass directly. In the open systems, instead, the nutritional factors are supplied to the cells from continuous adding fresh medium monitoring its components; this is possible by means of particular equipment named chemostate that is the one commonly used on industrial scale. In order to make some representative examples of the enormous potentialities of the plant biotechnologies, it can be underlined how investigators of many industries and university are studying protocols to obtain anticancer compounds as camptothecin, taxol, vinblastine and vincristine, through in vitro plant cell cultivations.[5] The camptothecin, naturally extracted from the trees *Camptotheca acuminata*, was investigated by the National Cancer Institute; after its anticancer activity was proven, it was demonstrated that it was toxic and that such toxicity could be removed chemically modifying the active principle; current methods are being studied to increase the in vitro yield and to eliminate the toxicity genetically, rather than chemically through metabolic engineering. Taxol, a compound against ovarian, lung, breast and colon cancers, is extracted in small amounts from the cork of mature trees of *Taxus brevifolia*. Approximately five mature 50 to 200 years old trees are required to take care of a single patient. According to these data, more than 125.000 trees would have to be uproot every year for dealing only this type of cancers.[5] Plant cell suspensions are able to produce taxol in much higher amounts than naturally found in the taxus bark; in this way the compound can be obtained in the purest form in two or three months, the time necessary to establish the cultivation. Moreover this technique offers the opportunity to select cell lines to begin other cell suspensions that produce alternative form of the taxol, that could turn out more beneficial.[5] At last, in vitro plant cultures can provide high yield of vindoline, a compound produced in cells of *Catharanthus roseus*; this substance can be subsequently converted to vincristine and vinblastine, through organic synthesis. These agents are particularly effective as a cure for cancers with rapid cell proliferation.

In Vitro Technologies for the Cultivation of Nutraceutical Plants

Different methodologies of in vitro culture can be employed, other than callus or cell cultures, to produce biomass of medicinal plant regardless the conservation of both phenotype and genotype of initial parent plants (adventitious regeneration of organs, organogenesis from callus, somatic embryogenesis, cultivation of genetic transformed plant material) whereas, to produce true-type plants, the micropropagation techniques are necessary to clone selected plants and are helpful to start field cultivation for a mass herbal production.

Adventitious Regeneration

Roots, buds and flowers are plant organs that can be directly formed de novo from one or few cells in plant tissue cultured in vitro. Roots and buds can be independently regenerated: usually the buds are formed at first, but in many cases induction of buds and development of adventitious roots need different culture medium compositions.[6] The factors which influence this process are still not well defined: in general we can say that regeneration of adventitious roots is frequently observed in plant tissue culture of several species and seems to depend on interaction between auxins, carbohydrates, light quality and photoperiod; shoot regeneration, instead, could be induced from an appropriated balance between auxins and cytokinins.

The new plantlets can also have origin indirectly from undifferentiated calli. Callus is a mass of undifferentiated cells that has been originated from any plant tissues. The induction and the formation of the callus, with the successive differentiation and organogenesis, are realized by the appropriate use of the plant growth regulators in the culture medium. In this field the research on the medicinal plants is rather vast; this is possible to verify from the examples in Table 1.

In general, it can be asserted that the induction of plantlets by callus formation can carry changed features and turn out plants extremely different from the mother plant from which the explant has been obtained (somaclonal variation). Such variability can lead to a secondary metabolite content of unexpected importance or, on the contrary, to useless metabolites.[3] Adventitious regeneration, due to the somaclonal variation, cannot be considered an appropriate method as concern clonal propagation of medicinal plants, because the main interest is to achieve, from in vitro cultures, standard chemical compounds characteristics of each species.

Somatic Embryogenesis

The somatic embryogenesis is a process through which groups of somatic plant cells or tissues leading to the formation of somatic embryos, which resemble the zygotic embryos of intact seeds, retaining the ability to develop into seedlings if placed in an appropriate culture environment. So, the regeneration of whole plants through somatic embryogenesis can offer, in comparison with the classic micropropagation through organogenesis, remarkable advantages thanks to the greater multiplication rate and the smaller costs supported to produce one plant.[7] Moreover embryogenic culture systems may be the basis for many biotechnological applications to plant improvement since they allow specific and directed changes to be introduced into elite individuals bypassing the undesirable consequences of sexual reproduction. Somatic embryos can be induced to proliferate directly on plant tissues cultured in vitro or in liquid cell suspension culture. The initiation of embryogenic calli needs the employment of a narrow range of plant growth regulators. The

Table 1. Medicinal plants regenerated from callus cultures (modified from Rout[8])

Plant	Explant	Response
Asparagus cooperi	st, in, r	Callus, shoot regeneration
Azasdirachta indica	an	Callus, shoot and root regeneration
Aegel marmelos	nu	Shoot regeneration
Catharanthus roseus	an	Callus
Cephaelis ipecacuanhal	s	Callus, cell differentiation
Centella asiatica	st, l	Callus, shoot regeneration
Cinchona pubescens	l	Callus, cell suspension
Datura innoxia	s	Aploid plantlet regeneration
Dioscorea deltoidea	hyp	Callus
Dioscorea deltoidea	ns	Callus, shoot regeneration
Hyoscyamus muticus	ct	Callus, shoot regeneration
Hyosciamus muticus	mp	Shoot regeneration
Lathyrus sativus	r, ld	Callus, shoot regeneration
Mentha arvensis var. pipe.	in	Callus, shoot regeneration
Plantago ovata	hyp	Callus, shoot regeneration
Psoralea coryfolia	l, st	Callus, shoot regeneration
Pinellia ternata	b, l, pt	Callus, shoot regeneration
Plumbago rosea	st	Callus, shoot regeneration
Plumbago zerylanica	l, st	Callus, shoot regeneration
Picrorhiza kurroa	l, st	Callus, shoot regeneration
Solanum melongena	hyp	Shoot regeneration
Solanum torvum	hyp	Shoot regeneration
Strychonus nuxvomica	hyp	Callus, shoot regeneration
Trigonella sp.	l	Callus
Trigonella elladona	l	Cell differentiation
Trigonella foenum-graec.	mp	Callus, shoot regeneration
Typhonium trilobatum	pt	Shoot regeneration
Vanilia planifolia	st	Callus, shoot regeneration

Abbreviations: an: anther; cs: cell suspension; in: internode; inf: flower stalk; ld: leaf disk; st: shoot tip; r: root; mp: mesophyll protoplast; nu: nucellar tissue; l: leaf; hyp: hypocotyl; ct: cotyledon; pt: petiole.

preponderance of protocols utilizes culture media with synthetic auxins, especially 2.4 dichlorophenoxyacetic acid. Cytokinins also are required to induce somatic embryo; the most commonly used are benzyladenine and thidiazuron. After a short pulse treatment with growth regulators the cells are placed on hormone free medium for embryo development. The somatic embryos mature in few weeks if transferred into suitable maturation media, obtaining, usually in few months, plants ready to the transfer in a greenhouse. In general, it has been noticed that reducing the concentration of the plant growth regulators, the development of embryos and their germination is promoted.[8] The germination of the somatic embryos has been obtained also in culture medium with low mineral content lacking of hormones.[9-11] On the contrary, the emergence of plants from somatic embryos is rather difficult and recently many researches have been carried out to increase plant survival. In particular, short-term treatments with several amino acids, osmotic compounds and growth regulators such as abscissic acid allowed to improve the embryo maturation and their germination ability.

Non zygotic embryo produced in bioreactors (but also vegetative propagated buds) can be encapsulated in synthetic seed coats to be practical used in commercial plant production as "clonal seeds". Many plant species, usually propagated by seeds, are attractive candidate for synthetic seed technology, since the per-plant cost might be reduced in comparison with the conventional vegetative propagation. The development of synthetic seed technology constitutes one of the most interest area of application for somatic embryogenesis studies.

Genetic Transformation

The direct manipulation of DNA sequences to alter gene expression in nutraceutical plant research is a booming business for the production of active ingredients to food components. For example some terpenes biosynthetic pathway of *Mentha* spp. were engineered to alter the essential oils in thricomes (cell organs specialized in producing and accumulating secondary metabolites) and increase resistance to fungal infections and abiotic stress.[12] Rather than insist on the creation of GM plants with productive and qualitative improvement, the role of genetic transformation seems to go towards the genetic improvement of cell culture and the establishment of hairy roots culture.[1] The processing methods can be the direct transfer of DNA or indirect taking advantage of the use of vectors (bacteria, viruses). The use of *Agrobacterium tumefaciens* Ti plasmid DNA or *Agrobacterium rhizogenes* RI plasmid for induction of hairy roots is among the indirect methods. These bacteria have the ability to induce adventitious roots (literally "hairy roots") to a large number of host plants, once integrated the T-DNA from Ri plasmid of *Agrobacterium rhizogenes*; Rol genes (or locus root) inserted in the plant DNA increase the sensitivity to auxin rather than codify for its biosynthesis. It allows transformed cells to develop hairy roots which can be cultured in bioreactor, in absence of growth regulators and can release secondary metabolites into medium or can be used to regenerate entire trasformed plants.

The Micropropagation

The micropropagation represents one of the numerous aspects of the vast field of the in vitro plant cultivation: it is a method of asexual reproduction (agamic propagation), alternative to the traditional ones, recently applied to several herbal and nutraceutical species. The last scope of this system of multiplication is to obtain, in relatively reduced times and area, enough number of plantlets identical to the mother plants in genotype and phenotype (Fig. 1). During the last few years numerous widespread species of nutraceutical interest have been micropropagated by explants deriving from the various organs of the mother plants (Table 2). The development of micropropagation protocols is the first step in order to select and to multiply plants with high productivity of active principle.

Conventional Methods of Micropropagation

The in vitro propagation, or micropropagation, has an enormous potential for the large-scale production of plants producing active compounds. Such technique, first of all, has two main advantages in comparison with the methods of in vivo traditional propagation: the multiplication rate, that it is greatly superior and the possibility to obtain biological material free from pathogens. For this reason regeneration in vitro is introduced as a fundamental passage to obtain an indefinite number of clones of the same plant, chosen for a determined characteristic, for example, in the case of medicinal species a high productive level of secondary metabolites.

Many things have changed since Murashige[13] proposed a three stages protocol to plant micropropagation: in the course of the years, in fact, at least five critical stages to obtain a satisfactory micropropagation process were recognized (Fig. 2).

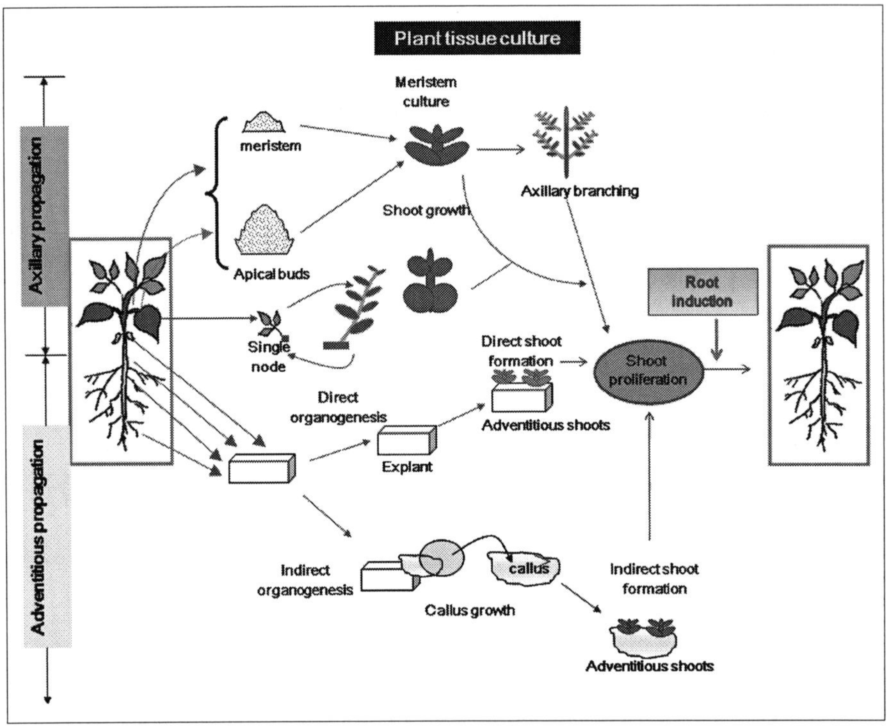

Figure 1. Plant tissue culture: adventitious and axillary propagation (modified from George et al[17]).

Table 2. In vitro multiplication of some important medicinal species (modified from Rout[8])

Plant Name	Explant Source	Response
Cinchona legderiana	st	Multiple shoots
Catharanthus roseus	st, ns	Multiple shoots
Dioscorea spp.	ns	Multiple shoots, plantlet formation
Gentiana kuroo	st, ns	Plantlet formation
Kaemferia galanga	ns, am	Multiple shoots, plantlet formation
Mentha spp.	st	Multiple shoots, plantlet formation
Ocimum basilicum	am	Multiple shoots, plantlet formation
Psoralea coryfolia	am	Multiple shoots, plantlet formation
Valeriana wallichii	apm	Multiple shoots, plantlet formation
Zingiber officinale	st	Multiple shoots, plantlet formation

Abbreviations: am: axillary meristem; apm: apical meristem; ns: nodal segment; st: shoot tip; ze: zygotic embryo.

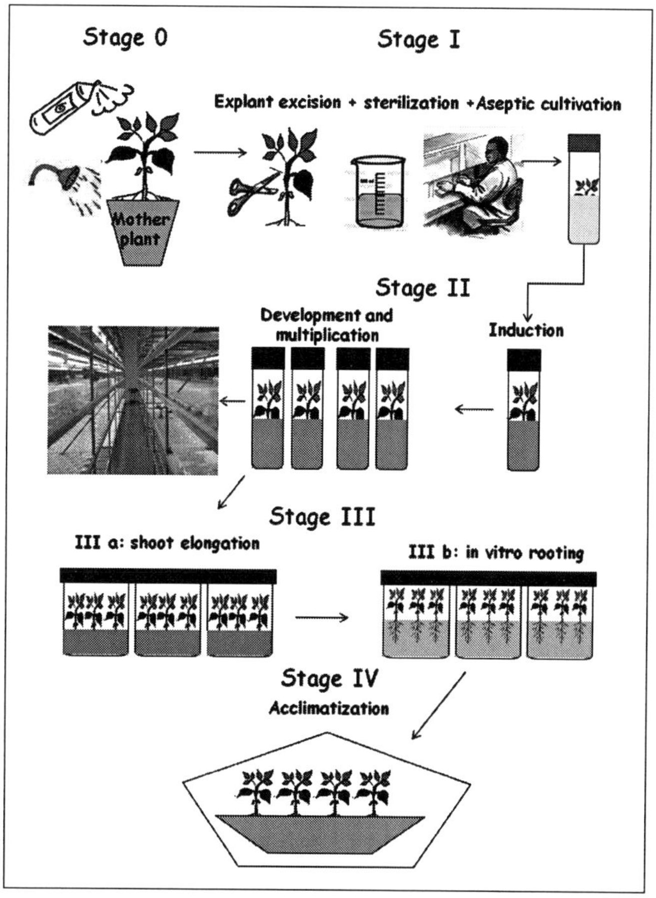

Figure 2. The five critical stages to obtain a micropropagation process.

The stages of the plant clonal propagation, according to Debergh and Maene[14] are:

Stage 0, preparation of the plant stock in health conditions;
Stage 1, beginning of the aseptic cultivation;
Stage 2, induction, development and multiplication;
Stage 3a, shoot elongation;
Stage 3b, rooting.
Stage 4, acclimatization;
Stages 1, 2 and 3 according to Debergh and Maene[13] coincide with stages I, II and III according to Murashige.[14]

In literature stage 0 is described as a remedy to the contamination problems, including all those that can improve the health of the mother plants, which are the basis of the in vitro propagation. It is useful also to remember the importance of the growth conditions of mother plants, which must be maintained in hygienic conditions with the control of temperature, light and relative humidity. The impact of stage 0 apparently is not only limited to the sanitary situation of the explant, but it seems to affect also the survival percentage of the successive stages of the micropropagation process. In vitro multiplication of plants occurs through cultivation of meristems, buds, petioles, roots, etc.; it needs some fundamental passages, first of all the achievement of aseptic conditions (Stage 1). Sterility is necessary for explants excised from plants grown in open field, which are contaminated from diverse microorganisms. Apart from external polluting agents, the inner contaminations can be expressed also after several years of in vitro sub-cultivation: endogenous bacteria or fungi could cause this phenomenon. The conventional methods of sterilization consisted of sodium hypochlorite (in variable percentages from the 5 to 15%), ethyl alcohol (from the 50 to 95%) and mercury chloride (0.01-0.1%). Often, the ethyl alcohol represents a preliminary sterilization step while the sodium hypochlorite is alternative to the mercury chloride use. Sterilization has variable time depending on the contamination state of the plant material, varying from 10 to 25 minutes. The absence of contaminations is fundamental for the in vitro culture maintenance, even if it seems that some species have a natural tendency to develop contaminations also in vitro without influence on the proliferation and/or the growth rate.

Beside contamination, many other factors must be considered in the selection of the plant explants to initiate an in vitro culture: physiological or ontogenic age of the organ that is the explant source, size and location of the explants, season in which the explant is excised.

The characteristics of the explants are decisive also for the successive stage of in vitro multiplication as demonstrated furthermore by the experiments on *Clerodendrum colebrookianu*[15] in which shoot apex, axillary shoot, leaf, petiole and root were compared as potential explants. In this species the optimum multiple shoot induction occurred subculturing axillary buds demonstrating how the nature and the physiological condition of the explant significantly influence the multiplication rate.

An other essential aspect of the micropropagation technique is the culture medium choice for the multiplication of a determined plant. Generally, the MS[16] medium is typically used for a wide range of species; it is composed of carbohydrates, vitamins, macro and microelements, necessary to the development of the plant material. Beyond the MS there are numerous culture media, differing from MS for mineral, amino acids, or for vitamin composition (Table 3).

In vitro multiplication (Stage 2) are based on the formation of shoots and roots induced on the explant by the plant growth regulators, therefore many studies are based on the amount and type of the plant hormone to employ for a certain plant species. In the middle 1980s numerous researches were focused on the influence and the interaction between exogenous plant growth regulators and the in vitro plant material.[8]

The effects of auxins and cytokinins on the shoot proliferation have been showed on many species,[17] the use of kinetin has been tested on *Picrorhiza kurroa*[18] and the use of thidiazuron (TDZ) for the bud proliferation has been tested on *Nothapodytes foetida.*[19] Moreover it was observed that in many genotypes the use of cytokinins in the shoot induction can be enhanced by the adoption of auxin low levels in the same medium.[20-22] According to Faria and Illg[23] the simultaneous use of BA, IAA and NAA induces in *Zingiber spectabile* a high bud multiplication rate, demonstrating that the number of shoot per explant depends on the concentration of the plant growth regulators and on the particular plant genotype.

The in vitro root induction (Stage 3) from the proliferated shoots is obtainable from culture media containing auxins, in various amounts depending on the type of auxin and the plant species. Marked differences are noticed among the plant species and the works performed to establish the optimal conditions for the in vitro root development are numerous.

Moderate or elevated concentrations of cytokinins inhibit the rooting.[8] Often the halved concentration of all the medium components (mainly of the sucrose amount) favours the rooting ability of the majority of micropropagated plants.[24]

The in vitro root induction is a phenomenon more difficult to obtain in comparison to the shoot growth; the excision that gives origin to a new shoot is a simple separation of a plant part from the source of water and minerals supplied by the roots; in turn, these elements can be supplied by the culture medium in which the explant is placed. On the contrary, the regeneration of a new rooting apparatus requires a deep modification in the shoot metabolism: the greater part of the plants is excised where there are not root meristems. For this reason, plants have difficulty to perform a new rooting apparatus without the aid of plant growth regulators.[24] The induction of the in vitro rhizogenesis in the micropropagated plants is fundamental for the successful in vivo

Table 3. Mineral and organic components of the principal culture media

Component	Chu[49]	Driver Kuniyuki[50]	Gamborg[51]	Hoagland[52]	Lloyd and McCown[53]	Murashige and Skoog[16]	Quoirin and Lepoivre[54]	Schenk and Hildebrandt[55]	White[56]
Minerals (mg/L)									
Ammonium nitrate		1416.0			400.0	1650.0	400.0		
Ammonium phosphate monobasic				115.03				300.0	
Ammonium sulfate	463.0		134.0						
Boric acid	1.6	4.8	3.0	2.86	6.2	6.2	6.2	5.0	1.5
Calcium chloride anhydrous	125.33	112.5	113.24		72.5	332.2		151.0	
Calcium nitrate		1367.0		656.4	386.0		833.77		200.0
Cobalt chloride $6H_2O$			0.025			0.025	0.025	0.1	
Cupric sulfate $5H_2O$		0.25	0.025	0.08	0.25	0.025	0.025	0.2	
EDTA disodium salt $2H_2O$	37.25	45.4	37.3		37.3	37.26	37.3	20.0	
Ferric sulfate									2.5
Ferric tartrate $2H_2O$				5.32					
Ferrous sulfate $7H_2O$	27.85	33.8	27.8			27.8	27.8	15.0	

Continued on next page

Table 3. Continued

Component	Chu[49]	Driver Kuniyuki[50]	Gamborg[51]	Hoagland[52]	Lloyd and McCown[53]	Murashige and Skoog[16]	Quoirin and Lepoivre[54]	Schenk and Hildebrandt[55]	White[56]
Magnesium sulfate	90.37	361.49	122.09	240.76	180.7	180.7	175.79	195.4	360.0
Manganese chloride $4H_2O$				1.81					
Manganese sulfate H_2O	3.33	33.5	10.0		22.3	16.9	0.76	10.0	5.04
Molybdenum trioxide				0.016					
Molybdic acid sodium salt H_2O		0.39	0.25		0.25	0.25	0.25	0.1	
Nickel sulfate $6H_2O$		0.005							
Potassium chloride									65.0
Potassium iodide	0.8		0.75			0.83	0.08	1.0	0.75
Potassium nitrate	2830.0		2500.0	606.6		1900.0	1800.0	2500.0	80.0
Potassium phosphate monobasic	400.0	265.0			170.0	170.0	270.0		
Potassium sulfate		1559.0			990.0				
Sodium phosphate monobasic			130.5						16.5
Sodium sulfate									200.0

Continued on next page

Table 3. Continued

Component	Chu[49]	Driver Kuniyuki[50]	Gamborg[51]	Hoagland[52]	Lloyd and McCown[53]	Murashige and Skoog[16]	Quoirin and Lepoivre[54]	Schenk and Hildebrandt[55]	White[56]
Zinc nitrate $6H_2O$		17.0							
Zinc sulfate $7H_2O$	1.5		2.0	0.22	8.6	8.6	8.6	1.0	2.67
Organics (mg/L)									
Ascorbic acid						5			
D-Biotin									
Glycine (free base)	2	2			2	2			
Myo-Inositol		100	100		100	100	100	1000	
Nictotinic acid (free acid)	1	1	1		0.5	0.5		5	
Pyridoxine HCl	0.5		1		0.5	0.5		0.5	
Thiamine HCl	0.5	2	10		1	0.1	0.4	5	

acclimatization; for many years rooting has been considered a process based on only one phase but, since the 1980s, it is subdivided in three main phases: induction (pre-initiation), initiation and expression (post-initiation).[17] The first phase is characterized by a lack of visible histological indications, the second phase corresponds to the first cytological modifications and cellular divisions, the third is characterized by the cellular development and by visible changes of the base of the explant. Other factors that more affected the in vitro rooting are: juvenility of the micropropagated plant material; dark treatments; conditions of autotrophy; peroxidase status at the base of the explant.

Another difficult phase of the conventional micropropagation is the acclimatization process (Stage 4) in which the plantlets or microcuttings grown in vitro are transferred outside, since they must be adapted to the new environmental conditions and, in particular to the sugar lack in the substrate and to the reduced relative humidity of the air. This term has been defined from Conover and Poole[25] exactly like "climatic adaptation of a living organism that it is transferred in a new atmosphere" and is a process that derives from the human direct action: when the same process happens naturally, the recommended terms are acclimation or adaptation. The low plantlet survival rate during the acclimatization process originates mainly from the insufficient control of loss of water caused from the conditions of previous elevated humidity in the cultivation vessel. The closed systems are essential for the maintenance of the in vitro sterility and the aseptic atmosphere avoids any pathogenic stress to the plantlets. These conditions make the in vitro plantlets little adapted to survive in greenhouse or open field. In fact, the leaves that are developed in vitro generally do not differentiate epicuticolar protections and their stomata don't close normally; moreover, their leaves have a little structured palisade layer and show a low photosynthetic activity: because plantlets show those physiological and anatomical characteristics, the micropropagate plantlets should have to go gradually towards the acclimatization processes.[26] In the conventional acclimatization the main objective is to maintain a high relative humidity and low lighting conditions simulating the previous in vitro phase (Fig. 3). It is important to reduce the relative humidity gradually and to increase the intensity of light to make the new leaves more similar to those developed in greenhouse or in open field. This sequence of events causes a "vicious circle" which leads to continuous losses of plant material, to remarkable losses of energy and to increases of the production costs for the maintenance of this environment.[27]

Two possible approaches exist in order to resolve these problems: one consists in controlling the atmosphere where the acclimatization takes place using sophisticated methodologies to monitor the culture and to adjust in real time the environment parameters ("control units of micro-computerized acclimatization"), the other consists in controlling the atmosphere during the last passages of the in vitro culture.

Unconventional Methods of Micropropagation

The autotrophy is the characteristic that distinguishes the plant from animal reign: the plants are able to produce the carbohydrates they need for their growth and the development of the primary or secondary metabolisms starts from the CO_2 in the air and from the photosynthesis.

The plants cultivated in vitro by the traditional techniques lose partially or completely such ability; instead of utilizing CO_2 in order to carry out the photosynthesis, they find the necessary energy for the metabolic cycles in the carbohydrates contained in the culture substrate. Because of the insufficient CO_2 in the containers, in fact, the in vitro plants develop heterotrophy or mixotrophy, absorbing sugars from the culture medium as the major carbon source. In particular in vitro cultures are defined heterotrophic, photoautotrophic and mixotrophic when the carbon source is represented respectively from the carbohydrates in the medium, from the CO_2 of the headspace of the vessel, or both.[28] In conventional micropropagation the main objective is to obtain as many "true-to type" plants as possible, rapidly and at low cost, without taking into account the quality of the environment in which they live. In the vessels used for the conventional micropropagation the photosynthesis (autotrophy) is hindered by the presence of carbohydrates in the medium and by the simultaneous limited gas exchange and insufficient light intensity. However it is possible to lead the explants towards the development of the photosynthetic activity using several strategies

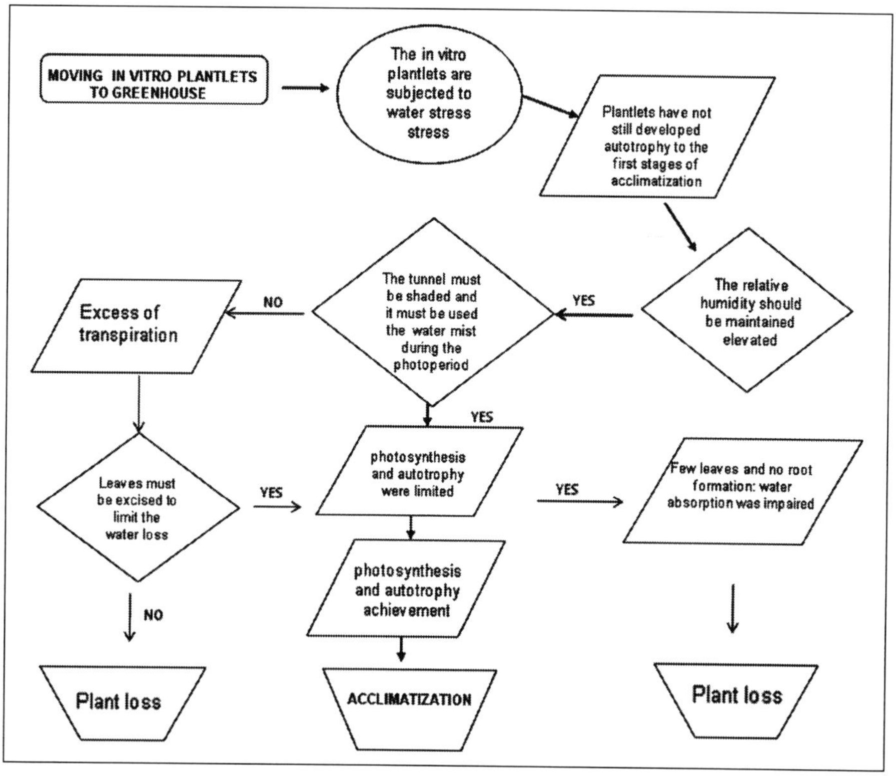

Figure 3. The crucial role of the autotrophy in the acclimatization process.

which come from simple systems of passive ventilation (filters ventilates can be placed on the vessel lids) which permit the recovery of the atmospheric level of CO_2 in the containers[29-32] to the most sophisticated techniques of carbonic enrichment.[33,34] The realization of the in vitro autotrophy helps to obtain therefore plantlets that can carry out a high photosynthetic activity with morphological characteristics more similar to those of the in vivo plants. The physical parameters that determine the micro-environment of the in vitro culture system which can promote a good photosynthetic efficiency are the light, the temperature, the gaseous atmosphere, the matrix of explant support and the type of container (Fig. 4).[35] The control of the micro- and macro-environmental factors are considered basic also from a practical point of view to reduce the production costs; it promotes in fact the growth and the development of the explant (increments in fresh and dry weight, in the leaf area, the in vitro rooting, the branching of the shoots), it reduces morphologic and physiologic disorders, it also reduces the loss of plants due to contaminations. The control of the in vitro culture environment increases the shoot uniformity and can lead to the reduction of the plant growth regulators and, at last, it helps the plants to grow more vigorous once they are transferred outside. In order to facilitate the achievement of good levels of autotrophy, in vitro cultivation must be carried on with increments in light intensity while the carbohydrates content in the culture medium must be reduced or completely eliminated. The plantlets, grown photo-mixotrophically in vitro gain an advantage in overcoming the in vivo stress in facing the acclimatization phase in comparison with the heterotrophic ones.[36,37] The autotrophic plantlets are suitable to realize "micro-greenhouses" cultures of greater dimensions than the traditional containers. Artificial substrates as agriperlite, vermiculite, polyurethane and cellulose filters (Fig. 5) are ideal to be employed for these cultural systems and they contribute to the development of photosynthetic activity.[38-40] In these culture

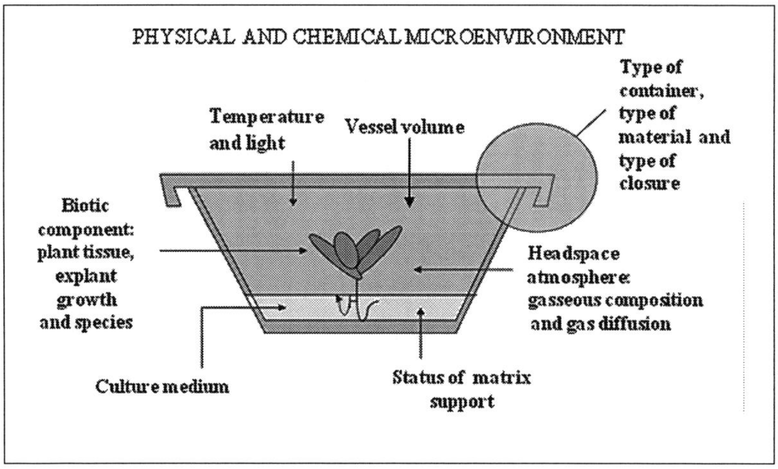

Figure 4. The vessel micro-environmental factors.

systems named "Microponic systems" the plantlets develop on inert supports in which the liquid medium moves like that of a miniaturized hydroponics culture.[41,42] The term "microponic" derives from the terms "micro-propagation" and "hydroponic", ideally a midway between these two techniques, circumventing the limitations of both and exploiting the advantages. The idea comes from the need of overcoming the limits of micropropagation, which are mainly due to the low growth and survival rates during acclimatization. Moreover, the high production cost of in vitro culture, could be resolved by the following actions: control of environmental factors for the promotion of autotrophic growth and for the reduction of physiological disorders; labor saving

Figure 5. *Echinacea angustifolia* L. rooting shoots on cellulose filters Sorbarod™ (Eaumgartner Papiers SA, Switzerland).

by introducing robotic systems; energy saving and recycling of resources; improving container efficiency; computerizing processes for mass propagation; introducing new materials for artificial substrates, containers and lamps.

The unconventional micropropagation could be employed in the secondary metabolites production: physical-chemical variables in the in vitro systems can be modulated[43] to affect positively the metabolites production. Secondary metabolites formation and plant growth are often inversely related: stress conditions[44] can restrain culture growth and can also start cell differentiation process or organogenesis inducing the enzymes of secondary metabolism.[8,45] Secondary metabolite compounds accumulated in vitro as much as the differentiated structures are comparable to those of the intact plant.[8] A functional photosynthetic activity in vitro might induce a stimulation of biosynthetic cell capacities.[46] Photosynthesis produces energy and molecules very important for the regulation of the secondary metabolism. An in vitro culture system with balanced environmental factors may generate a compromise between primary and secondary metabolism promoting active principles. Micropropagated plantlets of mint and thyme cultured with carbonic enrichment are able to produce high quantities of secondary metabolites.[47] To induce the formation of alkaloids in in vitro cultured shoots of *Catharanthus roseus* L., autotrophic cultures were established in liquid medium on cotton fibers as plant support.[48] Photoautotrophic cultures produced more alkaloids, allow longer subcultures of plantlets and required a supporting agent that included a 4-fold cost-saving and better diffusion of nutrients. Autotrophic cultures were more cost-effective towards the production of medicinal indole alkaloids than mixotrophic cultures. Below there are exemplified some miniaturized in vitro aerated systems (Figs. 6 and 7).

Figure 6. Microfloating system: a pump inflates air into the vessels under the liquid medium trough flexible silicon pipes.

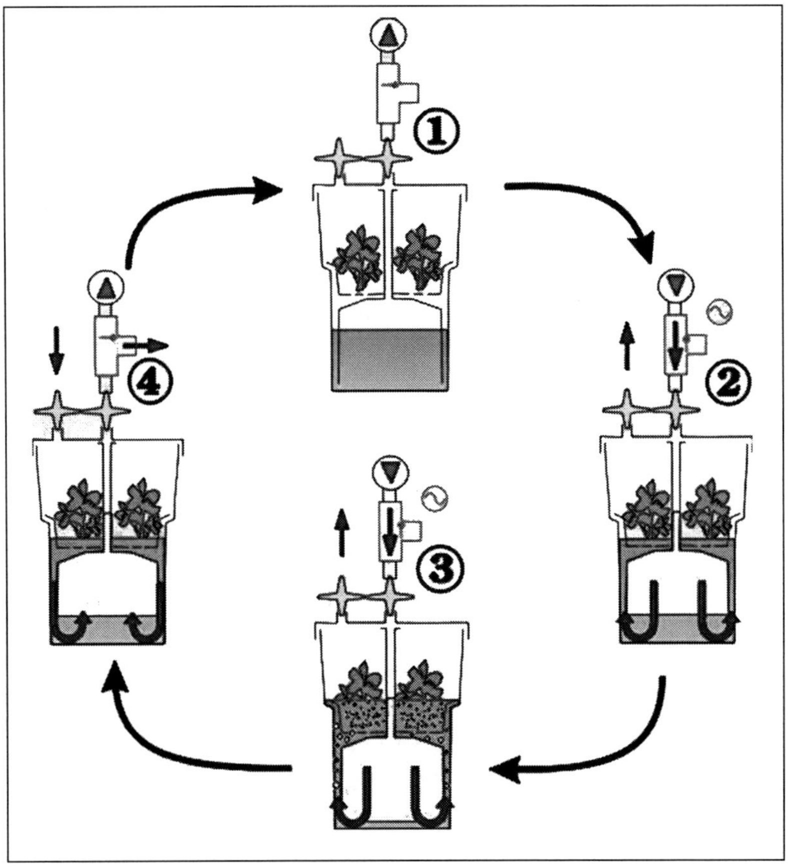

Figure 7. A system of temporary immersion in a container patented by CIRAD (Montpellier, France) called RITA® and distributed by Vitropic (*Saint Mathieu de Tréviers France*). Image available from http://www.vitropic.fr/rita.

Conclusion

The establishment of these protocols could allow the in vitro cultivation of plant genotypes selected for the production of active constituents. Our study outlines the possibility to obtain the in vitro biomass by direct and indirect organogenesis or cloning plantlets via micropropagation techniques. Phytochemical analysis demonstrated that in vitro developed shoots could be a suitable source of bioactive compounds characteristic of the plants of origin. On the other hand, in vitro culture could produce phytochemical profiles different from the mother plant species and guarantee varied yields of known or novel natural compounds.

The alternative use of unconventional micropropagation can induce autotrophy in the in vitro systems and drives the cultures toward the development of features comparable to those of in vivo growing plants. This could allow to obtain yields of good quality biomass. In particular, the adoption of autotrophic cultures permits to get uniformity of growth, consistent dry matter yields, speediness of growth increments, stable production of natural compounds and phytochemical profiles comparable to those characteristics of the mother plants. All these parameters are important to define the quality of in vitro plant biomass profitable to obtain natural compounds of nutraceutical interest.

References

1. Canter H, Thomas H, Ernst E. Bringing medicinal plants into cultivation: opportunities and challenges for biotechnology. Trends in Biotechnology 2005; 23:180-185.
2. Collin HA. Secondary product formation in plant tissue cultures. Plant Growth Regulation 2001; 34:119-134.
3. Bajaj YPS. Biotechnology for the improvement of medicinal plants. Acta Hort 1998; 457:37-45.
4. Jeong CS, Chakrabarty D, Hahna EJ et al. Effects of oxygen, carbon dioxide and ethylene on growth and bioactive compound production in bioreactor culture of ginseng adventitious roots. Biochemical Engineering Journal 2006; 27:252-263.
5. Fox S. Plant cell culture provides reservoir for beneficial drugs/rare plants. Genetic Engineering News 1992; 12:12, 13, 36.
6. Tripathi L, Tripathi JN. Role of biotechnology in medicinal plants. Tropical Journal of Pharmaceutical Research 2003; 2:243-253.
7. Jayasankar S, Gray DJ, Litz RE. High-efficiency somatic embryogenesis and plant regeneration from suspension cultures of grapevine. Plant Cell Reports 1999; 18:533-537.
8. Rout GR, Samantaray S, Das P. In vitro manipulation and propagation of medicinal plants. Biotechnology Advances 2000; 18:91-120.
9. Choi YE, Kim JW, Soh WY. Somatic embryogenesis and plant regeneration from suspension cultures of Acanthopanax koreanum Nakai. Plant Cell Reports 1997; 17:84-8.
10. Hirai G, Kasai N, Harada T. Somatic embryogenesis in mature zygotic embryo culture of Glehnia littoralis. Plant Cell Tissue and Organ Culture 1997; 48:175-80.
11. Rout GR, Samantaray S, Das P. Somatic embryogenesis and plant regeneration from callus culture of Acacia catechu: a multipurpose leguminous tree. Plant Cell Tissue and Organ Culture 1995; 42:283-5.
12. Veronese P, Li X, Niu X et al. Bioengineering mint crop improvement. Plant Cell Tisue Organ Culture 2001; 64:133-144.
13. Debergh PC, Maene LJ. A scheme for commercial propagation of ornamental plants by tissue culture. Scientia Hort 1981; 14:335-345.
14. Murashige T. Plant propagation through tissue culture. Annu Rev Plant Physiol 1974; 25:135-166.
15. Mao AH, Wetten A, Fay M et al. In vitro propagation of Clerodendrum colebrookianum Walp: a potential natural anti-hypertension medicinal plant. Plant Cell Rep 1995; 14:493-6.
16. Murashige T, Skoog F. A revised medium for rapid growth and bioassays with tobacco tissue cultures. Physiol Plant 1962; 15:473-493.
17. George EF, Hall MA, De Klerk GJ. Plant Propagation by Tissue Culture 3th Edition: Vol. 1, the Background. Springer, Berlin (Germany), 2007.
18. Lal N, Ahuja PS. Plantlet regeneration from callus in Picrorhiza kurroa Royle ex Benth. An endangered medicinal plant. Plant Tissue Cult 1996; 6:127-34.
19. Rai VR. Rapid clonal propagation of Nothapodytes foetida (wight) sleumer—a threatened medicinal tree. In-vitro Cell Dev Biol-Plant 2002; 38:347-51.
20. Rout GR, Das P. In vitro organogenesis in ginger (Zingiber officinale Rose). J Herbs, Spices and Med Plants 1997; 4:41-51.
21. Shasany AK, Khanuja SPS, Dhawan S et al. High regenerative nature of Mentha arvensis internodes. J Biosci 1998; 23:641-6.
22. Rout GR, Saxena C, Samantaray S et al. Rapid clonal propagation of Plumbago zeylanica L. Plant Growth Reg 1999; 28:1-4.
23. Faria RT, Illg RD. Micropropagation of Zingiber spectabile Griff. Sci Horticult 1995; 62:135-7.
24. Moncousin C. Rooting of in vitro cutting. In: Bajaj YPS, ed. Biotechnology in Agriculture and Forestry: High Tech and Micropropagation I. Berlin Heidelberg: Springer-Verlag, 1991;231-262.
25. Conover CA, Poole RT. Acclimatization of indoor foliage plants. Hort Rew 1984; 6:120-154.
26. Preece JE, Sutter EG. Acclimatization of micropropagated plants to the greenhouse and field. In: Debergh P, Zimmerman R, eds. Micropropagation of Horticultural crops Dordrecht: Kluwer Academic Publishers, 1990;71-93.
27. Kozai T. Acclimatization of micropropagated plants. In: Bajaj YPS, ed. Biotechnology in Agriculture and Forestry: High-tech and Micropropagation I. Berlin Heidelberg, Germany Springer Verlag, 1991;127-141.
28. Kozai T. Autotrophic micropropagation. In: Bajaj YPS, ed. Biotechnology in Agriculture and Forestry: High Tech and Micropropagation I. Berlin Heidelberg, Germany: Springer-Verlag, 1991;313-343.
29. Mensuali-Sodi A, Panizza M, Tognoni F. Quantification of ethylene losses in different container-seal systems and comparison of biotic and abiotic contributions to ethylene accumulation in cultured tissues. Physiol Plant 1992; 84:472-476.

30. Matthijs DG, Demeester JJ, Pascat B et al. Factors controlling the evolution of the gaseous atmophere during in vitro culture. In: Carre F, Chagvardieff P, eds. Ecophysiology and Photosynthetic in vitro Cultures Ecophysiology and Photosynthetic in vitro Cultures. Aix-en-Provence: CEA; 1995:129-140.
31. Marino G, Berardi G, Ancherani M. The effect of the type of closure on the gas composition of the headspace and the growth of GF677 peach x almond rootstock cell suspension cultures. In vitro cell dev Biol Plant 1995; 31:207-210.
32. Lucchesini M, Mensuali-Sodi A, Massai R et al. Development of autotrophy and tolerance to acclimatization of Myrtus communis L. transplants cultured in vitro under different aeration. Biol Plant 2001; 44:167-174.
33. Buddendorf-Joosten JMC, Woltering EJ. Components of the gaseous environment and their effects on plant growth and development in vitro. In: Lumsden PJ, Nicholas JR Davies WJ, eds. Physiology, Growth and Development of Plants in Culture. Dordrecht: Kluwer Academic Publishers, 1994;165-190.
34. Solarova J, Pospisilova J. Effects of carbon dioxide enrichment during in vitro cultivation and acclimation to ex vitro conditions. Biol Plant 1997; 9:23-30.
35. Desjardins Y. Photosynthesis in vitro. On the factors regulating CO_2 assimilation in micropropagation systems. Acta Hort 1995; 393:45-57.
36. Van Huylenbroeck J, Debergh P. Impact of sugar concentration in vitro on Photosynthesis and carbon metabolism during ex vitro acclimatization of Spathiphyllum plantlets. Phisiol Plant 1996; 96:298-304.
37. Lucchesini M, Monteforti G, Mensuali Sodi A et al. Leaf ultrastructure, photosynthetic rate and growth of myrtle plantlets under different in vitro culture conditions. Biologia plantarum 2006; 50:161-168.
38. Kirdmanee C, Kytaia Y, Kozai T. Effects of CO_2 enrichment and supporting material in vitro on photoautotrophic growth of Eucalyptus plantlets in vitro and ex vitro. In vitro cell dev Biol Plant 1995; 31:144-149.
39. Afreen-Zobayed F, Zobayed SMA, Kubota C et al. A combination of vermiculite and paper pulp supporting material for the photoautotrophic micropropagation of sweet potato. Plant Science 2000; 157:225-231.
40. Lucchesini M, Mensuali-Sodi A. Effects of vessel permeability to gas exchanges on the in vitro plantlets of Myrtus communis L. Agr Mediter 2000; 130:78-84.
41. Hempel M. From Micropropagation to "Microponics". Practical Hydroponics International 1993; 21-23.
42. Kozai T, Fujiwara K, Kitaja Y. Modelling, measurements and control in plant tissue culture. Acta Hort 1995; 393:63-73.
43. Lucchesini M, Mensuali-Sodi A. Influence of medium composition and vessel ventilation on in vitro propagation of Phillyrea latifolia L. Scientia Hort 2003; 100:117-125.
44. Mensuali Sodi A, Serra G, Vitagliano C et al. In vitro growth pattern of salt-stressed cells of Lavandin. Acta Hort 1990; 280:459-62.
45. Panizza M, Mensuali Sodi A, Tognoni F. Morphological differentiation in callus cultures of lavandin: a role of ethylene. Biologia Plantarum 1997; 39:481-89.
46. Baccou JC. Effect of photosynthesis on the secondary metabolism of cell cultures. In: Carre F, Chagvardieff P, eds. Ecophysiology and photosynthesis in vitro cultures. 1995 Aix-en Provance (France): CEA, 1995;1-3.
47. Tisserat B, Vaughn SF. Essential oils enhanced by ultra-high carbon dioxide levels from Lamiaceae species grown in vitro and in vivo. Plant Cell Rep 2001; 20:361-68.
48. Mitra A, Khan B, Rawal S. Photoautotrophic shoot culture: an economical alternative for the production of total alkaloid from Catharanthus roseus (L.) G. Don. Current Sci 1997; 73:608-9.
49. Chu CC. The N6 medium and its applications to anther culture of cereal crops. In: Proceedings of the Symposium on Plant Tissue Culture. Beijing, China: Science Press 1978;43-50.
50. Driver JA, Kuniyuki AH. In vitro propagation of Paradox walnut rootstock. Hort Science 1984; 19:507-9.
51. Gamborg OL, Miller RA, Ojima K. Nutrient requirements of suspension cultures of soybean root cells. Exp Cell Res 1968; 50:151-8.
52. Hoagland DR, Arnon DI. The water culture method for growing plants without soil. California Agr Expt Sta Circ 1938:347.
53. Lloyd G, McCown B. Commercially-feasible micropropagation of mountain laurel, Kalmia latifolia, by use of shoot tip culture. Combined Proceedings of the International Plant Propagators Society 1980; 30:421-27.
54. Quoirin M, Lepoivre P. Improved media for in vitro culture of Prunus sp. Acta Hort 1977; 78:437-42.
55. Schenk RU, Hildebrandt AC. Medium and techniques for induction and growth of monocotyledonous and dicotyledonous plant cell cultures. Can J Bot 1972; 50:199.
56. White PR. The cultivation of animal and plant cells. New York: Ronald Press, 1963.

Chapter 15

Plant Cell Cultures:
Bioreactors for Industrial Production

Barbara Ruffoni,* Laura Pistelli, Alessandra Bertoli and Luisa Pistelli

Abstract

The recent biotechnology boom has triggered increased interest in plant cell cultures, since a number of firms and academic institutions investigated intensively to rise the production of very promising bioactive compounds.

In alternative to wild collection or plant cultivation, the production of useful and valuable secondary metabolites in large bioreactors is an attractive proposal; it should contribute significantly to future attempts to preserve global biodiversity and alleviate associated ecological problems. The advantages of such processes include the controlled production according to demand and a reduced man work requirement.

Plant cells have been grown in different shape bioreactors, however, there are a variety of problems to be solved before this technology can be adopted on a wide scale for the production of useful plant secondary metabolites. There are different factors affecting the culture growth and secondary metabolite production in bioreactors: the gaseous atmosphere, oxygen supply and CO_2 exchange, pH, minerals, carbohydrates, growth regulators, the liquid medium rheology and cell density. Moreover agitation systems and sterilization conditions may negatively influence the whole process.

Many types of bioreactors have been successfully used for cultivating transformed root cultures, depending on both different aeration system and nutrient supply. Several examples of medicinal and aromatic plant cultures were here summarized for the scale up cultivation in bioreactors.

Introduction

Plants produce several different secondary metabolites, called phytochemicals mostly of them used as pharmaceuticals. In recent years biopharmaceutical/nutraceutical industry renewed increased attention in production of health-promoting secondary metabolites using plant cell and tissue cultures. Different efforts to improve their productivity had limited success, especially due to the lack of suitable technologies for such scale-up applications.

The production of in-vitro secondary metabolites can be possible through plant cell cultures.[1] This technology represents a good model to overcome many problems linked to the conventional agriculture such as variations in the crop quality due to environmental factors: drought, flooding and other abiotic stresses and/or biotic stresses as diseases or pest attacks. Moreover crop adulteration, losses in storage and handling may decline the secondary metabolites production, which cannot be prevented by inability of some authorities.

*Corresponding Author: Barbara Ruffoni—CRA- FSO Unità di Ricerca per la Floricoltura e le Specie Ornamentali, Sanremo, Italy. Email: b.ruffoni@istflori.it

Bio-Farms for Nutraceuticals: Functional Food and Safety Control by Biosensors
edited by Maria Teresa Giardi, Giuseppina Rea and Bruno Berra.

In order to optimize secondary metabolite production in in-vitro plant cultures the following strategies have been evaluated:

1. Establishment of cell suspension cultures of plant with a content of required phytochemicals,
2. Selection of highly productive clones,
3. Optimization of culture conditions.

Successful establishment of cell lines able to produce high yields of secondary compounds in cell suspension cultures has been reported first by Zenk.[2]

The accumulation of secondary products in plant cell cultures depends on the composition of the culture medium and on environmental conditions.[3,4]

For the production of secondary metabolites at commercial level some prerequisites have also to be considered, like: high demand, high product costs, uniform availability of raw material and technology cost inputs, otherwise the production will not be effective and commercially possible.

As consequence of these considerations, after the demonstration of desirable product presence, at reasonable levels, in plant cell cultures, we have to consider the problems of scale up from small (about 5 liters) to large vessels. Large volume automated culture vessels, called fermenters, have been successfully used to produce cultured plant cells and many bioreactor designs have been used for mass cell growth.[5]

The present chapter reviews various aspects of large scale plant cell and tissue cultures and the bioreactor systems, focusing the attention on some examples of medicinal and aromatic plant cultivation for secondary metabolite production.

Historical Background

For large scale cultures, the first success was obtained in 1960 by culturing cells of various plant species in a 134 liter bioreactor.[6]

In 1977, the culture of tobacco cells was obtained in a 20 L tank. In 1984, the first industrial production of shikonin (natural compound used as a dye and a medicinal compound) was performed in 750 L bioreactor by *Lithospermum erythrorhizon* cells in Japan. Later a German company started with the taxol production, an anticancer compound, using *Taxus* cell cultures with bioreactor capacities up to 75 L.[7]

However, this technology is still being developed and despite the advantages outlined above, there are a variety of problems to be overcome before it can be adopted on a wide scale for the production of useful plant secondary metabolites. The success of Mitsui Petrochemical Industry Co. Ltd. in Japan in producing shikonin on a commercial scale from *Lithospermum erythrorhizon* cultivations and that of Nitto Denko Co. Ltd. also in Japan in mass production of *Panax ginseng* or ginseng cells using 20 kL tanks and the other examples described above demonstrate that many of the problems can be overcome with perseverance. The economic feasibility of these processes is related to the value of the considered metabolite. It is possible to produce food additives, metabolites with pharmacological value, antioxidants, aromes, organic acids, aminoacids, vitamins and polisaccharides or secondary metabolites such as phenols, flavonoids, terpenoids, etc.[8]

There are known some significant differences between microbial and plant cell cultures that must be considered when attempting to apply plant cell cultures to the available technology. Table 1 shows a comparison of the characteristics of plant and microbial cultures during fermentation.

Large-scale cell culture production has been in part limited by the large size, rigid cell wall and extensive vacuole of plant cells, making them sensitive to shear stress.[9] In fact with normal blade impellers the cells may twist and can be broken triggering the cell death with the loss of entire cell culture. Thus, air-lift fermentors are recommended by some researchers. The low aeration requirement for plant cells is an advantage over microbial cultures in general. Furthermore the large size of the plant cells allows to extend their duplication time in comparison with the microbial ones, determining a prolonged period for a successful fermentation time. Other differences regard the product accumulation, since in plant cells they are stored in vacuoles, meanwhile microorganisms secrete the biosynthesised products in the medium. Thus, squeezing is necessary to obtain high metabolite yields from plant cell cultures, with the consequence of cells replenishment.

Table 1. Characteristics of microbial and plant cells during fermentation

Characteristics	Microorganisms	Plant Cells
Size	<2 μm	>10 μm
Shear stress	Insensitive	Sensitive
Water content	75%	>90%
Duplication time	<1 hour	Days
Aeration	1-2 vvm	0.3 vvm
Fermentation time	Days	Weeks
Product accumulation	Medium	Vacuole
Production phase	Uncoupled	Often growth-linked

Undifferentiated plant cells cultures often produce reduced quantities and different profiles of secondary metabolites in comparison with the intact plant. The poor production is attributed to a lack of differentiation in cultures. These features may change during the growth curve. On the other hand, there are cases of cultures that over-produce metabolites compared with the whole plant (Table 2).

The culture of undifferentiated plant biomass can be obtained in semisolid agarized culture (callus) and in liquid culture (cell suspensions). The non-organised cultures might show genetic instability such as poliploidy that generally increase during the culture time.

Table 2. Secondary metabolites produced in high levels by plant cell cultures

		Yields (% Dry Wt)		
Compound	**Plant Species**	**Culture**	**Plant**	**Culture Type***
Shikonin	*Lithospermum erythrorhizon*	20	1.5	s
Ginsenoside	*Panax ginseng*	27	4.5	c
Anthraquinones	*Morinda citrifolia*	18	0.3	s
Ajmalicine	*Catharanthus roseus*	1.0	0.3	s
Rosmarinic acid	*Coleus blumei*	15	3	s
Ubiquinone-10	*Nicotiana tabacum*	0.036	0.003	s
Diosgenin	*Dioscorea deltoides*	2	2	s
Benzylisoquinoline Alkaloids	*Coptis japonica*	11	5-10	s
Berberine	*Thalictrum minus*	10	0.01	s
Berberine	*Coptis japonica*	10	2-4	s
Anthraquinones	*Galium verum*	5.4	1.2	s
Anthraquinones	*Galium aparine*	3.8	0.2	s
Nicotine	*Nicotiana tabacum*	3.4	2.0	c
Bisoclaurine	*Stephania cepharantha*	2.3	0.8	s
Tripdiolide	*Tripterygium wilfordii*	0.05	0.001	s

*s: suspension; c: callus.

It is possible to induce the undifferentiated tissue (callus) by hormonal treatments from fragments of in vitro or in vivo plant tissues (leaf, petiole). Normally auxines as Naphthaleneacetic acid (NAA) or 2,4 dichlorophenoyacetic acid (2,4D) were used at different concentrations and combined with light supply and temperature.[10] The juvenility of the starting plant material has a great importance in the callus induction: the best callus can be produced from young tissues.

Callus must be friable for cell culture development, i.e., it can be easily disaggregated in liquid medium. In the first week of culture the suspension is constituted by cells aggregates with different size. Then a filtration of this material is necessary to select small clusters and single cells for starting a synchronised controlled growth.[11]

Protocol optimisation for scaling up the biomass is a precompetitive method for the extraction of suitable amounts of secondary metabolites and for industrial production, after an economical evaluation of the system.

The Plant Bioreactors

Type of Bioreactors

As the bioreactor is a physical/thermal system for maintenance of cells at the best culture conditions for a fast growth, several models can be considered as bioreactors, starting from simple close vessels that can be externally agitated up to complex aseptic systems controlled and regulated by appropriate software.[12]

Tulecke and Nickel[13] firstly developed a successful 10 L system in a sample carboy for the cultivation of plant cells. For use at laboratory scale, bioreactor is a large culture vessel made up of glass (up to 10 L) but large scale reactors are made up of stainless steel. These are fitted with a microprocessor control unit for the control of pH, dissolved oxygen, gas flow rate, agitation speed, nutrient factors, temperature inside the vessel and cell density for optimal growth and/or production. For sterilization procedures small reactors can be autoclaved, while commercial scale reactors are sterilized in situ by passing steam at appropriate pressure.

Secondary metabolites produced in plant cells are either released into medium or accumulated in the cells. Thus the spent medium or the biomass is harvested after suitable incubation period for the extraction of bioactive compounds. Sometimes different media for growth and production are used to obtain maximum secondary metabolite production.

Cell growth and secondary metabolite production represent the main factors for selecting the suitable process mode. In fact if growth and production occur simultaneously, a process mode which supports growth of cells over an extended period should be chosen; otherwise if product synthesis follows a period of rapid growth, an appropriate process mode is selected for its ability to maintain cultures in a slow-growing rate and to retain its productivity. It is imperative to note that a bioreactor is used when basic studies related to optimization of product yield have been completed.

Different process modes are used, such as batch culture, fed batch culture, rapid feed batch culture, two-stage batch culture and continuous culture (Fig. 1).

Factors Affecting the Growth in Bioreactor

In order to manage the biomass growth in bioreactors, various culture conditions must be controlled, i.e. the oxygen supply and CO_2 exchange, pH, minerals, carbohydrates, growth regulators, the liquid medium agitation and cell density.[14,15] A complete review on this topic[16] reported several aspects relating to the gaseous atmosphere first in the culture vessels and then related to the bioreactor environment. In vessels the aerial part is composed mainly of nitrogen (78%), oxygen (21%) and carbon dioxide (0.036%). The culture vessel gas composition is influenced by the volume of the vessel and the extent of ventilation. Plants evolve CO_2 and consume O_2 during respiration, while during photosynthesis CO_2 is used and O_2 is produced. If photoautotrophic conditions persisted in the plant cell culture, CO_2 levels increased during the dark period, while they decreased during the light period. Ethylene, ethanol, acetaldehyde and other hydrocarbons are additional components of the gaseous atmosphere in vitro. Most

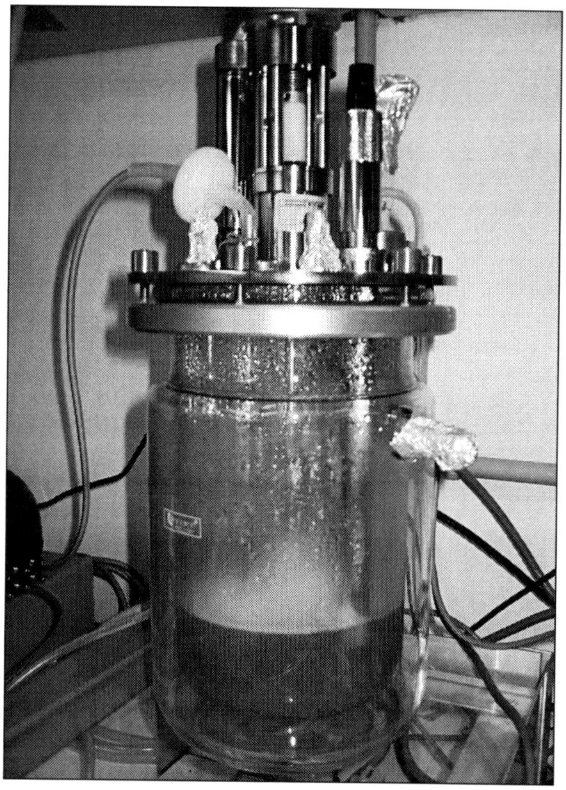

Figure 1. Scale up of *Salvia cinnabarina* hairy root culture in bioreactor (Applikon®, Germany).

of the effects of CO_2, O_2 and C_2H_4 on plant growth in vitro were reported for agar-gelled or cell suspension cultures.[17]

In bioreactors, the control of the gaseous phase depends on the gas flow and can be easily manipulated to provide the required levels of O_2, CO_2 and C_2H_4. The effect of aeration in *Linum album* in cell suspension culture in a 5 L stirred tank bioreactor equipped with low shear Setric impeller was reported to be important for particular compound production: 6-methoxypodophyllotoxin production was enhanced when cells were cultivated at 30% dissolved oxygen level.[18]

Oxygen. The amount of O_2 in bioreactor depends on the presence of O_2 in the gas phase above and in the air bubbles inside the medium, as well as on the dissolved O_2 in the medium. Air is released through a sparger located at the base of the bioreactor. The available oxygen for plant cells in liquid cultures, determined by oxygen transfer coefficient (kLa values), is the part that dissolves in water. Its depletion as a function of the metabolic activity of the growing cell biomass can affect the culture yield. Plant cells have a lower metabolic rate than microbial cells and a slow doubling time and therefore require a lower O_2 supply. In general, high aeration rates appear to reduce the biomass growth.[19]

The requirements for O_2 may vary from one plant species to another and must be supplied continuously to provide adequate aeration, since it affects metabolic activity and energy supply as well as anaerobic conditions. The level of O_2 in liquid cultures in bioreactors can be regulated by agitation or stirring and through aeration, gas flow and air bubble size. The use of a porous irrigation tube as a sparger generated fine bubbles, high kLa values, low mechanical stress and provided a high growth rate.[20]

High aeration rates were found to inhibit cell growth in cell suspensions cultured in airlift bioreactors. This result was explained to be due to an effect of "stripping" of the volatiles produced by the plant cells, which are apparently necessary for cell growth.[21]

Increasing O_2 levels from 21% to 80% in bioreactor cultures of Boston fern clusters enhanced growth values from 0.61 to 0.92. The growth value (GV) was calculated as the difference between the fresh weight at the end of the growth period (FW_1) and the initial fresh weight (FW_0) divided by the initial fresh weight $GV = (FW_1 - FW_0)/FW_0$. Reducing O_2 levels to 10% (v/v) affected cell differentiation in bioreactor cultures of carrot embryogenic tissue.[30]

CO_2. The effects of CO_2 were reported for both agarized cultures and cell suspension cultures used for secondary metabolite production.[17,22,23]

The contribution of CO_2 supply during the proliferation and multiplication stage in media supplied with sucrose in bioreactors is not clear. It is implicit that if photoautotrophic conditions do not prevail, CO_2 enrichment beyond the 0.36% in the air supply is unnecessary. It is reported that high aeration rates rather than excessive oxygen inhibit growth and this reduction could be due to depletion of CO_2 or to the removal of various culture volatiles including CO_2.[24,25] The requirement for CO_2 was not related to photosynthesis but to some other metabolic pathways involved in amino acid biosynthesis.

Ethylene. Ethylene is produced by cell suspension; generally the level of ethylene in the head-space in liquid cultures in flasks differs from that in continuously aerated bioreactor cultures. High rates of aeration, which are often required at high biomass densities, can cause "stripping" of volatiles that are apparently important for some plants grown in culture. In clusters of *Brodiaea* cultured in liquid medium, ethylene had no effect on growth, although its level was reduced in the presence of silver thiosulfate (an inhibitor of ethylene action). The level of ethylene was reduced from 0.38 to 0.12 ml/L in highly aerated bioreactor cultures of this plant material without affecting biomass growth.[26]

As another example, growth of *Thalictrum* cell suspension cultures was suitable in airlift system. At high cell density for berberine production gas-stripping also played a significant role and it was discovered that CO_2 and ethylene were important for product formation. By supplying a mixture of CO_2 and ethylene into the airlift system, the specific berberine content was increased two fold.[25]

Mineral Nutrients. Media with various modifications in the inorganic and organic constituents of Murashige and Skoog medium (MS).[27] are used for most plant species in agar-gelled or liquid cultures in vitro. The availability of mineral nutrients depends on the type of culture, whether agar-gelled or liquid, the type and size of the plant biomass and the physical properties of the culture. Factors such as pH, temperature, light, aeration, concentration of minerals, the medium volume and the viscosity of the medium determine the rate of absorption of the various nutritional constituents.[28,29] Plant cells growing in liquid cultures are better exposed to the medium components and the uptake and consumption are faster. In bioreactors, in which either humidified air or condensers are used to prevent dehydration, the level of the nutrients in the medium is affected mainly by the absorption rate and by cell lysis.[30] A decline in pH (lower values than 4.5) with subsequent increase to pH 5.5 was often attributed to the initial utilization of ammonium (NH_4^+), followed by a later uptake of nitrate. In several species the depletion of NH_4^+ is the first limiting factor of biomass growth.[16]

Other limitations of growth are due to the availability of phosphate, nitrogen and carbohydrates and to a lesser extent to calcium, magnesium and other ions.[31]

Mixing of dissolved nutrients of the culture medium is generally not a problem in suspension cultures. But in presence of aggregates serious problems can be found, since cells adhered to the surface of the tank at the bottom of the bioreactors.

Carbohydrates. Cultured plants require a constant supply of carbohydrates as their source of energy. Sucrose and to a lesser extent glucose, fructose, or sorbitol are the most commonly used carbohydrates in vitro. In general, sucrose is removed rather rapidly from the medium and after 10-15 days it can be completely depleted or reduced to 5-10 g/L from an initial level of 30 g/L in both agar-gelled and liquid cultures. At the same time, glucose and fructose (derived from sucrose hydrolysis) appear in the medium and can reach levels of 5-10 g/L *Catharanthus roseus* cell suspensions cultured in a column airlift bioreactor showed a lag phase of 5 days, during which there was a total hydrolysis of the sucrose to glucose and fructose.[21] In suspension cultures of alfalfa, sucrose also was hydrolyzed during the first 5 days. Most of the sugars uptake occurs after day 5 and glucose is taken up preferentially over fructose.[32]

pH. The initial pH in most plant cell cultures ranges between 5.5-5.9. Since most media are not buffered, changes occur during autoclaving and during the biomass growth. A rapid drop in pH to 4.0-4.5 took place within 24-48 h in cell suspension, organ and embryogenic cultures.[33-35] These changes were related to an initial ammonium uptake and acidification due to cell lysis. However, the pH increased after a few days and reached a stable level around pH 5.0-5.5, which was related to the uptake of nitrates. In spruce species cultured in liquid medium, the pH levels were shown to increase to 6.5-6.8 after 14 days in culture.[31]

Temperature. The control of the temperature in the liquid medium inside the bioreactor can be easily manipulated by using an heating element in the vessel or by circulating water in an enveloping jacket outside the vessel. There is, however, limited information on the effects of temperature in bioreactor cultures, which is usually kept constant at 25°C, with a short photoperiod.

Agitation Systems

Depending on the mode of agitation, bioreactors can be basically classified into following two types:

Mechanically Agitated Bioreactors. In these bioreactors medium is agitated with the help of a mechanically driven impeller and vary types of impellers have been used. One of these is flat-blade turbine impeller, in which high agitation breaks incoming air into small bubbles. Mechanically stirred bioreactors depend on impellers, including a helical ribbon impeller,[30] magnetic stirrers, or vibrating perforated plates.[19]

Pneumatically Agitated Bioreactors. These are classified in two types: bubble column and air-lift. These are tall and thin in comparison with agitation bioreactors.

In the bubble columns air is bubbled at the base of the column, thus medium is agitated. In air lift bioreactors, gas is sparged from the riser section to the top of the column and the medium flows downward in the down corner section. These two sections may be separated using a baffle, a concentric cylinder or an external loop. Circulation in the air-lift bioreactor promotes a better mixing and therefore offers advantages in uniformly suspending cells and clumps, although the oxygen transfer rate is low in the down-corner section. So the performance of an air-lift bioreactors is strongly dependent upon the geometry of the system. Mixing by gas sparging in bubble column or airlift bioreactors lacking impellers or blades is far less damaging for clusters than mechanical stirring, since they were shown to have a lower shearing stress.[26,34]

The main advantage of airlift bioreactors is their relatively simple construction, the lack of regions of high shear, reasonably high mass and heat transfer and relatively high yields at low input rates.[36]

Alternative Aeration: Silicon Tubing. A bubble free oxygen supply bioreactor with silicone tubing was found suitable for embryogenic cell suspensions and provided foam-free cultures.[37] A system of 8 independent units of bioreactor using silicon tubing as aeration system was established by Preil and Hvoslef-Heide, showing a good aeration percentage without damage cells.[38]

For hairy root culture, an acoustic mist bioreactor was found to increase root biomass significantly.[39]

Sterilization and Bioreactor Component

The sterilization routine is affected by the size of a bioreactor and its associated components. Hale et al[40] defined the following criteria for plant bioreactor design:

1. All components should be fully autoclavable and the sterility should be maintained for several weeks.
2. The growth chamber should be transparent for adequate light transmission and visibility.

3. Materials should selected to be more breakage resistant than glass.
4. Component should be easily assembled and disassembled for cleaning.
5. Reduce opening at minimum to avoid contamination.
6. The ventilation (oxygen) should be supplied to avoid damages of plant cells, tissues or organs.
7. Suitable mechanical agitation should be provided to protect the biomass from damage.
8. Glass or plastic bioreactor chambers must be strong to withstand the weight, the pressure and the turbulence.

Kinetics of the Cell Growth

The nutrient supply strictly conditioned the biomass growth into bioreactor.

The feeding system has implication on the growth pattern, therefore must be chosen considering the productivity objectives. A number of operating strategies can be applied in plant cell bioreactors:

Batch Culture (*Close System*): Inoculation of organisms in a fixed volume of liquid medium. Inside the vessel the growth conditions constantly change, with nutrient depletion and concomitantly a metabolites accumulation. The organisms perform a sigmoid growth curve reaching a plateau at the senescence. *Salvia officinalis* cells grown in batch culture exhibited this trend (Fig. 2).

Fed Batch Culture: Variation of the batch system. The culture medium is gradually added to the bioreactor with a consequent rise of plant biomass (linear relationship). An example of fed batch culture of *Cyclamen persicum* embryogenic cells is represented in Figure 3, in which dry weight biomass was measured during fermentations in fast or slow feeding.[41] A fast feeding fermentation was realised using 300 ml of 500 μm filtered suspension at minimum 80% of cell viability, put into the bioreactor with 200 ml of fresh medium; after 24 hours in constant conditions (batch phase) the fresh medium was added daily in increasing quantity of about 10% of the suspension total volume (feed-batch phase). The experiment stopped after 16 days when the volume reached 1.5 L.

A slow feeding fermentation in 500 ml suspension culture was performed with a growth cycle of 22 days; during that period the growth speed was lowered and 20-30 ml of fresh medium were added every 2 days and reaching 1.0 L of suspension at the end of the experiment.[41]

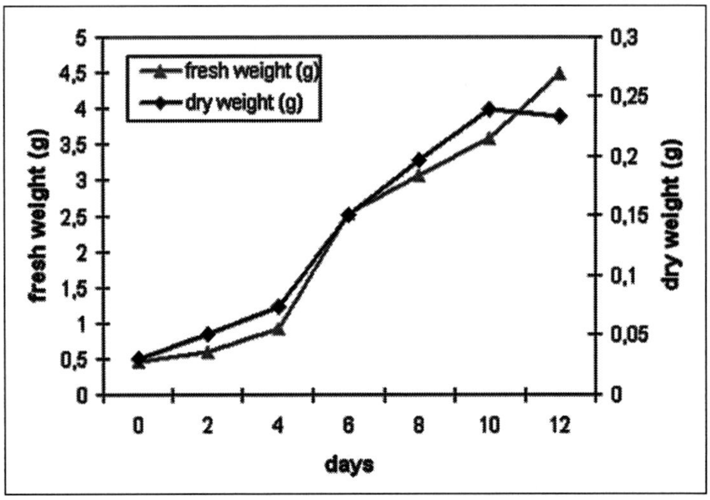

Figure 2. Growth curve of *Salvia officinalis* cells in batch culture. The typical sigmoid curve showed a logarithmic phase between the 4th and 10th day.

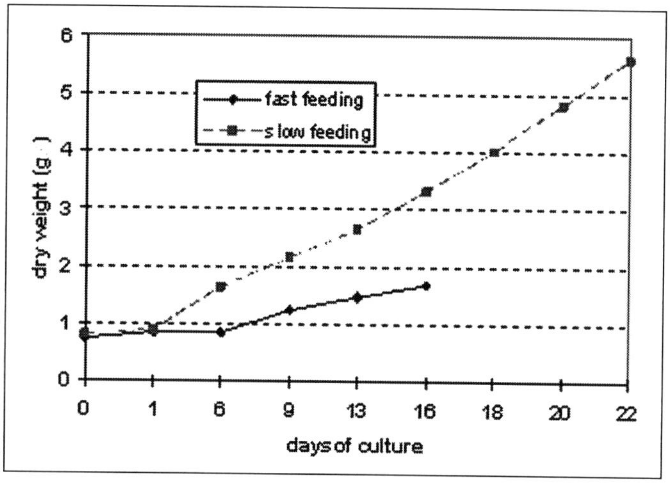

Figure 3. *Cyclamen persicum* embryogenic cell culture, scale up in bioreactor: dry weight biomass evaluation during fermentations in fast or slow feeding.[41]

Continuous Culture: During the exponential phase a volume of fresh medium is added and at the same time an equal volume of cell culture is discarded; a balanced growth can be obtained by using semi constant volumes of biomass quantity and nutrient and metabolite concentrations.

Bioreactors and Hairy Root Culture

Bioreactor cultivation represents the final step in the development of techniques for producing metabolites from plant in vitro systems.[42]

Many types of bioreactors have been successfully used for cultivating transformed root cultures such as conventional stirred tanks, stirred tanks with a separate impeller, bubble columns, mist reactors and balloon-type reactors.[43] Thus, it would be difficult, if not impossible, to select the "best" bioreactor design for cultivating transformed roots. However, for successful scale-up of hairy root-based processes, whatever type of bioreactor is used, several factors should be considered as morphology, unusual rheological properties and high stress sensitivity of hairy roots.[44]

However, the bioreactor cultivation of hairy roots has been considered in several reviews[43-45] and herein only recent advances (post-2002) will be considered.

The most traditional system to culture the transformed roots in liquid medium is the airlift bioreactors used for microorganisms or plant cells as reported for *Beta vulgaris* and *Artemisia annua.*[45,46]

A bubble bioreactor was found to efficiently support the scale up process for coculturing shoots and hairy roots from *Genista tinctoria* devoted to produce phytoestrogens.[47] To improve the homogenization of culture medium, an air lift mesh-draught reactor with wire helixes was designed for large scale culture of *Solanum chrysotrichum* hairy roots.[48] The principal advantage of this method is to reduce drammatically the volume of the culture medium and consequently to increase the concentration of the selected metabolites. This system was adopetd to ensure production of the antitumoral drug campthothecin from *Camptotheca* hairy roots.[49] Interestingly, in this study, coupled reactors of mean volume were used and the production capacity was enhanced by increasing the number of reactors. This represent a good alternative compared with the use of a single reactor of high volume and might reduce the risk occured in a bioreactor.

The development of disposable wave bioreactor systems represents a good advantage. The working principle of these systems is based on wave-induced agitation, which significantly reduces stress levels. Moreover, utilization of plastic disposable chambers minimizes the need

Figure 4. Temporary immersion system (RITA®, Vitpropic, Saint-Mathieu-de-Tréviers, France) for cultivating hairy root cultures.

for labor- and time-consuming cleaning and sterilization procedures and facilitates fulfillment of Good Manufacture Practice requirements.[50]

The ginsenoside production of *P. ginseng* hairy roots in 2 L wave bioreactors has been studied in detail.[51] The results showed that both biomass accumulation and ginsenoside production were significantly higher in 2 L wave systems than in shaken flasks. Large scale wave systems with capacities up to 600 L are now commercially available in Switzerland.

Temporary immersion systems (RITA®, Vitropic, Saint-Mathieu-de-Tréviers, France) (Fig. 4) have also been used for cultivating hairy root cultures from *B. vulgaris* and *Harpagophytum procumbens*.[52,53] Although the RITA® systems have been developed for plant in vitro propagation, their advantages (reduced hyperhydricity and lower consumable and labor costs) make them attractive for hairy root cultivation.[54] The scale of the RITA® systems is about 200 ml and the daily cumulative duration of the immersion stage (flooding) can vary from minutes to several hours. Pavlov and Bley[53] found that growth of *B. vulgaris* hairy roots is optimal with 15-min immersion/75-min standby cycles, while maximal amounts of betalain pigments are accumulated with 15-min flooding/60-min standby cycles.

Another significant problem occurring during the cultivation of hairy roots in bioreactor is associated with mass transfer limitations and root oxygen demands, greater in the meristems than the old tissues. The hairy roots typically grow in a "tuft-like" manner, which promotes the formation of oxygen and nutrient gradients in the tissue. Hairy roots can be grown in bioreactors at low tissue concentrations (<10-g dry weight/L) with virtually any configuration.[43] However, at high tissue concentrations (>10-g dry weight/L) in submerged bioreactors, several scale-up limitations may arise. For high tissue density cultivation, Ramakrishnan and Curtis[55] developed a 14 L pilot-scale reactor that operates in a "trickle-bed" mode. The dry mass of *H. muticus* cultivated in this bioreactor reached 36 g/L within 25 days and the calculated growth index (based on the dry weight) and doubling time were 180 and 3.3 days, respectively.

In addition, an attempt to cultivate transformed root cultures on a large-scale (500 L) has been reported using a procedure involving inoculum preparation in a 10 L seed vessel for 2 weeks, followed by aseptic transfer of the root inoculum to the main 500 L reactor.[56] Here the roots were immobilized on barbs, which further facilitate their harvesting. Although the final biomass yield (about 4-g dry weight/L) could not be considered high, this procedure addresses many potentially problematic issues (mainly technological) concerned with transferring the tissue inoculum from the seed reactor to a greater one(s).

Monitoring the roots' growth during the processes in reactor systems is highly important. In fact during the cultivation of organ cultures (e.g., hairy roots) it is impossible to obtain homogeneous samples of the tissue, which complicates measurements of growth and other process parameters. Therefore, conductivity measurements have been widely used to obtain indirect growth estimates.[53,57] Changes in conductivity are due to the cellular uptake of ionic nutrients (nitrate anions usually). Huang and Chou[58] found that redox potential changes during the cultivation of hairy root cultures in a mist trickling reactor reflect the assimilation of ammonium, nitrate, sucrose and growth phases.

It is worth noting that, in some cases, such linear relationships between reductions in conductivity and biomass growth increase do not exist, indicating that the relationships should be individually determined for every culture.[59]

The osmolarity of the culture medium offers another possibility for indirect measurements of the root growth. Osmolarity measurements have several advantages over conductivity measurements, as osmolarity takes into account the total number of moles of all solutes present in the medium.[60] Use of such indicators could also be considered for monitoring hairy roots growth.

Scale-Up of Medicinal and Aromatic Plants for Secondary Metabolite Production

Zedoary (Curcuma zedoaria)

Zedoary (*Curcuma zedoaria* Roscoe), a member of the Zingiberaceae family, is a species, which grows wild in the eastern Himalayas and is cultivated in India, Sri Lanka, China, Japan, Thailand and Vietnam. Essential oils, curcumin and terpenoids are the main secondary metabolites of zedoary and show the principal pharmacological activities associated with the plant. The constituents of zedoary rhizome oil have been investigated extensively and zedoary has been recognized as a rich source of terpenoids.[61] Zedoary essential oils evidence antimicrobial and antimutagenic activities. Curcumin is well-known for its antitumor, antioxidant, antiamyloid and anti-inflammatory properties. Curcumin shows a free radical scavenger and antioxidant activity and inhibits lipid peroxidation and oxidative DNA damage.[62,63]

Cell suspension cultures for zedoary are generally considered the most suitable system for large-scale applications in the biotechnology industry; therefore zedoary was utilized for monitoring the possible use of these cells to produce of essential oil and curcumin.

A Vietnamese researcher team has established a model batch culture system in a small bioreactor, comparing the production of fresh biomass and metabolites in bioreactors with that obtained from usual in vitro protocols.[64] After the stabilization of the shaking liquid cell culture, the cell biomass were then collected and transferred into a bioreactor (Biotron, Inc. Korea) with a 5 L working volume and three impellers and then propagated at an agitation rate of 150 rpm and an aeration rate of 2.5 L/min for 14 days. Mixing and aeration were achieved using sterile gas from an air pump through a flow meter and an air filter. Determinations of the inoculum sizes, agitation rates and aeration rates suitable for cell biomass production were already known.[65] The

Table 3. Curcumin content and specific activities of POD, SOD and CAT in culture of zedoary cell (from Loch et al 2008). Different letters indicate significantly different means using Duncan's test ($P < 0.05$).

Culture Time (Days)	Content of Curcumin (% of Dry Cell Weight)	POD (U/mg Protein)	SOD (U/mg Protein)	CAT (U/mg Protein)
2	8.20^{ab}	0.01^{e}	13.12^{cd}	6.22^{de}
4	8.31^{ab}	0.02^{e}	13.57^{cd}	7.73^{d}
6	8.34^{ab}	0.13^{cd}	14.66^{c}	10.37^{c}
8	8.52^{ab}	0.15^{cd}	14.76^{c}	16.11^{bc}
10	8.55^{ab}	0.25^{c}	14.84^{c}	17.57^{b}
12	8.63^{ab}	0.41^{b}	14.96^{c}	17.96^{b}
14	9.69^{a}	0.63^{a}	16.60^{b}	19.59^{b}
16	8.63^{ab}	0.08^{d}	15.76^{bc}	12.84^{c}
18	8.31^{ab}	0.04^{de}	14.77^{c}	10.41^{c}
20	8.22^{ab}	0.01^{e}	12.17^{d}	7.77^{d}
Control	5.88^{b}	0.07^{d}	18.58^{a}	16.12^{bc}

online monitoring of the pH and temperature in the cultures (5.8 and 25°C, respectively) was measured by connecting the pH electrode and temperature sensor to the bioreactor. The biomass concentration reached a maximum value of 67.73 g fresh weight/L (approximately 6.77 g dry weight/L) after 14 days of culture.

The highest cell biomass was further used for the extraction of secondary metabolites (essential oil and curcumin) and for the determination of the activities of antioxidant enzymes peroxidase (POD), superoxide dismutase (SOD) and catalase (CAT).[64] In general, the essential oil contents (percentage of dry cell weight) increased between days 2 and 14 and reached maximum values at the end of the log phase (14th day of culture) in both extractions via steam distillation (1.78%) and petroleum ether (0.69%). See Table 3 (from Loch et al 2008).

These results may prove useful in the development of a large-scale production protocol. Scale-up protocols will need to be designed and tested in the future.

The data presented in Table 3 suggest that the accumulation of curcumin in zedoary cells can be associated with the increased levels of activities of POD, SOD and CAT showing the highest amounts at the 14th day of culturing.[64] The curcumin induction of such detoxifying enzymes proved the potential value of curcumin as a protective agent against oxidative stress, as reported in other detailed studies.[66]

*Coneflower (*Echinacea sp*)*

Echinacea sp is a traditional perennial herb plant native of North America and is widely distributed in the world for commercial purposes. Secondary metabolites obtained from roots and aerial parts of three species, *Echinacea purpurea* (called purple coneflower), *Echinacea angustifolia* and *Echniacea pallida,* are of particular importance.[67]

Roots from *E. angustifolia* were historically used in phytomedicine, but also *E. purpurea* extracts have antioxidative, antibacterial, antiviral, antifungal properties and are used for treating common cold, respiratory and urinary diseases.[68] The most important potential active compounds in *E. purpurea* are caffeic acid derivatives namely caftaric acid, chlorogenic acid, cynarin, echinacoside and chichoric acid. Of these, chichoric acid has immunostimulatory property and can promote phagocyte activity in vitro and in vivo. It also has anti-hyaluronidase activity, a protective effect

Table 4. Growth, productivity and bioactive substances in adventitious roots of* Echinacea purpurea *cultured in different conditions

	Biomass Yield		Caffeic Acid Derivatives (mg/g DW)			
Type of Culture	Fresh Weight (Kg)	Dry Weight (Kg)	Caftaric Acid	Chlorogenic Acid	Chicoric Acid	Total
500-L balloon-type bioreactors	26.3 ± 0.5	3.6 ± 0.1	2.8 ± 0.1	4.4 ± 0.4	20.2 ± 0.8	27.4 ± 0.5
1000-L drum-type bioreactors	40.5 ± 0.5	5.1 ± 0.1	3.9 ± 0.1	4.9 ± 0.1	22.5 ± 0.6	31.5 ± 0.6
Natural adven-tious roots	-	-	2.4 ± 0.2	-	6.2 ± 0.9	8.6 ± 1.1

on the free radical-induced degradation of collagen as well as antiviral activity[68] where it inhibits HIV-1 integrase and replication.[69]

Because commercial preparations are commonly made from root tissues, in vitro protocols for their proliferation could improve the commercial availability. Therefore many attention have been focused on the efficient biomass production for fast growth rates and stable metabolite productivity.[70,71]

Recently research efforts were focused on developing bioreactor methodologies for the efficient production of caffeic acid derivatives from adventitious *Echinacea* spp. root cultures.[67] Culture systems in airlift bioreactors (20 L, 500 L balloon-type, bubble bioreactors and 1000 L drumtype bubble bioreactor) were then developed for the production of chichoric acid, chlorogenic acid and caftaric acid. In the 20 L balloon type bubble bioreactors a maximum yield of 11 g dry biomass/L was achieved after 60 days.

Pilot scale balloon-type bubble bioreactors (500 L working capacity) and horizontal drum bioreactors (1000 L working capacity) were also used for the cultivation of adventitious roots. These bioreactors have sparger positioned at the bottom, used to generate air bubbles less then 0.5 mm in diameter. Aeration rate was controlled at 0.1 vvm.[67] 3.6 kg and 5.1 kg dry biomass were achieved in the 500 L and 1000 L bioreactors, respectively (Table 4).

Adventitious roots grown in pilot scale bioreactors were also efficient in accumulation of caffeic acid derivatives and the total caffeic acids contents were about 27 mg/g dry weight and 31 mg/g dry weight with adventitious roots grown in the 500 L balloon type bubble bioreactor and 1000 L drum bioreactor, respectively. The accumulation of 5 mg/g dry weight chlorogenic acid, 22 mg/g dry weight chichoric acid and 4 mg/g dry weight caftaric acids were achieved with adventitious roots grown in 1000 L bioreactors.

Sometimes during the scale up of plant cell and organ cultures, a decrease in productivity may occur.[72,73] However, the scale up of adventitious root cultures of *E. purpurea* showed no decrease in biomass production and caffeic acid productivity. Comparison of caffeic acid derivative contents in adventitious roots and field grown plants revealed that the contents of caftaric acid (1.6-fold), chichoric acid (3.6-fold) were higher in the adventitious roots than in the roots of field grown plants.

These results may be useful for biotechnological application of *E. purpurea* adventitious root cultures for the production of caffeic acid derivatives on a large scale.

Basil (Ocimum basilicum)

Ocimum basilicum L., a popular Lamiaceous plant known as sweet basil, is used as a kitchen herb in the production of "pesto", a typical Italian sauce known for its unmistakable aroma and as an ornamental in house gardens. The aromatic characters of each type of basil is determined by genotype and depends on their major chemical constituents in the essential oils. Basil aromatic leaves and essential oils are widely used as antioxidants, flavouring agents in foods, confectionary products, beverages as well as in perfumery.[74] Basil is known to contain the antioxidant phenolic compound, rosmarinic acid, one of the most common caffeic acid esters occurring in Lamiaceae family.

Hairy roots and cultured cells of sweet basil (*Ocimum basilicum* L.) are able to produce rosmarinic acid.[75,76]

The use of bioreactors for the growth of sweet basil cell suspensions and for the regeneration from nodal explants can both improve and scale up rosmarinic acid accumulation and plant micropropagation of this medicinal plant species.[5] Thus, proliferating callus tissue from nodal cultures were transferred for three weeks in an 5 L disposable presterilized plastic airlift bioreactor (Osmotek Lifereactors) (ml/reactor, 75 explants/reactor).[77,78]

During that incubation period, suspension cultures grew faster in the airlift bioreactor than in a 250 ml flask and the average biomass increased even more remarkable (1457%).[5]

In bioreactors, enhanced growth was highly positive correlated with rosmarinic acid accumulation ($r^2 = 0.99$), producing higher levels of metabolites than in 250 ml flasks suspension cultures. Further observations indicated that the culture fresh biomass growth was also positively correlated with radical oxygen species (ROS) production; ROS are by-products of normal cellular metabolism produced by mitochondria, chloroplasts and peroxisomes, so their increase in concentration might be associated with the rapid turn-over of primary metabolites.[5] In fact rosmarinic acid biosynthesis was previously associated with primary metabolic processes, leading to cellular growth.[76]

On the other hand, dry weight accumulation as well as soluble protein and carbohydrate concentrations were negatively associated with biomass growth. This culture growth was mainly due to cellular enlargement, a process associated with increased ROS formation.[79]

In conclusion both cell suspensions and nodal explants of *O. basilicum* represent appropriate sources of good rosmarinic acid biosynthesis under scale-up conditions.

Lavender (Lavandula vera)

Lavandula species, belonging to Lamiaceae plants, are mainly grown for their essential oils, which are used in perfumery, cosmetics, food processing and aromatherapy products.[80] Lamiaceae plants are well known producers of phenolic compounds and cell suspensions of *Lavandula vera* MM specie represented a promising producer of rosmarinic acid.[81]

Many investigations were focused on transferring the suspension cell process in a laboratory bioreactor and on the optimization of cultivation conditions. Experiments were performed in a 3 L propeller-stirred bioreactor (BioFlo 110, New Brunswick) using 1,8 L cultivation nutrient medium, inoculated with 7-day-old shake-flask suspension.[52] During cultivation of plant cell suspensions were selected different nutrient media and the dissolved oxygen and agitation speed were separately optimized.

Significant production of rosmarinic acid was achieved up to 3489.4 mg/L in the 3 L bioreactor cultivation although yields of biomass were relatively invariable in the different culture used (Table 5). The obtained data confirmed the proposed algorithm for the optimization of rosmarinic acid biosynthesis by *L. vera*, which is of technological significance and revealed possibilities for the next scale-up of the process.[52] (Table 5).

Ginseng (Panax ginseng)

The use of modified nutrient medium, as well as elicitors and air lift bioreactors are common methodology to improve production of the active ginsenosides (saponins) from Ginseng (*Panax ginseng* C.A. Meyer), a worldwide important medicinal plants, whose active components have

Table 5. Biomass, RA yield and productivity detected in Lavandula in different bioreactor conditions (adapted from ref. 52)

Parameter	Shake-Flask Culture in Standard LS Nutrient Medium	Shake-Flask Culture in Modified LS Medium	3-L Bioreactor Culture in Modified LS Medium
RA Yield, mg/L	68.0	1786.7	3489.4
Biomass, g Dry Weight/g sucrose	0.54	0.72	0.59
RA productivity, mg/L day	8.5	148.9	268.4

attributed cardio-protective, immunomodulatory, antifatigue, hepato-protective physiological and pharmacological effects.[82]

It is well known that synthesis of secondary metabolites and enzymes in plants is usually associated with plant defense responses to different stress conditions.[83,84]

The effects of methyl jasmonate and salicylic acid on changes of the activities of major antioxidant enzymes and ginsenoside accumulation were investigated in ginseng roots (*Panax ginseng* L.) in 4 L air lift bioreactors.[85]

Selected adventitious roots were collected at different time of growth after methyl jasmonate and salicylic acid elicitation in airlift bubble type bioreactors. These elicitors considerably increased the saponin accumulation without changing biomass until seven days, however, biomass decreased in salicylic acid-treated roots compared to methyl jasmonate. Moreover both treatments induced an oxidative stress in *P. ginseng* roots, increasing superoxide anion (O_2^-) formation and conseguently lipid peroxidation.[86] The results suggested that methyl jasmonate and salicylic acid act as signalling molecules inducing saponin accumulation and O_2^- may function as a signal for the induction of defence genes and could enhance the ginsenoside production.[87]

The finding that methyl jasmonate and salicylic acid enhance both the antioxidant defence systems and the secondary metabolite formation without affecting biomass accumulation of *P. ginseng* roots increases the usefulness of this culture system for production of pharmacologically ginsenosides.[85]

Sage (Salvia miltiorrhiza)

Salvia miltiorrhiza Bunge (Lamiaceae) roots or Danshen in Chinese, is a well-known Chinese herb, which is widely used in modern and traditional medicine for the treatment of menstrual disorders and blood circulation diseases and for the prevention of inflammation. A major class of bioactive ingredients of Danshen is ascribed to the lipophilic diterpene pigments generally known as tanshinones.[88] The hairy root culture of *S. miltiorrhiza* has been established as an alternative more efficient production of tanshinones then the whole plant growth in farms.[87,89]

Elicitation is one of the most effective means for improving secondary metabolite production in plant tissue and cell cultures including hairy root cultures;[84] the effect of elicitation is a general productivity which depends on the biomass growth rate and biomass concentration in the culture.[90]

The most common elicitors used are fungal carbohydrates or polysaccharides, jasmonic acid or methyl jasmonate, chitosan and heavy metal ions. In addition to these agents, hyperosmotic stress had an effective stimulation for the production of various secondary metabolites in plant cell cultures.[90] Although single elicitors have been mostly used, the combined dose of two different elicitors has been shown more effective due to a synergistic effect.[91,92]

Wu and Shi[90] observed an increase (three to four fold) in the total tanshinone content of *S. miltiorrhiza* roots by osmotic stress and yeast elicitor treatment, separately, but more significantly by their combination (about eight fold).

The application of three different means to *S. miltiorrhiza* hairy root cultures, such as a semi-continuous culture process, a yeast elicitor addition and in situ adsorption of tanshinones (with a hydrophobic polymeric resin, X-5) applied at the late exponential growth phase, increased the root biomass to 30.5 g dry weight/L (versus 8-10 g dry weight/L in batch mode). Moreover the volumetric tanshinone yield reached 87.5 mg/L (about 15-fold increase), with 76.5% adsorbed to the resin.[93]

These results demonstrated that the integration of multiple elicitation, in situ adsorption and semi-continuous operation can synergistically enhance tanshinone production in *S. miltiorrhiza* hairy root cultures.

Conclusion

Bioreactors offer a great hope for the large scale synthesis of bioactive compounds in medicinal and aromatic plants. Since the biosynthetic efficiency of population varies a high yielding variety is recommended as a starting material. There are also several difficulties associated with large scale bioreactor technology, mainly focused on product yield cost, optimising growth rate and product release, expression of desirable traits and culture stability. Genetic transformation may provide increased and efficient system for in vitro production of secondary metabolites. Recent progress in the scaling up of hairy root cultures is making this system an attractive tool for industrial processes.

References

1. Tripathi L, Tripathi JN. Role of biotechnology in medicinal plants Tropic. J Pharm Res 2003; 2(2):243-53.
2. Zenk MH. The impact of plant cell culture on industry. In: Thorpe TA, ed. Frontiers of Plant Tissue Culture. Calgary: Int Ass for Plant Tissue Culture 1978:1-13.
3. Stafford A, Morris P, Fowler MW. Plant cell biotechnology: A perspective. Enz Microbial Technol 1986; 8:578-97.
4. Robins RI. Secondary products from cultured cells and organs: Molecular and cellular approaches. In: Dixon RA, Gonzales RA, eds. Plant Cell Culture. Oxford: IRL Press, 1994:169-97.
5. Kintzios S, Kollias H, Straitouris E et al. Scale-up micropropagation of sweet basil (Ocimum basilicum L.) in an airlift bioreactor and accumulation of rosmarinic acid. Biotechnol Lett 2004; 26(6):521-23.
6. Merillon JM. Large-scale production in bioreactors. In: Ramawat KG, Merillon JM, eds. Biotechnology Secondary Metabolites. Enfield: Science publishers Inc, 1999:331-50.
7. Mühlbach H. Use of plant cell cultures in biotechnology. Biotech Ann Rev 1998; 4:113-71.
8. Deus B, Zenk MH. Exploitation of plant cells for the production of natural compounds. Biotechnol Bioeng 1982; 24:1965-74.
9. Sun X, Linden JC. Shear stress effects on plant cell suspension cultures in a rotating wall vessel bioreactor. J Ind Microbiol biotechnol 1999; 22:44-47.
10. George EF. Plant propagation by tissue culture. Edington: Exegetics Limited Publications, 1993.
11. Ruffoni B, Giovannini A. Produzione di biomassa in vitro: induzione e scale-up. In: Tognoni F, Pardossi A, Mensuali Sodi A, eds. Colture Artificiali di Piante Medicinali. Roma: Aracne 2007:147-65.
12. Suri SS, Sharma R, Ramawat KG. Production of secondary metabolites in bioreactors. In: Khan IA, Khanum A, eds. Role of Biotechnology in Medicinal and Aromatic Plants Vol II. Hyderabad: UKAAZ Publications, 2000:437-51.
13. Tulecke W, Nickel LG. Production of large amounts of plant tissue by submerged cultures. Science 1959; 130:863-64.
14. Heyerdahl PHO, Olsen AS, Hvoslef-Eide AK. Engineering aspects of plant propagation in bioreactors. In: Aitken-Christie J, Kozai T, Smith MAL, eds. Automation and Environment Control in Plant Tissue Culture. Dordrecht: Kluwer Acad Publ, 1995:87-123.
15. Kieran PM, MacLoughlin PF, Malone DM. Plant cell suspension cultures: some engineering considerations. J Biotechnol 1997; 59:39-52.
16. Ziv M. Bioreactor technology for plant micropropagation. Hort Rev 2000; 24:1-30.
17. Buddendorf-Joosten JMC, Woltering EJ. Components of the gaseous environment and their effects on plant growth and development in vitro. In: Lumsden PJ, Nicholas JR and Davies WJ, eds. Physiology Growth and Development of Plants in Culture. Dordrecht: Kluwer Acad Publ, 1994:165-90.
18. Baldi A, Srivastava AK, Bisaria VS. Effect of aeration on production of anticancer lignans by cell suspension cultures of Linum album. Appl Biochem Biotechnol 2008; 151(2-3):547-55.
19. Cazzulino D, Pederson H, Chin CK. Bioreactors and image analysis for scale-up and plant propagation. In: Vasil IK, ed. Cell Culture and Somatic Cell Genetics of Plants. San Diego: Academic Press, 1991:147-77.

20. Takayama S, Akita M. Bioreactor techniques for large-scale culture of plant propagules. Adv Hort Sci 1998; 12:93-100.
21. Smart NJ, Fowler MW. An airlift column bioreactor suitable for large-scale cultivation of plant cell suspensions. J Exp Bot 1984; 35:531-37.
22. Scragg AH. Fermentation systems for plant cells. In: Charlwood BV, Rhodes MJC, eds. Secondary Products from Plant Tissue Culture. London: Clarendon Press, 1990:243-63.
23. Scragg AH. Large-scale plant cell culture: methods, applications and products. Curr Opin Biotech 1992; 3:105-09.
24. Hegarty PK, Smart NJ, Scragg H et al. The aeration of Catharanthus roseus L.G. Don suspension cultures in airlift bioreactors: the inhibitory effect at high aeration rates on culture growth. J Exp Bot 1986; 7:1911-20.
25. Kim DL, Pederson H, Chin CK. Cultivation of Thalictrum rugosum cell suspension in an improved airlift bioreactor: stimulatory effects of CO_2 and C_2H_4 on alkaloid production. Biotech Bioengin 1991; 38:331-39.
26. Ilan A, Ziv M, Halevy AA. Propagation and corm development of Brodiaea in liquid cultures. Scientia Hort 1995; 63:101-12.
27. Murashige T, Skoog F. Revised medium for rapid growth and bioassay with tobacco tissue cultures. Physiol Plant 1962; 15:473-497.
28. Williams RR. Towards a model of mineral nutrition in vitro. In: Kurata K, Kozai T, eds. Transplant Production Systems. Dordrecht: Kluwer Acad Publ, 1992:213-29.
29. Debergh PC, De Riek J, Matthys D. Nutrient supply and growth of plants in culture. In: Lumsden PJ, Nicholas RJ, Davies WJ, eds. Physiology, Growth and Development of Plants in Culture. Dordrecht: Kluwer Acad Publ, 1994:58-68.
30. Archambault J, Lavoie L, Williams RD et al. Nutritional aspects of Daucus carota somatic embryo cultures performed in bioreactors. In: Terzi M, Cella R, Falavigna A, eds. Current Issues in Plant Molecular and Cellular Biology. Dordrecht: Kluwer Acad Publ, 1995:681-87.
31. Tautorus TE, Lulsdorf MM, Kikcio SI et al. Nutrient utilization during bioreactor culture and maturation of somatic embryo culture of Picea marianna and Picea glauca-engelmannii. In Vitro Cell Dev Biol Plant 1994; 30:58-63.
32. McDonald KA, Jackman AP. Bioreactor studies of growth and nutrient utilization in alfalfa suspension cultures. Plant Cell Rep 1989; 8:455-58.
33. Stuart DA, Strickland SG, Walker KA. Bioreactor production of alfalfa somatic embryos. HortScience 1987; 22:800-03.
34. Ziv M, Hadar A. Morphogenic pattern of Nephrolepsis exaltata Bostoniensis in agar-gelled or liquid culture. Implication for mass propagation. Israel J Bot 1991; 40:7-16.
35. Jay V, Genestier S, Courduroux JC. Bioreactor studies on the effect of medium pH on carrot (Daucus carota L.) somatic embryogenesis. Plant Cell Tis Org Cult 1994; 36:205-209.
36. Kawase Y. Liquid circulation in external-loop airlift bioreactors. Biotech Bioengin 1989; 33:540-46.
37. Luttman R, Florek P, Preil W. Silicone-tubing aerated bioreactors for somatic embryo production. Plant Cell Tiss Org Cult 1994; 39(2):157-70.
38. Hvoslef-Eide AK, Hohe A, Ruffoni B et al. COST experiences with bioreactors. Acta Hort 2006; 725:549-59.
39. Chatterjee C, Corell MJ, Weathers PJ et al. Simplified acoustic window mist bioreactor. Biotechnol Tech 1997; 11:155-58.
40. Hale SA, Young RE, Adelberg JW et al. Bioreactor development for continual-flow liquid plant tissue culture. Acta Hort 1992; 319:107-12.
41. Ruffoni B, Savona M. Somatic embryogenesis in floricultural crops: experiences of massive propagation of Lisianthus, Genista and Cyclamen. In: Teixeira da Silva JA, ed. Floriculture, Ornamental and Plant Biotechnology Vol II. London: Global science Books, 2006:305-13.
42. Bourgaud F, Gravot A, Milesi S et al. Production of plant secondary metabolites: a historical perspective. Plant Sci 2001; 161:839-51.
43. Curtis WR. Hairy roots, bioreactor growth. In: Spier RE, ed. Encyclopedia of Cell Technology. New York: Wiley, 2008; 2:27-841.
44. Wysokinska H, Chmiel A. Transformed root cultures for biotechnology. Acta Biotechnol 1997; 17:131-59.
45. Kim Y, Wyslouzil BE, Weathers PJ. Secondary metabolism of hairy root cultures in bioreactors. In Vitro Cell Dev Biol Plant 2002; 38:1-10.
46. Thimmaraju R, Bhagyalakshmi N, Ravishankar GA. In situ and ex situ adsorption and recovery of betalaines from hairy root cultures of Beta vulgaris. Biotechnol Progr 2004; 20(3):777-85.
47. Luczkiewicz M, Kokotkiewicz A. Cocultures of shoots and hairy roots of Genista tinctoria L. for synthesis and biotransformation of large amounts of phytoestrogens. Plant Sci 2005; 169(5):862-71.

48. Caspeta L, Quintero R, Villareal ML. Novel airlift reactor fitting for hairy root cultures: developmental and performance studies. Biotechnol Prog 2005; 21(3):735-40.
49. Lorence A, Medina-Bolivar F, Nessler CL. Camptothecin and 10-hydroxycamptothecin from Camptotheca acuminata hairy roots. Plant Cell Rep 2004; 22(6):437-41.
50. Eibl R, Eibl D. Design and use of the wave bioreactor for plant cell culture. In: Gupta SD, Ibaraki Y, eds. Plant Tissue Culture Engineering. Focus on Biotechnology. Berlin: Springer Verlag, 2006; 6:203-27.
51. Palazon J, Mallol A, Eibl R et al. Growth and ginsenoside production in hairy root cultures of Panax ginseng using a novel bioreactor. Planta Med 2003; 69:344-49.
52. Pavlov AI, Georgiev MI, Panchev MP. Optimization of rosmarinic acid production by Lavandula vera MM plant cell suspension in a laboratory bioreactor. Biotechnol Prog 2005; 21(2):394-46.
53. Pavlov A, Bley T. Betalains biosynthesis by Beta vulgaris L. hairy root culture in a temporary immersion cultivation system. Process Biochem 2006; 41:848-52.
54. McAlister B, Finnie J, Watt MP et al. Use of the temporary immersion bioreactor system (RITA®) for the production of commercial Eucalyptus clones in Mondi Forests (SA). Plant Cell Tiss Org Cult 2005; 81:347-58.
55. Ramakrishnan D, Curtis WR. Trickle-bed root culture bioreactor design and scale-up: growth, fluid-dynamics and oxygen mass transfer. Biotechnol Bioeng 2004; 88:248-60.
56. Wilson PDG. The pilot-scale cultivation of transformed roots. In: Doran PM, ed. Hairy Roots and Culture Applications. Amsterdam: Harwood Academic Publishers, 1997:179-90.
57. Heyon KJI, Yoo YJE. Adaptive estimation of hairy root mass using conductometry. J Microbiol Biotechnol 2003; 13:641-46.
58. Huang S-Y, Chou S-N. Elucidation of the effects of nitrogen source on proliferation transformed hairy roots and secondary metabolite productivity in a mist trickling reactor by redox potential measurement. Enz Microb Technol 2006; 38:803-13.
59. Georgiev M, Heinrich M, Kerns G et al. Production of iridoids and phenolics by transformed Harpagophytum procumbens root cultures. Eng Life Sci 2006; 6:593-596.
60. Suresh B, Rajasekaran T, Rao SR et al. Studies on osmolarity, conductivity and mass transfer or selection of a bioreactor for Tagetes patula L. hairy roots. Process Biochem 2001; 36:987-93.
61. Garg SN, Naquvi AA, Bansal RP et al. Chemical composition of the essential oil from the leaves of Curcuma zedoaria Rosc. of Indian origin. J Essential Oil Res 2005; 17:29-31.
62. Braga ME, Leal PF, Carvalho JE et al. Comparison of yield, composition and antioxidant activity of turmeric (Curcuma longa L.) extracts obtained using various techniques. J Agric Food Chem 2003; 51:6604-11.
63. Park Y, Patek R, Mayersohn M. Sensitive and rapid isocratic liquid chromatography method for the quantisation of curcumin in plasma. J Chromatogr Bull 2003; 796:339-46.
64. Loch NH, Diem DT, Binh DH et al. Isolation and characterization of antioxidation enzymes from cells of zedoary (Curcuma zedoaria Roscoe) Cultured in a 5 L bioreactor. Mol Biotechnol 2008; 38:81-87.
65. Loch NH, Ha TTT, Hirata Y. Effect of several factors on cell biomass production of zedoary (Curcuma zedoaria Roscoe) in bioreactor. Vietnam J Biotechnol 2006; 4:213-20.
66. Okada K, Wangpoengtrakul C, Tanaka T et al. Curcumin and especially tetrahydrocurcumin ameliorate oxidative stress-induced renal injury in mice. J Nutrition 2002; 131:2090-95.
67. Wu CH, Murthy HN, Hahn EJ et al. Large-scale cultivation of adventitious roots of Echinacea purpurea in airlift bioreactors for the production of chichoric acid, chlorogenic acid and caftaric acid. Biotechnol Lett 2007; 29(8):1179-82.
68. Barrett B. Medicinal properties of Echinacea: a critical review. Phytomed 2003; 10:66-86.
69. Lin Z, Neamati N, Zhao H et al. Chichoric acid analogues as HIV-1 integrase inhibitors. J Med Chem 1999; 42:1401-14.
70. Choi JW, Cho GH, Byun SY. Integrated bioprocessing for plant cell cultures. Adv Biochem Eng Biotechnol 2001; 72:63-102.
71. Kim SK, Hahn EJ, Murthy HN et al. Adventitious root growth and ginsenoside accumulation in Panax ginseng cultures as affected by methyl jasmonate. Biotechnol Lett 2004; 26:1619-22.
72. Kwok KH, Doron PM. Kinetic and stoichiometric analysis of hairy roots in a segmented bubble column reactor. Biotechnol Prog 1995; 11:429-435.
73. Scragg AH, Morris P, Allan EJ et al. Effect of scale-up on serpentine formation by Catharanthus roseus suspension cultures. Enz Microb Technol 1987; 9:619-24.
74. Chang SS, Ostric-Matijasevic B, Hsieh OAL et al. Natural antioxidant from Rosemary and Sage. J Food Sci 1977; 42:1102-07.
75. Tada H, Murakami Y, Omoto T et al. Rosmarinic acid and related phenolics in hairy root cultures of Ocimum basilicum. Phytochem 1996; 42:431-34.
76. Kintzios S, Makri O, Panagiotopoulos EM et al. In vitro rosmarinic acid accumulation in sweet basil (Ocimum basilicum L.). Biotechnol Lett 2003; 25:405-08.

77. Ziv M, Ronen G, Raviv M. Proliferation of meristematic clusters in disposable presterilized plastic bioreactors for large scale micropropagation of plants. In Vitro Cell Dev Biol Plant 1998; 34:152-58.
78. Konstas J, Kintzios S. Developing a scale-up system for the micropropagation of cucumber (Cucumis sativus L.): the effect of growth retardants, liquid culture and vessel size. Plant Cell Rep 2003; 21:538-48.
79. Schopfer P, Liszkay A, Bechtold M et al. Evidence that hydroxyl radicals mediate auxin-induced extension growth. Planta 2002; 214:821-28.
80. Harborne J, Williams C. Phytochemistry of the genus Lavandula. In LaVender, The genus LaVandula; Lis-Balchin, M, eds. Taylor and Francis: London, 2002; 29:86-99.
81. Ilieva M, Pavlov A. Rosmarinic acid production by Lavandula vera MM cell suspension culture. Appl Microbiol Biotechnol 1997; 47:683-88.
82. Attele AS, Wu JA, Yuan CS. Ginseng pharmacology: multiple constituents and multiple actions. Biochem Pharmacol 1999; 58:1685-93.
83. Ebel J, Mithoefer A. Early events in the elicitation of plant defense. Planta 1998; 206:335-48.
84. Zhao J, Lawrence CD, Verpoorte R. Elicitor signal transduction leading to production of plant secondary metabolites. Res Rev Paper Biotechnol Adv 2005; 23:283-33.
85. Ali MB, Hahn EJ, Paek KY. Copper-induced changes in the growth, oxidative metabolism and saponin production in suspension culture roots of Panax ginseng in bioreactors. Plant Cell Rep 2006; 25(10):1122-32.
86. Ali MB, Yu KW, Hahn EJ et al. Methyl jasmonate and salicylic acid elicitation induces ginsenosides accumulation, enzymatic and non-enzymatic antioxidant in suspension culture Panax ginseng roots in bioreactors. Plant Cell Rep 2006b; 25(6):613-20.
87. Chen THH, Chen HBG, Chen F et al. Production of lithospermic acid B and rosmarinic acid in hairy root cultures of Salvia miltiorrhiza. J Ind Microbiol Biotechnol 1999; 22:133-38.
88. Tang W, Eisenbrand G. Chinese drugs of plant origin: chemistry, pharmacology and use in traditional and modern medicine. Berlin: Springer-Verlag, 1992:891-902.
89. Hu ZB, Alfermann AW. Diterpenoid production in hairy root cultures of Salvia miltiorrhiza. Phytochem 1993; 32:699-703.
90. Wu JY, Shi M. Ultrahigh diterpenoid tanshinone production through repeated osmotic stress and elicitor stimulation in fed-batch culture of Salvia miltiorrhiza hairy roots. Appl Microbiol Biotechnol 2008; 78(3):441-48.
91. Linden JC, Phisalaphong M. Oligosaccharides potentiate methyl jasmonate-induced production of paclitaxel in Taxus canadensis. Plant Sci 2000; 158:41-5.
92. Zhao J, Hu Q, Zhu WH. Enhanced catharanthine production in Catharanthus roseus cell cultures by combined elicitor treatment in shake flasks and bioreactors. Enz Microb Technol 2001; 28:673-81.
93. Yan Q, Hu Z, Tan RX et al. Efficient production and recovery of diterpenoid tanshinones in Salvia miltiorrhiza hairy root cultures with in situ adsorption, elicitation and semi-continous operation. J Biotechnol 2005; 119:416-24.

Chapter 16

Determination of the Antioxidants' Ability to Scavenge Free Radicals Using Biosensors

Montserrat Cortina-Puig,* Beatriz Prieto-Simón, Mónica Campàs, Carole Calas-Blanchard and Jean-Louis Marty

Abstract

Free radicals are highly reactive molecules generated during cellular metabolism. However, their overproduction results in oxidative stress, a deleterious process that can damage cell structures, including lipids and membranes, proteins and DNA. Antioxidants respond to this problem, scavenging free radicals. This chapter critically reviews the electrochemical biosensors developed for the evaluation of the antioxidant capacity of specific compounds. Due to the ability of these devices to perform simple, fast and reliable analysis, they are promising biotools for the assessment of antioxidant properties.

Introduction

Free radicals, such as reactive oxygen species (ROS), are highly unstable molecules containing unpaired electrons, generated in vivo during metabolic processes. These molecules are neutralised by antioxidants, naturally produced by the body. However, environmental or behavioural stressors (pollution, sunlight exposure, cigarette smoking, excessive alcohol consumption, etc.) or simply a malfunction of the antioxidant production may lead to a free radical excess, resulting in oxidative stress. Oxidative stress can damage lipids, proteins, enzymes, carbohydrates and DNA, impeding normal cell functioning. These biochemical alterations construct the molecular basis in the development of cancer, neurodegenerative and autoimmune disorders, cardiovascular diseases and diabetes. In such conditions, external supply of antioxidants is essential to countervail the deleterious consequences of oxidative stress. Since antioxidants are naturally present in vegetables, a balanced diet helps the body to prevent these diseases.

The determination of free radicals and antioxidants has been widely investigated in the food technology and human health fields. Traditional techniques such as spectrophotometry, fluorescence and gas or liquid chromatography,[1,2] are being replaced by other innovating technologies. In this direction, electrochemical biosensors are promising tools, suitable for fast analyses, based on inexpensive instrumentation and simple operation protocols. Whereas in the medical field the main objective is the evaluation of the ability of some compounds to scavenge free radicals, in food science research aims to detect and quantify them. In this sense, two different kinds of biosensors are reported in the antioxidant domain. On one hand, several amperometric biosensors

*Corresponding Author: M. Cortina-Puig—BIOMEM group, Université de Perpignan, 52 Avenue Paul Alduy, 66860 Perpignan Cedex, France. Email: mcortina12@gmail.com

Bio-Farms for Nutraceuticals: Functional Food and Safety Control by Biosensors edited by Maria Teresa Giardi, Giuseppina Rea and Bruno Berra.

for the detection of mono and polyphenols (the main antioxidant compounds in food) have been developed on the basis of enzymes such as tyrosinase, laccase or peroxidase.[3] These configurations allow the evaluation of the usually named "total phenol content". On the other hand, biosensors for the assessment of the antioxidant capacity are based on the free radical scavenging activity. This review is focused on the biosensors for measuring the antioxidant capacity. All biosensors developed for this purpose are electrochemical and use ROS in their configurations.

In the following sections, the different biosensors developed for the assessment of the antioxidant capacity are described according to the radical of interest. As ROS are not commercially available because of their highly reactive nature and their very short lifetime, the biochemical, chemical and physicochemical processes to produce them are also summarised.

Monitoring Superoxide Radical ($O_2^{\bullet -}$)

Superoxide radical ($O_2^{\bullet -}$) is mainly produced by the oxidation of xanthine (or hypoxanthine) to uric acid in the presence of the enzyme xanthine oxidase (XOD); $O_2^{\bullet -}$ is formed as an intermediate of this reaction:[4]

$$\text{xanthine} + O_2 + H_2O \xrightarrow{\text{XOD}} \text{uric acid} + 2H^+ + O_2^{\bullet -} \qquad \text{(Eq. 1)}$$

The addition of NaOH to dimethylsulphoxide (DMSO) also generates $O_2^{\bullet -}$.[5] This $O_2^{\bullet -}$ production is inversely proportional to the water concentration in DMSO and solutions obtained are stable for up to three days.

The simple injection of KO_2 in aprotic organic solvents, especially DMSO, also results in $O_2^{\bullet -}$ generation by means of the following reaction:[6,7]

$$KO_2 \longrightarrow K^+ + O_2^{\bullet -} \qquad \text{(Eq. 2)}$$

With the aim to assess antioxidant capacity based on the measurement of $O_2^{\bullet -}$ concentration, two main types of biosensors have been developed, using (cyt *c*) or (SOD) enzyme. Both cyt *c*- and SOD-based sensors usually incorporate XOD as radical generator. $O_2^{\bullet -}$ determination using a cyt *c*-based sensor lacks selectivity, since this heme protein is not specific for $O_2^{\bullet -}$. Moreover, its inherent property as a peroxidase, able to reduce H_2O_2 endogenously coexisting in biological systems, greatly limits its application for the detection of $O_2^{\bullet -}$ in real samples. SOD-based biosensors, on the contrary, use to be much more specific and sensitive.

Cyt c-Based Biosensors

The detection principle of these biosensors is based on the redox reaction of cyt *c* (Fig. 1). The immobilised cyt *c* is reduced by $O_2^{\bullet -}$ and immediately regenerated at the surface of the electrode polarised at the oxidation potential. The current generated due to the electron transfer from the radical, via cyt *c*, to the electrode is proportional to the radical concentration.[8] In order to avoid interference from H_2O_2, generated by spontaneous dismutation of $O_2^{\bullet -}$ (Eq. 3), catalase enzyme is sometimes added to the reaction media. The addition of antioxidants reduces the radical concentration and thus the oxidation current, allowing the quantification of the antioxidant capacity.

$$2O_2^{\bullet -} + 2H^+ \longrightarrow O_2 + H_2O_2 \qquad \text{(Eq. 3)}$$

Usually, the electron transfer between the electrode and the redox protein is extremely slow. Consequently, traditional electrochemical methods cannot detect it.[9] An elegant way to modify gold electrodes, the most commonly used in this area, is the formation of self-assembled monolayers (SAMs). Short-chain alkanethiols show a high efficiency of communication between cyt *c* and the electrode.[8,10] However, they cannot form dense films and thus, do not effectively block the electrode from interfering substances. In order to eliminate electroactive interferences, e.g., H_2O_2

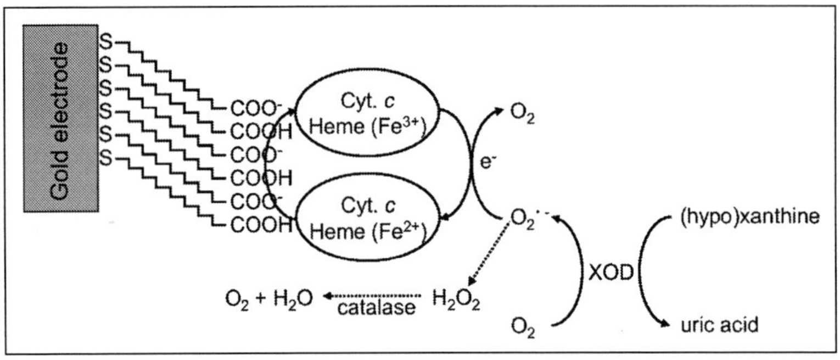

Figure 1. Scheme of the detection principle of $O_2^{\bullet-}$ produced by xanthine-XOD system, using a cyt *c*-modified electrode.

and uric acid, products of the XOD-catalysed reaction, cyt *c* has been immobilised on long-chain thiol (mercaptoundecanoic acid)-modified electrodes. These cyt *c*-based biosensors can be used for $O_2^{\bullet-}$ detection and for the analysis of antioxidant activities (Fig. 1). Following this strategy, Lisdat et al[11,12] analysed the ability of SOD to dismutate $O_2^{\bullet-}$ radicals. Besides, Ignatov et al[13] evaluated the antioxidant capacity of flavonoids with the same cyt *c*-modified electrode. They established the following trend: flavanols > flavonols > flavones > flavonones > isoflavonones.

In order to increase the electron transfer rate of cyt *c*, Ge and Lisdat[7] employed mercaptoundecanoic acid/mercaptoundecanol mixed SAM-modified gold electrodes. This procedure had already been described and applied by Gobi and Mizutani[14,15] to the $O_2^{\bullet-}$ sensing. However, they still used short-chain modifiers, which limited the selectivity. These mixed SAM-modified biosensors showed more sensitivity to $O_2^{\bullet-}$ and were applied to the study of the antioxidant capacity of hydrophobic antioxidants. Beissenhirtz et al[16] tested the antioxidant capacity of ascorbic acid and Biochanin A in a mixture of 40% DMSO and 60% phosphate buffer. Results concluded that Biochanin A was less effective than ascorbic acid as an antioxidant. They also studied the antioxidative properties of some cosmetic creams in the same medium.

From the Ge and Lisdat[7] approach, some authors tried to simultaneously detect both $O_2^{\bullet-}$ and H_2O_2 produced in the course of its spontaneous dismutation.[17,18] Following the concept of "lab-on-a-chip", Krylov et al[19] developed a fluidic chip which combined the generation of $O_2^{\bullet-}$ and H_2O_2 with a two-electrode biosensor detection system (Fig. 2). In this way, the antioxidant capacity of different potential scavengers of the respective reactive species was quantified in a flow-injection mode. The antioxidant capacity of ascorbic acid found using this strategy was very close to previous works.[7,16] The same group has recently applied this fluidic chip to the determination of the

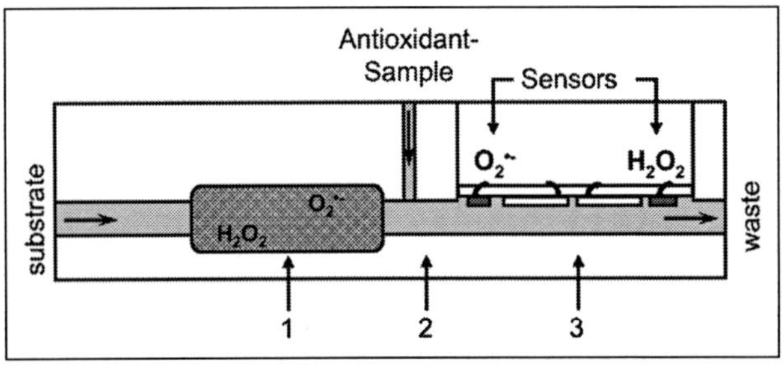

Figure 2. Schematic view of the fluidic chip. The main compartments are: ROS-generation chamber (1), mixing section (2) and detection chamber with the electrode chip (3). (Reprinted from ref. 20 with permission from The Royal Society of Chemistry.)

antioxidative capacity of complex lipophilic mixtures, such as cosmetic creams.[20] The antioxidative properties of a model cosmetic cream doped with green tea extract were also characterised.

The study of such SAM-modified electrodes showed that the sensitivity of the sensor was directly proportional to the amount of immobilised protein.[7] With the aim of increasing the amount of immobilised biomolecules, Beissenhirtz et al[21] constructed multilayer structures of cyt *c* and poly(anilinesulfonic acid) (PASA) on long-chain mixed SAM-modified electrodes. These multilayer electrodes were successfully applied for the quantitative detection of $O_2^{\bullet-}$. Besides, they were much more sensitive than monolayer electrodes.[22,23]

Apart from in vitro analysis, cyt *c*-based biosensors allow the measurement of the $O_2^{\bullet-}$ produced in vivo, e.g., during ischemia and reperfusion injury.[24-26] Beissenhirtz et al[27] compared the in vitro $O_2^{\bullet-}$ scavenging activity and the in vivo antioxidant potential of methanolic extracts prepared from 10 Chinese tonifying herbs. Results did not show quantitative correlation. However, for 8 out of 10 samples a similar tendency was found.

Superoxide Dismutase-Based Biosensors

SOD biosensors are shown as a promising alternative to cyt *c* biosensors for the evaluation of the antioxidant capacity. SOD enzyme is involved in cell protection mechanisms against oxidative damage from ROS. It specifically catalyses the dismutation of the $O_2^{\bullet-}$ producing O_2 and H_2O_2 via a cyclic oxidation/reduction electron transfer mechanism.

$$SOD(Cu^{2+}) + O_2^{\bullet-} \longrightarrow SOD(Cu^{2+}) + O_2 \qquad \text{(Eq. 4)}$$

$$SOD(Cu^{+}) + O_2^{\bullet-} + 2H^{+} \longrightarrow SOD(Cu^{2+}) + H_2O_2 \qquad \text{(Eq. 5)}$$

The high reactivity of $O_2^{\bullet-}$ limits its direct detection, suggesting a more accurate measurement based on the determination of one of the enzymatically generated species, O_2 or H_2O_2 which can be easily detected using amperometric transducers. That is the case of the so-called first-generation SOD-based sensors. Although some of them are based on the measurement of O_2, most SOD biosensors are usually based on the H_2O_2 oxidation at the electrode surface. Nevertheless, the high potential (>0.5 V vs Ag/AgCl) often required for the electrochemical reaction results in interference problems, limiting the samples where these biosensors can be applied. Some approaches to improve the selectivity have been proposed, based on the use of a H_2O_2-impermeable Teflon membrane[28] or on the simultaneous detection of $O_2^{\bullet-}$ and H_2O_2 with a two-channel sensor.[6] The latter sensor system is based on two glassy carbon working microelectrodes, one covered by an electrodeposited polypyrrole/horseradish peroxidase (PPy/HRP) membrane acting as a H_2O_2 sensor and another covered by a composite membrane composed of an inside layer of PPy/HRP and an outside layer of SOD used as a working electrode for $O_2^{\bullet-}$ detection. Although these biosensors have not yet been applied to the assessment of the antioxidant capacity, this application can be envisaged, since the addition of a sample with antioxidant properties will produce a decrease in the signal, due to their reaction with $O_2^{\bullet-}$ and consequent decrease in concentration.

Campanella and coworkers[29-36] have been working on the development of different SOD-based biosensors for assessing the antioxidant capacity of several compounds. Most of their work is based on the immobilisation of SOD in a κ-carrageenan gel and the amperometric detection of H_2O_2.[37] The gel is sandwiched between an internal cellulose acetate membrane, which improves the selectivity of the electrode by blocking the access to possible electroactive interferences and an external dialysis membrane. This biosensor has been used to evaluate: red and white wines,[29] fresh aromatic herbs, olives and fresh fruit, bulbs and vegetables, plant products sold by herbalists and/or pharmacies, tea,[30-32] dry spices,[33] algae,[34] phytotherapeutic diet integrators[35] and drugs containing as mean component acetylsalicylic acid,[36] ascorbic acid, cysteine, melatonin and β-carotene.[37] The developed biosensor also enables the measurement of the antioxidant activity of healthy and diseased human kidney tissues in vitro.[37]

Due to the low solubility in water of most antioxidant compounds, the same authors have modified their biosensor to ensure its performance in non-aqueous solvents.[35] The developed $O_2^{\bullet-}$ biosensor, successfully used in a DMSO solution, is based on SOD entrapped in a cellulose triacetate layer (sandwiched between two gas-permeable membranes) or in a κ-carrageenan gel layer (sandwiched between an external gas permeable membrane and an internal cellulose acetate membrane), coupled to an O_2 amperometric transducer.[38] This is an interesting approach because antioxidants are not a homogeneous group of compounds (vitamins, inorganic compounds, essential amino acids, polyphenols, etc.), thus biosensors should be able to work in both aqueous and organic solutions.

These biosensors provide a reliable method to measure the antioxidant capacity, as it has been demonstrated by comparing results with other detection methods, e.g., cyclic and pulse voltammetry, spectrophotometry (*N,N*-dimethyl-*p*-phenylenediamine (DMPD)-$FeCl_3$ method) and fluorimetry (Oxygen Radical Absorbance Capacity (ORAC) method), the last one often considered as the reference method. Moreover, the developed biosensors are robust, easily miniaturisable, can perform analyses in situ and do not require expensive or sophisticated equipment.

Radical species are generated in a large number of enzymatic reactions. Due to the possible influence of these species on the enzymatic activity of SOD, the same authors have studied the inhibitor or activation effect of other radicals on the enzyme sensor.[39] Experimental results show that $NO^{\bullet}$, used as a model radical, actively modulates the biosensor response. However, the effect of $NO^{\bullet}$ modulation on the SOD enzymatic sensor was not the expected inhibition, but rather an activation effect, since $NO^{\bullet}$ reduces the SOD metallic centre Cu^{2+} to Cu^{+}, thus, reactivating the enzyme more rapidly. This modulating effect must be taken into account when measuring the antioxidant capacity of real samples.

Emregül[40] developed another biosensor, successfully applied to determine the antioxidant properties of acetylsalicylic acid-based drugs and the antioxidant activity of healthy and cancerous human brain tissues. In this case, SOD immobilisation on a platinum electrode surface is carried out within gelatine, which provides a biocompatible microenvironment around the enzyme and efficiently stabilises its activity.

The previously described $O_2^{\bullet-}$ biosensors are the only sensors used for the assessment of the antioxidant capacity of specific compounds. In what follows, other biosensors, not yet used for this purpose, are reviewed due to their potential applicability.

Mesáros and coworkers[41,42] developed an amperometric enzyme electrode for $O_2^{\bullet-}$ determination by anodic polymerisation of pyrrole and concomitant incorporation of SOD on a platinum wire. The overoxidation of the polymer resulted in an insulating film. The generated H_2O_2 diffused into the film until the electrode surface, where was oxidised at +0.7 V (vs. SCE). The selectivity of this biosensor was studied by Descroix and Bedioui,[43] who concluded that the biosensor was unaffected by the presence of interfering substances and therefore, the biosensor could be used in a blood matrix. Moreover, the microsize of the sensor enables its use for measurements in vivo and in a single cell.

An interesting approach is that based on the electron transfer from SOD instead of the measurement of the enzymatic products. In this direction, some second-generation SOD-based sensors have been developed. Ohsaka et al[44] demonstrated the ability of methyl viologen to efficiently mediate the electron transfer between polyethylene oxide-modified SOD and the electrode in DMSO. Also with the aim to mediate oxidation between SOD and a platinum electrode, Endo et al[45] used an electron carrier which is composed of ferrocene-carboxyaldehyde, cyanamide and bovine serum albumine cross-linked with glutaraldehyde, obtaining a SOD biosensor which shows similar sensitivity to that developed by Campanella and coworkers,[29-36] enabling real-time measurement of $O_2^{\bullet-}$ concentration in various sample solutions and tissues.

Nevertheless, the third-generation biosensor, based on rapid and direct electron transfer of SOD, without any mediator, is the most attractive due to the simple sensor design, the high sensitivity and the selectivity. Like for cyt *c*-based biosensors, the required potentials are much lower than those involved in the H_2O_2 oxidation, avoiding the interference of this molecule generated by

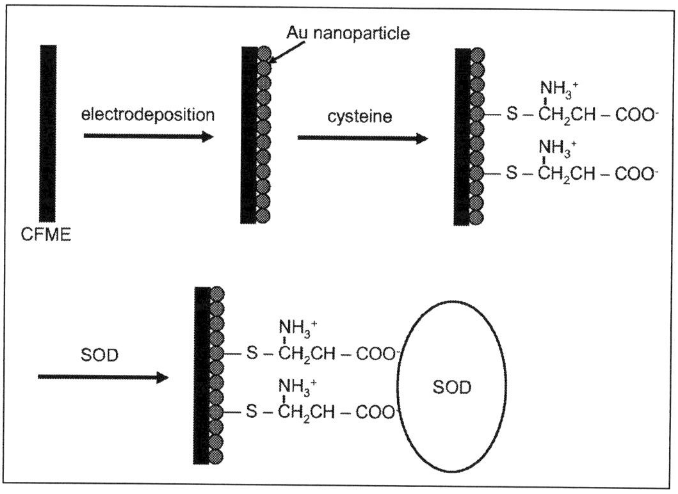

Figure 3. Immobilisation process of SOD enzyme onto CFMEs. (Reprinted from ref. 51 with permission from Elsevier.)

spontaneous dismutation. Using the same principle, Tian and coworkers[46-50] suggested that electron transfer could be efficiently promoted by thiol SAMs formed on gold electrodes. Cysteine SAMs act as effective promoters for the direct electron transfer of the SOD, favouring both the oxidation of $O_2^{\bullet-}$ to O_2 and its reduction to H_2O_2. Thus, potentials of +0.3 V and −0.2 V (vs Ag/AgCl) can be applied, being a great advantage especially in biological systems, because the operating potential can be suitably chosen by taking into account the potential interferences. In more recent studies, the same authors[51] have demonstrated the efficient electron transfer between SOD and carbon fiber microelectrodes (CFMEs) modified with cysteine-SAM gold nanoparticles (Fig. 3).

Di and coworkers[52] developed another third-generation biosensor for $O_2^{\bullet-}$, based on the entrapment of SOD in a thin silica-poly(vinyl alcohol) (silica-PVA) sol-gel film deposited on a gold electrode. The uniform porous structure of the film not only acts as a stabilising matrix, but also provides a fast response rate. Moreover, the low reduction potential applied (−0.15 V vs SCE) minimises possible interferences.

Slightly beyond the scope of this review, a hybrid ultra-sensitive electrophysiological $O_2^{\bullet-}$ sensor has been developed, based on "membrane-engineered" mammalian cells immobilised in an alginate matrix.[53] SOD molecules are electroinserted in cells, which act as catalytic units able to convert $O_2^{\bullet-}$ to H_2O_2. This dismutation process induces changes to the cell membrane potential, which are measured by appropriate microelectrodes according to the principle of the bioelectric recognition assay (BERA). It is interesting to remark the sensitivity improvement (one hundred-fold higher) achieved with this novel biosensor when compared to the previously described SOD biosensors.

Monitoring Hydroxyl Radical ($OH^\bullet$)

Hydroxyl radicals ($OH^\bullet$) can be generated by the Fenton reaction, where reduced transition metal ions, such as Fe(II), Cu(I) or Cr (II), react with H_2O_2 in a one-electron redox reaction, producing $OH^\bullet$ and hydroxide anion:

$$Me^{n} + H_2O_2 \longrightarrow Me^{n+1} + OH^{-} + OH^{\bullet} \qquad \text{(Eq. 6)}$$

The addition of a reducing agent increases the radical generation rate. Alternatively, the transition metal can be reduced by the application of an appropriate electrode potential.[54]

OH• can also be produced trough the photocatalytic oxidation of water by using TiO_2. In this case, the radical generation starts with the absorption of light of a wavelength higher than its band gap by TiO_2, which results in the transition of an electron from the valence band (VB) to the conduction band (CB), leaving a hole behind (Eq. 7). Then, adsorbed water or hydroxide ions are trapped by holes to produce OH• (Eq. 8 and 9). Subsequently, electrons are trapped by the reaction with adsorbed O_2 to produce $O_2^{\bullet -}$ (Eq. 10), which then forms more OH• (Eq. 11):[55]

$$TiO_2 + h\nu \longrightarrow h^+_{VB} + e^-_{CB} \quad \text{(Eq. 7)}$$

$$h^+_{VB} + H_2O_{(ads)} \longrightarrow OH^{\bullet} + H^+ \quad \text{(Eq. 8)}$$

$$h^+_{VB} + 2OH^-_{(ads)} \longrightarrow OH^- + OH^{\bullet} \quad \text{(Eq. 9)}$$

$$e^-_{CB} + O2 \longrightarrow O_2^{\bullet -} \quad \text{(Eq. 10)}$$

$$O_2^{\bullet -} + 2H_2O \longrightarrow 2OH^{\bullet} + 2OH^- + O_2 \quad \text{(Eq. 11)}$$

OH• radicals are known to damage DNA by oxidation of the bases, which results in their destruction and release, or attack of the deoxyribose moieties, which results in strand breaks.[56-58] The lesions produced in the DNA seem to be related with several diseases, such as cancer.[58] In this case, the presence of antioxidants involves a decrease in DNA alterations.

Taking advantage of this property, several DNA-based sensors have been developed for the measurement of the antioxidant capacity of different compounds. In many cases the Fenton reaction is used as the inducing method to damage the biomolecule. The simplest strategy is based on the immobilisation of double stranded DNA (dsDNA), usually from calf thymus, on screen-printed carbon electrodes by simple adsorption and the detection of the guanine oxidation peak between +0.8 and +1.0 V (vs Ag/AgCl) by square wave voltammetry (SWV).[59,60] Since the peak current intensity is proportional to the guanine concentration, the immersion of the DNA-modified electrode into a Fenton solution produces a signal decrease in the peak current intensity (Fig. 4). The introduction of antioxidants into the Fenton solution results in the scavenging of the OH•. Consequently, after immersion of the electrode into this solution, the peak current intensity is very close to the original one, which demonstrates the DNA integrity. Using this strategy, Mello et al[60] evaluated the antioxidant capacity of different plant extracts and concluded the following trend: *Baccharis genstelloides* > *Peumus boldus* > *Foeniculum vulgare* > *Cymbopogom citrates* > *Camellia sinensis* > *Mentha piperita*. When comparing with the spectrophotometric test using 1,1-diphenyl-2-picrylhydrazyl (DPPH), the trend was not the same, certainly due to the different involved radical. Although the authors did not identify the responsible compounds, the scavenging effect is likely to be due to the polyphenols present in plants.

Another way to detect the DNA damage is using an electrochemical label able to interact with dsDNA. This interaction is based on an intercalation phenomenon (predominantly at high ionic strength) or on an electrostatic interaction (predominantly at low ionic strength). The amount of redox label bound to the DNA layer, which can be measured by differential pulse voltammetry (DPV), decreases proportionally to the concentration of radicals present into the sample.[61,62] As before, the presence of antioxidants recovers the electrochemical signal. Using tris-1,10-phenanthroline cobalt (III), $(Co(phen)_3)^{3+}$ as a redox marker, Bučková et al[63] evaluated the antioxidant capacity of different yeast polysaccharides and found the following trend: mannan (*Candida krusei*) > extracellular mannan (*Candida utilis*) > mannan (*Candida albicans*) > glucomannan (*Candida utilis*). The same strategy was used by Labuda et al[64] to assess the antioxidant activity of plant extracts

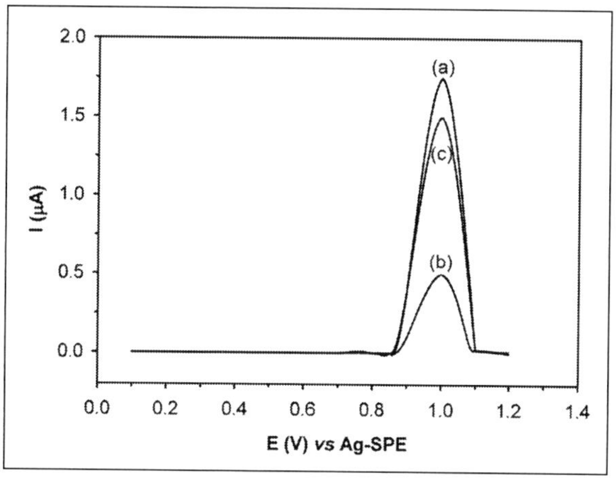

Figure 4. SWV curves corresponding to the: a) DNA-based biosensor signal (blank signal); b) DNA-based biosensor signal after immersion in a Fenton solution; c) DNA-based biosensor signal after immersing in a Fenton solution with *Baccharis genstelloides* extracts. (Reprinted from ref. 60 with permission from Elsevier.)

containing rosmarinic acid and/or caffeic acid (phenolic compounds). In this case, the trend was: lemon balm > oregano > thyme > agrimony. When studying flavonols and flavanols, the following antioxidant trend was observed: quercetin > rutin > epigallocatechin gallate > catechin.[65] Liu et al[66] used methylene blue (MB) as intercalating probe and, instead of using a Fenton solution, DNA was photooxidised on TiO_2-modified indium-tin oxide (ITO) electrodes. SWV measurements allowed them to conclude that gallic acid is ~16 times more efficient as antioxidant than glutathione. The same group has recently developed a kinetic model to study the protecting role of antioxidants and has evaluated the redox antioxidant capacity of gallic acid, glutathione, trolox, uric acid, ascorbic acid and BSA using tris-2,2'-bipyridine ruthenium (II), $(Ru(bpy)_3)^{2+}$ and SWV.[67] In this case, the screen-printed carbon was doped with TiO_2 nanoparticles, creating a porous surface structure on which dsDNA adsorbed better because of specific phosphate-TiO_2 interactions. The oxidative damage was here produced by $(Ru(bpy)_3)^{3+}$, an efficient oxidant of guanine and adenine.

DNA sensors are promising devices to perform simple tests for the routine evaluation of the antioxidant capacity of samples in an easy way. Moreover, the choice of screen-printed carbon as electrode material leads to disposable and low-cost analysis tools.

Monitoring Nitric Oxide Radical ($NO^{\bullet}$)

Nitric oxide radicals ($NO^{\bullet}$) can be generated in an enzymatic or non-enzymatic way. On one hand, $NO^{\bullet}$ is synthesised by the action of the different forms of the enzyme nitric oxide synthase (NOS) on L-arginine in the presence of many cofactors.[68] On the other hand, $NO^{\bullet}$ can also be produced by the dismutation reaction of nitrite in acidic solution:[69]

$$3HNO_2 \longrightarrow H^+ + NO_3^- + 2NO^{\bullet} + H_2O \qquad \text{(Eq. 12)}$$

Most of the techniques for assaying $NO^{\bullet}$ release use indirect methods relying on measurements of secondary species such as nitrites (a $NO^{\bullet}$ oxidation product). The most widely used techniques are spectroscopic and some bioassays based on physiological effects of $NO^{\bullet}$.[70]

Electrochemical techniques, especially amperometry, appear as an attractive alternative for monitoring this radical. Biosensors for the determination of $NO^{\bullet}$ have been mainly focused on two different approaches. The first one is based on the direct oxidation of $NO^{\bullet}$ on a platinum electrode coated by a gas-permeable membrane,[71] which allows the permeation of gases while excluding other materials. On the basis of this pseudo-type Clark's electrode, several $NO^{\bullet}$-electrochemical (micro) sensors have been developed by using carbon, glassy carbon, platinum or gold electrodes covered by selected types of membranes, such as Nafion,[72] Nafion and cellulose acetate[73] or polycarbazole.[74] These organised layers are employed to enhance the selectivity for $NO^{\bullet}$.

The second amperometric approach is based on the electrocatalytic oxidation of $NO^{\bullet}$ on the electrode surface modified with electropolymerised films of metalloporphyrin[75] and metallophthalocyanine,[76] modified with carbon nanotubes[77,78] or by the deposition of *o*-phenylenediamine layers.[79,80]

On the other hand, some chemically modified electrodes for the reductive detection of $NO^{\bullet}$ have also been developed. These biosensors combine facilitated $NO^{\bullet}$ reduction by iron porphyrins (hemes) with $NO^{\bullet}$'s high affinity to heme proteins such as peroxidase[81] and hemoglobin.[82,83] However, a problem encountered with the reductive detection of $NO^{\bullet}$ is the severe interference of dissolved oxygen, whose electroreduction is thermodynamically more favourable than that of $NO^{\bullet}$.

Many researchers have been working on the development of biosensors for the $NO^{\bullet}$ determination; however, no applications to the antioxidant capacity assessment have been published to date.

Conclusion

There are two types of biosensors for the detection of antioxidants: those based on the total phenol content determination and those based on the antioxidant capacity measurement. The latter deserve special attention, since they provide information about the real antioxidant capacity of a compound, enabling the exploitation of its beneficial properties. This review describes in detail the electrochemical biosensors for the assessment of the antioxidant capacity, based on the determination of the free radical scavenging activity. Although many biosensors focused on the $NO^{\bullet}$ determination have been developed, no applications to the antioxidant capacity assessment can be found. On the contrary, many biosensors based on the determination of both $O_2^{\bullet-}$ and $OH^{\bullet}$ have been applied to evaluate the antioxidant capacity of different compounds. For the determination of $O_2^{\bullet-}$, usually generated through the xanthine/XOD enzymatic system, cyt *c* and SOD biosensors are commonly used. Whereas all cyt *c* biosensors are based on the direct electron transfer between the immobilised redox protein and the electrode surface, promoted by SAMs, only first-generation SOD sensors have been applied to evaluate the antioxidant capacity. Nevertheless, SOD-based biosensors are more sensitive and, in principle, more selective. In order to determine $OH^{\bullet}$, mainly generated by the Fenton reaction, DNA-based sensors have been developed. These sensors are based on the damage induced to DNA by free radicals and they are shown to be an attractive alternative to assess antioxidant capacity.

The applicability of electrochemical biosensors to the analysis of compounds believed to have antioxidant capacity is demonstrated. There is a growing interest in the development of such devices, as demonstrate the numerous configurations recently appeared in the literature. Their advantages in terms of simplicity, rapidity and low cost respect to traditional techniques promote their investigation and exploitation. Nevertheless, further work is required to avoid the interferences problem and to consolidate them as practical and current antioxidant assessment tools.

References

1. Prior RL, Wu X, Schaich K. Standardized methods for the determination of antioxidant capacity and phenolics in foods and dietary supplements. J Agric Food Chem 2005; 53:4290-4302.
2. Roginsky V, Lissi EA. Review of methods to determine chain-breaking antioxidant activity in food. Food Chem 2005; 92:235-254.
3. Mello LD, Kubota LT. Review of the use of biosensors as analytical tools in the food and drink industries. Food Chem 2002; 77:237-256.
4. Fridovich I. The biology of oxygen radicals. Science 1978; 201:875-880.
5. Hyland K, Auclair C. The formation of superoxide radical anions by a reaction between O_2, OH^- and dimethyl sulfoxide. Biochem Biophys Res Commun 1981; 102:531-537.

6. Lvovich V, Scheeline A. Amperometric sensors for simultaneous superoxide and hydrogen peroxide detection. Anal Chem 1997; 69:454-462.
7. Ge B, Lisdat F. Superoxide sensor based on cytochrome c immobilized on mixed-thiol SAM with a new calibration method. Anal Chim Acta 2002; 454:53-64.
8. Tammeveski K, Tenno TT, Mashirin AA et al. Superoxide electrode based on covalently immobilized cytochrome c: modelling studies. Free Radical Biol Med 1998; 25:973-978.
9. Krylov AV, Pfeil W, Lisdat F. Denaturation and renaturation of cytochrome c immobilized on gold electrodes in DMSO-containing buffers. J Electroanal Chem 2004; 569:225-231.
10. Manning P, McNeil CJ, Cooper JM et al. Direct, real-time sensing of free radical production by activated human glioblastoma cells. Free Radical Biol Med 1998; 24:1304-1309.
11. Lisdat F, Ge B, Ehrentreich-Forster E et al. Superoxide dismutase activity measurement using cytochrome c-modified electrode. Anal Chem 1999; 71:1359-1365.
12. Lisdat F, Ge B, Reszka R et al. An electrochemical method for quantification of the radical scavenging activity of SOD. Fresenius J Anal Chem 1999; 365:494-498.
13. Ignatov S, Shishniashvili D, Ge B et al. Amperometric biosensor based on a functionalized gold electrode for the detection of antioxidants. Biosens Bioelectron 2002; 17:191-199.
14. Gobi KV, Mizutani F. Efficient mediatorless superoxide sensors using cytochrome c-modified electrodes: surface nano-organization for selectivity and controlled peroxidase activity. J Electroanal Chem 2000; 484:172-181.
15. Gobi KV, Mizutani F. Amperometric detection of superoxide dismutase at cytochrome c-immobilized electrodes: xanthine oxidase and ascorbate oxidase incorporated biopolymer membrane for in-vivo analysis. Anal Sci 2001; 17:11-15.
16. Beissenhirtz M, Scheller F, Lisdat F. Immobilized cytochrome c sensor in organic/aqueous media for the characterization of hydrophilic and hydrophobic antioxidants. Electroanalysis 2003; 15:1425-1435.
17. Shipovskov S, Ferapontova EE, Gazaryan I et al. Recombinant horseradish peroxidase—and cytochrome c-based two-electrode system for detection of superoxide radicals. Bioelectrochem 2004; 63:277-280.
18. Krylov AV, Beissenhirtz M, Adamzig H et al. Thick-film electrodes for measurement of superoxide and hydrogen peroxide based on direct protein–electrode contacts. Anal Bioanal Chem 2004; 378:1327-1330.
19. Krylov AV, Adamzig H, Walter AD et al. Parallel generation and detection of superoxide and hydrogen peroxide in a fluidic chip. Sens Actuators B 2006; 119:118-126.
20. Krylov AV, Sczech R, Lisdat F. Characterization of antioxidants using a fluidic chip in aqueous/organic media. The Analyst 2007; 132:135-141.
21. Beissenhirtz MK, Scheller FW, Lisdat F. A superoxide sensor based on a multilayer cytochrome c electrode. Anal Chem 2004; 76:4665-4671.
22. Guo Z, Chen J, Liu H et al. Electrochemical determination of superoxide based on cytochrome c immobilized on DDAB-modified powder microelectrode. Anal Lett 2005; 38:2033-2043.
23. Dronov R, Kurth DG, Möhwald H et al. A self-assembled cytochrome c/xanthine oxidase multilayer arrangement on gold. Electrochim Acta 2007; 53:1107-1113.
24. Scheller W, Jin W, Ehrentreich-Förster E et al. Cytochrome c based superoxide sensor for in vivo application. Electroanalysis 1999; 11:703-706.
25. Büttemeyer R, Philipp AW, Mall JW et al. In vivo measurement of oxygen-derived free radicals during reperfusion injury. Microsurg 2002; 22:108-113.
26. Büttemeyer R, Philipp AW, Schlenzka L et al. Epigallocatechin gallate can significantly decrease free oxygen radicals in the reperfusion injury in vivo. Transplant P 2003; 35:3116-3120.
27. Beissenhirtz MK, Kwan RCH, Ko KM et al. Comparing an in vitro electrochemical measurement of superoxide scavenging activity with an in vivo assessment of antioxidant potential in Chinese tonifying herbs. Phytother Res 2004; 18:149-153.
28. Song MI, Bier FF, Scheller FW. A method to detect superoxide radicals using Teflon membrane and superoxide dismutase. Bioelectrochem Bioenerg 1995; 38:419-422.
29. Campanella L, Bonanni A, Finotti E et al. Biosensors for determination of total and natural antioxidant capacity of red and white wines: comparison with other spectrophotometric and fluorimetric methods. Biosens Bioelectron 2004; 19:641-651.
30. Campanella L, Favero G, Persi L et al. Evaluation of radical scavenging properties of several plants, fresh or from a herbalist's, using a superoxide dismutase biosensor. J Pharm Biomed Anal 2001; 24:1055-1064.
31. Campanella L, Bonanni A, Tomassetti M. Determination of the antioxidant capacity of samples of different types of tea, or of beverages based on tea or other herbal products, using a superoxide dismutase biosensor. J Pharm Biomed Anal 2003; 32:725-736.
32. Campanella L, Bonanni A, Favero G et al. Determination of antioxidant properties of aromatic herbs, olives and fresh fruit using an enzymatic sensor. Anal Bioanal Chem 2003; 375:1011-1016.

33. Bonanni A, Campanella L, Gatta T et al. Evaluation of the antioxidant and prooxidant properties of several commercial dry spices by different analytical methods. Food Chem 2007; 102:751-758.
34. Campanella L, Martini E, Tomassetti M. Antioxidant capacity of the algae using a biosensor method. Talanta 2005; 66:902-911.
35. Campanella L, Bonanni A, Bellantoni D et al. Biosensors for determination of total antioxidant capacity of phytotherapeutic integrators: comparison with other spectrophotometric, fluorimetric and voltammetric methods. J Pharm Biomed Anal 2004; 35:303-320.
36. Campanella L, Bonanni A, Bellantoni D et al. Comparison of fluorimetric, voltammetric and biosensor methods for the determination of total antioxidant capacity of drug products containing acetylsalicylic acid. J Pharm Biomed Anal 2004; 36:91-99.
37. Campanella L, Favero G, Persi L et al. New biosensor for superoxide radical used to evidence molecules of biomedical and pharmaceutical interest having radical scavenging properties. J Pharm Biomed Anal 2000; 23:69-76.
38. Campanella L, De Luca S, Favero G et al. Superoxide dismutase biosensors working in non-aqueous solvent. Fresenius J Anal Chem 2001; V369:594-600.
39. Campanella L, Persi L, Tomassetti M. A new tool for superoxide and nitric oxide radicals determination using suitable enzymatic sensors. Sens Actuators B 2000; 68:351-359.
40. Emregül E. Development of a new biosensor for superoxide radicals. Anal Bioanal Chem 2005; 383:947-954.
41. Mesáros S, Vanková Z, Grunfeld S et al. Preparation and optimization of superoxide microbiosensor. Anal Chim Acta 1998; 358:27-33.
42. Mesáros S, Vanková Z, Mesárosová A et al. Electrochemical determination of superoxide and nitric oxide generated from biological samples. Bioelectrochem Bioenerg 1998; 46:33-37.
43. Descroix S, Bedioui F. Evaluation of the selectivity of overoxidized polypyrrole/superoxide dismutase based microsensor for the electrochemical measurement of superoxide anion in solution. Electroanalysis 2001; 13:524-528.
44. Ohsaka T, Shintani Y, Matsumoto F et al. Mediated electron transfer of polyethylene oxide-modified superoxide dismutase by methyl viologen. Bioelectrochem Bioenerg 1995; 37:73-76.
45. Endo K, Miyasaka T, Mochizuki S et al. Development of a superoxide sensor by immobilization of superoxide dismutase. Sens Actuators B 2002; 83:30-34.
46. Tian Y, Mao L, Okajima T et al. Superoxide dismutase-based third-generation biosensor for superoxide anion. Anal Chem 2002; 74:2428-2434.
47. Tian Y, Shioda M, Kasahara S et al. A facilitated electron transfer of copper-zinc superoxide dismutase (SOD) based on a cysteine-bridged SOD electrode. Biochim Biophys Acta 2002; 1569:151-158.
48. Tian Y, Mao L, Okajima T et al. Electrochemistry and electrocatalytic activities of superoxide dismutases at gold electrodes modified with a self-assembled monolayer. Anal Chem 2004; 76:4162-4168.
49. Ohsaka T, Tian Y, Shioda M et al. A superoxide dismutase-modified electrode that detects superoxide ion. Chem Commun 2002; 990-991.
50. Tian Y, Ariga T, Takashima N et al. Self-assembled monolayers suitable for electron-transfer promotion of copper, zinc-superoxide dismutase. Electrochem Commun 2004; 6:609-614.
51. Tian Y, Mao L, Okajima T et al. A carbon fiber microelectrode-based third-generation biosensor for superoxide anion. Biosens Bioelectron 2005; 21:557-564.
52. Di J, Bi S, Zhang M. Third-generation superoxide anion sensor based on superoxide dismutase directly immobilized by sol-gel thin film on gold electrode. Biosens Bioelectron 2004; 19:1479-1486.
53. Moschopoulou G, Kintzios S. Application of "membrane-engineering" to bioelectric recognition cell sensors for the ultra-sensitive detection of superoxide radical: A novel biosensor principle. Anal Chim Acta 2006; 573-574:90-96.
54. Fojta M, Kubicarova T, Palecek E. Electrode potential-modulated cleavage of surface-confined DNA by hydroxyl radicals detected by an electrochemical biosensor. Biosens Bioelectron 2000; 15:107-115.
55. Nagaveni K, Hegde MS, Ravishankar N et al. Synthesis and structure of nanocrystalline TiO_2 with lower band gap showing high photocatalytic activity. Langmuir 2004; 20:2900-2907.
56. Portugal J, Waring MJ. Hydroxyl radical footprinting of the sequence-selective binding of netropsin and distamycin to DNA. FEBS Lett 1987; 225:195-200.
57. Jaruga P, Dizdaroglu M. Repair of products of oxidative DNA base damage in human cells. Nucl Acids Res 1996; 24:1389-1394.
58. Evans MD, Cooke MS. Factors contributing to the outcome of oxidative damage to nucleic acids. Bioessays 2004; 26:533-542.
59. Mascini M, Palchetti I, Marrazza G. DNA electrochemical biosensors. Fresenius J Anal Chem 2001; 369:15-22.
60. Mello LD, Hernandez S, Marrazza G et al. Investigations of the antioxidant properties of plant extracts using a DNA-electrochemical biosensor. Biosens Bioelectron 2006; 21:1374-1382.

61. Labuda J, Bučková M, Vanícková M et al. Voltammetric detection of the DNA interaction with copper complex compounds and damage to DNA. Electroanalysis 1999; 11:101-107.
62. Korbut O, Buckova M, Tarapcik P et al. Damage to DNA indicated by an electrically heated DNA-modified carbon paste electrode. J Electroanal Chem 2001; 506:143-148.
63. Bučková M, Labuda J, Sandula J et al. Detection of damage to DNA and antioxidative activity of yeast polysaccharides at the DNA-modified screen-printed electrode. Talanta 2002; 56:939-947.
64. Labuda J, Bučková M, Heilerová L et al. Detection of antioxidative activity of plant extracts at the DNA-modified screen-printed electrode. Sensors 2002; 2:1-10.
65. Labuda J, Bučková M, Heilerová Ľ et al. Evaluation of the redox properties and anti/pro-oxidant effects of selected flavonoids by means of a DNA-based electrochemical biosensor. Anal Bioanal Chem 2003; 376:168-173.
66. Liu J, Roussel C, Lagger G et al. Antioxidant sensors based on DNA-modified electrodes. Anal Chem 2005; 77:7687-7694.
67. Liu J, Su B, Lagger G et al. Antioxidant redox sensors based on DNA modified carbon screen-printed electrodes. Anal Chem 2006; 78:6879-6884.
68. Porasuphatana S, Tsai P, Rosen GM. The generation of free radicals by nitric oxide synthase. Comp Biochem Phys C 2003; 134:281-289.
69. Younathan JN, Wood KS, Meyer TJ. Electrocatalytic reduction of nitrite and nitrosyl by iron(III) protoporphyrin IX dimethyl ester immobilized in an electropolymerized film. Inorg Chem 1992; 31:3280-3285.
70. Archer S. Measurement of nitric oxide in biological models. FASEB J 1993; 7:349-360.
71. Shibuki K, Okada D. Endogenous nitric oxide release required for long-term synaptic depression in the cerebellum. Nature 1991; 349:326-328.
72. Bedioui F, Trevin S, Devynck J. The use of gold electrodes in the electrochemical detection of nitric oxide in aqueous solution. J Electroanal Chem 1994; 377:295-298.
73. Pariente F, Alonso JL, Abruña HD. Chemically modified electrode for the selective and sensitive determination of nitric oxide (NO) in vitro and in biological systems. J Electroanal Chem 1994; 379:191-197.
74. Prakash R, Srivastava RC, Seth PK. Polycarbazole modified electrode; nitric oxide sensor. Polym Bull 2001; 46:487-490.
75. Malinski T, Taha Z. Nitric oxide release from a single cell measured in situ by a porphyrinic-based microsensor. Nature 1992; 358:676-678.
76. Pontié M, Lecture H, Bedioui F. Improvement in the performance of a nickel complex-based electrochemical sensor for the detection of nitric oxide in solution. Sens Actuators B 1999; 56:1-5.
77. Wang Y, Li Q, Hu S. A multiwall carbon nanotubes film-modified carbon fiber ultramicroelectrode for the determination of nitric oxide radical in liver mitochondria. Bioelectrochem 2005; 65:135-142.
78. Zhang L, Zhao G-C, Wei X-W et al. A nitric oxide biosensor based on myoglobin adsorbed on multiwalled carbon nanotubes. Electroanalysis 2005; 17:630-634.
79. Friedemann MN, Robinson SW, Gerhardt GA. o-phenylenediamine-modified carbon fiber electrodes for the detection of nitric oxide. Anal Chem 1996; 68:2621-2628.
80. Park J-K, Tran PH, Chao JKT et al. In vivo nitric oxide sensor using nonconducting polymer-modified carbon fiber. Biosens Bioelectron 1998; 13:1187-1195.
81. Casero E, Darder M, Pariente F et al. Peroxidase enzyme electrodes as nitric oxide biosensors. Anal Chim Acta 2000; 403:1-9.
82. Fan C, Li G, Zhu J et al. A reagentless nitric oxide biosensor based on hemoglobin-DNA films. Anal Chim Acta 2000; 423:95-100.
83. Fan C, Liu X, Pang J et al. Highly sensitive voltammetric biosensor for nitric oxide based on its high affinity with hemoglobin. Anal Chim Acta 2004; 523:225-228.

Chapter 17

Biosensors for the Determination of Phenolic Metabolites

Simona Carmen Litescu,* Sandra Eremia and Gabriel Lucian Radu

Abstract

Antioxidants are groups of chemical substances, the most abundant being polyphenols, mainly found in plants, fruits and vegetables. They include flavonoids, flavonoid derivatives, polyphenols, carotenoids and anthocyanins. Currently, the nutritional quality of many foodstuffs is guaranteed by the presence of antioxidant compounds. The importance of these chemicals as indicators and preservatives of nutritional quality makes necessary the development of accurate, versatile and rapid analytical tools necessary to detect their presence in many foodstuffs and to assess their antioxidant efficacy. In this chapter, enzyme-based biosensors such as monophenol monooxygenase (tyrosinase), catechol oxidase (laccase) and horseradish peroxidase (HRP) are reviewed. Actually, these biosensors are the most commonly used for the detection of polyphenols and flavonoids content.

Introduction

Biosensors are flexible, specific and accurate analytical tools for the detection of antioxidant compounds. An antioxidant is defined as any compound that, when present in low concentration compared to that of an oxidizable substrate, significantly delays or prevents the substrate oxidation. The antioxidants present in the human body act as a defence against highly damaging chemical species such as the free radicals generated during cellular metabolism. The ability of polyphenols, flavonoids and other molecules to scavenge free radicals is connected to the specificity of their chemical structure resulting from the presence of aromatic rings. When these compounds react with a free radical the unpaired electron stemming from the radical is neutralized through the delocalization over the aromatic ring. The captured electron is stabilized by the resonance effect of the aromatic nucleus causing termination of the free radical chain reaction. In this context it should be underlined that polyphenolic compounds inhibit oxidation by means of a variety of mechanisms.[1-4]

The formal redox potential of polyphenols allows them to act towards free radicals both as electron donors and as hydrogen donors.[5] This behaviour is due to the fact that they can be oxidized to phenoxyl radical releasing an electron or loosing a hydrogen atom from the OH group. The phenoxyl radical stabilization is possible by means of H intramolecular bonds.[6] The H donation is possible thanks to the low bonding OH dissociation energy required to complete the reaction pathway.[7] Up to now, the highest reported antioxidant efficacy was obtained for polyphenols 1,2

*Corresponding Author: Simona Carmen Litescu—Centre of Bioanalysis, National Institute for Biological Sciences Bucharest, 296 Splaiul Independentei, Bucharest 6, 060031, Romania. Email: slitescu@gmail.com

Bio-Farms for Nutraceuticals: Functional Food and Safety Control by Biosensors
edited by Maria Teresa Giardi, Giuseppina Rea and Bruno Berra.

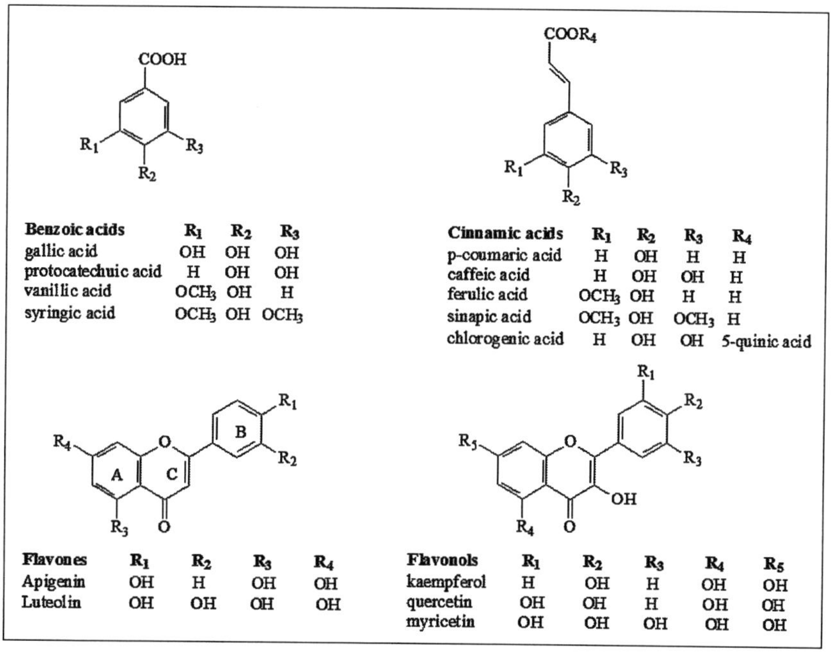

Figure 1. Chemical structures of main antioxidant polyphenols.

dihydroxy substituted on the aromatic ring. As mentioned above, this behaviour finds an explanation by presupposing the stabilization of the radical by intramolecular H bonds.[6]

The term "polyphenols" comprises different subgroups of phenolic acids and flavonoids and their derivatives. Assessing the polyphenols content involves a very important sampling step, which consists in extracting the secondary metabolites from raw sources.

The extraction is performed using various plant sources such as leaves, roots, citrus fruits, grapes, berries-seeds etc. either as fresh or dried material. In the last decade, the use of in vitro cultivation has also made it possible to increase a plant's capability to biosynthesize polyphenols. In particular, it was discovered that, under specific conditions such as oxidative stress induced by high light illumination or temperature, plants increase the production of secondary metabolites among which are polyphenols. This allows researchers to control and raise the antioxidant production in in vitro cultivation permitting the application of molecular biology techniques for large-scale production of antioxidants and their use as food additives.

The main subclasses of flavonoids are the flavonols which include quercetin, kaempferol and myricetin; the flavones which include apigenin, luteolin and tangeritin; the flavanones which include catechins, catechin gallates, naringenin and hesperetin; the isoflavones which include genistein, daidzen and glycitein.[3,8] Flavonoids and related structures are shown in Figure 1.

Biosensors Used in the Determination of Polyphenols

The development of accurate, sensitive and specific methods for the quantitative determination of polyphenols is a challenging task for researchers. To increase the simplicity and performance level of polyphenols determination, highly sensitive, fast and selective methods able to replace classical methods such as high-performance liquid chromatography, up to now the most successful and accurate method, are required.

Biosensors are a sub-group of chemical sensors capable of operating directly in complex matrices to detect analytes and to ensure, at the same time, the requirements of accuracy, sensitivity and selectivity which are the reason for their extensive use in recent years. Biosensors allow quantitative and semi-quantitative analysis based on the use of a biological recognition element (biochemical receptor), which is in direct and spatial contact with a transductor element.

Biosensors can be classified according to the type of active biological component involved in the mechanism, the mode of signal transduction or a combination of these two aspects. The choice of the biological material and the transducer depends on the sample properties and the type of physical variable to be measured.

The type of bio-component determines the degree of selectivity or specificity of the biosensor. The recognition elements are divided in three groups: bio-catalytic, bio-affinity and hybrid receptors.

Bio-Catalytic Receptors

Bio-catalytic receptors can be mono or multi-enzyme systems, whole cells systems (using microorganisms such as bacteria, fungi, eukaryotic cells and yeasts) and cells organelles and plant or animal tissue slices-based systems. The biosensors using microorganisms, plant or animal tissue as bio-components offer the advantage of not requiring extraction and purification procedures that are time-expensive and very laborious, however, because of their high selectivity,[9] enzyme- based biosensors are the most commonly used sensing agents.

Bio-Affinity Receptors

The affinity-based biosensors may be based on antibodies or nucleic acids chemo-receptors. These species provide highly selective interactions with specific ligands leading to thermodynamically very stable complexes. Antigen-antibody complexes may be coupled to every type of transductor element, but, generally different active substances are used as labelling compounds to increase the detectable signal. Among these are enzymes, fluorescent compounds, electrochemically active substances, radionuclides and avidin-biotin complexes.[10,11]

Hybrid Receptors

Hybrid receptors are based on nucleic acid chemo-receptors. The deoxyribonucleic acid (DNA) structure is a double helix consisting of two polynucleotide strands, each strand being constituted of a polymeric chain containing the nucleobases adenine, thymine, cytosine and guanine. Hybrid receptors are made using a unique well-defined sequence of nucleic acid bases. The detection, in this way, becomes highly specific and selective because the hybridization and the recognition occurs only in presence of the complementary DNA fragment.

Most biosensors devoted to the determination of phenolic metabolite content use biocatalytic receptors and electrochemical-based transducers, the measurements being performed by an amperometric system. In this section some of these biosensors are presented emphasizing the performance criteria achieved during biosensor development. The general criteria to evaluate biosensors performance are based on the IUPAC[11] requirements and are based on the calibration characteristics -as sensitivity, working at linear concentration range, detection and quantitative determination limits- and selectivity, reliability, steady-state, transient response times, reproducibility, stability and lifetime.

When the determination of the phenols content in a sample is performed, the total amount of polyphenols is generally detected rather than each of them individually, since the overall response is the most important.[12]

The term 'total phenolics' refers to all phenols that are responsible for the total antioxidant capacity of a specific sample.

The electrochemical reaction used in the amperometric detection of polyphenols is based on two subsequent steps: first, at the electrode surface, the substrate (polyphenol) is oxidised by means of a bio-catalyst (enzyme), in the presence of oxygen. Successively, the regeneration of the enzyme to its original oxidation state occurs, carried out by the electron transfer from a suitable compound (for example, phenols or flavonoids in their reduced form).

The most commonly used biocatalysts in the determination of phenolics content belong mainly to two enzymes classes, phenoloxidases and peroxidases.

Since phenolic-derivatives are suitable substrates for oxidases, electrodes modified with tyrosinase, laccase, peroxidase and cellobiose dehydrogenase have been developed to detect the phenolic compounds. All these proteins belong to the class of 'oxidase' enzymes in which phenols work as electron donors. Laccase and tyrosinase are the most commonly used biological recognition elements in polyphenolics biosensors. The main difference between the two enzymes consists in their specificity: while tyrosinase catalyses only ortho-substituted phenols (polyphenols), laccase is able to act as efficient bio-catalyst for a wider class of polyphenols.[13]

Laccase-Based Biosensors

Laccase is a cuproprotein belonging to the small group of enzymes named blue oxidases[14] able to catalyze the oxidation of various aromatic compounds.

This behaviour can be explained on the basis of the peculiar affinity of laccase towards oxygen that makes it an efficient catalyst when acting as electron acceptor.[15] Laccase is also able to oxidise many nonphenolic compounds.[16] Moreover, it catalyzes the hydrogen atom removal from hydroxyl group of either *ortho-* or *para-*substituted mono- and poly-polyphenolic substrates.[17,18] The rationale supporting the use of laccase as biomediator in phenols determination is linked to the extremely high sensitivity of this enzyme towards phenolic compounds.

A phenolic substrate, in the presence of laccase, undergoes one-electron oxidation leading to the formation of an aryl-oxy radical that can be converted to a quinone in a second step of the oxidation. The principle of polyphenol detection which uses a laccase-based amperometric biosensor is briefly represented in Figure 2.

The catalytic features of laccase are, as expected, highly dependent on the laccase source. The highest catalytic activity has been found in laccase from fungal sources.

Immobilisation of laccase for use as bio-recognition element has been attempted on different solid supports like graphite,[19] redox hydrogel on glassy carbon,[20] carbon paste,[21] carbon fibres,[22] polyethersulphone membranes[23] or platinum.[24]

A biosensor based on the immobilisation of laccase derived from the fungus *Coriolus Versicolor* on polyethersulphone membranes fixed on Pt–Ag supports was reported by Gomes[24] and applied to the determination of polyphenols, flavonoids (caffeic acid, gallic acid, catechin, rutin, *trans*-resveratrol, quercetin) and anthocyanidins (malvidin) from complex samples. The developed

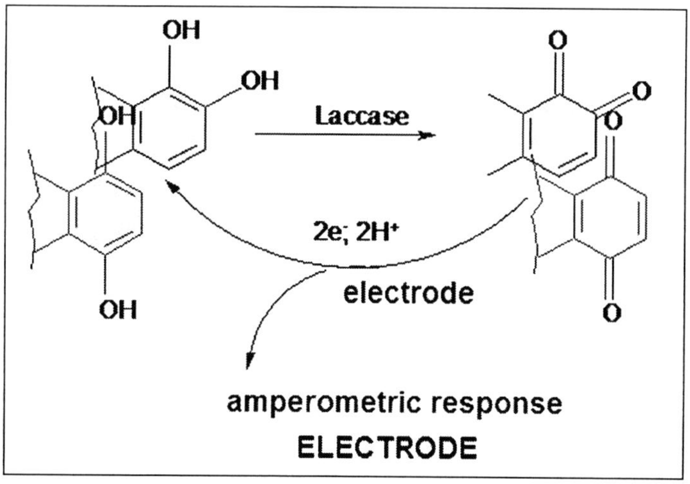

Figure 2. Basis of the detection of polyphenols using laccase–based amperometric biosensors.

biosensor achieved a limit of detection of 1.0×10^{-6} mol L^{-1}, a linearity range from 2.0 to 14.0×10^{-6} mol L^{-1}, high sensitivity (0.0566 mA/mol L^{-1}) and reproducibility (R.S.D. < 10%) when catechin and caffeic acid were used as substrates.

Wilkołazka et al[13] reported a laccase-based biosensor for the detection of catechin hydrate, epicatechin, epicatechin gallate, prodelphinidin and caffeic acid. The enzyme was immobilised by physical adsorption on the surface of a graphite electrode. The electrodes were inserted in a flow-injection cell and the obtained sensitivities were between 57.92 nA/μmol L^{-1} and 7.81 nA/μmol L^{-1} depending on the substrate used, with limits of detection ranging between 0.56 μmol L^{-1} and 2.44 μmol L^{-1}.

Gamella et al[25] have developed a laccase-based biosensor for the determination of polyphenol index in wines. The estimation of the polyphenol index was performed both in batch and flow injection conditions. The enzyme was immobilised by cross-linking with glutaraldehyde onto a glassy carbon electrode, while caffeic acid and gallic acid were used as standard compounds.

Tyrosinase-Based Biosensors

The tyrosinase biosensors are restricted to the monitoring of phenolic compounds with at least one free *ortho*-position. Tyrosinase catalyses two different oxygen-dependent reactions that occur consecutively: the *o*-hydroxylation of monophenols to *o*-diphenols (cresolase activity) and the subsequent oxidation of *o*-diphenols to *o*-quinones (catecholase activity).[26,27]

Carralero Sanz et al[28] have reported the development of a tyrosinase biosensor based on the immobilisation of the enzyme onto a glassy carbon electrode (GCE) modified with electrodeposited gold nanoparticles. The GCE was modified with gold nanocrystals after previous polishing and rinsing. The modification was performed by immersion of the electrode into $HAuCl_4$ and applying a potential of −200 mV for 1 min. Then tyrosinase was added onto the modified electrode surface. The developed biosensor was applied for the estimation of the content of phenolic compounds in beverages.

For the detection of polyphenols in tea, Abhijith[29] constructed a tyrosinase biosensor based on the immobilisation of the enzyme onto a Clark-oxygen electrode membrane cross-linking the protein with glutaraldehyde. The principle of the biosensor was the enzymatic transformation of polyphenols during oxygen consumption. The change in the dissolved oxygen amount depended on the concentration of catechins in the sample. The oxygen consumption at the electrode consumes electrons, resulting in an electrochemical signal that is proportional to the concentration of polyphenols in the sample.

A multilayer tyrosinase based-biosensor sensor was described by Schuhmann.[30] A redox dye was covalently bound to an electrogenerated poly-ω-carboxyalkylpyrrole layer which was covered by a second layer of polypyrrole incorporating the enzyme. This multilayer configuration was able to prevent electrode fouling caused by the polymerization of quinone derivatives.

A tyrosinase biosensor was also developed by Liu et al[31] based on the immobilisation of tyrosinase in a positively charged Al_2O_3 sol-gel membrane onto a glassy carbon electrode. It was found that Al_2O_3 sol-gel has two functions: the hydrophilic, porous and positively matrix provides a friendly microenvironment for the enzyme to retain its functional activity and also acts as an effective promoter for the electron transfer between *o*-quinone and the electrode.

A mediator-free phenol biosensor was developed by Li and coworkers.[32] Tyrosinase, was adsorbed on the surface of the ZnO nanoparticles by electrostatic interactions and subsequently immobilized onto a glassy carbon electrode surface via a chitosan film. The phenolic compounds (catechol, p-cresol and phenol) were determined by the direct reduction of bio-catalytically generated quinones at −200 mV.

Recently, amperometric phenol-oxidases sonogel carbon based biosensors were developed for the determination of polyphenols in complex samples.[33] The detection limit for caffeic acid in the case of the laccase biosensor was of 0.06 μmol L^{-1} and the linear range 0.04-2 μmol L^{-1}. The enzymatic solution was mixed with glutaraldehyde and then modified with 0.5% bovine serum albumin (BSA).

A biosensor based on horseradish peroxidase (HRP) and DNA immobilized onto silica–titanium has also been reported by Mello and Kubota[34] and applied to determination of polyphenolic compounds in vegetable samples. The biosensor performance characteristics reported against chlorogenic acid exhibited a linear response range from 1 to 50 mmol L^{-1}, applied a potential −50 mV versus Ag/AgCl, biosensor sensitivity—expressed as current density on concentration- is about 181 nA/mmol L^{-1} cm^{-2} and the obtained detection limit of 0.7 mmol L^{-1}. The biosensor response compared to the Folin–Ciocalteau method proved the suitability of the biosensor for the quantitative analysis of the total polyphenol in the tested plant extract samples.

Conclusion

Biosensors designed and developed by immobilisation or co-immobilization of one or two phenol-oxidase enzymes on the surface of solid supports—gels, graphite, printed inks, conductive metals etc.—exhibit very encouraging performances when applied to the determination of polyphenols. Depending on their construction, the devices show reasonable stability and working lifetime even in complex sample matrices (beverages—beer, wines, food and food raw materials). As exemplified above, biosensors for the determination of polyphenols based on laccase and tyrosinase are versatile and work well both in batch and in flow analysis. When flow injection analysis is performed the determination sensitivity improves, reaching detection limits of 560 nmol L^{-1}. This level of sensitivity is comparable to chromatographic analysis (high-performance liquid chromatography or gas-chromatography).

Moreover, the use of polyphenol-oxidases based biosensors to assess 'total phenol content' from plant extracts and foodstuff ensures a higher selectivity compared to the classical Folin Ciocalteu method. The former method, unlike the second is, in fact, exempt from the interferences caused by other compounds (e.g., sugars, ascorbic acid) occurring in plant material.[35]

References

1. Steinmetz KA, Potter JD. s.l. Vegetables, fruit and cancer. II. Mechanisms. Cancer Causes Control 1991; 2(6):427-442.
2. Cuvelier ME, Richard H, Berset C. Comparison of the antioxidative activity of some acid-phenols: Structure—activity relationship. Biosci Biotech Biochem 1992; 56:324-325.
3. Kahkonen MP, Hopia A, Heinonen M. Berry phenolics and their antioxidant activity. J Agric Food Chem 2001; 49:4076-4082.
4. Stich HF, Rosin MP. Naturally occurring phenolics as antimutagenic and anticarcinogenic agents. Adv Exp Med 1984; 177:1-29.
5. Tsao R, Deng Z. Separation procedures for naturally occurring antioxidant phytochemicals. J Chromatogr B 2004; 812:85-99.
6. Halliwell B. How to characterize a biological antioxidant. Free Radic Res Commun 1990; 9(1):1-32.
7. Laguerre M, Lecomte J, Villeneuve P. Evaluation of the ability of antioxidants to counteract lipid oxidation: existing methods, new trends and challenges. Progress in Lipid Research 2007; 46(5):244-282.
8. Larson RA. Naturally occurring antioxidants. New York: Lewis Publishers, 1997.
9. Ogawa A, Arai H, Tanizawa H et al. On-line screening method for antioxidants by liquid-chromatography with chemiluminescence detection. Anal Chim Acta 1999; 383:221-230.
10. Fitzpatrick J, Fanning L, Hearty S et al. Applications and recent developments in the use of antibodies for analysis. Anal Lett 2000; 33(13):2563-2609.
11. Thévenot DR, Toth K, Durst R et al. Electrochemical biosensors: recommended definitions and classification. Biosens Bioelectron 2001; 16(1-2):121-131.
12. Blasco AJ, Rogerio MC, Gonzalez MC et al. "Electrochemical Index" as a screening method to determine "total polyphenolics" in foods: A proposal. Anal Chim Acta 2005; 539:237-244.
13. Jarosz-Wilkołazka A, Ruzgas T, Gorton L. Use of laccase-modified electrode for amperometric detection of plant flavonoids. Enzyme Microb Tech 2004; 35:238-241.
14. Thurston CF. The structure and function of fungal laccases. Microbiology 1994; 140:19-26.
15. Xu F. Oxidation of phenols, anilines and benzenethiols by fungal laccases: correlation between activity and redox potentials as well as halide inhibition. Biochemistry 1996; 35:7608-7614.
16. Yaropolov AI, Skorobogat'ko OV, Vartanov SS et al. Laccase—properties, catalytic mechanism and applicability. Appl Biochem Biotechnol 1994; 49:257-80.
17. Leonowicz A, Cho NS, Luterek J et al. Fungal laccase: properties and activity on lignin. J Basic Microbiol 2001; 41:183-225.

18. Decker H, Tuczek F. Tyrosinase/catecholoxidase activity of hemocyanins: structural basis and molecular mechanism. Trends Biochem Sci TIBS 2000; 25:392-397.
19. Decker H, Dillinger R, Tuczek F. How does tyrosinase work? Recent insights from model chemistry and structural biology. Angew Chem Int Ed 2000; 39:1591-1595.
20. Yaropolov AI, Kharybin AN, Emneus et al. Flow injection analysis of phenols at a graphite electrode modified with co-immobilised laccase and tyrosinase. Anal Chim Acta 1995; 308:137-144.
21. Leech D, Daigle F. Optimisation of a reagentless laccase electrode for the detection of the inhibitor azide. Analyst 1998; 123:1971-1974.
22. Freire RS, Thongngnamdee S, Duran N et al. Mixed enzyme (laccase/tyrosinase)-based remote electrochemical biosensor for monitoring phenolic compounds. Analyst 2002; 127:258-261.
23. Leite OD, Lupetti KO, Fatibello-Filho O et al. Synergic effect studies of the bienzymatic system laccase–peroxidase in a voltammetric biosensor for catecholamines. Talanta 2003; 59(5):889-896.
24. Gomes SASS, Rebelo MJF. A new Laccase biosensor for polyphenols determination. Sensors 2003; 3:166-175.
25. Gamella M, Campuzano S, Reviego AJ et al. Electrochemical estimation of the polyphenol index in wines using a laccase biosensor. J Agric Food Chem 2006; 54(21):7960-7967.
26. El Kaoutit M, Naranjo-Rodriguez I, Temsamani KR et al. Investigation of biosensor signal bioamplification: Comparison of direct electrochemistry phenomena of individual Laccase and dual Laccase-Tyrosinase copper enzymes, at a Sonogel-Carbon electrode. Talanta 2008; 75:1348-1355.
27. Quan D, Kim Y, Shin W. Characterization of an amperometric laccase electrode covalently immobilized on platinum surface. J Electroanal Chem 2004; 561:181-189.
28. Carralero Sanz V, Luz Mena M, Gonzalez-Costes A et al. Development of a tyrosinase biosensor based on gold nanoparticles-modified glassy carbon electrodes: Application to the measurement of a bioelectrochemical polyphenols index in wines. Anal Chim Acta 2005; 528(1):1-8.
29. Abhijith KS, Kumar Sujith PV, Kumar MA et al. Immobilised tyrosinase-based biosensor for the detection of tea polyphenols. Anal Bioanal Chem 2007; 389:2227-2234.
30. Schmidt HL, Schuhmann W. Reagentless oxidoreductase sensors. Biosens Bioelectron 1995; 11:127-135.
31. Liu J, Su B, Lagger G et al. Antioxidant redox sensors based on DNA modified carbon screen-printed electrodes. Anal Chem 2006; 78:6879-6884.
32. Li YF, Liu ZM, Liu YL et al. A mediator-free phenol biosensor based on immobilizing tyrosinase to ZnO nanoparticles. Analyt Biochem 2006; 349:33-40.
33. El Kaoutit M, Naranjo-Rodriguez I, Temsamani KR et al. Dual Laccase–Tyrosinase based sonogel–carbon biosensor for monitoring polyphenols in beers. J Agric Food Chem 2007; 55:8011-8018.
34. Mello LD, Sotomayor MDPT, Kubota LT et al. HRP-based amperometric biosensor for the polyphenols determination in vegetables extracts. Sensors and Actuators B 2003; 96:636-645.
35. Prior RL, Wu X, Schaich K. Standardized methods for the determination of antioxidant capacity and phenolics in foods and dietary supplements. J Agric Food Chem 2005; 55(11):4028-4041.

Chapter 18

Methods for the Determination of Antioxidant Capacity in Food and Raw Materials

Simona Carmen Litescu,* Sandra Eremia and Gabriel Lucian Radu

Abstract

A comprehensive description of the most frequently used methods to determine the antioxidant activity in food and raw materials is given. The methods are classified into two categories, depending on the type of the assessment carried out. Several methods for the assessment of antioxidant efficacy using free radical scavenging such as Oxygen Radical Absorbance Capacity Assay (ORAC), Total Radical Trapping Antioxidant Parameter assay (TEAC), Ferric reducing antioxidant power assay (FRAP) and 2,2′-diphenyl-1-picrylhydrazyl (DPPH) assay are described. An example of methods based on the assessment of antioxidant efficacy using significant biological substrates is also presented. Critical opinions concerning the proposed methods are presented.

Introduction

Free radicals are chemical species with one or more unpaired electrons in the valence shell.[1] The formation of radicals takes place via the homolytic scission of covalent bonds and occurs most frequently between two atoms of similar electronegativity. In organic chemistry, this is often the O-O bond in peroxide species or O-N bonds. Inside the human body some radicals are produced as a natural byproduct of oxygen metabolism and have important roles in cell signaling. They are known as reactive oxygen species (ROS). Free radicals are generally very unstable and reactive and to reach a stable electronic configuration, tend to capture an electron from other molecules. As a consequence of this action, oxidative damages can occur in the human body when these species and other pro-radical reactive species such as hydrogen peroxide or nitrogen monoxide are present in high concentrations. The first damage occurs at the level of cell organelles.[2]

An organism's defence against the attack of ROS is ensured by antioxidants. A lack of equilibrium between free radicals and antioxidants leads to a condition called 'oxidative stress' that has been defined by Sies[3] as a disturbance in the prooxidant-antioxidant balance in favour of the pro-oxidants, leading to potential damage.

Usually, the term 'antioxidant' is referred to any compound able to block or delay the reaction of a substrate with molecular oxygen or reactive oxygen species but a fundamental issue to decide if a certain compound could be considered antioxidant comes from the evaluation of the concentration ratio between the oxidizable substrate and the presumed antioxidant. To emphasize the definition

*Corresponding Author: Simona Carmen Litescu—Centre of Bioanalysis, National Institute for Biological Sciences Bucharest, 296 Splaiul Independentei, Bucharest 6, 060031, Romania. Email: slitescu@gmail.com

Bio-Farms for Nutraceuticals: Functional Food and Safety Control by Biosensors edited by Maria Teresa Giardi, Giuseppina Rea and Bruno Berra.

by Halliwell and Gutteridge,[4] an antioxidant is 'any substance that is present in low concentrations compared to those of an oxidizable substrate and significantly delays or prevents oxidation of that substance'. Normally to report a compound as an effective antioxidant the concentration ratio FR : antioxidant has to be in the limit of 100:1.

Frequently, some 'retarder' compounds—molecules able to diminish the oxidation rate—are often wrongly reported as antioxidants. The confusion originates from the ability of these compounds to diminish the oxidation rate when present in very large amounts.

As a consequence, when addressing the determination of antioxidant effect, several arguments have to be taken into account:

a. In the first instance the difference between 'antioxidant capacity' and 'antioxidant activity' has to be considered.
   While 'antioxidant activity' regards the reaction between a single antioxidant species and the free radical, the 'antioxidant capacity' regards the reaction between an antioxidant solution, eventually containing a mixture of antioxidant compounds and the radical.[5]
   While 'antioxidant activity' is defined as the rate constant of the reaction between a unique antioxidant and a given free radical, in fact, 'antioxidant capacity' is defined as the number of moles of free radical scavenged by an antioxidant testing solution that could lead to a different result for the same radical.
b. The potential "own-reactivity" of radicals derived from parent antioxidants, versus the reaction products.
c. The specificity of the antioxidant compound towards the free radical as there are no universal antioxidants able to efficiently quench any type of reactive oxygen species.

In this chapter several models for the in vitro assessment of the antioxidant capacity of food and raw materials are presented.

Models for Antioxidant Activity Determination in Food and Raw Materials

Plants and vegetable materials are very important sources of antioxidant compounds. Antioxidant phytochemicals such polyphenols, flavonoids and isoflavones, anthocyanins etc have been the subject of many recent studies in addition to their use as food additives.

The most widely employed antioxidant phyto-derivatives in food industry are compounds derived from hydroxybenzene (phenol) and are commonly known as polyphenols. The antioxidant effect of these compounds is connected with their chemical structure and their capability to delocalize electrons over the aromatic ring. When these compounds react with a free radical the captured electron is delocalized and stabilized by the resonance effect of the aromatic nucleus preventing free radical chain reactions. This is often called radical scavenging, but polyphenolic compounds inhibit oxidation by various mechanisms, mainly depending on the source material and possible presence of synergists or antagonists.

Several methods have been developed for the evaluation of the antioxidant effect of molecules though, as many variables have to be taken into account when measuring the antioxidant characteristics of a compound, the results have to be treated with caution.

There is no universal system able to provide information about the 'true' antioxidant power or capacity of a single antioxidant or complex mixture of antioxidant phytochemicals[8,9] and a comparative evaluation of antioxidant efficacy is difficult to perform because the activity depends on the substrate, the reaction medium, the oxidation conditions, interfacial phenomena and the antioxidant partitioning properties between phases.

The need to use standardized methods has resulted in a sort of bio-analytical protocol to assess the antioxidant efficacy of phenolic compounds from food and raw materials that includes some common basic features:[10]

1. The quantification and, if possible, the identification of the phenolic compounds;
2. The quantification of the radical scavenging activity and the determination of the formal reduction potential;
3. The evaluation of the inhibition or ending point of lipid oxidation in biological model systems;

4. The study of the efficiency against relevant oxidative markers.

Namely, a good radical scavenging activity does not necessarily correlate with a good antioxidant activity and thus not all the compounds showing a high radical scavenger effect show good antioxidant properties.

To ascribe antioxidant properties to a compound, it is also necessary to determine the efficacy in preventing the oxidation of relevant substrates such as lipids, lipoproteins, DNA etc... against relevant free radical species such as the peroxyl-, superoxide or hydroxyl radical.

Depending on these assumptions, the mechanism involved and the type of assessment, antioxidant capacity assays can be divided in two main categories. The first category is an "assessment of antioxidant efficacy in relation to free radical species". This category includes different reaction mechanisms models such as:

1. Hydrogen atoms transfer reactions model (HAT) based on the transfer of hydrogen atoms;
2. Single electron transfer reactions model (SET) based on the transfer of a single electron;
3. Hydrogen-electron transfer reactions model combining the two mechanisms HAT and SET.[11]

The second category is an "assessment of antioxidant efficacy using biological significant markers and significant substrates".[12-14] This category involves the determination of antioxidant efficacy via evaluation of the damaging effects on a biological substrate produced by reactive species of oxygen (ROS) or related nitrogen oxide species (RNOS) when reacting lipids, lipoproteins, DNA etc.

Examples of antioxidant assays pertaining to both the categories are described in this chapter.

Different assays based on HAT and SET mechanisms are detailed. One example of evaluating antioxidant capacity via the damage to lipids is also given.

Models Based on HAT Mechanisms

According to published data, the HAT reaction is a key step promoting radical chain reactions. Nevertheless, as reported by Prior and coworkers,[15] monitoring the reaction pathway, it is very difficult to distinguish between the hydrogen atom and the electron transfer reactions. The two reactions may take place simultaneously and the mechanism of the reaction is determined by the antioxidant's structure and solubility, the partition coefficient and the solvent polarity.

Hydrogen atom transfer reactions also depend on the pH of the medium. These reactions are very fast and generally occur through the intermediacy of peroxyl radicals. Hydrogen atom transfer reaction can be depicted as follows:

$$X^{\bullet} + AH \rightarrow XH + A^{\bullet}$$

$X^{\bullet}$—Free Radical; AH—any presumed antioxidant compound able to act as hydrogen atom donor.

Oxygen Radical Absorbance Capacity Assay (ORAC)

Oxygen radical absorbance capacity assay (ORAC) is a method using β-phycoerythrin (β-PE) as a fluorescent probe. β-phycoerythrin is a hydrosoluble protein isolated from *Porphyridium cruentum*, which absorbs visible light, presenting high fluorescence yield and sensibility to ROS.[16] The assay is based on the production of a free radical generated in situ which is able to react with the fluorophore causing a decrease in the fluorescence intensity. In the presence of an antioxidant compound the fluorescence decay is inhibited. The result is expressed by calculating the area under the fluorescence curve. The ORAC method has been frequently utilised to evaluate the antioxidant capacity of water-soluble phytochemicals, but is affected by several drawbacks: firstly, β-PE is an irreproducible protein whose exact composition varies between different production lots. Furthermore, it is photo-sensitive so that it loses its fluorescence even in the absence of the free radicals and moreover it interacts with antioxidant molecules such as polyphenols as consequence of a nonspecific protein binding.

A way to minimise these drawbacks is to use different fluorescent probes having high molar absorption coefficients, quantum yields and photochemical stabilities such as fluorescein and 6-carboxyfluorescein.[17,18]

Fluorescein and fluorescein-derivatives are excited at 490 nm and the emission is measured at 514 nm, according to the characteristics of the fluorescent probe used. In the absence of antioxidants, as the reaction occurs, the fluorescence intensity decreases. In the presence of antioxidant species, the fluorescence decay is inhibited. The inhibition is calculated respect to 6-hydroxy-2,5,7,8-tetramethylchroman-2-carboxylic acid (trolox). Trolox is generally used as reference compound because it is a water-soluble vitamin with proven antioxidant activity against recognised reactive oxygen species.

Further improvements have been carried out on the ORAC method by Huang et al[19] using a configuration with a plate reader and an automatic sampling system. These changes allow the time of analysis to be shortened and eliminate inadvertent errors occurring in sample preparation.

Total Radical-Trapping Antioxidant Parameter Assay (TRAP)

The total radical-trapping antioxidant parameter assay (TRAP) is one of the first methods used to determine the total antioxidant capacity of blood plasma or serum.[20] The species involved in a TRAP assay are the peroxyl-radicals generated from the thermolysis of 2,2′-azo-bis(2-amidinopropane) hydrochloride (AAPH) and the peroxidizable materials contained in the plasma or other biological fluids.[5] The AAPH is added to a plasma in which oxidizable components are monitored through the oxygen consumption at the surface of an oxygen electrode. The inhibition of oxidation by the presence of antioxidant species is the measuring principle of the method. The time interval of the reaction induction (lag phase) is compared to the interval time generated by the reference compound, Trolox. The major drawback of this assay is the lack of stability of the oxygen electrode.[21] This inconvenience has been solved using chemiluminescence to detect the reaction end-point. Plasma oxidation, mediated by peroxyl radicals derived from 2,2′-azo-bis(2-amidinopropane) dihydrochloride, in fact, is accompanied by a significant visible chemiluminescence (CL). The chemiluminesce, probably caused by hydroperoxide-like reaction intermediates, is quenched when the antioxidant is added to the system. The degree of chemiluminescence quenching is proportional to the radical trapping ability of the antioxidant sample.[22]

Crocin Bleaching Assay

The crocin bleaching assay determines the antioxidant capacity of a sample using crocin as fluorescent marker.

Crocin is a natural carotenoid present in some flowers genus such as *Crocus* and *Gardenia.* Dissolved in water it forms an orange solution. Crocin is often extracted from *Crocus sativus* (saffron) and can be used to determine the antioxidant capacity of a sample, monitoring the absorbance of the solution when the free radical and the antioxidant are added. If AAPH radical is added to the crocin solution, this undergoes bleaching.[23] In the presence of an antioxidant species the bleaching rate decreases and the antioxidant efficacy can be calculated as a function of the bleaching inhibition. Quantitatively, the antioxidant capacity is the ratio between the crocin bleaching rate in absence and in presence of antioxidants.

Models Based on SET Mechanisms

SET-based assays are used to assess the capability of an antioxidant to reduce a specific oxidant.

SET reactions are usually slow, require a long period and are multi-step processes that can be schematically presented as follows:

$$X^{\bullet} + AH \rightarrow X^{-} + AH^{\bullet +}$$

$$AH^{\bullet +} + H_2O \rightarrow A^{\bullet} + H_3O^{+}$$

$$X^{-} + H_3O^{+} \rightarrow XH + H_2O$$

$$M(III) + AH \rightarrow AH^{+} + M(II)$$

$X^{\bullet}$—Free Radical; *AH*—any presumed antioxidant able to transfer an electron; *M*—a 3*d* metal able to initiate a chain reaction.

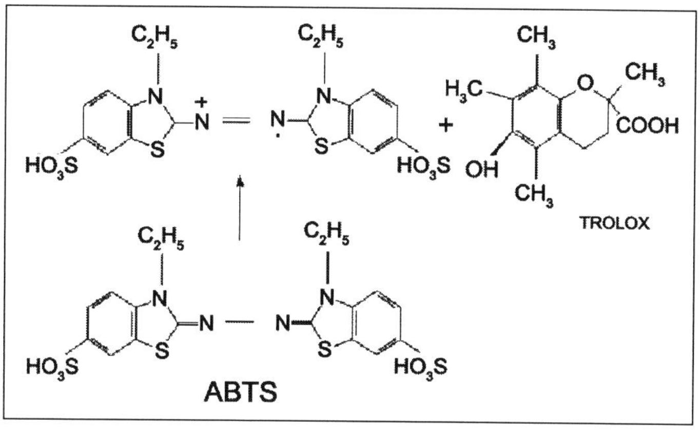

Figure 1. Reference reaction ABTS model (ABTS and trolox).

Trolox Equivalent Antioxidant Capacity Assay (TEAC)

The TEAC assay determines the antioxidant capacity of a sample studying the changes in the specific absorbency of the 2,2′-azino-bis(3-ethylbenzthiazoline-6-sulphonic acid) (ABTS). radical cation. The first method using the ABTS$^{\cdot+}$ cation was developed and commercialized by Randox Laboratories (San Francisco, USA) in an attempt to provide a standardized system to evaluate the antioxidant status.

In this procedure, the radical is formed by means of the reaction between ABTS acid and the ferrylmyoglobin radical generated by the activation of meth-myoglobin with hydrogen peroxide.

The sample to be tested is added before the ABTS$^{\cdot+}$ formation and, when the radical promoters are added, the solution reduces the formed radicals inducing a lag phase. The lag phase duration can be correlated to the antioxidant properties of the tested solution and is used to calculate the antioxidant capacity of the sample.

This method was widely used till it was discovered that a competitive reaction occurs: while scavenging the radical, the antioxidant also scavenges the ferrylmyoglobin radicals. To avoid this drawback, an improved technique to generate ABTS$^{\cdot+}$ has been developed by Re and later optimised by Miller and coworkers.[24] In this method the cation is obtained through the reaction between ABTS and potassium persulfate.

In order to produce ABTS$^{\cdot+}$, ABTS acid is dissolved in water and treated with potassium persulphate. The mixture is allowed to stand at room temperature in the dark for 12-16 h until it turns dark-blue. The solution is diluted before use until the absorbance at 734 nm reaches ca. 0.7. The ABTS$^{\cdot+}$ radical formed shows absorption maxima at various wavelengths: 415 nm, 645 nm, 734 nm and 815 nm. The reaction with the antioxidant solution is monitored at 415 nm, this being the most stable absorption maximum over time. When the antioxidant compound is added to the reacting mixture, the radical is reduced in proportion to the compound concentration (see Fig. 1) causing a colour variation, which is spectrophotometrically determined. The variation is calculated with respect to the 6-hydroxy-2,5,7,8-tetramethylchroman-2-carboxylic acid (trolox) and expressed as Trolox equivalents.[25] TEAC assay is versatile enough to be applied to both water-soluble compounds and lipid-soluble antioxidants. A further improvement to the method has been provided by Pinto[26] using an automatic sequential injection analysis system (SIA) for evaluation of the antioxidant capacity of white and red wines.

2,2-Diphenyl-1-Picryhydrazyl Radical Assay (DPPH)

The DPPH assay measures the antioxidant properties of compounds in reference to their ability to scavenge the radical anion 2,2-diphenyl-1-picryhydrazyl (DPPH$^{\cdot-}$). DPPH$^{\cdot-}$ is a very stable and commercially available free radical, able to accept an electron or a hydrogen atom creating the

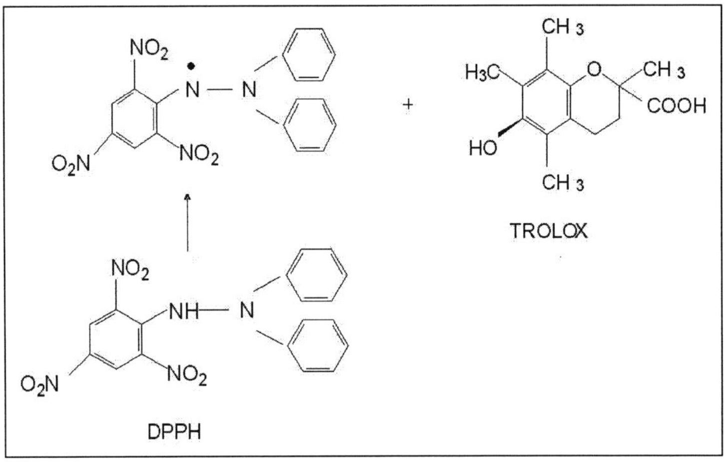

Figure 2. Reference reaction DPPH model (DPPH and trolox).

diamagnetic molecule DPPH (nonradical). The radical absorbs visible light at 515 nm appearing red-violet coloured but when is mixed with a protic solution the reduced form (nonradical) is generated with loss of the violet colour and the appearance of pale yellow colour.[27]

During early work on DPPH, researchers concluded that the reaction takes place through the transfer of a hydrogen atom between the radical and the solution. In fact, the reaction starts with an electron transfer while hydrogen atom abstraction is a slow secondary reaction only occurring in strong hydrogen bond accepting solvents such as methanol and ethanol.[28] As in other electron transfer-based assays, the scavenging activity is strongly influenced by the pH and the solvent properties. A good choice to analyze both lipophilic and hydrophilic antioxidants is the mixture 50% (v/v) water/ethanol.

Antioxidant compounds are able to donate electrons and hydrogen atoms, therefore, the antioxidant capacity can be evaluated using both TEAC and DPPH methods. The colour variation undergone by DPPH is schematically presented in Figure 2. The variation is generally expressed using the parameter of efficient concentration (EC50). EC50 is the antioxidant concentration that produces a 50% reduction of the initial DPPH· concentration. The main drawback of EC_{50} determination is its dependence on the initial concentration of DPPH˙ radical. Higher accuracy is obtained by expressing the result as the absorbance variation respect to the dose-response curve of a standard antioxidant such as Trolox.

DPPH˙ assay is a valid and easy method to evaluate the scavenging activity of antioxidant compounds because the radical is a stable molecule and does not need to be generated 'in situ'. The results are highly reproducible and comparable to other antioxidant detecting methods as ABTS. A drawback is that the solvent effect has to be quantified carefully. In the case of a protic solvent, in fact, the competition of hydrogen abstraction between the antioxidant and the solvent can lead to false positive responses, which, obviously, invalidate the determination.

Ferric Reducing Antioxidant Power Assay (FRAP)

FRAP assay measures the antioxidant capacity by studying the reduction of the complex ferric tripyridyl triazine (Fe^{III}-TPTZ) at low pH.

The reduction of ferric ions to ferrous ions, leads to an intense blue coloured ferrous-tripyridyl-triazine complex[9] the formation of which can be followed spectrophotometrically.

In the presence of antioxidant species in the solution, these act as reductants for Fe (III). The difference of absorbance with respect to a reaction mixture containing ferrous ions of known concentration is directly related to the total ferric reducing power of the antioxidant in the sample.

The FRAP assay provides fast and reliable results for plasma, single antioxidants in pure solution and for mixtures of antioxidants in aqueous solutions.

Moreover, the FRAP assay is simple and inexpensive. The only drawback of this method is that it cannot be used to determine antioxidants containing oxidizable groups such as –SH or which react with Fe (II).

Assessment of Antioxidant Efficacy Using Significant Biological Substrates

The determination of antioxidant capacity using biological substrates is related to the ability of a compound or mixture of compounds to inhibit an oxidative damaging process. The measurement is generally performed using significant biological markers such as DNA strands, RNA strands and lipids.

Most of the methods for this type of assay are based on lipid peroxidation, which is the oxidative process of lipid damage caused by the presence of free radicals.

Assays Based on the Oxidation of Lipids

Lipids peroxidation can be initiated in three ways:

1. Using a thermal method:
   A lipid solution/suspension is heated in the presence of dissolved oxygen to promote the formation of lipoperoxide radicals (LOO·)
2. Using an azo-initiator:
   The initiator generates peroxyl radicals[29] that react with lipid or low density lipoproteins[30] leading to lipo-peroxides
3. Using hydroxyl radical (HO·) generating systems:
   The radical is created using UV irradiation, titanium dioxide (TiO_2) or the Fenton reaction (Fe(II)-H_2O_2).

The lipoperoxides formed are trapped or quenched by the presumed antioxidant, which acts as the terminator of the radical chain reaction. A schematic representation of the lipid peroxidation using the last two reactions is given in Figure 3.

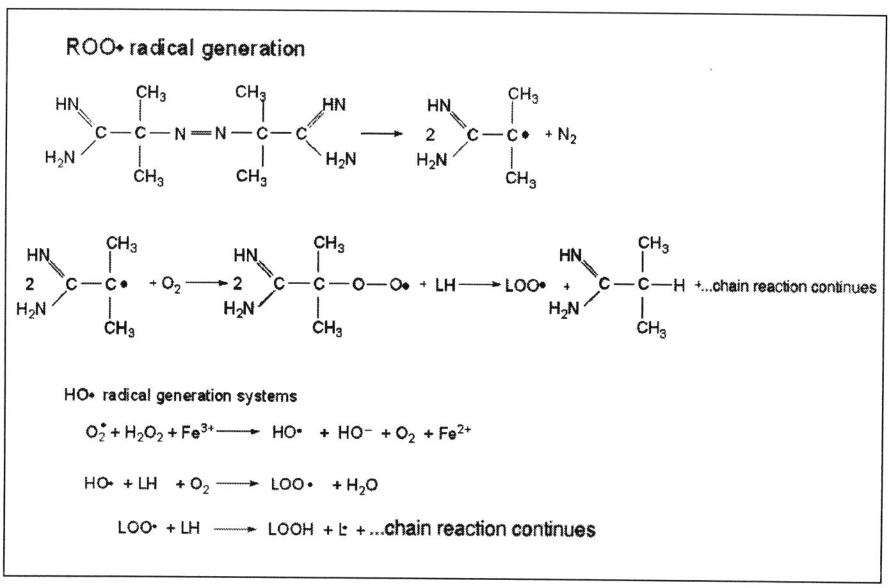

Figure 3. Induction of lipid peroxidation.

Lipid oxidation can be monitored in different ways: one approach consists of measuring the reaction induction period- the so called lag-time reaction- but generally it involves calculations via graphical methods that are inaccurate and poorly reproducible.[31] Other approaches are based on oxygen consumption[32] or inhibition concentration value (IC_{50}) i.e., the antioxidant concentration that results in a 50% of inhibition of the reaction. The calculation of IC_{50} is a very useful and successful method because is the only one which provides directly comparable data.

The drawback is that it can be applied easily only in those cases in which long-lived free radicals are involved or when DNA damage has to be monitored. While monitoring chain reactions, in fact, the results do not depend only on the antioxidant activity but are affected by other parameters such as the chain length, the time measurement and the rate rate determining steps.

Conclusion

The commonly accepted methods for evaluating antioxidant capacity rely on the inhibition of radical chain reactions caused by a presumed antioxidant.

Most of the methods are based on the decrease of specific absorbancy of a long-lived free radical in the presence of the antioxidant. The decrease of the 2,2′-diphenyl-1-picrylhydrazyl radical (DPPH) and of the radical cation derived from 2,2′-azino-bis(3-ethylbenzthiazoline-6-sulphonic acid) (ABTS) are the stable radicals mainly used in 'in vitro' assays as they provide easily comparable results.

Despite the possibility of determining antioxidant capacity using important biological markers and substrates, the most frequently used methods are still based on the assessment of antioxidant efficacy as radical scavengers. This choice is due to the fact that methods using biological significant markers are affected by several drawbacks such as:

1. The lack of reproducibility when using the oxygen consumption method that is connected with the general lack of reproducibility of oxygen electrodes.
2. The lack of reproducibility when using lipoproteins which are obtained by extraction from tissues.
3. The complexity of the procedures used to follow the reaction(s).
4. The difficulties of automating these kinds of methods.

Finally, the ORAC, TEAC or DPPH assays are not strictly related to a compound's efficacy against ROS and consequently not strictly related to the antioxidant activity. These methods are more frequently used to assess the radical scavenging behaviour of food and raw materials.

References

1. Halliwell B. Reactive species and antioxidants. Redox biology is a fundamental theme of aerobic life. Plant Physiol 2006; 141:312-322.
2. Somogyi A, Rosta K, Pusztai P et al. Antioxidant measurements Physiol Meas 2007; 28(4):R41-R55.
3. Sies H. Oxidative stress: from basic research to clinical application. Am J Med 1991; 91(3C):31S-38S.
4. Halliwell B, Gutteridge JMC. Antioxidant defences: endogenous and diet derived in Free Radicals in Biology and Medicine. 4th Ed. Oxford University Press: Clarendon, 2007; 79-81.
5. Prior RL, Cao G. In vivo total antioxidant capacity: comparison of different analytical methods. Free Rad Biol and Med 1999; 27:1173-1181.
6. Meyer AS, Isaksen A. Application of enzymes as food antioxidants. Trends in Food Sci and Technol 1995; 6:300-304.
7. Thomas C. Approaches and rationale for the design of synthetic antioxidants as therapeutic agents. In: Packer L, Cadenas E. eds. Handbook of Synthetic Antioxidants 1st ed. New York: Marcel Dekker, 1997; 3-25.
8. Frankel EN. In search of better methods to evaluate natural antioxidants and oxidative stability in food lipids. Trends Food Sci Technol 1993; 4:220-225.
9. Ou B, Huang D, Hampsch-Woodill M et al. Analysis of antioxidant activities of common vegetables employing oxygen radical absorbance capacity (ORAC) and ferric reducing antioxidant power (FRAP) assays: a comparative study. J Agric Food Chem 2002; 50:3122-3128.
10. Becker EM, Nissen LR, Skibsted LH. Antioxidant evaluation protocols: Food quality or health effects. Eur Food Res Technol 2004; 219:561-571.
11. Huang B, Ou B, Prior RL. The chemistry behind antioxidant capacity assays. J Agric Food Chem 2005; 53(6):1841-1856.

12. Pryor WA. Bio-assay for oxidative stress status (BOSS), Amsterdam: Elsevier Science 2001.
13. Griffiths SW, Cooney CL. Development of a peptide mapping procedure to identify and quantify methionine oxidation in recombinant human α1-antitrypsin. J Chromatogr A 2002; 942:133-143.
14. Roginsky V, Lissi E. Review of methods to determine chain-breaking antioxidant activity in food. Food Chemistry 2005; 92(2):235-254.
15. Prior RL, Wu X, Schaich K. Standardized methods for the determination of antioxidant capacity and phenolics in foods and dietary supplements. J Agric Food Chem 2005; 55(11):4028-4041.
16. Cao G H, Alessio HM, Cutler RG. Oxygen radical absorbency capacity assay for antioxidants. Free Radical Biol Med 1993; 14:303-311.
17. Naguib YMA. A Fluorometric method for measurement of oxygen radical-scavenging activity of water-soluble antioxidants. Anal Biochem 2000; 284:93-98.
18. Ou B, Hampsch-Woodill M, Prior RL. Development and validation of an improved oxygen radical absorbance capacity (ORAC) assay using fluorescein as the fluorescent probe. J Agric Food Chem 2001; 49:4619-4628.
19. Huang D, Ou B, Hampsch-Woodill M et al. Development and validation of oxygen radical absorbance capacity assay for lipophilic antioxidants using randomly methylated β-cyclodextrin as the solubility enhancer. J Agric Food Chem 2002; 50:1815-1821.
20. Wayner DDM, Burton GW, Ingold KU et al. Quantitative measurement of the total peroxyl radical trapping antioxidant capability of human blood plasma by controlled peroxidation. The important contribution made by plasma proteins. FEBS Lett 1985; 187:33-41.
21. Rice-Evans C, Miller NJ. Total antioxidant status in plasma and body fluids. Meth Enzymol 1994; 234:279-293.
22. Alho H, Leinonen J. Total antioxidant activity measured by chemiluminescence methods. Meth Enzymol 1999; 299:3-15.
23. Bors W, Michel C, Saran M et al. The involvement of oxygen radicals during the autoxidation of adrenalin. Biochim Biophys Acta 1978; 540:162-172.
24. Miller N J, Rice-Evans CA. Antioxidant activity of resveratrol in red wine. Clin Sci 1995; 41:1789-1793.
25. Re R, Pellegrini N, Proteggente A et al. Antioxidant activity applying an improved ABTS radical cation decolorization assay. Free Rad Biol Med 1999; 26:1231-1237.
26. Pinto PCAG, Saraiva MFSL, Reis S et al. Automatic sequential determination of the hydrogen peroxide scavenging activity and evaluation of the antioxidant potential by the 2,2′-azinobis(3-ethylbenzothiazoline-6-sulfonic acid) radical cation assay in wines by sequential injection analysis. Anal Chim Acta 2005; 531:25-32.
27. Sanchez-Moreno C. Methods used to evaluate the free radical scavenging activity in foods and biological systems. Food Sci Technol Int 2002; 8:121-137.
28. Foti MC, Daquino C, Geraci C. Electron-Transfer Reaction of Cinnamic Acids and Their Methyl Esters with the DPPH• Radical in Alcoholic Solutions. J Org Chem 2004; 69:2309-2314.
29. Niki E. Antioxidants in relation to lipid peroxidation. Chemistry and Physic of lipids 1987; 44:227-253.
30. Noguchi N, Gotoh N, Niki E. Dynamics of the oxidation of low density lipoprotein induced by free radicals. Biochim Biophys Acta 1993; 1168:348-357.
31. Loshadkin D, Roginsky V, Pliss E. Substituted p-hydroquinones as a chain-breaking antioxidant during the oxidation of styrene. Int J Chem Kin 2002; 34:162-171.
32. Roginsky V. Kinetics of oxidation of polyunsaturated fatty acid esters inhibited by substituted phenols. Kinetics and Catalysis 1990; 31:475-481.

CHAPTER 19

Analytical Methods for the Extraction and Identification of Secondary Metabolite Production in 'In Vitro' Plant Cell Cultures

Alessandra Bertoli,* Barbara Ruffoni, Laura Pistelli and Luisa Pistelli

Abstract

The production of plant secondary metabolites by in vitro culture is one of the most challenging and thrilling field of recent scientific researches. In the few last years, pharmaceutical and food industry demand in phytochemicals has increased steadily. Therefore, the establishment of in vitro plant protocols has to be monitored by phytochemical investigation of their selected extracts in order to supply standardized raw material. In this chapter, the advantages and disadvantages of some modern techniques have been described for the sampling, extraction and analysis of the in vitro plants and derivatives. Depending on the volatile or nonvolatile substances produced by in vitro plant raw material, different kinds of laboratory facilities are needed for the extraction and quali-quantitative analysis. Recent extraction technology such as Accelerated Solvent Extraction or Microwave Assisted Extraction in combination with hyphenated techniques such as Gas Chromathography-Mass Spectrometry (GC-MS) and Liquid Chromatography-Mass Spectrometry (LC-MS) represent a modern approach to perform fast and reproducible analytical methods for the quality control of secondary metabolite production in 'in vitro' plant material.

Introduction

Although the production of many secondary metabolites by plant tissue cultures has not been feasible, future developments may hold promise both for polar and apolar plant constituents.[1,2] The current inability to understand the chemical complexity of all plant metabolites and the limitations of most instrumental techniques represent the major difficulty in plant metabolomic approaches. Selective extraction methods combined with hyphenated technologies such as LC-MS and GC-MS have been introduced recently to obtain the most exhaustive visualization of the plant metabolome.[3,4] Many responses, due to environmental and seasonal stimuli, involving altered plant gene expressions, result in large variations in the plant metabolite pool. At the same time, these different responses may result only in temporal or spatial metabolite variations. In addition, it is important to point out that the presence of some excessive plant metabolites can cause significant chemical interferences in the method performance. For example, high levels of primary metabolites such as sugars often interfer with the ability to profile flavonoids in plant extracts. In vitro protocols are currently available to

*Corresponding Author: Alessandra Bertoli—Dipartimento di Scienze Farmaceutiche, University of Pisa, via Bonanno, 33, Pisa, Italy. Email: bertoli@farm.unipi.it

Bio-Farms for Nutraceuticals: Functional Food and Safety Control by Biosensors
edited by Maria Teresa Giardi, Giuseppina Rea and Bruno Berra.

obtain high regeneration of many plant species and specific important secondary metabolites.[5,6] These natural compounds have sometimes such a structural complexity that chemical synthesis is too difficult or not currently possible. Nowadays, food, cosmetic and pharmaceutical companies depend largely upon raw materials derived from naturally occurring supplies, which are risking to be severely depleted. Therefore, plant cell, shoot and root cultures or transgenic roots may be a promising raw plant material alternative to wild or cultivated species.[6-8]

Plant cells are biosynthetically totipotent, which means that each cell in culture retains complete genetic information and hence is able to produce the range of chemicals found generally in the parent plant. Furthermore, some Plant Cell and Tissue Cultures (PCTC) sometimes produce higher amounts of specific secondary metabolites in comparison with the cultivated or wild plants.[9,10] In this context, phytochemical investigations have to be performed to guarantee the standardization of the final PCTC products. This is extremely important especially when enriched in vitro raw plant material is requested for industrial uses.[11-12] Recent developments in the analytical technologies provide more information in obtaining the phytochemical profiles of plant extracts. Very fast and high-resolution separation systems have been carried out by ultra-performance liquid chromatography (UPLC) in tandem with high-accuracy mass instruments for the on-line identification of analytes. Furthermore, experimental spectrometry and spectroscopy based on specific databases of plant constituents narrow the gap between the detected signals and the metabolite identification. Although many modern analytical techniques, are available, the comprehensive quali-quantitative analysis of all metabolites in a plant tissue is a very ambitious goal and is still far off.[13–14] As the standardization of the raw plant material is generally a critical step, the PCTCs could allow to control better the variations in the secondary metabolite yields. To achieve this task, it is extremely important to select the PCTCs protocols by monitoring biomass and metabolite production simultaneously.

The term 'metabolomics', originated from metabolite profiling, refers generally to quali-quantitative analyses of complex mixtures with a physiological origin.[13,14] Regarding the relationship between the in vitro plant material and its parent plant, it is well known that the PCTC processing technologies may cause different secondary metabolite profiles and yields. In addition, it is quite difficult to guarantee PCTCs with the same large variety of phytochemicals, which generally exists in the correspondent mother plants. Due to the very limited understanding of plant secondary metabolites, it is currently not possible to predict how a product profile will be modified by PCTC process nor can we control these metabolomic modifications completely. Furthermore, the modern analytical techniques play a crucial role in defining not only a safe phytochemical composition, but also contamination traces of additives, generally added during the PCTC establishment.[15] Many countries have developed policies and guidelines to regulate food ingredients produced by PCTCs and they request more and more fast and reproducible analytical control.[16] It is generally agreed that the safety of food products derived from new technologies should not depend on the process system. Therefore, it is important to develop fast, reproducible and sensitive analytical methods in order to control not only the production of the desidered bioactive compounds but also the traces of toxic metabolites or additives in the PCTC extracts.

Sampling and Storage

Unlike the production of cultivated plants is seasonally limited, cell cultures provide a system that can be used year-round and that is independent of the seasons, geographic location and political situations. Plant cells can be induced to produce metabolites of interest and these can be located in the nutrient medium or in cells.[17] In plants, the large diversity in secondary metabolites, generally greatly exceeds the variety of primary metabolites. The biosynthesis of secondary metabolites can occur in all tissues and cells. However, it seems to be rule that biosynthesis is restricted to special tissues or even special cells and it is correlated with differentiation and development.[18,19] Even at intracellular level, the differences are evident: some products are formed in the cytosol, others in organelles, such as mitochondria, chloroplasts, vesicles, or in membrane systems.[20,21] The accumulation occurs in many or even all plant organs and the intercellular transport (via phloem or xylem) must be taken into account for plants.[20] The secondary metabolites need to be stored in plants at

Table 1. The main types of plant matrices from in vitro cultures

Type	Description	Profile	Features in SMP Parent Plant
Callus or cell suspension cultures growth	-Rapid growth -Similarity to microbial cultures -Lack of differentiation and organization	–, AP	-Useful for large scale-production -Lack consistency in cell and secondary metabolite production
Organized tissue cultures	-Differentiation and organization -Shoots, root, other organ cultures -Stable growth	–/+, SMP	-Consistent SMP in shoots, roots, similar, higher or different than in intact plant
Transformed tissue cultures	-Infection of wounded plants -Agrobacterium rhizogenes HR disease -Integration of T-DNA of Ri-plasmid into the plant genome -Stability at genetic and biochemical levels -Rapidly growing and highly branched roots -Adventitious shoots	–/+, SMP	Many concerns about safety issues of final products

(–): differ from that of parent plant; (+): similar to parent plant, AP: accumulation of precursors (disrupt regular metabolic pathways), SMP: secondary metabolites production.

appropriate (usually high) concentrations and in the correct cellular or intracellular sites in order to be effective as defence molecules. In the case of PCTCs, which are characterized by rapid cell division, greatly reduced cell-cell interaction, as well as a lack of tissue-specific morphological and cytological differentiation, other regulatory factors have to be considered for understanding the secondary product metabolism. Furthermore, the nutrients or phytohormones added to the growth medium influence greatly the cell metabolism. The metabolic compartment such as chloroplasts and the storage sites (vacuoles, cell wall and growth medium) are the main compartment involved in secondary metabolism of plant cells.[21] Furthermore, the turnover of a secondary metabolite may occur by several kinds of alternative reactions depending on the availability of the necessary enzymes in the cultured cells and the cell culture growth phase.[22]

The quali-quantitative analysis of secondary metabolites can be carried out on these main classes of PCTC products:

a. *Raw materials.*

The PCTC technologies supply generally three main categories of raw plant matrices for the extraction and analysis of secondary metabolites. They have different features in physiological structures and in the secondary metabolite production (Table 1).

It is important to point out that the chemical composition of cultured plants is greatly influenced by the medium components. Synthetic plant hormones such as IBA, NAA and 2,4-D, 2-isopentenyladenine are generally used in the PCTC establishment and normally are not found in the natural raw plants.[23,24] Suitable analytical methods have to be performed to determine their residue levels even if their working concentrations are generally at a level of several parts per million or lower. Hairy root (HR) is a particular kind of raw in vitro plant material induced by *Agrobacterium* species, which are common in agricultural field too. It is well known that the gene of the bacteria is present in the plant cell and several amino acids, which are not typical of humans are produced by *Agrobacterium* transformed HR. More studies are necessary to establish whether there are any toxic substances of concern from *Agrobacterium* spp. as well as it is important to define the phytochemical composition of their extracts.[10,25,26]

b. *Extracts or purified fractions.*
   Quali-quantitative analysis is necessary to define the phytochemical composition in order to guarantee the quality control of extracts or their purified fractions. It is assumed that the main constituents of the PCTC extracts are not so different from the mother plants, as the production of secondary metabolites is dependent upon the plant genes. However, some types of phytochemicals are not produced by undifferentiated cell cultures and the bioactive constituents yields depend on the medium components and cell line selection. In addition, any contamination derived from the culture medium additives, must be determined as they are generally considered toxic.
c. *Phytochemicals.*
   After purification steps by preparative chromatographic methods, NMR and mass spectrometry experiments are generally use to define the chemical structure of the isolated phytochemicals. The physi-chemical equality between the phytochemical derived from mother plants and that produced by PCTCs is required. Fast and reproducible analytical methods by LC-MS and GC-MS are useful to guarantee the purity grade of a specific secondary metabolite produced from PCTC.
   In conclusion, sampling of the in vitro plant material is a crucial 'pre-analytical' step in each of these cases. Furthermore, the conditions of the biological material should be as homogeneous as possible in order to have reproducible results. After sampling, the in vitro plant material have to be properly transported to the laboratory and stocked as frozen or freezing-dried. Otherwise, the analytical results can be invalidated since chemical reactions, microbial decomposition, photochemical reactions can modify the sample composition greatly. In particular, the sampling of in vitro plant cultures producing flavouring and fragrances represents a fundamental step in the analysis of their aromatic constituents. Adequate storage temperatures as well as proper materials for the containers should be taken in account in order to preserve sample integrity and to avoid any contamination.

Sample Preparation and Extraction

After sampling step, sub-samples ideally representative of the whole sampled raw material will be collected for the qualitative and quantitative analysis of its secondary metabolites. However, several other chemical compounds existing in the matrix as well as contaminants may interfer with the determination of these secondary metabolites.

To overcome these problems, a proper sample preparation and extraction steps are included in the development of a quali-quantitative analytical method.

Sample preparation is perhaps the most underestimated part of plant metabolomic analyses. In any biological system, metabolites of a wide chemical diversity are present in a dynamic range of concentrations. The different classes of secondary metabolites in PCTCs matrix (Table 2) greatly influenced the choice of sample preparation procedures, chromatographic and detection methods. According to Krishnan,[27] a typical cell may contain 5000 metabolites which are expected to be different in concentrations and chemical properties in the plant material. The challenge is to perform a sample preparation method "to capture" the most plant constituents. More steps are included in sample preparation to select a particular class of compounds in the desidered extract. In this context, an important role is played by novel sample preparation techniques which have shown significant advantages over conventional methods, such as a reduction in organic solvent consumption as well as in sample degradation, multiple clean-up and concentration steps before chromatographic analysis. For metabolomics applications, a fast, reproducible, unselective extraction method is preferred to detect the wide range of metabolites that occur in a plant, avoiding unforeseen chemical modifications.[28] Performing appropriate parameters of the extraction may mean to simplify or avoid altogether to clean up the analytical sample. The nature of plant material and its bioactive components should be considered simultaneously in order to achieve good extraction efficiency, reproducibility and specificity. The clean-up step, the removal of chlorophyll or inert substances present in the raw plant material is an important contribution to optimize the

Table 2. The main classes of secondary metabolites detected in tissue and suspension cultures of higher plants

Phenylpropanoids	Alkaloids	Terpenoids	Quinones	Steroids
Anthocyanins	Acridines	Carotenes	Anthroquinones	Cardiac glycosides
Proanthocyanidins	Indoles	Triterpenes	Benzoquinones	Pregnenolone derivatives
Coumarins	Betalaines	Monoterpenes	Naphthoquinones	
Flavonoids	Quinolizidines	Sesquiterpenes		
Hydroxycinnamoyl derivatives	Furnoquinones	Diterpenes		
Isoflavonoids	Isoquinolines			
Lignans	Purines			
Phenolenones	Pyridines			
Stilbenes	Tropane alkaloids			
Tanins				

ratio signal-to-noise in the chromatographic profile.[28] Semi-polar compounds (e.g., phenolic acids, flavonoids, alkaloids and glycosylated sterols) are successfully extracted by methanol/water solutions, while the apolar carotenoids or aromatic constituents are better extracted in chloroform or *n*-hexane. The choice of solvent extraction has to be also compatible with chromatographic step. In fact, for reversed phases in LC-MS analyses, solvents such as ethyl acetate or chloroform are not advisable, as these do not dissolve in the mobile phase nor do they produce an efficient spray ionization. In the case of GC-MS analysis of essential oils, they are generally diluted in an apolar solvent such as *n*-hexane suitable for the chromatographic separations based on a gas flow.

Most of the common solid-liquid extraction techniques for plants such as maceration, reflux, Soxhlet extraction and sonication are taken into consideration also for the extraction of in PCTCs. However, these extraction methods use large volumes of solvent and show long extraction time, low selectivity, low extraction yields and scarce reproducibility. Moreover, the large amounts of toxic solvents arise lot problems in the respect environmental regulamentation and operator's safety. There are several new modern techniques such as accelerated solvent extraction, supercritical fluid extraction, microwave assisted extraction, solid-phase extraction and solid-phase microextraction, which can be automated to monitor the secondary metabolite production in PCTC batches in a short time.

Accelerated Solvent Extraction (ASE)

This technique is one of the most promising pressure liquid chromatography automated extraction process, successfully employed for various pollutants, pharmaceuticals and lipids (Fig. 1).[29-31] ASE is an innovative sample preparation technique that combines elevated temperature and pressures on liquid solvents to achieve a faster and more efficient removal of analytes from various matrices. In the ASE system, the extraction process is carried out at temperatures exceeding the boiling point of a solvent. The extraction of analytes from plant material is based on the analyte solubility, mass transfer effects and disruption of surface equilibria by using temperature and pressure simultaneously. In particular, pressure facilitates extractions from plant material in which the analytes have been trapped in matrix pores or in water-sealed pores or in air-bubble-sealed pores. ASE equipment represents a good solution in savings in time, solvents and laboratory costs. The ASE automated system may be suitable for the monitoring secondary metabolite production during the in vitro establishment because of the reduced solvent consumption (from five to two) and a good reproducibility (probably due to the minimal handly of sample). However, there are not many data available on the solubility of natural compounds in the solvents at the

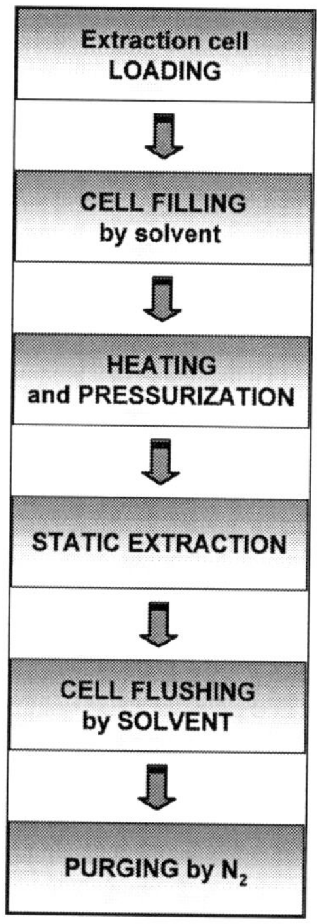

Figure 1. Operating mode by accelerated solvent extraction (ASE) technique.

pressures and temperatures employed in pressure liquid chromatography. Previous studies on this technology evaluated a range of chemical structures (curcuminoids, saponins, flavonoids, polyphenols, terpenes) present in different vegetal matrixes such as roots, leaves, fruits, herbs and rhizomes.[31-34] PCTCs provide an alternative production of food flavors and fragrance ingredients. The essential oils can be extracted by official procedures using Clevenger apparatus (Pharmacopoeia), but ASE extractor using *n*-hexane or dichloromethane in turn is suitable to obtain the same characteristic constituents of aromatic mother plant.[31,32] This technology has only recently been used also for the extraction of plant constituents such as antioxidants.[33] At higher temperatures, although most phenolic antioxidants are stable, catechin and epicatechin are degradated. In extracting polyphenols such as catechin and epicatechin from tea and grape seed, it was found that among water, methanol, ethanol and ethyl acetate, the solvent methanol had the highest yield.[35] ASE was also used for the extraction of saponins, phloroglucinols and procyanidins.[31,36] Bertoli et al[37] investigated the accelerated solvent extraction of chlorogenic acid, flavonoids, hypericin and hyperforin from different rigenarated lines of *Hypericum perforatum* hairy roots.

Supercritical Fluid Extraction (SFE)

This technique uses the properties of gases above their critical points to extract selective soluble components from a plant material. Carbon dioxide is an ideal solvent for the extraction of natural products because it is nontoxic, nonexplosive, readily available and easy to remove from the extracted products (Fig. 2). Recently there has been an increasing interest in the SFE with carbon dioxide as solvent for the extraction of antioxidants from cultivated and in vitro plants.[38-41]

SFE has the ability to use low temperatures leading to less deterioration of the thermally labile components in the extract.[42,43] The main disadvantage of SFE is that it is often difficult to extract poly-oxigenated aglycons and their glycosides from plant matrices . To enhance the

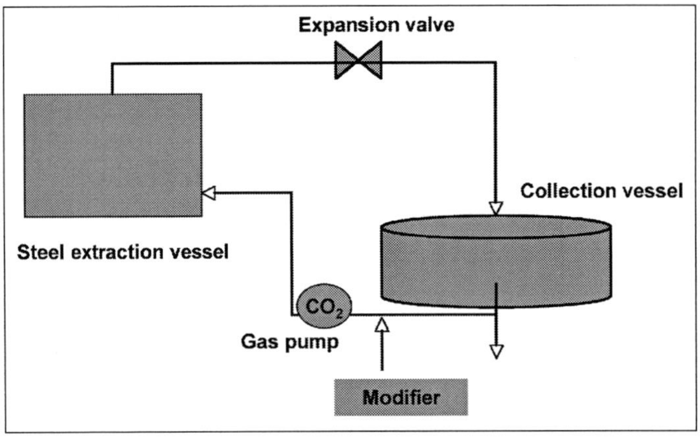

Figure 2. Scheme of supercritical fluid extraction (SFE) technique.

recovery of these metabolites, it is important to find a proper organic cosolvent called 'modifier' such as methanol to add in suitable and small amounts to the plant material before the extraction. However, the recovery of the most polar analytes has not always satisfied. In the case of SFE extracts of *Rosmarinus officinalis,* the most active antioxidant constituents are phenolic diterpenes such as carnosic acid, carnosol, rosmanol, epi- and iso-rosmanol together with rosmarinic acid. On the other hand, rosmanol and epi-, iso-rosmanol are considered minor constituents resulting from the degradation of carnosic acid.[44,45] There is a considerable interest in replacing the traditional methods with SFE technology, but the high costs of this equipment and the difficulties in the extraction of the more polar plant constituents have made it to be no so widespread yet. Nowadays, SFE technique is used especially to improve the quality of essential oils avoiding any thermal stress to the terpenes component.[46,47]

Microwave Assisted Extraction (MAE)

The use of MAE results in a significant reduction in the extraction time and solvent consumption in contrast to conventional liquid-solid extraction methods such as Soxhlet or Clevenger apparatus of different chemical classes of secondary metabolites.[48-50] MAE was used also for the automatized extraction of important polar plant constituents such as isoflavones from soy or taxanes from Taxus.[51-52]A fast, sensitive and selective procedure employing a combination of microwave-assisted extraction and solid phase extraction was applied also to phenolic compounds in plant materials.[53] However, MAE is expecially used for the laboratory scale extraction of essential oils and the microwave distillation is an alternative technique, which combines the microwave heating and the dry distillation at atmospheric pressure.[54]

Solid Phase Micro Extraction (SPME)

The aroma of in vitro plant material can be studied by sampling directly the static or dynamic headspace by SPME technique (Fig. 3). Apolar and polar fibers, inserted in a special holder, are put in contact with the headspace developed over the PCTC sample to adsorb the volatile compounds emitted spontaneously.[55-57] As many stress factors influenced the volatile organic compounds emissions of plants in their habitat, producing plants under in vitro conditions offers important advantages in this respect as it implies exclusion of external contamination.[58] A reproducible in vitro growth involves control of important external factors such as temperature, light characteristic and relative humidity. Sudden changes in headspace composition can be explained and repeated sufficiently for statistical treatment under in vitro conditions.[59] Although in vitro plants can grow successfully when the culture container is properly sealed to avoid any external contamination, for the headspace analysis the conditions required are even more stringent in

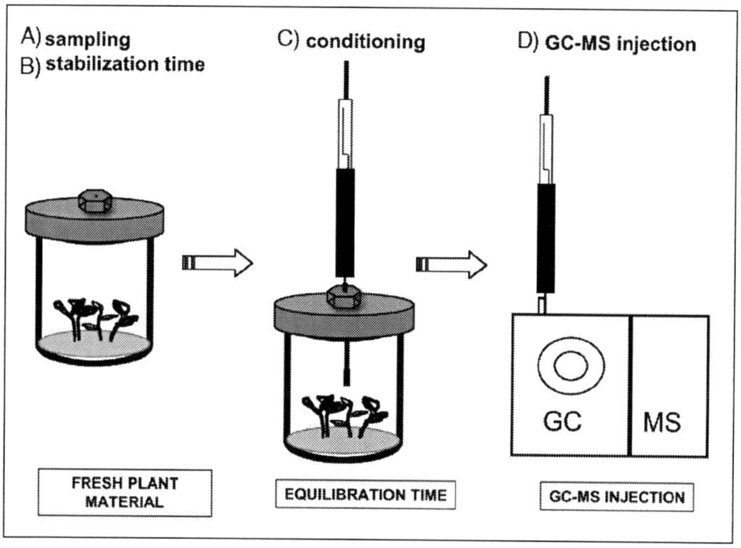

Figure 3. SPME-GC-MS analysis scheme for fresh plant material.

order to equilibrate the HS over plant sample before its adsorbtion. Furthermore, studies on the adsorptive characteristics of the culture medium, septa and/or the glassware were carried out in order to define potential interferences in the extraction yields.[60] If the stabilizing time between fiber and headspace is too short, sensitivity is restricted; on the other hand, when it is increased too much, the total analysis time becomes excessively long. For practical reasons, a sampling time of 30 min was generally used as a compromise with respect to the equilibration requirements. Regarding the recovery of in vitro aroma by SPME, apolar monoterpenes together with some sesquiterpenes were studied. The headspace profile of in vitro plant material is sometimes really different from that obtained from the mother plant. Differences (up to 40%) in absolute amounts of each volatile component in in vitro plant samples can be caused by differences in the sample amounts and differences in the growth capacity. The dissimilarity in emission of mono- and sesquiterpenes between parent and in vitro plants suggests a different compartmentalization in the synthesis of monoterpenes (molecular formula: $C_{10}H_{16}$) compared with sesquiterpenes (molecular formula: $C_{15}H_{24}$). In addition, the SPME-GC-MS profiles of PCTC, generally showed compounds containing 6 carbon atoms which are originated from the degradation of the cell wall fatty acids.[60]

Chromatographic Separation and Detection of Phytochemicals

The wide chemical variety of plant constituents requires suitable quali-quantitative methods to evaluate one or more marker compounds after the extraction step (Fig. 4). It is generally accepted that a single analytical technique will not provide sufficient detection of the plant metabolomic profile which is generally a compromise between speed, selectivity and sensitivity of the analytical method.[61]

Thin Layer Chromatography (TLC) and High Performance-Thin Layer Chromatography (HP-TLC)

TLC is the oldest and widespread analytical chromatographic technique for the screening of plant extracts. The separation process involves a suitable adsorbent (stationary phase) and a solvent or solvent mixture (mobile phase). [62] By TLC methods, a broad range of substances dissolved in all solvents even aggressive reagents can be tested. However, the separation efficiency

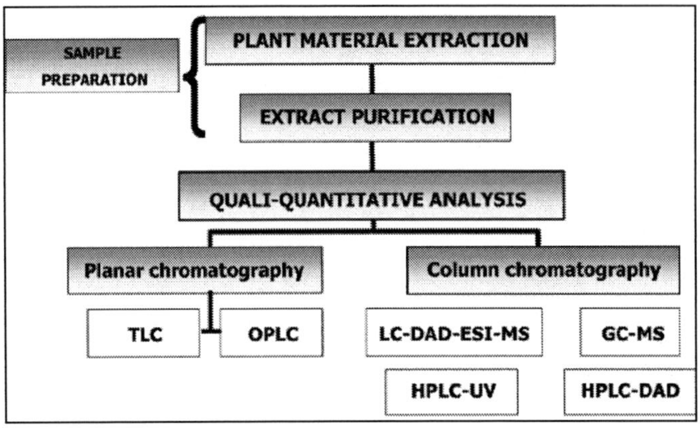

Figure 4. Flow-chart of analytical steps in the phytochemical investigation.

of HPLC and capillary GC are considerably higher than TLC. On the other hand, analytical TLC techniques require a very simple and low cost-equipment as well as reduced amounts of samples. The UV detection and densitometry also allow quantitative determination.[62] Recently HP-TLC technique, allows faster and authomatic development.[63,64] A preliminary phytochemical screening of PCTCs by thin layer chromatography is generally considered useful to perform quali-quantitative HPLC methods. A general inspection of the methanolic extracts obtained from cell suspension cultures, calli, regenerated shoots and roots of *Hypericum perforatum* was performed by analytical TLC.[65] The ginsenoside content in different sources (field-grown roots, in vitro cultures, calluses, liquid cultures) of some ginseng species (*Panax ginseng, P. quinquefolium* and *P. vietnamensis*) were evaluated by HP-TLC technique combining an automatic TLC sampler and a scanner. In this study, HP-TLC was faster and simpler than HPLC.[66] The steviol-glycosides production in intact plants of *Stevia rebaudiana* was compared with different types of PCTCs by HP-TLC. For quantitative analysis, densitometric quantification of five glycosides (stevioside, rebaudiosides A, B and C and steviolbioside) was performed.[67]

Over Pressured Layer Chromatography (OPLC)

It is a relatively recent equipment, not yet generalized although it gives efficient separations in the analysis of essential oils and plant extracts.[63,68] The OPLC technique is a unique liquid chromatography technology which made some authors even talk about 'flat columns'. In fact, OPLC system uses a programmable pump to deliver the mobile phase to the 'flat column' and the resulting forced flow leads to a faster separation and improved efficiency than capillary flow in the TLC system.[69] OPLC is a planar chromatographic method which uses a pressured chamber where the vapor phase above the sorbent is practically eliminated. The eluent is pushed through the sorbent layer and a pump can perform the continuous development. OPLC technique like TLC can be used with various stationary phases. OPLC system which is described as a bridge between thin layer chromatography and high performace liquid chromatography, is more rapid and reliable than TLC procedures for the analysis of a large number of complex plant extracts.[69]

Two modes of detection are commonly used with OPLC: a) on-line mode, where the separated compounds are detected simultaneously as in HPLC and the complete range of HPLC detectors can be connected (e.g., UV, fluorometer); b) off-line mode, where the developed plate is removed from the purification unit and views the progress of the analyte separation by spray reagents or a densitometer. OPLC is a very flexible technique since the flat column can be reconditioned for further elutions and two-dimensional chromatographic separations can be performed with off-line mode detection. Bioautography is a particular on-planar column detection method that can be used in combination with OPLC to screen bioactive or toxic constituents in plant extracts. In this case, the OPLC column serves a double purpose: as the separation medium and mechanical support for a specific cell culture. In fact, the biological activity of plant extract constituents is directly tested on specific fungi or bacteria sprayed on the chromatographic plate. The evaluation of interaction zones on the adsorbent bed between the specific spot of a plant constituent and a particular microorganism can greatly help with the fast identification of potential plant active principles.[70,71]

High-Performance Liquid Chromatography (HPLC)

It is a proven technique that has been used in laboratories worldwide over the past 30-plus years.[28,72] Chromatograms of plant extracts are used as fingerprints and compared with standard compounds in order to identify the plant material and its constituents. HPLC is thus one of the best suited technique for an efficient separation of the crude plant extracts, as shown by Sakakibara (2003) who claim to have found a method capable of quantifying every polyphenol in vegetables, fruits and teas.[73] The reversed-phase columns may be considered the most popular columns used in the analytical separation of plant secondary metabolites, even if new stationary phases have been exploited.[74-75] The versatility of HPLC system was showed in the analysis of rosmarinic acid produced in hairy roots.[75] An in vitro propagation protocol was developed to obtain shoot and root cultures from *Sanicula graveolens* (Apiaceae). Their content of chlorogenic acid and quercetin 3-O-glucoside was assessed by HPLC-diode array detector (HPLC-DAD).[77] The research of plant material with phytoestrogenic activity was carried out also in PCTCs: the isoflavone accumulation in vitro cultures of *Genista tinctoria* and *Pueraria lobata* was used to investigate by HPLC-DAD.[78,79] In vitro cultures of St. John's wort were found to contain hyperforin and three related polyprenylated acylphloroglucinol derivatives. The accumulation of these compounds after different stimulation was detected by HPLC-DAD system.[80] Changes in phenolic metabolism after elicitation with *Colletotrichum gloeosporioides* were monitored by HPLC-DAD in cell suspension cultures of *Hypericum perforatum* L.[81]

Photodiode array detector is generally coupled with HPLC system as it allows to collect spectra in the established wavelength range and to define the spectral homogeneity (purity) of the analytes. This system is one of the most versatile tool in the screening of metabolic profiles of plant extracts. In addition, the analytical high performance liquid chromatography 'piloted' the preparative isolation of camptothecin and triterpenoids from in vitro cultures by the optimization of the experimental separations and checking the different fractions.[82,83] However, the baseline separation using HPLC normally requires complex solvent gradient programs and long analysis times. In addition, unequivocal identification of flavonoids, one of the largest and widespread plant secondary metabolite classes, which have similar UV spectra and elution times, cannot be guaranteed. Furthermore, UV detection is unable to fulfill the phytochemical task, since lots of plant constituents have not chromophoric groups.

Liquid Chromatography and Mass Spectrometry (LC-MS)

It combines the high separation power of high-performance liquid chromatography with the structural information of mass spectrometry. The sensitivity and specificity of LC-MS methods are drastically improved relatively to the traditional UV detection and allows the use of very fast chromatographic separations with high peak purity value.[4,84-86] A key development of this technique is the interface Electron Spray Ionization that transfers analyte molecules from solution to the gas phase, suitable for mass analysis. The mass detection of a molecule is conditioned by the capacity of the analyte to ionize while being part of a complex mixture. Apart from the chemical properties of the molecule itself, eluent flow, composition of sample matrix as well as ionization source, influence ionization greatly. The use of ionization enhancers, sample clean-up methods and different ionization sources are some of the possibilities that can improve ionization of analytes in the positive or negative modes.[87] LC-MS mostly uses as soft-ionization sources the atmospheric pressure ionization, with the electrospray ionization or atmospheric pressure chemical ionization. The performance of each soft-ionization mass spectrometer can be described by means of several intrinsic parameters: mass resolving power (or resolution), mass accuracy, linear dynamic range and, sensitivity. Improvement of these parameters enables more effective identification of the molecular mass of the injected analyte. However, drawbacks are noise in the typical LC-MS raw data and retention time shifts which depend on the complexity of the matrix and ion suppression effects. For general applications in plant metabolomics, the isolation of one ion and the tandem mass spectrometry spectrometry experiments to obtained daughter fragments can be highly informative in order to elucidate metabolite structures directly in the non purified plant extract sample.[87,88] Based on these main features, LC-MS is at present a widely applied technique for the fast and sensitive quali-quantitative analysis of plant metabolites in order to compare wild, cultivated and in vitro plant material.[3,4,85-88] Recent studies were carried out for the simultaneous analysis of salidroside and other main constituents in callus and plant extracts of *Rhodiola spp.* In these cases, the developed LC-MS method showed an extremely versatility for the simultaneous analysis of the transformation of cinnamyl alcohol precursor into rosavin or L-tyrosine as well as tyrosol precursors into salidroside in callus.[89,90]

Ultra Performance Liquid Chromatography (UPLC)

It is an emerging technique for carrying out rapid and highly efficient quali-quantitative analysis. In fact, it makes possible to perform very high-resolution separations in short periods of time with little solvent consumption utilizing very small solid phase particles. Furthermore, the hyphenation of UPLC to mass spectrometry can be extremely advantageous for a very fast quali-quantitative analysis of complex plant matrices.[91,92] Robustness and reproducibility (retention time and mass accuracy) as well as efficient ionization of the analytes are essential for obtaining consistent data by UPLC-MS system. In conventional HPLC the choice of particle size must be a compromise as the smaller is the particle size, the higher column back-pressure is generating in system. On the contrast, the UPLC technology takes full advantage using columns packed with smaller particles and/or higher flow rates to perform superior resolution and sensitivity with shorter running time. UPLC removed the barrier of traditional chromatographic packing material by development of new, highly efficient, mechanically strong, 1.7 mm bridge hybrid particles that are stable over a broad pH operating range. Considering the shorter column wash-out time, UPLC methods can be regarded as "green" procedures because of the negligible solvent consumption for each chromatographic run.[93] Due to very narrow and sharp peaks, more number of them may appear in less time which may facilitate in analysis of complex mixtures and that may give more qualitative information regarding plant metabolomic profile. Furthermore, the UPLC approach enables the detection of analytes at very low concentrations because of the improve signal-to-noise ratio. Quali-quantitative methods have been reported for the fast simultaneous determination of flavonoids and saponins in many plant extracts.[94-96] This technology is quite new, but it has been applied also to the secondary metabolites produced in in vitro plant tissues. Some alkaloid fractions were analysed by UPLC in a study to describe the genetic engineering and expression of the terminal step of vindoline biosynthesis in *Catharanthus roseus* hairy root cultures.[97]

Gas-Chromatography Mass Spectrometry (GC-MS)

It is the most popular and useful analytical tool in the research field of volatile plant secondary metabolites.[98] The advantages of GC clearly lie in its high sensitivity of detection for almost all the volatile chemical compounds present in the apolar plant extracts or essential oils. Furthermore, the high selectivity of capillary columns guarantees high resolution of many volatile compounds simultaneously within comparatively short times. This fact is extremely important in the case of essential oils, generally contain a hundred of terpenes and derivatives. In addition, the unambiguous identification of plant metabolite is possible by computer matching with commercial and experimental databases of mass spectra and retention indices. In fact, the method of ionization generally used in conjunction with gas chromatography is Electron Impact, a hard-ionization method which provides reproducible mass spectra and allows library searching.[99] In the case of gas chromatography, it is well known that the volatility and thermal stability of the analytes represent sometimes a great limit. Another limitation of GC-MS is its inability to handle polar or high molecular plant metabolites and cofactors are often a problem too. However, GC-MS is the elective technique to study the aroma produced by in vitro PCTCs which is generally due to many volatile compounds like in the mother plants.[100,101] One of the most commercialized aromas is strawberry, where 143 different constituents were identified and even the slightest change in the composition of these compounds can cause significant flavor modification.[102] Changes in the composition of the volatile constituents may also be used as indicators of oxidation, enzymatic changes and microbial fermentation of the analysed in vitro plant material. The analysis of volatile biotransformation products in plant cell cultures has been widely reported.[102,103] Micro-propagated plant material can produce the same characteristic volatile constituents of the parent plants both in their fresh aroma than in their essential oil. On the other hand, different levels of aromatic constituents and complete absence of the typical volatiles derived from the counterpart adult plants have been reported for in vitro cultures.[55,104]

A study on *Peganum harmala* showed that loss terpenes by evaporation severely limited the ability of cultures to accumulate these volatile components. However, the incorporation of controlled release polymers like substrate storage produced a dramatic increase in volatile constituents, particularly of geraniol and linalool.[106]

Direct Infusion (DI)

The direct infusion using electrospray ionisation mass spectrometry (DI-ESI-MS), that is to say MS detection without prior chromatographic separation of the extract, is able to produce rich information on mass spectra of plant constituents.[107]

For a preliminary qualitative screening, it is possible to inject directly into the mass spectrometer the unpurified plant extracts without the conventional chromatographic separation by liquid chromathography column. In fact, the 'molecular' ions in a complex sample may be sufficiently distinguished by their m/z values by ESI-MS direct infusion. Mass spectra are generally acquired in scan mode detection and ESI-MS conditions were optimized using the available standards for the different class of plant metabolites. However, it is important to point out that direct infusion technique has limited utility for the quantitative analysis of plant metabolites due to undesirable effects of ion suppression.[108]

Direct infusion by ESI-MS has found recent application for the rapid characterization of vincristine and vinblastine in *Catharanthus roseus* as well as for the secondary metabolites in microbial extracts.[109-110] The same technology was used to identify chemical differences that occurred in the expression of secondary metabolites by 44 actinomycetes cultivated under six different fermentation conditions.[111]

The reproducibility of ESI-MS experiments is generally guaranteed by the combinaton with chemometrics analysis (Cluster analysis, Principal Component Analysis). ESI-MS is considered competitive in terms of speed with other detection methods, especially in conjunction with MS/MS analysis, but appropriate caution is recommended as it must be considered a potential valuable compromise between speed and information.[112]

The described approach is specific, sensitive, rapid and does not require prepurification steps. It provides the fingerprint of plant extracts, since it permits to detect simultaneously different constituents and it may be suggested for the discrimination of productive batches of in vitro plant material.

Conclusion

Nowadays, the homogenous tissues of PCTCs represent a valuable alternative to wild or cultivated plants as the growth stages, environmental conditions and sampling should be less critical for the standardization of their metabolomic profile. However, this fact shows obviously some restrictions as many of the highly expressed metabolites are often unique and can provide exclusive bases for the differentiation of cell states, organs, tissues, varieties and organisms. Due to the very limited understanding of plant secondary metabolism, it is currently not possible to predict how PCTC processes will modify the final metabolomic profile nor can we control any modification. Most PCTC systems are clearly not able to produce exactly the same phytochemical profile of their parent plant. Therefore, any phytochemical variations have to be characterized in order to produce standardized raw PCTC plant material as ingredients of food, cosmetics or phytomedicines. In addition, the establishment and propagation of plant cell cultures require chemicals which are usually not a part of the human diet and which will have to be removed during the final extract. The levels of natural or synthetic dangerous constituents in PCTC products have to be assure lower than toxic concentration.

The extraction and identification of selected analytes by the classical analytical methods is generally a time and solvent—consuming process which not guarantees reproducible results. Therefore, modern extraction techniques such as ASE, MAE combined with chromatographic systems (OPLC, HPLC-DAD, GC-MS, LC-MS) or direct infusion (DI-ESI-MS) allow to develop rapid and effective analytical methods in order to define the quality and safety of PCTC products.

Acknowledgement

This work was supported by the NUTRA-SNACKS project funded within the EU Framework Programme 6.

References

1. Oksman-Caldentey K-M, Inze D. Plant cell factories in the post-genomic era: new ways to produce designer secondary metabolites. Trends Plant Sci 2004; 9(9):433-40.
2. Verpoorte R, Memelink J. Engineering secondary metabolite production in plants. Curr Opin Biotechnol 2002; 13(2):181-7.
3. Sumner LW, Mendes P, Dixon RA. Plant metabolomics: large-scale phytochemistry in the functional genomics era. Phytochemistry 2003; 62(6):817-836.
4. Halket JM, Waterman D, Przyborowska AM et al. Chemical derivatization and mass spectral libraries in metabolic profiling by GC/MS and LC/MS/MS. J Exp Bot 2005; 56:219-43.
5. Kurz W, Constabel F. Production of secondary metabolites. In: Altman A, Colwell R, ed. Agricultural Biotechnology, 1997. New York: CRC Press, 1998:183-213.
6. Rout GR, Samantaray S, Das P. In vitro manipulation and propagation of medicinal plants. Biotechnol Adv 2000; 18(2):91-120.
7. Bourgaud F. Plant cell and tissue culture for the production of food ingredients. Plant Sci 2001; 160(3):571-572.
8. Wu J, Zhong JJ. Production of ginseng and its bioactive components in plant cell culture: Current technological and applied aspects. J Biotechnol 1999; 68(2):89-99.
9. Zhong JJ. Plant cell culture for production of paclitaxel and other taxanes. J Biosci Bioeng 2002; 94(6):591-9.
10. Ramachandra RS, Ravishankar GA. Plant cell cultures: Chemical factories of secondary metabolites. Biotechnol Adv 2002; 20(2):101-53.
11. Charchoglyan A, Abrahamyan A, Fujii I et al. Differential accumulation of hyperforin and secohyperforin in Hypericum perforatum tissue cultures. Phytochemistry 2007; 68(21):2670-7.
12. Zhao J, Davis LC, Verpoorte R. Elicitor signal transduction leading to production of plant secondary metabolites. Biotechnol Adv 2005; 23(4):283-333.
13. Trethewey RN. Metabolite profiling as an aid to metabolic engineering in plants. Curr Opin Plant Biol 2004; 7(2):196-201.
14. Weckwerth W, Fiehn O. Can we discover novel pathways using metabolomic analysis? Curr Opin Biotechnol 2002; 13(2):156-160.
15. Ton-Jen Fu. Plant cell and tissue culture for food ingredient production, safety considerations. In: Fu TJ, ed. Plant Cell and Tissue Culture for the Production of Food Ingredients. New York: Kluwer Accademic Plenum Publishers, 1999:237.
16. Beru N. Food Ingredients from plant cell and tissue culture: regulatory considerations. In: Fu TJ, ed. Plant Cell and Tissue Culture for the Production of Food Ingredients. New York: Kluwer Accademic Plenum Publishers, 1999:265.
17. Murashige T, Skoog F. A revised medium for rapid growth and bioassays with tobacco tissue culture. Physiol Plant 1962; 15:472-5.
18. Wiermann R. Secondary products and cell tissue differentiation. In: Conn EE, ed. The Biochemistry of Plants. New York: Academic Press 1981; 7:85-115.
19. Wink M. Physiology of accumulation of secondary metabolites with special reference to alkaloids. In: Constabel F, ed. Cell Culture and Somatic Cell Genetics of Plants. London: Academic Press, 1987; 4:17-42.

20. Wink M. Production of secondary metabolites by plant cell cultures in relation to the site and mechanism of their accumulation. In: Marin B, ed. Plant Vacuoles. NATO Advanced Institute, Series 134. 1987:477-484.
21. Guern J, Renaudin JP, Brown SC. The compartmentation of secondary metabolites in plant cell cultures. Constabel F, Vasil IK, eds. In: Cell Culture and Somatic Cell Genetics. New York: Academic Press, 1987; 9:43-76.
22. Barz W, Beimen A, Drager B et al. Turnover and storage of secondary products in cell cultures. In: Charlwood B, Rhodes M, eds. Secondary Products from Plant Tissue Culture. Oxford: Claredon Press, 1990:79-102.
23. San-Francisco S, Houdusse F, Zamarreno AM et al. Effects of IAA and IAA precursors on the development, mineral nutrition, IAA content and free polyamine content. Scientia Horticulturae 2005; 106(1):38-52.
24. Thorpe TA. The current status of plant tissue culture. In: Bhojwani, ed. Handbook of Plant Cell Culture, Techniques and Applications. New York: Macmillan, 1990; 4:1-33.
25. Srivastava S, Srivastava AK. Hairy root culture for mass-production of high-value secondary metabolites. Crit Rev Biotechnol 2007; 27(1):29-43.
26. Saito K, Yamazaki M, Murakoshi I. Transgenic medicinal plants: Agrobacterium-mediated foreign gene transfer and production of secondary metabolites. J Nat Prod 1992; 55(2):149-162.
27. Krishnan P, Kruger NJ, Ratcliffe RG. Metabolite fingerprinting and profiling in plants using NMR. J Exp Bot 2005; 56(410):255-265.
28. Huie W. A review of modern sample-preparation techniques for the extraction and analysis of medicinal plants. Anal Bioanal Chem 2002; 373:23.
29. Carabias-Martinez R, Rodriguez-Gonzalo E, Revilla-Ruiz P et al. Pressurized liquid extraction in the analysis of food and biological samples. J Chromatogr A 2005; 1089(1):1-17.
30. Smelcerovic A, Spiteller M, Zuehlke S. Comparison of methods for the exhaustive extraction of hypericins, flavonoids and hyperforin from Hypericum perforatum L. J Agric Food Chem 2006; 54(7):2750-53.
31. Benthin B, Danz H, Hamburger M. Pressurized liquid extraction of medicinal plants. J Chromatogr A 1999; 837(1):211-219.
32. Choa SK, Abd El-Atya AM, Choia J-H et al. Optimized conditions for the extraction of secondary volatile metabolites in Angelica roots by accelerated solvent extraction. J Pharm Biomed Anal 2007; 44(5):1154-8.
33. Mendiola JA, Rodríguez-Meizoso I, Señoráns FJ et al. Antioxidants in plant foods and microalgae extracted using compressed fluids. JEAF 2008; 7(10):3279-87.
34. Chena J, Li W, Yang B et al. Determination of four major saponins in the seeds of Aesculus chinensis Bunge using accelerated solvent extraction followed by high-performance liquid chromatography and electrospray-time of flight mass spectrometry. Anal Chim Acta 2007; 596:273-280.
35. Alonso-Salces RM, Korta E, Barranco A et al. Pressurized liquid extraction for the determination of polyphenols in apple. J Chromatogr A 2001; 933(1):37-43.
36. Anand R, Verma N, Gupta DK et al. Comparison of extraction techniques for extraction of bioactive molecules from Hypericum perforatum L. plant. J Chromatogr Sci 2005; 43(10):530-1.
37. Bertoli A, Giovannini A, Ruffoni B et al. Bioactive constituent production in St. John's Wort in vitro hairy roots regenerated plant lines, J Agric Food Chem 2008; 56:5078-5082.
38. Cavero S, Garćıa-Risco MR, Marın FRJ et al. Supercritical fluid extraction of antioxidant compounds from oregano, chemical and functional characterization via LC-MS and in vitro assays. J Supercrit Fluids 2006; 38:62-6.
39. Wu SJ, Tsai JY, Chang SP et al. Supercritical carbon dioxide extract exhibits enhanced antioxidant and anti-inflammatory activities of Physalis peruviana. J Ethnopharmacol 2006; 108:407-413.
40. Celiktas OY, Bedir E, Sukan FV. In vitro antioxidant activities of Rosmarinus officinalis extracts treated with supercritical carbon dioxide. Food Chem 2007; 101(4):1457-64.
41. Caruso JL, Callahan J. Carnosic acid in green callus and regenerated shoots of Rosmarinus officinalis. Plant Cell Rep 2000; 19(5):500-3.
42. Modey WK, Mulholand DA, Raynor MW. Analytical supercritical fluid extraction of natural products? A review. Phytochem Anal 1996; 7:1-15.
43. Jarvis AP, Morgan ED. Isolation of plant products by supercritical-fluid extraction. Phytochem Anal 1997; 8:217-222.
44. Senorans FJ, Ibanez E. Liquid chromatographic-mass spectrometric analysis of supercritical-fluid extracts of rosemary plants. J Chromatogr A 2000; 870(1-2):491-499.
45. Carvalho RN, Moura LS, Rosa PTV et al. Supercritical fluid extraction from rosemary (Rosmarinus officinalis): kinetic data, extract's global yield. J Supercrit Fluids 2005; 35(3):197-204.
46. Reverchon E, Daghero J, Marrone C et al. Supercritical fractionation extraction of phenel seed oil and essential oil: experiments and mathematical modelling. Ind Eng Chem Res (381999) 3069-3075.

47. Reverchon E, Della Porta G, Taddeo R. Extraction of sage essential oil by supercritical CO_2: influence of some process parameters. J Supercrit Fluids 8 1995;302-309.
48. Carro N, García CM, Cela R. Terpenic compounds, responsible for a variety of aromas in musts and wines can be extracted with good recoveries using MAE. Analyst 1997; 122:325.
49. Chen SS, Spiro M. Study of microwave extraction of essential oil constituents from plant materials. J Micro Power Electromagn Energy 1994; 29:231-241.
50. Cavero S, García-Risco MR, Marína FR et al. Supercritical fluid extraction of antioxidant compounds from oregano: Chemical and functional characterization via LC-MS and in vitro assays, J Supercrit Fluids 2006; 38(1):62-69.
51. Rostagno MA, Palma M, Barroso CG. Microwave assisted extraction of soy isoflavones. Anal Chim Acta 2007; 588(2):274-282.
52. Mattina MJI, Berger WAI, Denson CL. Microwave assisted extraction of taxanes from Taxus biomass. J Agric Food Chem 1997; 45:4691-4696.
53. Trbová D, Matjíek D, Vlek J et al. Combined microwave-assisted isolation and solid-phase purification procedures prior to the chromatographic determination of phenolic compounds in plant materials. Anal Chim Acta 2004; 513(2):435-444.
54. Ferhata MA, Meklatia BY, Smadjab J et al. An improved microwave Clevenger apparatus for distillation of essential oils from orange peel, J Chromatogr A 2006; 1112(1-2):121-126.
55. Bertoli A, Pistelli L, Morelli I et al. Volatile constituents of micropropagated plants of Bupleurum fruticosum L. Plant Sci 2004; 167(4):807-810.
56. Banthorpe DV, Branch SA. Ability of plant callus cultures to synthesize and accumulate lower terpenoids. Phytochemistry 1986; 25(3):629-636.
57. Jain M, Banerji R. In vitro production of essential oil from proliferating shoots of Rosmarinus officinalis, Planta Medica 1991; 57(2):122-124.
58. Charron CS, Cantliffe DJ, Heath RR. Volatile emissions from plants. Hort Rev 1995; 17:43-72.
59. Maes K, Debergh PC. Volatiles emitted from in vitro grown tomato shoots during abiotic and biotic stress. Plant Cell Tissue Organ Cult 2003; 75(1):73-8.
60. Maes K, Vercammen J, Pham-Tuan H et al. Critical aspects for the reliable headspace analysis of plants cultivated in vitro. Phytochemical Analysis 2001; 12(3):153-8.
61. Exarchou V, Fiamegos YC, van Beek TA. Hyphenated chromatographic techniques for the rapid screening and identification of antioxidants in methanolic extracts of pharmaceutically used plants. J Chromatogr A 2006; 1112(1-2):293-302.
62. Stahl E. Thin Layer Chromatography: A Laboratory Handbook. San Diego: Academic Press, 1965:485-502.
63. Pothier J, Galand N, Ouali M et al. Comparison of planar chromatographic methods (TLC, OPLC, AMD) applied to essential oils of wild thyme and seven chemotypes of thyme. Il Farmaco 2001; 56:505-511.
64. Galand N, Pothier J, Dollet J et al. OPLC and AMD, recent techniques of planar chromatography: their interest for separation and characterization of extractive and synthetic compounds. Fitoterapia 2002; 73(2):121-34.
65. Pasqua G, Avato P, Monacelli B et al. Metabolites in cell suspension cultures, calli and in vitro regenerated organs of Hypericum perforatum cv. Topas. Plant Sci 2003; 165:977-982.
66. Kevers C, Jacques P, Gaspar T et al. Comparative titration of ginsenosides by different techniques in commercial ginseng products and callus cultures. J Chromatogr Sci 2004; 42(10):554-8.
67. Bondarev N, Reshetnyak O, Nosov A. Peculiarities of diterpenoid steviol glycoside production in in vitro cultures of Stevia rebaudiana Bertoni. Plant Sci 2001; 161:155-163.
68. Botz L, Nagy S, Kocsis B. Planar Chromatography: A Retrospective View for the Third Millennium. Budapest: Springer, 2001:103.
69. Nyiredy S. The bridge between TLC and HPLC: overpressured layer chromatography (OPLC). Trends Anal Chem 2001; 20:91-101.
70. Tabanca N, Demirci B, Baser KHC et al. Characterization of volatile constituents of Scaligeria tripartita and studies on the antifungal activity against phytopathogenic fungi. J Chromatogr. J Chromatogr B Analyt Technol Biomed Life Sci 2007; 850(1-2):221-229.
71. Cheel J, Schmeda-Hirschmann G, Jordan M et al. Zeitschrift für Naturforschung. C, Free radical scavenging activity and secondary metabolites from in vitro cultures of Sanicula graveolens. J Biosci 2007; 62:555-62.
72. Meloan CE. Chemical Separations: Principles, Techniques and Experiments. Canada: Wiley and Sons, 1999.
73. Sakakibara H et al. Simultaneous determination of all polyphenols in vegetables, fruits and teas. J Agric Food Chem 2003; 51:571-581.

74. Tanaka N, Kobayashi H, Ishizuka N et al. Monolithic silica columns for high-efficiency chromatographic separations. J Chromatogr A 2002; 965:35-49.
75. Tanaka N, Kimura H, Tokuda D et al. Simple and comprehensive two-dimensional reversed-phase HPLC using monolithic silica columns. Anal Chem 2004; 76:1273-1281.
76. Sook YL, Hui X, Yong KK et al. Rosmarinic acid production in hairy root cultures of Agastache rugosa Kuntze. World J Microbiol Biotechnol 2008; 24:969-972.
77. Cheel J, Schmeda-Hirschmann G, Jordan M et al. Free radical scavenging activity and secondary metabolites from in vitro cultures of Sanicula graveolens. Z Naturforsch C 2007; 62(7-8):555-62.
78. Łuczkiewicz M, Głod D. Morphogenesis-dependent accumulation of phytoestrogenes in Genista tinctoria in vitro cultures. Plant Sci 2005; 168:967-979.
79. Thiem B. In vitro propagation of isoflavone-producing Pueraria lobata (Willd.) Ohwi. Plant Sci 2003; 165(5):1123-8.
80. Charchoglyana A, Abrahamyana A, Fujiib I et al. Differential accumulation of hyperforin and secohyperforin in Hypericum perforatum tissue cultures. Phytochemistry 2007; 68:2670-2677.
81. Conceição LFR, Ferreres F, Tavares RM et al. Induction of phenolic compounds in Hypericum perforatum L. cells by Colletotrichum gloeosporioides elicitation. Phytochemistry 2006; 67(2):149-155.
82. Pasqua G, Monacelli B, Valletta A. Cellular localization of the anti-cancer drug camptothecin in Camptotheca acuminata. Eur J Histochem 2004; 48:321-328.
83. Pasqua G, Silvestrini A, Monacelli B et al. Triterpenoids and ellagic acid derivatives from in vitro cultures of Camptotheca acuminata Decaisne. Plant Physiol Biochem 2006; 44(4):220-225.
84. Tolstikov VV, Fiehn O. Analysis of highly polar compounds of plant origin: Combination of hydrophilic interaction chromatography and electrospray ion trap mass spectrometry. Anal Biochem 2002; 301:298-307.
85. Tolstikov VV, Lommen A, Nakanishi K et al. Monolithic silica-based capillary reversed-phase liquid chromatography/electrospray mass spectrometry for plant metabolomics. Anal Chem 2003; 75:6737-40.
86. Wolfender JL, Rodriguez S, Hostettmann K. Liquid chromatography coupled to mass spectrometry and nuclear magnetic resonance spectroscopy for the screening of plant constituents. J Chromatogr A 1998; 794:299-316.
87. Cole RB. Cole RB. Electrospray Ionization Mass Spectrometry-Fundamentals. Instrumentation and Applications. New York: Wiley, 1997.
88. Lee JS, Kim DH, Liu KH et al. Identification of flavonoids using liquid chromatography with electrospray ionization and ion trap tandem mass spectrometry with an MS/MS library. Rapid Commun Mass Spectrom 2005; 19(23):3539-3548.
89. Tolonen A, György Z, Jalonen J et al. LC/MS/MS identification of glycosides produced by biotransformation of cinnamyl alcohol in Rhodiola rosea compact callus aggregates. Biomed Chromatogr 2004; 18(8):550-8.
90. Gyorgy Z, Tolonen A, Pakonen M et al. Enhancing the production of cinnamyl glycosides in compact callus aggregate cultures of Rhodiola rosea. Plant Sci 2004; 166(1):229-36.
91. Novakova L, Matysova L, Solich P. Advantages of application of UPLC in pharmaceutical analysis. Talanta 2006; 68(3):908-18.
92. Gruza J, Novák O, Strnad M. Rapid analysis of phenolic acids in beverages by UPLC-MS/MS. Food Chem 2008; 111(3):789-794.
93. De Villiers A, Lestremau F, Szucs R et al. Evaluation of ultra performance liquid chromatography. J Chromatogr A 2006; 1127(1):60-9.
94. Guan J, Lai CM, Li SP. A rapid method for the simultaneous determination of 11 saponins in Panax notoginseng using ultra performance liquid chromatography J Pharm Biomed Anal 2007; 44:996-1000.
95. Chena XJ, Jic H, Zhang QW et al. A rapid method for simultaneous determination of 15 flavonoids in Epimedium using pressurized liquid extraction and ultra-performance liquid chromatography. J Pharm Biomed Anal 2008; 46:226-235.
96. Xiangyu D, Guihua G, Shuning Z et al. Qualitative and quantitative analysis of flavonoids in the leaves of Isatis indigatica Fort. by ultra-performance liquid chromatography with PDA and electrospray ionization tandem mass spectrometry detection. J Pharm Biomed Anal 2008; 48:562-567.
97. Magnotta M, Murata J, Chen J et al. Expression of deacetylvindoline-4-O-acetyltransferase in Catharanthus roseus hairy roots. Phytochemistry 2007; 68:1922-31.
98. Adams RP. Identification of Essential Oil Components by Gas Chromatography-Mass Spectroscopy. Carol Stream: Allured Publ Corp, 1995.
99. Schauera N, Steinhausera D, Strelkovb S et al. GC-MS libraries for the rapid identification of metabolites in complex biological samples. FEBS Lett 2005; 579:1332-1337.
100. Schauera N, Steinhausera D, Strelkovb S et al. Strawberry flavour: analysis and biosynthesis. J Sci Food Agric 1997; 74:421-434.

101. Khaled MS, Salem A, Charlwood BV. Accumulation of essential oils by Agrobacterium tumefaciens-transformed shoot cultures of Pimpinella anisum. Plant Cell Tissue Organ Cult 1995; 40(3):75-93.
102. Gbolade AA, Lockwood GB. Volatile constituents from parsley cultures. Flavour and Fragrance Journal 2006; 4(2):69-71.
103. Figueiredo AC, Almendra MJ, Barroso JG et al. Biotransformation of monoterpenes and sesquiterpenes by cell suspension cultures of Achillea millefolium L. ssp. Millefolium. Biotechnol Lett 1996; 18:8.
104. Gbolade AA, Lockwood GB. XIX Petroselinum crispum (Mill.) Nyman (Parsley), in vitro culture, production and metabolism of volatile constituents. In: Bajaj YPS, ed. Biotechnology in Agriculture and Forestry-Medicinal and Aromatic Plants. Berlin: Springer-Verlag, 1999; 43:324-336.
105. Nogueira JF, Romano A. Essential oils from micropropagated plants of Lavandula viridis. Phytochem Anal 2002; 13(1):4-7.
106. Zhu W, Asghari G, Lockwood GB. Factors affecting volatile terpene and nonterpene biotransformation products in plant cell cultures. Fitoterapia 2000; 71(5):501-6.
107. Mauri P, Pietta P. Electrospray characterization of selected medicinal plant extracts. J Pharm Biomed Anal 2000; 23:61-68.
108. Sterner JL, Johnston MV, Nicol GR et al. Signal suppression in electrospray ionization Fourier transform mass spectrometry of multi-component samples. J Mass Spectrom 2000; 35:385-391.
109. Favretto D, Piovan A, Filippini R et al. Monitoring the production yields of vincristine and vinblastine in Catharanthus roseus from somatic embryogenesis. Semiquantitative determination by flow-injection electrospray ionization mass spectrometry. Rapid Comm Mass Spec 2001; 15:364-369.
110. Higgs RE, Zahn JA, Gygi JD et al. Rapid method to estimate the presence of secondary metabolites in microbial extracts. Appl Environ Microbiol 2001; 67:371-376.
111. Zahn JA, Higgs RE, Hilton MD. Use of direct-infusion electrospray mass spectrometry to guide empirical development of improved conditions for expression of secondary metabolites from actinomycetes, Appl Environ Microbiol 2001; 67:377-386.
112. Goodacrea R, York EV, Heald JK et al. Chemometric discrimination of unfractionated plant extracts analyzed by electrospray mass spectrometry. Phytochemistry 2003; 62(6):859-863.

Chapter 20

Biosensors for Functional Food Safety and Analysis

Teresa Lavecchia,* Arianna Tibuzzi and Maria Teresa Giardi

Abstract

The importance of safety and functionality analysis of foodstuffs and raw materials is supported by national legislations and European Union (EU) directives concerning not only the amount of residues of pollutants and pathogens but also the activity and content of food additives and the health claims stated on their labels. In addition, consumers' awareness of the impact of 'functional foods' on their well-being and their desire for daily healthcare without the intake pharmaceuticals has immensely in recent years. Within this picture, the availability of fast, reliable, low cost control systems to measure the content and the quality of food additives and nutrients with health claims becomes mandatory, to be used by producers, consumers and the governmental bodies in charge of the legal supervision of such matters. This review aims at describing the most important methods and tools used for food analysis, starting with the classical methods (e.g., gas-chromatography GC, high performance liquid chromatography HPLC) and moving to the use of biosensors—novel biological material-based equipments. Four types of biosensors, among others, the novel photosynthetic proteins-based devices which are more promising and common in food analysis applications, are reviewed. A particular highlight on biosensors for the emerging market of functional foods is given and the most widely applied functional components are reviewed with a comprehensive analysis of papers published in the last three years; this report discusses recent trends for sensitive, fast, repeatable and cheap measurements, focused on the detection of vitamins, folate (folic acid), zinc (Zn), iron (Fe), calcium (Ca), fatty acids (in particular Omega 3), phytosterols and phytochemicals. A final market overview emphasizes some practical aspects of biosensor applications.

Introduction to Safety and Functionality Analysis in Food

The term 'functional food' was first used in Japan to indicate food having a high content of natural metabolites with recognized health benefits such as, for instance, anticancer, antilipidemic, anticholesterol, antimicrobial, antibacterial, antifungal, antiviral, antihypertensive, antiinflammatory and antioxidant properties. From Japan, the concept of "functional food" spread to Europe and the United States (US) and, in 1999, the European Commission's Concerted Action on Functional Food Science in Europe, defined that: "a food product can only be considered functional if together with the basic nutritional impact it has beneficial effects on one or more functions of the human organism thus either improving the general and physical conditions or/and decreasing the risk of the evolution of diseases". The amount of intake and form of the functional food should be as normally expected for dietary purposes. Therefore, it should not be in the form of pill or capsule but

*Corresponding Author: Teresa Lavecchia—Biosensor S.R.L, Via le Tivoli, 00018 Palombara Sabina, Rome, Italy. Email: t.lavecchia@biosensor.it

Bio-Farms for Nutraceuticals: Functional Food and Safety Control by Biosensors edited by Maria Teresa Giardi, Giuseppina Rea and Bruno Berra.

as normal food.[1] For many years, companies attempting to launch functional food in Europe have faced a variety of legislative frameworks regulating the approval of products. With the European Parliament's second reading of 2003 this situation changed. According to the new regulation on nutrition and health claims made on foods (EC No. 1924/2006), a list of authorised claims will be published for all member states and nutrient profile will be required for food making health claims. These new trends on the market along with the requirements and regulations in the fields of environmental protection, control of biotechnological processes and certification of food and water quality are the reason for the growing demand on rapid, sensitive and inexpensive measurements for foodstuffs analysis.

The automation of industrial manufacture of products used in large quantities in everyday life, such as human and veterinary pharmaceuticals, plasticizers and various industrial additives have stimulated governments, researchers, health professionals and food industries to avoid the use of substances hazardous to human health. Some contaminants are formed naturally and by human activities carried over from water, air or soil into the food chain or created as by-products of the food production process itself. For example, acrylamide has been found in potato crisps resulting from cooking practices and aflatoxin, a toxin produced by fungi, has been found sometimes in nuts. Chemicals are, sometimes, improperly added to foodstuffs to alter their characteristics and encourage their purchase. An extraordinary example is the recent scandal in China, in 2008, concerning the addition of melamin, an industrial chemical used to make plastic receptacles to milk and infant formulas. Due to China's wide range of export food products, this problem affected countries on all continents with an estimated number of 94,000 victims claimed. The chemical was added to milk in order to mimic a higher content of proteins. In the same period, the Italian government prohibited the sale of "mozzarella" cheese because of dioxin contamination.[2] Dioxins are fat soluble compounds that tend to accumulate in higher animals, including humans, with effects on health such as impairment of the immune system, nervous system, hormonal systems and reproductive functions; these chemicals are also suspected of causing cancer. On the subject of food contamination, EU legislation ratifies that food containing a level of contaminant that is unacceptable from a public health viewpoint, in particular at the toxicological level, cannot enter the market. Therefore, maximum levels have been established for the contaminants of greatest concern to consumers, either due to their toxicity or their potential prevalence in the food chain. These include toxins, heavy metals, dioxins, nitrates and pesticides.

Related to the problem of food contamination is the new topic of "functional food". Generally, in the food industry, the quality of a product is evaluated through periodic chemical and microbiological analyses. The adopted procedures conventionally use physicochemical, biological and chemical and serological techniques (i.e., chromatography, spectrophotometry, electrophoresis, titration and others) to identify contaminants. Although these methods assure excellent results, they do not allow easy continuous monitoring, because they are expensive, slow, need well trained operators and in some cases, require several extraction steps or sample pretreatment, increasing the analysis time. In the subsequent sections, the classical techniques used for food analysis are briefly described, followed by a thorough description of biosensor technology and detection methods, with a particular focus on biosensors for functional foods analysis. A final overview of the biosensor market demonstrates their wide industrial applicability and potential business volume.

Classical Methods for Food Analysis: Chemical and Microbiological Methods

The physicochemical, biological and serological tests techniques (i.e., chromatography, spectrophotometry, electrophoresis, titration and others) have often been used to identify contaminants. Chemicals are generally analysed using GC or HPLC. The methods aim at the separation of the components of a complex sample and their identification through specific detectors: the most commonly used of which are flame ionization (FID) and thermal conductivity (TCD) for GC; ultraviolet light (UV), fluorescence (FL) or mass spectrometry (MS) for HPLC. The Polymerase Chain Reaction (PCR) and immunology-based methods are the most common tools used for pathogen detection, involving DNA analysis and counting of bacteria and antigen–antibody interactions, respectively.[3] PCR is based on the amplification of short DNA sequences generating millions or more copies of the DNA fragments, subsequently detected by gel electrophoresis.

Direct counts of microorganisms by microscope are possible but because of the difficulty arising from their small size, an easier method consists of spreading the sample over a wide area (i.e., nutrient agar plate) and counting the number of colonies that grow. Nevertheless, alternative techniques are necessary because culturing methods are excessively time-consuming. This is an obvious inconvenience in many industrial applications, particularly in the foods sector. The field of immunology-based methods for bacteria detection provides a very powerful method: the enzyme-linked immunosorbent assay test (ELISA)[4] is the most established technique combining the specificity of antibodies and the sensitivity of simple enzyme assays by using antibodies or antigens coupled to an easily assayed enzyme. Older ELISAs use chromogenic substrates, though newer assays employ fluorogenic substrates enabling much higher sensitivity.

Biosensor Technology

Biosensor technology offers a useful alternative to facilitate routine analysis of industrial products. A wide range of applications is targeted by the producer industry such as the agricultural sector, veterinary analysis, industry for drinkable products, fermentation industry, waste water management, environmental pollution monitoring, microbial contamination, clinical diagnosis, drug monitoring and analysis, mining, military and defence, aerospace personnel safety and many others.

A biosensor is an analytical device composed of a sensitive biological element in intimate contact with a physicochemical transducer (Fig. 1).

Biosensors can be classified according to the type of the involved biocomponent, mechanism or mode of signal transduction: sometimes, all these features are considered. The biological elements can be systems containing enzyme(s) (mono or multienzyme), plant or animal tissues, microorganisms, organelles, cell receptors, antibodies, nucleic acids, etc. The transduction elements are generally electrochemical, optical or piezoelectric. The electrical signals can be based on a change in the measured current at a fixed applied bias (amperometric), a change in the measured voltage between the electrodes (potentiometric), a change in the ability to transport charge (conductometric). The

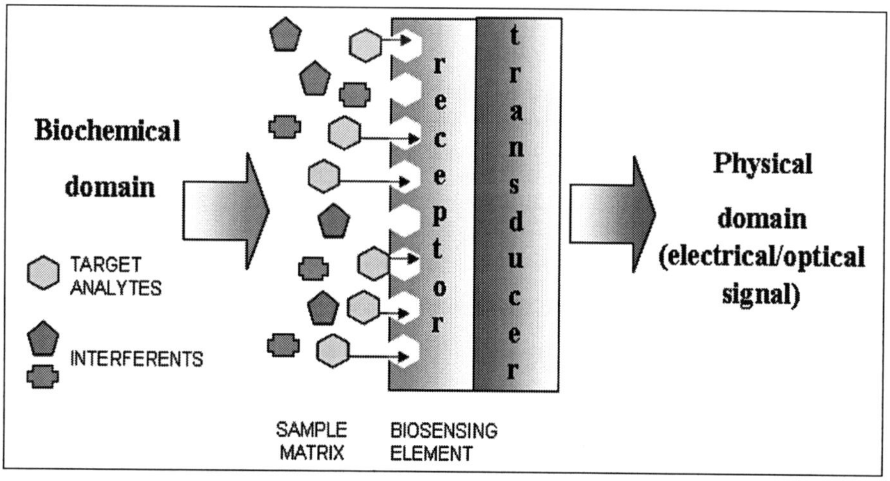

Figure 1. Scheme of a biosensoristic device. The biological sensing element is in contact with the surface of a physicochemical transductor. When the device is exposed to a sample containing the target analyte, a biochemical reaction occurs producing a signal converted by the transductor and detected as an output in the physical domain.

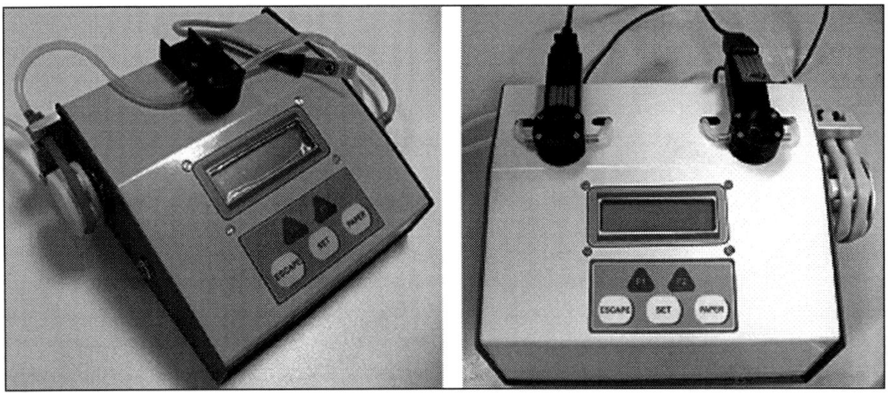

Figure 2. Pictures of the compact amperometric biosystem (single cell and two parallel cells model) that integrates biologic material sample holder and measurement chamber with electronics and automatic fluidic. The instrument is produced by the company Biosensor Srl (Palombara Sabina-Rome, Italy).

optical sensors generally measure absorbance, fluorescence, chemiluminescence, surface plasmon resonance or changes in light reflectivity. Mass sensors produce a signal based on the mass of chemicals that interact with the sensing film. Arrays of many different detector molecules have been applied in so-called electronic nose devices, where the pattern of response from the detectors is used to fingerprint a volatile complex mixture. Current commercial electronic noses, however, use organic receptors more than biological material.

The recent scientific literature reports that the enzyme-based biosensors dominating the market in the last three years in the food sector are related to electrochemical transduction systems. Most of the cited articles[5-9] concern biosensor devices for detection of pesticides, toxins and microorganisms. Frequent applications are also documented for glucose.

Despite the enormous diversity of research involving biosensors, applications in the food industry are still sparing: since 1998 no more than 100 articles have been published containing the word 'biosensors' and 'food', in 2007 only four appeared. Nevertheless, amperometric biosensors based on tyrosinase and laccase enzymes have been reported for the detection of polyphenol content in foodstuffs and beverages, such as olive oil,[10] wine,[11] tea[12] and biscuits.[13] Novel portable and versatile amperometric biosystems have been developed for phenol analysis and launched on the market to provide an instrument based on the screen printed electrode (SPE) technology able to work with different samples and experimental protocols[14] (Fig. 2).

Recent trends in biosensor technology show that four types of biosensors are going to dominate the scene in the field of foodstuffs analysis: (1) antibody or antigen-based biosensors, (2) Deoxyribonucleic acid (DNA)-based sensors, (3) whole cell-based biosensors and (4) photosynthetic protein-based biosensors.

Antibody or Antigen-Based Biosensors

Antibody or antigen-based biosensors are also known as immunosensors. They can be considered a modified version of the enzyme-linked immunosorbent assay (ELISA) test. The latter is source of inspiration for many biosensor applications. Immunosensors detect the tiny changes that occur when an antibody binds to an antigen. The biosensor merely replaces the traditional colorimetric detection system of the ELISA test, in order to increase the working range, speed and sensitivity. In a simple immunosensor the transducer surface is coated or immobilized with antigen or antibody. An excess of specific antibody-enzyme conjugate or antigen-enzyme is spread on the surface and allowed to bind. Later, the analyte solution interacts with the surface and binds only to the target molecules. Unbound material is washed off and discarded. The amount of

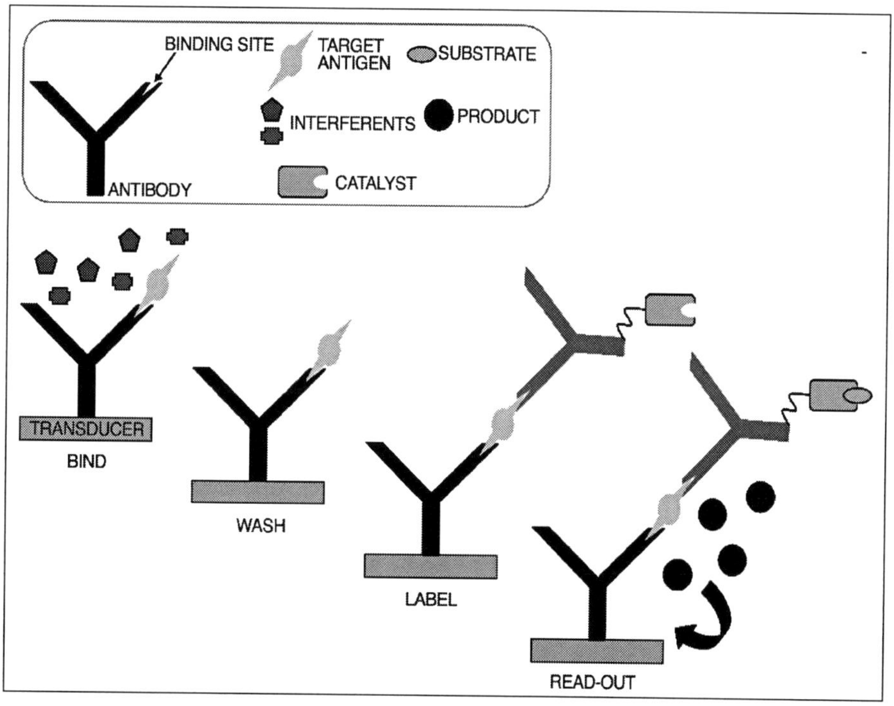

Figure 3. Schematic of an immunosensor: the immobilized antibody is able to specifically recognize the target antigen in a complex sample; after washing out the interferents, a specific antibody-enzyme conjugate complex binds to the target molecule to enable the biological event transduction. The enzyme is responsible to initiate the reaction from an added substrate which produces biomolecules to be directly detected by the electrochemical/optical transducer.

antibody-enzyme conjugate released or antigen-enzyme conjugate bound is determined directly from the transduced signal (Fig. 3).

Antigen coated surfaces are generally preferred because a disadvantageous orientation of antibody molecules onto the surface can cause loss of activity in the antibody. Frequently, the antigens are conjugated with a protein which favors the immobilized antigen and the interaction with the specific antibody-enzyme conjugate. Antigen-antibody biosensors have been reported in the detection of pesticides,[15] veterinary drugs,[16] steroids,[16-17] pathogenic bacteria and their toxins.[16,18]

Whole Cell-Based Biosensors

Cell-Based Biosensors (CBBs) utilize the physiological response of living cells (changes in behavior, metabolism, or induction of cell death) to detect biologically active agents. As detectors, CBBs exploit the natural sensitivity of cells to a wide range of biochemical stimuli making them well-suited for applications requiring functional screening of unknown agents such as environmental monitoring or drug discovery. In contrast to molecule-based biosensing technologies which tend to identify specific target analyte(s) in an unknown matrix, whole cells-based biosensors aim at investigating specific or global biological effects induced by analytes such as gene expression, metabolic activity, viability, bioavailability, toxicity and genotoxicity.[19] At the moment, CBB application in the fields of foodstuffs and nutraceutical analysis is still limited: a few articles report examples in which yeast cells are used as biomediators to detect ethanol,[20-21] lactic acid[22] and carbohydrates.[23]

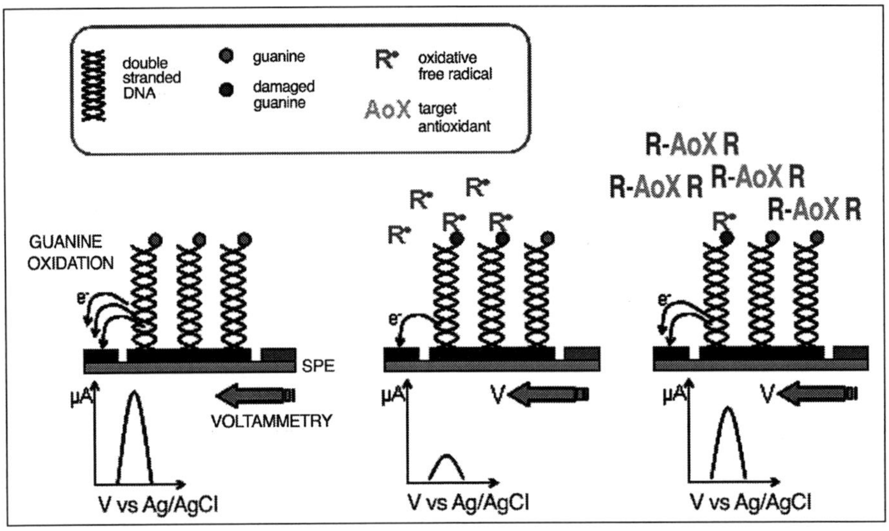

Figure 4. Schematic of a DNA-based biosensor for antioxidant content measurement: left) DNA strands are immobilized on a Screen-Printed Electrode (SPE), that collects the current generated by the oxidation of guanine bases when a linear voltage gradient is applied; centre) guanine bases are damaged by oxidative agents, thus reducing the current signal; right) a sample containing antioxidant agents scavenge the oxidant molecules preserving the guanine bases and (partially) restoring the current output.

DNA-Based Biosensors

DNA-based biosensors are obtained by immobilizing well-defined sequences of DNA single strands onto an electrode surface. The DNA structure is a double helix consisting of two strains made from repeating units called nucleotides. The nucleotides contain both the segment of the backbone of the molecule, which holds the chain together and a base, which interacts with the other DNA strand in the helix. Each type of base on one strand forms a bond with just one type of base on the other strand, which is called complementary base-pairing. The base-pairing gives the ability of one single strand to recognize its complementary strand and underlies the principle of DNA-based biosensors detection. If a well-defined DNA sequence (probe) is immobilized onto a surface and an unknown DNA sequence is added, the formation of a thermodynamically stable complex can occur. If the unknown sequence combines with the probe, for the principle of base-pairing, recognition occurs and the detection and identification are possible. Because of these properties, DNA-based biosensors are very promising for food analysis, particularly for the detection of pathogenic microorganisms. Examples concerning DNA-based biosensor applied to food samples have been reported to detect bacteria and pathogenic fungi such as *Escherichia coli,*[24-26] *Salmonella spp,*[27] *Listeria monocytogenes,*[28] *Fusarium culmorum*[29] and to evaluate the antioxidant power of different plant extracts[30,31] (Fig. 4). In some cases DNA strands have simply been co-immobilized with the biomediator on the surface of electrochemical devices to improve stability and sensitivity.[32]

Photosynthetic Protein-Based Biosensors

Promising biotools in sensor applications are novel devices using intact cells of photosynthetic microorganisms (cyanobacteria and algae) and plant components (thylakoids or isolated Photosystem II-enriched particles) as biorecognition elements. Cyanobacteria, algae and thylakoid membranes of many higher plants contain the photosynthetic complex Photosystem II (PSII). PSII is a light-driven oxidoreductase protein complex enabling photosynthesis by means of a photoinduced electron transfer from water to plastoquinones. In the photosynthetic process, solar photons are captured by the antenna chlorophyll pigments surrounding the PSII reaction centers and start a cascade of oxido-reduction reactions ending with the reduction of the terminal acceptor quinone B (Q_B), located within the protein subunit D1 of PSII, in the Q_B pocket. The photosynthetic process can be monitored by immobilizing cyanobacteria, algae or thylakoid membranes onto an electrode surface and registering the current that flows out after photostimulating the biomediator by means of a Light Emitting Diode (LED) light. This is the set-up of a photosynthetic protein-based biosensor. These devices are particularly useful for pesticides and heavy metal detection and many applications are available especially for water monitoring. A wide variety of photosynthetic biosensors in the recent scientific literature has been reported by Campàs et al[33] and Rouillon et al.[34] Giardi and coworkers[14,35] showed that recent advances in the molecular biology of the green alga *Chlamydomonas reinhardtii*, by means of direct site mutagenesis, enable punctual amino acid substitutions in the sequence of the reaction center D1 protein, producing particularly specific and sensitive mutants. Depending on the position and type of amino acid substitution, chemically different herbicides showed differential affinity for the D1 binding niche. This result allowed the recognition of different subclasses of pesticides offering useful developments for analysis of effluent, irrigation and drinking water as well as foodstuffs and functional food.

Functional Foods

The most common functional foods are those fortified with vitamins and/or minerals such as retinol (vitamin A), tocopherol (vitamin E), folic acid, zinc, iron and calcium.[36] More recently, however, attention has moved to food fortified with various nutrients such as omega-3 fatty acids, phytosterols and phytochemical substances having antioxidant properties such as carotenoids, phenolic compounds (flavonoids, phenolic acids, catechins, stilbenes, curcumin etc.), phytoestrogens (isoflavones, lignans) or anticarcinogenic agents such as limonene, allicin, glucosinolates and capsaicinoids.[37-38] All these substances are able to promote good health, but, if in excess, in some cases, they can result in the opposite effect: vitamins C and E protect cells against damage by free radicals thanks to their antioxidant properties, but, vitamin E excess may increase the risk of bleeding, particularly for adults who are also taking an anticoagulant. Occasionally, adults who utilized very high doses developed muscle weakness, fatigue and nausea.

Table 1 summarizes the main substances or elements added to foodstuff for nutraceutical purposes, their effect on the human body and the methods used for their detection.

Examples of Biosensors for Functional Food Analysis

In the matter of 'functional food' scarce progress has been observed in the field of biosensors. Nevertheless, the most significant examples are reported in the following section, listed according to the functional ingredient or active principle under detection.

Biosensors for Vitamins, Folic Acid and Fatty Acids

Vitamin content in food products derives from naturally available vitamins and their supplementary addition. The labeled content of vitamins on food products generally refers to the total level of vitamins, including natural and added vitamins. Commonly used methods for vitamin detection such as microbiology or HPLC, in many cases, are used to determine the fortified level of vitamins where the natural level of vitamins is negligible. Sample preparation is time-consuming and a common protocol for different vitamins in complex matrices is not yet available. Few articles report the determination of vitamins through biosensor-based methods, in contrast, many articles deal with applications for water-soluble vitamins, because fat-soluble vitamins (such as vitamin A and E) analysis is more complicated. For such analyses, organic solvents are required to dissolve the sample, often being incompatible with the biological mediator. Moreover, biospecific recognition elements for fat-soluble vitamins are not available. Concerning the scientific literature, some articles report about RBP detection and the study of its active site[39-42] and in only one paper this protein is

Table 1. Nutraceutical elements added to foodstuffs and detection method

Nutraceutics and other Compounds	Effects on Human Body			
	Deficiency	**Excess**	**Common Detection Methods**	**Biosensor Methods**
Retinol	Irreversible blindness	High brain pressure	HPLC-UV	RBP-based optical biosensors for retinol content evaluation
Tocopherol	Degeneration of neurons.	Risk of stroke.	HPLC-UV	DNA-based electrochemical biosensors for tocopherol content
Folate	Dementia	No toxic effects	Folate conjugase microbiological assay, HPLC and LC/MS	Biosensor-based immunoassay and SPR optical biosensors for folate content
Metal ions: zinc (Zn), iron (Fe), calcium (Ca)	Zn: impairment of the body's immune system	Zn: damage of the immune system	Zn, Fe: AA (cold vapour technique) and ICP-MS	Fe: Lactoferrin and phytochelatin-based electrochemical, fluorescent siderophore bacteria-based optical biosensors
	Fe: anemia.	Fe: damage to the intestine and coronary arteries	Ca: AA	Zn: Alkaline phosphatase, metallothionein and phytochelatin-based electrochemical biosensors,
	Ca: abnormal heart rhythms	Ca: abnormal heart rhythms and kidney irreversible damage		Ca: for human tissues and organelles detection, phosphotyrosine-based, electrochemical biosensors. FRET chelating fluorescent protein-based and SPR albumin-based optical biosensors
Omega-3 fatty acids	Heart problems	Brain artery leakage	GC analysis	NF
Phytosterols	Hypercholester olemia and cardiovascular disease	No toxic effects	GC and HPLC	NF
Carotenoids	Exposure to DNA damage and cancer	Toxic effects for smokers.[35]	HPLC-UV/VIS, HPLC-MS/MS, LC-MS/MS	NF

continued on next page

Table 1. Continued

Nutraceutics and other Compounds	Effects on Human Body		Common Detection Methods	Biosensor Methods
	Deficiency	Excess		
Polyphenols	Exposure to oxidative stress and DNA damage	Reduction of iron absorption.	LC-UV, HPLC-UV/VIS, HPLC-CL, HPLC-MS/MS, CE-CL	Superoxide dismutase, peroxidase, cytochrome C and DNA-based electrochemical biosensors for polyphenol antioxidant power Tyrosinase, laccase, peroxidase and composite peroxidase/DNA-based electrochemical biosensors for polyphenol content
Phytoestrogens	Risk of cardiovascular problems, cancer and osteoporosis	Hormonal imbalance	Common methods: GC-MS, HPLC-UV, HPLC-FL, HPLC-ED, HPLC-MS, CE-UV, CE-FL, CE-ED, CE-MS, immunoassays.	NF
Limonene	Risk of Cancer	No toxic effects	GC-MS, GC–FID, GC-O, PTR-MS	NF
Allicin	Risk of cardiovascular problems and cancer.	No toxic effects	HPLC-UV, HPLC-FL, HPLC-MS, HPLC-ED, HPLC-ICP-MS, HPLC-ESI-MS/MS	NF
Glucosinolates	Risk of carcinomas and cancer.	Thyroid gland disorders	LC-MS/MS, CE-LIF, HPLC-ESI-MS/MS, LC-MS/MS, LC-APCI-MS, HPLC-MS, LC/MS.	Composite myrosinase/glucoseoxidase-based electrochemical biosensors and myrosinase-based optical biosensors
Capsaicinoids	Risk of carcinomas	No toxic effects for ingestion. Irritating action on skin, eyes and lungs.	HPLC-FL, LC/MS.	NF

RBP: retinol binding protein; LC: liquid chromatography; SPR: surface plasmon resonance; AA: atomic absorption (spectrophotometry); ICP: inductively coupled plasma; FRET: fluorescence resonance energy transfer; UV-VIS: ultraviolet-visible light (spectroscopy); MS/MS: tandem mass spectrometry; CE: capillary electrophoresis; CL: chemiluminescence; ED: electrochemical detection; O: olfactometry; PTR: proton transfer reaction; ESI: electrospray ionization; LIF: laser-induced fluorescence); APCI: atmospheric pressure chemical ionization; NF: not found.

used as the biomediator for vitamin A recognition[43] Concerning vitamin E detection, one article deals with tocopherol recognition by means of a DNA-modified screen-printed carbon electrode.[44]

Folic acid (also known as vitamin M and folacin) and folate (the anion) are forms of the water-soluble Vitamin B9. Folate and its derivatives occur in nature as polyglutamate with 5-methyltetrahydrofolate being the most abundant. The traditional folate extraction method involves two steps including heat treatment to release folate from its binding proteins and folate conjugase treatment, to hydrolyze polyglutamyl folate. In the 1990s a new method known as 'trienzyme-extraction' was developed for folate extraction from food to monoglutamyl folate. The method involves the use of α-amylase, protease and folate conjugase for a more complete extraction of folate trapped in the carbohydrate or protein matrices in food than the traditional method.[45] Current clinical methods for folate, however, give different results and cannot distinguish its forms. With regard to biosensor detection of folate, some articles discuss the use of optical biosensors to analyze folate in breakfast cereals, drinks, milk and soy-based infant formulas. At any rate, the signal transduction is obtained applying SPR technique. The method is based on the specific interaction between folate-binding protein (FBP) and pteroyl-L-glutamic acid (PGA). The difference between free FBP and associated FBP allows the PGA concentration to be determined.[46-48]

Biosensors for Zn, Fe and Ca

Conventional analytical techniques for heavy metals such as Zn and Fe are cold vapour AA spectrophotometry and ICP-MS. Light elements such as Ca are generally determined by simple AA spectrophotometry. These techniques are precise but mostly laboratory bound and suffer from the disadvantages of high cost and the need for trained personnel.[49] Extensive work has been conducted on biosensors for heavy metals detection; devices for Zn and Fe analysis have been developed using both electrochemical and optical transduction. Among the electrochemical biosensors the following were developed: for Zn, a conductometric alkaline phosphatase, for Fe, a potentiometric lactoferrin, for Zn, a voltammetric metallothionein and for Zn and Fe, a voltammetric phytochelatin biosensors.[50-52] Among the optical biosensors the following have been reported: devices using fluorescence microscopy and fluorescence quenching applied to siderophore bacteria biomediators[53,54] (for Fe) or using FRET and SPR applied to chelating fluorescent protein biomediators and albumin-based biosensors.[55,56] Concerning calcium detection, few biosensor applications consider foodstuffs analysis, most of them concerning detection in the form of calcium pantothenate[57] (vitamin B5). The reason is that calcium is added to nutraceutical foods in the form of bioavailable compounds such as calcium pantothenate (vitamin B5), calcium carbonate, calcium citrate, calcium lactate, calcium gluconate and others.[58-60] Because of the numerous forms in which calcium is added to food, it is easier for researchers to develop biosensors for calcium detection after its absorption in the human body[61-63] rather than to develop biosensors for the detection of each calcium compound.

Biosensors for Phytochemicals

Several phytochemical compounds are known in the field of nutraceutics for their health-supporting properties. Up to now, various biosensors have been developed for the detection of these compounds with the great majority of devices manufactured to quantitatively analyze specific molecules in unknown samples and/or their 'antioxidant power'. The major fraction of phytochemical biosensors production is aimed at the concentration of polyphenolic compounds and their use in biosensoristic applications as antioxidant shield. Few articles report on the total concentration of glucosinolates. The content of polyphenols has been evaluated using different kinds of biomediators such as tyrosinase, laccase, horseradish peroxidase (HRP) and horseradish peroxidase/DNA mixtures. Tyrosinase and laccase belong to the protein class of phenol oxidases; they can be used to detect phenolic compounds by catalyzing their oxidation using molecular oxygen. The 'quinone' species formed and liberated into the solution can be electrochemically rereduced leading to the possible development of an electrochemical sensor. HRP and HRP/DNA-based electrodes mixture belong to the protein class of peroxidase and detect phenolic compounds catalyzing their oxidation by means of hydrogen peroxide. DNA is added to improve biosensor stability and sensitivity. Also in this case quinone species are formed that can be rereduced on the electrode surface[64] and resulting in an amperometric device formed. In relation to the analysis of glucosinolates, biosensors are reported involving the use of myrosinase and glucose oxidase enzymes both for electrochemical and optical biosensors. The electrochemical device is created by immobilizing a mixture of myrosinase and glucose oxidase onto a Clark-type oxygen electrode. In this bienzymatic system, myrosinase catalyzes the hydrolysis of glucosinolates to glucose. Subsequently, the glucose is oxidized by glucose oxidase with the consumption of oxygen. The oxygen level is monitored and related to glucosinolates concentration.[65] The optical device utilizes analogous chemical reactions but the glucose is determined by using an oxygen-sensitive optode membrane.[66] With regard to biosensors for the determination of the antioxidant power of phytochemicals, recent papers show that most of the developed products measure the antioxidant power as the activity derived only from phenolic compounds or measured as the total content of all species able to scavenge free radicals. The reason is that antioxidant compounds are frequently present at the same time in foodstuffs, with polyphenols being the dominant species and they cannot be separated easily. Antioxidants are also naturally present inside the human body, since they act as a defense against free radicals generated during cellular metabolism. The overproduction of radicals leads to damage of the human cell structures, including lipids and membranes, proteins and DNA known as 'oxidative stress'. Biosensors measuring the antioxidant power of phytochemical compounds are based on Cytochrome c (Cyt c), DNA and superoxide dismutase protein[67] (SOD). These biosensors are electrochemical and use reactive oxygen species (ROS) including oxygen ions, radicals and peroxides, both inorganic and organic. Since ROS are highly reactive species and are characterized by a very short life-time they need to be produced 'in situ'. Cyt c and SOD-based sensors usually contain the xanthine oxidase enzyme (XOD) as the generator of the superoxide radical anion ($O_2\bullet-$). $O_2\bullet-$ reduces the Cyt c immobilized on the surface of an electrode that is suddenly re-oxidized when an oxidation bias is applied to the electrode. The current flow is proportional to the radical concentration and decreases in presence of antioxidant species. This allows the quantification of the antioxidant capacity.

DNA-based biosensors use a different mechanism of action. In this case, DNA strands are immobilized on an electrode surface and radical production is generally obtained by oxidizing a transition metal cation such as iron Fe(II), copper Cu(I) or chromium Cr(II) with hydrogen peroxide (H_2O_2) resulting in the formation of hydroxyl radicals ($\bullet OH$). This species reacts directly with DNA destroying the guanine bases present in the double helix. The antioxidant activity is followed by studying the variation of the oxidation peak of guanines in square wave voltammetry (SWV). The use of superoxide and DNA-based biosensors to detect the antioxidant capacity has been reported by Campanella et al[68,69] following the xanthine–xanthine oxidase system to analyze wines, various teas and herbal products such as camomille, dog rose and ginseng. Mello et al,[64,70] analyzed plant extracts from *Baccharis genstelloides*, *Peumus boldus*, *Foeniculum vulgare*, *Cymbopogom citrates*, *Camellia sinensis* and *Mentha piperita* based on the generation of OH radicals. The polyphenol, sulfite and ascorbic acid content of different wine samples was also examined using a tyrosinase biosensor, a sulfite oxidase biosensor and an ascorbate oxidase biosensor, respectively. Other articles by Cortina et al[26] and Litescu et al[71] report SOD, Cyt c and DNA-based antioxidant biosensors applied for the determination of selected radical species and for the development of possible models for "in vitro" assessment of antioxidant or radical-scavenger capacity of foodstuffs and raw materials. The reported biosensor applications pertain to the monitoring of $O_2\bullet-$, $\bullet OH$, nitric oxide radical ($NO\bullet$), adding knowledge and results to the numerous configurations recently appeared in the literature.

Up to now, no biosensors for fatty acids and phytosterols have yet been developed[72] and these coumpounds are still detected through the classical physicochemical methods. Fatty acids and phytosterols, for the most part, are detected using GC analysis.[73-74]

The Market for Biosensors

There are many advantages to using biosensors as analytical tools: high selectivity, specificity, relatively low cost of construction and storage, potential for miniaturization. Moreover, they enable easy automation, simple and portable equipment construction for fast analysis and can be used in quality control laboratories, in industrial facilities of raw material delivery or at critical stages along the food process chain. In the food industry, biosensor applications range from pesticides and heavy metal detection (such as atrazine, diuron, linuron, Pb, Cd, Hg) to veterinary drugs (such as hormones, antibiotics), pathogenic bacteria (*E. coli, L. monocytogenes*) and toxins detection (such as aflatoxins, ochratoxins). In environmental monitoring, water treatment is a critical means of pathogen defense, without which communities can suffer rapid large scale exposure if contamination occurs. To illustrate the urgency for increased food control, we note that each year about 5000 people die from *Salmonella* and/or *E. coli* induced food poisoning in the USA.[75] At the present, commercialization of biosensors is in stark contrast to the promise indicated by the research results and remain in the laboratories. The involvement of industry end-users in such technology transfer activities is crucial to reduce the time-to-market of the biosensors. The world biosensor market was $7.3 (US) billion in 2003 with the glucose biosensor being the most widely commercialized of all biosensors.[76] During the last four years, the direction of biosensor R and D has significantly changed in response to new biotechnological innovations and some market reports indicate that the global market for biosensors and other bioelectronics is expected to grow to 8.2 billion in 2009, with the total market potential for detection of pathogens in the USA having an annual growth rate of 4.5%.[74] Nowadays, many biosensors companies are involved in biosensor fabrication/marketing whereas others only provide the pertinent materials and instruments for biosensor fabrication. The most significant change in the market climate is due to biosensor technology penetration into nonmedical applications. The business and regulatory challenges of biosensor commercialization have pushed the market in unexpected directions. Despite medical applications—especially the "killer app" (glucose sensor)—has been the backbone of the biosensor industry, accounting for the majority of revenues; it can take five years and cost more than $40 million to get a medical sensor to the marketplace. Poorly capitalized sensor developers can go out of business before achieving commercialization. These circumstances have led some companies to divert some or all of their resources to developing biosensors for the industrial or environmental markets. The substantial overlap between medical and nonmedical sensing technology makes this possible. This new focus for commercial biosensors also includes the food industry and related applications on functional food analysis, which could become a profitable market niche.

Conclusion

The growing interest of consumers in the role of nutrition in health is the primary driver behind the success of the food market. Its increasing desire to be more proactive in optimising personal well-being is an additional driving force for the functional food market. Consumers' awareness of nutrition and healthy eating has resulted in the emergence of a wide variety of fortified foods in recent years, but, while much attention has been directed to this targeted nutrition in its attempts to create more specific health claims, only minor emphasis has been placed on safety and scientific analysis aspects. In this regard, specific, new technologies have been developed to examine functional food components and additives, the most appealing being biosensors. In the last three years, while a wide variety of biosensors has dominated the market to detect pesticides, heavy metals, pollutants and toxic compounds, no progress has been observed in the field of functional foods analysis. Despite technological progress, no biosensors have been yet developed for the easy detection of fatty acids and phytosterols; few attempts have been made to determine vitamins (frequently dealing only with applications for water-soluble vitamins) and no specific biosensors have been developed for phytochemical nutrients. The most promising input, in this sense, is shown by the biosensors developed for the determination of antioxidant power, since, ideally all plant bioactive compounds should be tested against the damaging effects of free radicals; moreover, the use of biosensors to evaluate the antioxidant status could provide an early, non-invasive indication of some diseases linked to oxidative stress or their progression. In this field, various biosensors have already been manufactured using superoxide dismutase protein (SOD), horseradish peroxidase protein (HRP), cytochrome C and DNA fragments as biomediators. The stability and sensitivity of these biosensors have been demonstrated in several systems. One future development probably with a significant impact in biosensor technology is the ability to detect radicals in the nanomolar range since often the radical species can be involved in the beginning of inflammatory and/or vascular diseases. Thus, more sensitive biosensors may have a distinguished future as diagnostic tools in medicine.

Acknowledgement

This work was supported by the NUTRA-SNACKS project funded within the EU Framework Programme 6.

References

1. Diplock AT et al. Scientific concepts of functional foods in Europe-consensus document. Br J Nutr 1999; 81(1):1-27.
2. Lyons PJ. The New York Times 2008.
3. Lazcka O, Del Campo FJ, Muñoz FX. Pathogen detection: A perspective of traditional methods and biosensors. Biosens and Bioelectron 2007; 22(7):1205-1217.
4. Crowther JR. Elisa: Theory and Practice, 1st ed., Humana Press, 1995:42.
5. Zhang S, Wang N, Yu H et al. Covalent attachment of glucose oxidase to an Au electrode modified with gold nanoparticles for use as glucose biosensor. Bioelectrochemistry 2005; 67(1):15-22.
6. Azevedo AM, Prazeres DM, Cabral JM et al. Ethanol biosensors based on alcohol oxidase. Biosens and Bioelectron 2005; 21(2):235-247.
7. Liu Y, Qu X, Guo H et al. Facile preparation of amperometric laccase biosensor with multifunction based on the matrix of carbon nanotubes–chitosan composite. Biosens and Bioelectron 2006; 21:2195-2201.
8. Prieto-Simón B, Fàbregas E. New redox mediator-modified polysulfone composite films for the development of dehydrogenase-based biosensors. Biosens and Bioelectron 2006; 22(1):131-137.
9. Mita DG, Attanasio A, Arduini F et al. Enzymatic determination of BPA by means of tyrosinase immobilized on different carbon carriers. Biosens and Bioelectron 2007; 23(1):60-65.
10. Capannesi C et al. Electrochemical sensor and biosensor for polyphenols detection in olive oil. Food Chemistry 2000; 71:553-562.
11. Junior ARS, Rebelo MJF. Biosensor for the polyphenolic content of wine determination. Portugaliae Electrochimica Acta 2008; 26:117-124.
12. Abhijith KS, Kumar PV, Kumar MA et al. Immobilised tyrosinase-based biosensor for detection of tea polyphenols. Anal Bioanal Chem 2007; 389:2227-2234.
13. Morales MD et al. A composite amperometric tyrosinase biosensor for the determination of the additive propyl gallate in foodstuffs. Microchemical Journal 2005; 80:71-78.
14. Tibuzzi A, Pezzotti G, Lavecchia T et al. A portable light-excitation equipped bio-amperometer for electrogenic biomaterials to support the technical development of most biosensors. Sensors and Transducers 2008; 88(2):9-20.
15. Jiang X, Li D, Xu X et al. Immunosensors for detection of pesticide residues. Biosens Bioelectron 2008; 23(11):1577-1587.
16. Ricci F, Volpe G, Micheli L et al. A review on novel developments and applications of immunosensors in food analysis. Anal Chim Acta 2007; 605(2):111-29.
17. Liu X, Wong DKY. Picogram-detection of estradiol at an electrochemical immunosensor with a gold nanoparticle|Protein G-(LC-SPDP)-scaffold. Talanta 2009; 77(4):1437-1443.
18. Wang XH, Wang S. Sensors and biosensors for the determination of small molecule biological toxins. Sensors 2008; 8(9):6045-6054.
19. Elad T, Lee JH, Belkin S et al. Microbial whole-cell arrays. Microb Biotechnol 2008; 1(2):137-148.
20. Rotariu L, Bala C, Magearu V. New potentiometric microbial biosensor for ethanol determination in alcoholic beverages. Anal Chim Acta 2004; 513(1):119-123.
21. Svensson K, Bülow L, Kriz D et al. Investigation and evaluation of a method for determination of ethanol with the SIRE® Biosensor P100, using alcohol dehydrogenase as recognition element. Biosens and Bioelectron 2005; 21(5):705-711.
22. Garjonyte R, Melvydas V, Malinauskas A. Amperometric biosensors for lactic acid based on baker's and wine yeast. Microchimica Acta 2008; preview.
23. Svitel J, Tkác J, Vostiar I et al. Gluconobacter in biosensors: applications of whole cells and enzymes isolated from gluconobacter and acetobacter to biosensor construction. Biotechnol Lett 2006; 28(24):2003-2010.
24. H. Sun H, Zhang Y, Fung Y. Flow analysis coupled with PQC/DNA biosensor for assay of E. coli based on detecting DNA products from PCR amplification. Biosens Bioelectron 2006; 22(4):506-512.
25. Rodriguez MI, Alocilja EC. Embedded DNA-polypyrrole biosensor for rapid detection of Escherichia Coli. Sensors Journal 2005; 5(4):733-736.
26. Wu VC, Chen SH, Lin CS. Real-time detection of Escherichia coli O157:H7 sequences using a circulating-flow system of quartz crystal microbalance. Biosens and Bioelectron 2007; 22(12):2967-2975.
27. Lermo A, Campoy S, Barbé J et al. In situ DNA amplification with magnetic primers for the electrochemical detection of food pathogens. Biosens Bioelectron 2007; 22 (9-10):2010-2017.
28. Wu L et al. The 2nd International Conference on Bioinformatics and Biomedical Engineering 2008.

29. Zezza F, Pascale M, Mulè G et al. Detection of Fusarium culmorum in wheat by a surface plasmon resonance-based DNA sensor. J Microbiol Methods 2006; 66(3):529-537.
30. Mello LD, Hernandez S, Marrazza G et al. Investigations of the antioxidant properties of plant extracts using a DNA-electrochemical biosensor. Biosens Bioelectron 2006; 21(7):1374-1382.
31. Cortina M et al. Determination of the antioxidants' ability to scavenge free radicals using biosensors, in press.
32. Mello LD, Kubota L, Sotomayor MPT. HRP-based amperometric biosensor for the polyphenols determination in vegetables extract. Sens Actuators B Chem 2003; 96(3):636-645.
33. Campàs M, Carpentier R, Rouillon R. Plant tissue-and photosynhesis-based biosensors. Biotechnol Adv 2008; 26:370-378.
34. Rouillon R, Piletsky S, Breton F et al. Photosystem II biosensors for heavy metals monitoring. In: Giardi MT, Piletska EV, eds. Biotechnological Applications of Photosynthetic Proteins: Biochips, Biosensors and Biodevices. Austin: Landes Bioscience 2006.
35. Tibuzzi A, Pezzotti G, Lavecchia T et al. A portable light-excitation equipped bio-amperometer for electrogenic biomaterials to support the technical development of most biosensors. Sensors and Transducers 2008; 88(2):9-20.
36. Giardi MT et al. Clamydomonas reinhardtii genetic variants as probes for fluorescence multi-biomediator sensing system in pollutants detection, in press.
37. Sloan AE. The top ten functional food trends. Food Technology 2000; 54:33-62.
38. Kapusta I, Janda B, Szajwaj B et al. Flavonoids in horse chestnut (Aesculus hippocastanum) seeds and powdered waste water byproducts. J Agric Food Chem 2007; 55(21):8485.
39. Krzyzanowska J, Czubacka A, Oleszek W. Dietary phytochemicals and human health, in press.
40. Goodman GE, Thornquist MD, Balmes J et al. The beta-carotene and retinol efficacy trial: incidence of lung cancer and cardiovascular disease mortality during 6-year follow-up after stopping beta-carotene and retinol supplements. J Natl Cancer Inst 2004; 96(23):1729-31.
41. Lee SJ, Youn BS, Park JW et al. ssDNA aptamer-based surface plasmon resonance biosensor for the detection of retinol binding protein 4 for the early diagnosis of type 2 diabetes. Anal Chem 2008; 80(8):2867-73.
42. Redondo C, Vouropoulou M, Evans J et al. Identification of the retinol-binding protein (RBP) interaction site and functional state of RBPs for the membrane receptor. FASEB J 2008; 22:1043-1054.
43. Yao , Han WW, Zhou YH et al. Catalytic reaction mechanism of human photoreceptor retinol dehydrogenase: a theoretical study. Journal of Theoretical and Computational Chemistry 2008; 7(4):565-578.
44. Malpeli G, Folli C, Cavazzini D et al. Purification and fluorescent titration of cellular retinol-binding protein. Springer Protocols. Methods Mol Bio 1998; 89:111-122.
45. Ramanathan K, Svitel J, Dzgoev A et al. Biomaterials for molecular electronics development of optical biosensor for retinol. Appl Biochem Biotechnol 2001; 96(1-3):287-301.
46. Ferancová A et al. Anti/pro-oxidative properties of selected standard chemicals and tea extracts investigated by DNA-based electrochemical biosensor. European Food Res Technol 2004; 219:416-420.
47. Puwastien P et al. International inter-laboratory analyses of food folate. J Food Compost Anal 2005; 18(5):387-397.
48. Nollet LML. Handbook of food analysis, 2nd ed., Marcel Dekker Inc., 2004(3):2138.
49. Indyk HE, Evans EA, Bostrom Caselunghe MC et al. Determination of biotin and folate in infant formula and milk by optical biosensor-based immunoassay. J AOAC Int 2000; 83:1141-1148.
50. O'Kane AA, McGrath T, Ferguson J et al. Quantification of water-soluble B-vitamins using an SPR optical biosensor. IFT Annual Meeting 2003. Chicago.
51. Verma N, Singh M. Biosensors for heavy metals. BioMetals 2005; 18:121-129.
52. Berezhetskyy AL, Sosovska OF, Durrieu C et al. Alkaline phosphatase conductometric biosensor for heavy-metal ions determination. Biosens Bioelectron 2003; 18(5-6):547-553.
53. Adam V, Petrovia J, Potesil D et al. Study of metallothionein modified electrode surface behavior in the presence of heavy metal ions-biosensor. Electroanalysis 2005; 17(18):1649-1657.
54. Adam V et al. Phytochelatin modified electrode surface as a sensitive heavy-metal ion biosensor. Sensors 2005; 5:70-84.
55. Chung Chun Lam CK, Jickells TD, Richardson DJ et al. Fluorescence-based siderophore biosensor for the determination of bioavailable iron in oceanic waters. Anal Chem 2006; 78(14):5040-5045.
56. Gupta V, Saharan K, Kumar L et al. Spectrophotometric ferric ion biosensor from pseudomonas fluorescens culture. Biotechnol Bioeng 2008; 100(2):284-296.
57. Evers TH, Appelhof MA, de Graaf-Heuvelmans PT et al. Ratiometric detection of Zn(II) using chelating fluorescent protein chimeras. J Mol Biol 2007; 374(2):411-425.
58. Wu CM, Lin LY. Utilization of albumin-based sensor chips for the detection of metal content and characterization of metal–protein interaction by surface plasmon resonance. Sens Actuators B Chem 2005; 110(2):231-238.

59. Gao Y, Guo F, Gokavi S et al. Quantification of water-soluble vitamins in milk-based infant formulae using biosensor-based assays. Food Chemistry 2008; 110(3):769-776.
60. Straub DA. Calcium supplementation in clinical practice: a review of forms, doses and indications. Nutrition in Clinical Practice 2007; 22(3):286-296.
61. To-O K, Kamasaka H, Nishimura T et al. Absorbability of calcium from calcium-bound phosphoryl oligosaccharides in comparison with that from various calcium compounds in the rat ligated jejunum loop. Biosci Biotechnol Biochem 2003; 67(8):1713-1718.
62. Haro JF, Martinez C, Ros G et al. Stability of calcium bioaccessibility and sensory parameters during the storage of fortified juices. Food Science and Technology International 2006; 12(4):281-285.
63. Pham E, Chiang J, Li I et al. A computational tool for designing FRET protein biosensors by rigid-body sampling of their conformational space. Structure 2007; 15(5):515-523.
64. Palmer AE, Tsien RY. Measuring calcium signaling using genetically targetable fluorescent indicators. Nature Protoc 2006; 1(3):1057-1065.
65. Bi X, Wong WL, Ji W et al. Development of electrochemical calcium sensors by using silicon nanowires modified with phosphotyrosine. Biosens Bioelectron 2008; 23(10):1442-1448.
66. Mello LD, Kubota L, Sotomayor MPT. HRP-based amperometric biosensor for the polyphenols determination in vegetables extract. Sens Actuators B Chem 2003; 96(3):636-645.
67. Wu B, Zhang G, Shuang S et al. A biosensor with myrosinase and glucose oxidase bienzyme system for determination of glucosinolates in seeds of commonly consumed vegetables. Sens Actuators B Chem 2005; 106(2):700-707.
68. Choi MMF, Liang MMK, Lee AWM. A biosensing method with enzyme-immobilized eggshell membranes for determination of total glucosinolates in vegetables. Enzyme Microb Technol 2005; 36(1):91-99.
69. Prieto-Simón B, Cortina M, Campas M et al. Electrochemical biosensors as a tool for antioxidant capacity assessment. Sens Actuators B Chem 2008; 129:459-466.
70. Campanella L, Bonanni A, Tomassetti M. Determination of the antioxidant capacity of samples of different types of tea, or of beverages based on tea or other herbal products, using a superoxide dismutase biosensor. J Pharm Biomed Anal 2003; 32(4-5):725-736.
71. Campanella L, Bonanni A, Finotti E et al. Biosensors for determination of total and natural antioxidant capacity of red and white wines: comparison with other spectrophotometric and fluorimetric methods. Biosens Bioelectron 2004; 19(7):641-651.
72. Mello LD, Hernandez S, Marrazza G et al. Investigations of the antioxidant properties of plant extracts using a DNA-electrochemical biosensor. Biosens Bioelectron 2006; 21(7):1374-1382.
73. Cortina M, Prieto-Simon B, Capas M et al. Determination of the antioxidants' ability to scavenge free radicals using biosensors. In: Giard M, Rea G, Berra B, eds. Bio-Farms for Nutraceuticals: Functional Food and Safety Control by Biosensors. Austin: Landes Bioscience, e-pub ahead of print.
74. Litescu SC, Eremia S. Methods for antioxidant capacity determination from food and raw materials, in press.
75. Hounsome N, Hounsome B, Tomos D et al. Plant metabolites and nutritional quality of vegetables. Journal of Food Science 2008; 73(4):R48-R65.
76. Masood A, Stark KD, Salem N Jr. A simplified and efficient method for the analysis of fatty acid methyl esters suitable for large clinical studies. J Lipid Res 2005; 46: 2299-2305.
77. Antolín EJM, Canavaciolo VLG, PéRez RS. Gas chromatographic determination of high molecular weight alcohols from policosanol in omega-3 fish oil by acylation with acetyl chloride. J AOAC Int 2008; 91(5):1013-1019.
78. Frankel EN. Lipid oxidation, 2nd ed., Bridgewater:The Oily Press, 2005.
79. Kitase A, Hino K, Furutani T et al. In situ detection of oxidized n-3 polyunsaturated fatty acids in chronic hepatitis C: correlation with hepatic steatosis. J Gastroenterol 2005; 40(6):617-624.
80. Newman JD, Turner APF. Home blood glucose biosensors: a commercial perspective. Biosens Bioelectron 2005; 20(12):2435-2453.
81. Alocilja EC, Radke SM. Market analysis of biosensors for food safety. Biosens Bioelectron 2003; 18(5-6):841-846.

Chapter 21

Biosensors for Secondary Metabolites, Two Case Studies: Ochratoxin A and Microcystin

Monica Campàs,* Beatriz Prieto-Simón and Régis Rouillon

Abstract

Secondary metabolites are chemical compounds that are not directly involved in the normal growth, development or reproduction of organisms. Due to the toxicity shown by some of these compounds, their presence can represent a threat to human health. Reliable detection systems able to control their presence are required, as a tool to ensure public health. This chapter offers an overview of different techniques developed for the detection of toxic secondary metabolites, taking ochratoxin A and microcystins as two representative examples. While ochratoxin A is a mycotoxin produced by several species of fungi, microcystins are cyanotoxins released by certain strains of cyanobacteria. Biosensor-based strategies are emphasized as powerful screening tools.

Introduction

While primary metabolism is common to all organisms, showing only slight differences in the synthetic pathways, secondary metabolism is often dependent on the species. Secondary metabolites are chemical compounds not directly involved in the normal growth, development or reproduction of organisms. Lack of secondary metabolites causes a mild impairment, lowered survivability/fecundity and aesthetic differences. Secondary metabolites are usually produced to act as defenses against predators, parasites and diseases, and to facilitate the reproductive processes.

The toxicity shown by some secondary metabolites leads to the necessity to develop detection systems to control their levels in various samples. This chapter summarizes the different detection techniques for toxic secondary metabolites, taking ochratoxin A and microcystins as examples, and focuses on the development of biosensor-based strategies as powerful screening tools, from a review (ochratoxin A) and a more experimental (microcystins) point of view. Ochratoxin A, produced by several species of fungi, has been commonly found in various food products and microcystins, released by certain strains of cyanobacteria, have been observed in fresh, brackish and marine water all over the world.

A Mycotoxin: Ochratoxin A

Ochratoxin A: Generalities

Mycotoxins are considered as the most important chronic dietary risk factor over synthetic contaminants, food additives or pesticides residues,[1] being a major matter of concern for human health. Their ubiquity in nature is due to the facility of the mycotoxin-producing mould species to

*Corresponding Author: Monica Campàs—IMAGES, EA4218, Centre de Phytopharmacie, Université de Perpignan Via Domitia 52 avenue Paul Alduy 66860 Perpignan cedex, France. Email: campas@univ-perp.fr

Bio-Farms for Nutraceuticals: Functional Food and Safety Control by Biosensors edited by Maria Teresa Giardi, Giuseppina Rea and Bruno Berra.

grow on a wide range of substrates under different conditions (e.g., moisture, oxygen and temperature). Nowadays, the main problem caused by mycotoxins is the spoilage of agricultural products, which depends on environmental and storage conditions. It is estimated that approximately 25% of the world's crops are contaminated to some extent with mycotoxins.[2,3]

Ochratoxin A (OTA) (3-methyl-5-chloro-8-hydroxy-3,4-dihydroisocoumarin linked to phenylalanine) is a naturally-occurring mycotoxin produced by several species of *Aspergillus* and *Penicillium*. It has been commonly found in cereals, coffee beans, cocoa, nuts, dried fruits, wine, beer, spices and animal organs. Toxicological studies have showed that OTA is a neurotoxic, immunotoxic, genotoxic, teratogenic, myelotoxic and carcinogenic (in mice and rats) agent, which is generally adsorbed from the gastrointestinal tract in animals and has strong toxic effects on their livers and kidneys.[4] OTA has been directly related to the Balkan Endemic Nephropathy, kidney being the main affected organ.[5] Moreover, due to the fact that OTA is also associated with an increased incidence of tumors of the upper urinary tract in humans, the International Agency for Research on Cancer (IARC) has classified OTA as possible human carcinogen. Although the European Union is currently considering the guidelines and the tolerance levels to be set for some products (green and roasted coffee and coffee products, wine, beer, grape juice, cocoa and cocoa products, and spices), there are already some regulatory limits for OTA levels in raw cereal grains (5 μg/kg), products derived from cereals (3 μg/kg) and dried vine fruits (10 μg/kg).[6] With the aim of minimizing the risk of exposure to OTA, many research groups have devoted their efforts to develop reliable, simple, cost-effective and fast analytical methods for OTA monitoring and control in food and feed samples. These methods must be highly sensitive in order to enable trace level monitoring and OTA quantification fulfilling the requirements of the stringent legislation. Moreover, the wide range of matrices where OTA can be found poses a great challenge to analytical chemists, above all when they are contaminated with other mycotoxins, which can induce synergistic effects. Thus, a high selectivity is also a crucial requirement.

Traditional Detection Methods

OTA analysis is usually performed by traditional methods, such as liquid chromatography (LC), high-performance chromatography (HPLC), thin-layer chromatography (TLC) and gas chromatography (GC), coupled to UV/Vis, fluorescence or mass spectrometry (MS).[7] Most of these methods involve expensive equipment, highly qualified personnel and time-consuming steps. Furthermore, before the detection, some steps are required with the aim of removing matrix interfering components and enhance sensitivity (e.g., extraction with organic solvents, sample clean-up, pre-concentration and sometimes analyte derivatization). Clean-up and pre-concentration steps are often performed by liquid-liquid partition[8] or solid-phase extraction (SPE).

The most widely used method for OTA detection is LC with fluorescence detection. Aresta and co-workers[9] determined OTA in wine samples with a detection limit of 0.07 μg/L, using a solid-phase microextraction (SPME) method, based on a polydimethylsiloxane/divinylbenzene (PDMS/DVB) fiber, interfaced with LC-fluorescence detection. Results greatly improved those previously obtained by the same authors[10] using an equivalent system with a UV detector. A very low detection limit for OTA (0.8 ng/L) was achieved by Medina et al[11] with a new method using zinc acetate, as a precipitating agent, and SPE silica cartridges for the clean-up step, coupled to LC-MS. Other researchers found particularly efficient the use of immunoaffinity columns (IACs) for sample extraction and purification steps. As an example, Garcia-Villanova et al[12] combined IACs with HPLC for a highly sensitive OTA detection. The molecular recognition principle of IACs, based on antigen-antibody interactions, leads to fast and highly selective separations of OTA from other interfering compounds. However, protocols are time-consuming, require trained personnel and may affect the accuracy of the analysis. Reversed-phase HPLC coupled to fluorescence detection[13] or to MS using electrospray ionisation[14] have also been used, the latter achieving low limits of detection (1 ng/L). Another chromatographic technique is capillary electrophoresis (CE), which bases the separation on the whole charge of each molecule when an electrical potential is applied. Good detection limits were achieved when diode array detection (DAD) was combined with CE for the determination of OTA in wine.[15] Fluorescence polarization (FP) spectroscopy must be also mentioned as a solution-phase assay widely used in clinical studies.[16] Although not very common, this method involves shorter times of analysis, since it is not necessary to separate free and bound OTA.

Enzyme-linked immunosorbent assays (ELISAs) deserve special attention, as consolidated methods able to be applied to the development of new emerging techniques. ELISAs are characterized by the inherent high selectivity coming from the affinity interactions between antibodies and antigens. They allow parallel analysis of multiple samples, which makes them useful as simple screening tools for routine surveillance programs. Although most ELISA protocols require no sample clean-up other than filtration and dilution, they still involve time-consuming washing steps. Moreover, cross-reactivity and matrix dependence are their major drawbacks. ELISA tests have been widely used for OTA detection.[17-19] Some of the proposed trade kits for OTA purification and detection are: Ridascreen Ochratoxin A (Biopharm/Diffchamp), Ochratoxin A (Neogen), Easyscreen ochratoxin (RP Diag/Rhone Poulenc), Ochratest (Vicam/AES) and Ochratoxin A Assay Kit (Biosystems).

Bio/Sensors

The need to develop high performing methods for OTA analysis able to deal with the current drawbacks involved in traditional methods is evident. In the last years, new approaches have appeared, including bio/sensors as powerful screening tools. The use of bio/sensors as detection devices involves many advantages, such as relative low cost of construction, potential miniaturization and easy automation, characteristics very useful in the development of a portable instrument for fast analysis and monitoring. In spite of being at an early stage of development, preliminary results are promising.

Immunosensors

Among the new emerging techniques developed for OTA detection, immunochemical methods are the most commonly found. Affinity interactions between antigen-antibody make those methods highly selective. Immunosensors have appeared with the aim of simplifying the immunoassay protocols and developing robust and portable analysis tools. Immunosensors present several advantages compared to traditional ELISAs, such as the ability to perform faster analysis and continuous monitoring, and the possibility to regenerate and reuse the biological sensing element. Moreover, when combined with the appropriate detection technique, such as electrochemistry, they can also be portable. Nevertheless, the need of labels may be a limitation and efforts are now focused on the development of alternative label-free strategies with convenient detection techniques.

Electrochemical Immunosensors

The combination of the high selectivity of ELISAs with the high sensitivity of the electrochemical techniques provides promising devices for OTA detection. Alarcón et al[20] developed two OTA immunosensors based on indirect and direct competitive ELISAs (Figs. 1A and 1D). Differential pulse voltammetry (DPV) was used as electrochemical technique to measure the amount of 1-naphthol produced by the alkaline phosphatase (AP) label. Although a better sensitivity was achieved with the direct format, both approaches attained limits of detection below the minimum limit legally established of 375 ng/L, corresponding to 3 μg/kg in all cereal products (LOD_{direct} = 60 ng/L and $LOD_{indirect}$ = 100 ng/L). The immunosensor was applied to the analysis of wheat samples, obtained after a one-step extraction procedure, and was proved to be a rapid, selective and sensitive OTA detection device, comparable to HPLC.

Our group is currently studying the behavior of different indirect strategies for the development of an electrochemical OTA immunosensor, and comparing the performance of horseradish peroxidase (HRP)- and AP-IgG labels (Fig. 1). The need of labels for the electrochemical transduction is still the main limitation of immunosensors. With the aim of circumventing this drawback,

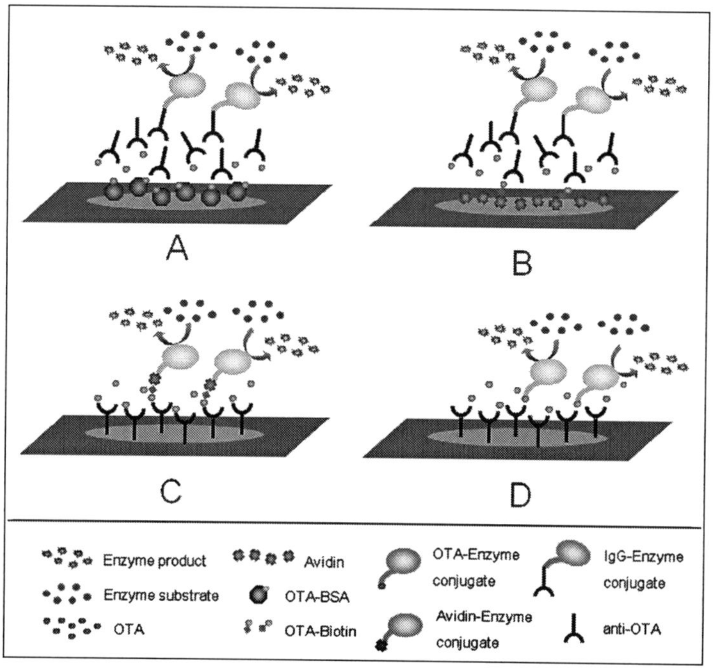

Figure 1. Indirect (A,B,C) and direct (D) competitive ELISA configurations for the determination of OTA.

impedimetric measurements are envisaged as an alternative transduction principle that, moreover, will allow much faster analysis.

Piezoelectric Immunosensors

Piezoelectric devices, such as quartz crystal microbalance (QCM), present the advantage of being label-free. However, the small size of OTA poses a problem for their direct detection. This problem can be solved by the use of indirect formats; however, sensitivity will probably be compromised. This is the case of the work currently performed in our research group for the development of an OTA immunosensor based on QCM measurements. In order to enhance the sensitivity, some additional steps incorporating labeled nanoparticles are envisaged. Moreover, the possibility to work in flow systems and to regenerate the surface of the quartz crystal by passing a 1% sodium dodecyl sulphate (SDS) solution during 2 min, makes the immunosensor strategy attractive for continuous control measurements.

Optical Immunosensors

Different optical transduction methods have been used in the development of immunosensors for mycotoxins. While some of them, like fluorescence, require a label molecule, others, such as surface plasmon resonance (SPR) only rely upon changes of mass concentration at the sensor surface. However, up to now only an optical waveguide lightmode spectroscopy (OWLS)-based biosensor has been developed for OTA determination in grain samples.[21] OWLS technique uses an evanescent field to measure variations of the refractive index due to the change of the layer thickness. Since, as expected, the direct non-competitive strategy does not provide sensitive enough detection limits, a direct competitive approach has been developed. In this case, a sensitive detection range between 0.5 and 10 μg/L was obtained and results for the analysis of real samples correlated with those measured with ELISA.

A special case of optical immunosensors for OTA detection is that of the array biosensor developed by the Naval Research Laboratory (NRL).[22] The main advantage of biosensor arrays is their ability to detect multiple target analytes simultaneously, which results in a saving of time. This is especially useful for the analysis of multi-component samples, since the presence of several mycotoxin types in the same solution may involve synergistic effects that affect the individual determination of each mycotoxin. Moreover, the ability to perform parallel analysis enables the assessment and quantification of matrix effects, allowing to obtain more accurate results. The OTA developed NRL array biosensor is based on a competitive immunoassay format and fluorescently labeled anti-OTA antibodies on a microscope slide that acts as a waveguide support. Although fluorescence requires the use of labels, limiting their usefulness, this transduction method is undoubtedly the most powerful one. The developed array is a rapid, simple and sample pre-treatment-free analysis technique, useful for the analysis of cereals and other food samples, where limits of detection in the μg/g range have been attained. Although a pre-concentration treatment would even lower these limits of detection, this extra step would increase the analysis time and the required expertise, directly affecting the advantages of these devices. The same authors proposed a NRL array biosensor for the simultaneous determination of OTA, deoxynivalenol (DON), aflatoxin B_1 (AFB_1) and fumonisins in the same sample.[23]

In order to avoid the use of label molecules, we have also been working on the development of an OTA immunosensor with SPR transduction. Nevertheless, preliminary results, like those obtained by QCM, are not enough sensitive to detect the direct OTA binding, involving the need of indirect protocols that usually affect sensitivity. Furthermore, the use of the common labels often is not enough to achieve the detection limits set by stringent regulations. As a result, nanotechnology, through the use of nanoparticles as labels, is envisaged as a powerful tool for improving the sensitivity and achieving more reliable systems.

MIP-Based Sensors

As an alternative to highly selective biomolecules, novel recognition molecules, such as molecularly imprinted polymers (MIPs) have been developed. MIPs are synthetic receptors with several inherent advantages when compared to biochemical/biological recognition systems, such as low cost, robustness and stability under long storage periods. These polymers bear high affinity sites able to selectively recognize the analyte, based on its shape, size or functional group distribution. These characteristics have led to consider MIPs as an excellent alternative for the clean-up and preconcentration of samples containing mycotoxins. Some MIPs have been already developed for OTA, as promising SPE adsorbents for clean-up and preconcentration.[24-26] Nevertheless, the high cost and toxicity of OTA usually involves the use of a synthetic mimic as imprinted template, which greatly limits their affinity for the analyte. Future research is required, as currently available MIPs are not selective enough to compete with natural biorecognition receptors. Up to now, there is only one reference to the use of MIPs recognizing OTA in real samples.[27]

A further step will be their integration into a sensing device, such as a biosensor, taking advantage of their tolerance to different solvents and extreme pH or ionic strength, without suffering from bioactivity degradation. Preliminary results with a zearalenone-MIP-based biosensor[28] enable to envisage the development of MIP-based biosensors for OTA.

A Cyanotoxin: Microcystin

Microcystins: Generalities

Cyanobacteria, also known as blue-green algae, are one of the oldest life forms on earth. Their presence has been observed in fresh, brackish and marine water all over the world. Certain strains of cyanobacteria produce toxins. Whereas it is known that cyanobacteria growth is associated to several environmental factors, such as high nutrient concentrations, low turbulence flow regimes, high light intensity and quality, warm temperatures and presence of trace metals,[29,30] the conditions that cause toxin production by certain strains of cyanobacteria are still poorly understood. Cyanobacterial toxins can be classified according to the chemical structure or the toxicity. Depending on the chemical structure, there are cyclic peptides, alkaloids and lipopolysaccharides. According to the toxic effects, cyanotoxins are classified as hepatotoxins, neurotoxins, cytotoxins, dermatoxins or irritant toxins.

Microcystins (MCs) are monocyclic heptapeptides, cyclo(D-Ala^1-L-X^2-D-MeAsp^3-L-Y^4-Adda^5 -D-Glu^6-Mdha^7), containing 5 constant amino acids and 2 variable ones (X and Y), with a molecular weight between 900 and 1100 Da.[31] To date, nearly 80 variants of MCs have been identified, each one showing different polarity, lipophilicity and toxicity.[32] Among them, microcystin-LR (MC-LR), with leucine (L) and arginine (R) as variable amino acids, is the most frequent and toxic congener. The hydrophobic Adda (3-amino-9-methoxy-2,6,8-trimethyl-10-phenyldeca-4,6-dienoic acid) chain allows MCs to penetrate the hepatocytes.[33] There, they irreversibly inhibit the serine/threonine protein phosphatases type 2A (PP2A) and 1 (PP1) by the formation of covalent bonds, which results in hyperphosphorylation of cytoskeletal filaments, hepatocyte deformation, tumor promotion and liver damage.[34]

Due to their cyclic structure, MCs are extremely stable in water, which explains why they are difficult to degrade or remove. Their hepatotoxicity produces adverse health effects on animals and humans.[34] The potential risk for the public health has led the World Health Organization (WHO) to define a provisional guideline value of 1 µg/L for MC-LR in drinking water that, in principle, should be enough to protect the consumer.[35]

Traditional Detection Methods

In order to guarantee the water quality and to preserve the human health, several detection methods have been developed for the detection of MCs. The mouse bioassay is the simplest one. This screening assay gives an indication of the global toxicity of a sample.[36] However, due to the low sensitivity and reliability, as well as the ethical problems, this bioassay is being replaced by other detection methods.

Chromatographic techniques, such as HPLC usually coupled to a UV detector, are the most widespread detection methods.[37] These techniques are characterized by their high sensitivity and selective detection, which allows the identification of MC variants. However, only a few MC variant standards are commercially available, which limits their applicability.

Taking advantage of the PP inhibiting power of MCs, an enzymatic inhibition assay has been developed for their detection. This assay, usually colorimetric although also radiometric and fluorimetric, informs about the toxicity of the sample.[38-40] PPs, however, may also be inhibited by other compounds, which makes the inhibition assay not specific.

ELISAs for MCs have also been described. They are based on monoclonal (MAbs) or polyclonal (PAbs) antibodies and have the advantage of being highly specific. However, as the antibodies usually recognize the Adda chain, common to all MC and nodularin (pentacyclic cyanobacterial hepatotoxin) congeners, misinterpretations associated to the cross-reactivity may arise.[41,42] Moreover, as the recognition events are based on the structure (not on the toxicity), immunoassays may indicate false positives from a toxicological point of view.

None of the traditional detection methods is ideal. The choice of the detection method depends on the purpose and, since they provide different and complementary information, their combination will provide a total and complete MC analysis.

Biosensors

With the purpose to improve the applicability of the screening methods, our group has developed an immunosensor[43] and an enzyme sensor[44,45] for the electrochemical detection of MCs.

Immunosensors

Colorimetric MC Detection on Wells and Electrodes

A competitive ELISA-based immunosensor was developed. To demonstrate the viability of the approach and to optimize the experimental parameters, preliminary colorimetric experiments were performed. Assays were firstly carried out on microtiter wells and afterwards on screen-printed carbon electrodes. In a first step, MAbs or PAbs were immobilized on the support by simple adsorption. Then, a direct competition between free MC-leucine arginine (MC-LR) variant and MC-LR-HRP conjugate was performed, the enzyme label providing the colorimetric (or electrochemical, in the electrochemical strategy) signal.

Colorimetric checkerboard titrations demonstrated that microtiter wells were completely coated using 1 μg/mL and 1:2,750 (dilution from stock) solutions of MAb and PAb, respectively. The concentrations of MC-LR-HRP were also optimized to 0.219 (MAb) and 0.235 μg/mL (PAb), and at these concentrations the non-specific adsorption of the conjugate on bare wells was less than 8%. Direct competitive ELISAs were then performed on wells and experimental parameters optimized. The concentrations of free MC-LR causing 50% inhibition of binding of the MC-LR-HRP conjugate, named inhibition coefficient 50 (IC_{50}), in the optimized ELISAs were 0.14 (MAb) and 1.60 μg/L (PAb). The lower value for the MAb demonstrates its higher sensitivity towards MC-LR when compared to the PAb. However, reproducibility was better with the PAb (17% maximum relative standard deviation in front of 26%, n = 3).

When screen-printed carbon electrodes were used as supports, higher MC-LR-HRP concentrations were required to obtain absorbance values similar than those obtained in microtiter wells (1.950 and 2.775 μg/mL for MAb and for PAb, respectively). This effect was probably due to the more difficult accessibility of the antibody, now immobilized on a rougher surface. Although checkerboards showed that under these conditions the non-specific adsorption of the conjugate on bare electrodes was not significantly important (less than 3%), the use of such high concentrations in the competitive assay impeded the free MC-LR to replace the MC-LR-HRP conjugate. Consequently, the optimal conjugate concentrations found in the assays on wells were used in subsequent experiments. Direct competitive ELISAs showed IC_{50} values slightly higher than those obtained with the antibodies immobilized on wells: 0.28 (MAb) and 1.81 μg/L (PAb). It seems that this effect is due to the more difficult accessibility, since higher free MC-LR concentrations are needed to replace the bound conjugate. As before, values were more reproducible with PAb (11% maximum relative standard variation in front of 22%, n = 3).

Electrochemical MC Detection with the Immunosensor

Since HRP does not provide direct electron transfer with the transducer, 5-methyl-phenazinium methyl sulfate (MPMS) was chosen as redox mediator between the enzyme label and the electrode. Chronoamperometric measurements were performed at a working potential of −200 mV (vs. Ag/AgCl).

When the antibody/conjugate configurations without MC-LR (i.e., maximum signals) were tested, the systems provided intensity currents of the same order of magnitude: 1041 and 1298 nA (background currents subtracted) with MAb and PAb, respectively. The slight differences were not significant and were certainly due to the relative standard deviation (10% for MAb and 12% for PAb, n = 5). Controls without antibody showed less than 19% of non-specific adsorption of the MC-LR-HRP on bare electrodes. This non-specific adsorption did not interfere in our measurements, as competition electrode assays with high toxin concentrations attained current values similar to those obtained with the controls without conjugate. The standard curves for MC-LR provided IC_{50} values slightly lower than those obtained by colorimetry on electrodes (0.10 and 1.73 μg/L for MAb and PAb, respectively), demonstrating the higher sensitivity of the electrochemical technique. Once again, the MAb system provided lower limits of detection. The maximum relative standard deviation values were 17 (MAb) and 16% (PAb) (n = 4), and working ranges 10^{-4}-10^{2} (MAb) and 10^{-1}-10^{2} μg/L (PAb) were obtained. Taking into account that our purpose was to develop a fast and simple screening tool to asses the potential danger of a sample and that our immunosensors are able to detect MC-LR at concentrations below the WHO recommendations, the possibility to apply them to the analysis of real samples was demonstrated.

Enzyme Sensors

Enzyme Immobilization

The entrapment into a polymeric matrix was used as enzyme immobilization method, since this technique does not modify the enzyme structure neither its recognition capacity. PP2A and the polymer (Poly(Vinyl Alcohol)-Stilbazolium Quaternary (PVA-SbQ) or Poly(Vinyl Alcohol) Azide-unit pendant Water-soluble Photopolymer (PVA-AWP)) were mixed, spread on the screen-printed carbon electrodes and let them polymerize under neon light.

Colorimetric enzymatic assays were performed to characterize the immobilization yield and to optimize the enzyme:polymer ratio. Results demonstrated that after 30-min incubation of the enzyme-modified electrodes in buffer (time required for MC incubation), the absorbance values decreased to 62% respect to non-incubated electrodes. The decrease in the activity could be due to the partial enzyme inactivation and/or to the leakage of the enzyme from the net. Nevertheless, the response was high enough to continue with the electrochemical biosensor development. Regarding the enzyme:polymer ratio, high polymer amounts may provide higher immobilization yields but may restrict the enzyme flexibility and functionality and high enzyme amounts may provide higher responses but may limit the sensitivity of the biosensor. The choice depends on the own particular interests. The 1:2 enzyme:polymer ratio provided higher absorbance values (0.259; c.v. = 17%) than those obtained with the 2:1 (0.041; c.v. = 9%) and the 1:3 (0.189; c.v. = 15%) ratios. The 1:2 ratio also provided the highest immobilization yields after 30-min incubation in buffer (62% in front of 16% and 41% for the 2:1 and the 1:3 ratios, respectively). Results clearly demonstrated that the 1:2 enzyme:polymer ratio was the optimum one for the biosensor construction.

Stability studies demonstrated that the PVA entrapment technique not only immobilizes the enzyme but also retains its activity (in fact, this enzyme is rapidly inactivated at room temperature and even at 4°C).

Colorimetric MC Detection on Wells and Electrodes

In order to demonstrate the viability of the approach, PP2A inhibition with MC-LR standards was firstly performed with the enzyme in solution and detected by colorimetry. The 50% inhibition coefficient (IC_{50}) towards MC-LR was 3.06 μg/L.

When the assay was performed with immobilized PP2A, the enzyme also recognized the MC presence, but the matrix limited the mass transport of both the toxin and the enzyme substrate. This phenomenon decreased the sensitivity of the assay. The diffusion barrier also decreased the reproducibility between electrodes, the coefficients of variation increasing from 15 (with the enzyme in solution) to 25%. Nevertheless, results were satisfactory enough to continue with the biosensor development.

Electrochemical MC Detection with the Enzyme Sensor

Since PP2A is not electrochemically active, the electrochemical transduction was achieved using an enzyme substrate electroactive only after dephosphorylation by the enzyme. Among several phosphorylated compounds, the laboratory-made catechyl phosphate was recognized by the enzyme and showed the lowest electrode fouling (effect due to the cyclic structure). Catechyl phosphate allowed us to work at + 450 mV (vs. Ag/AgCl), oxidation peak corresponding to the generated catechol (Fig. 2). The electrochemical standard curve for the inhibition of PP2A immobilised on screen-printed carbon electrodes by MC-LR showed an IC_{50} of 83.08 μg/L. Despite the lower sensitivity respect to the colorimetric approach with the enzyme in solution, the amperometric biosensor is useful as screening tool to discriminate between toxic and nontoxic samples. Work is in progress to improve the reproducibility and the sensitivity of the electrochemical strategy by the use of novel electrode materials, signal amplification systems and molecularly engineered supersensitive enzymes.

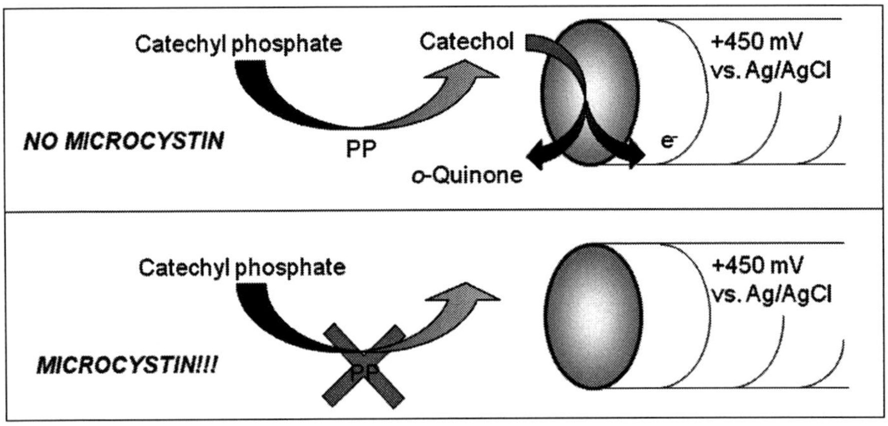

Figure 2. PP2A inhibition-based electrochemical biosensor for the determination of MCs.

Conclusion

Secondary metabolism of cyanobacteria, fungi, plants and animals is responsible for the production of a wide range of chemical compounds. These metabolites are widely studied due to their interest in numerous domains: toxicology, environment, pharmacology, cosmetics, medicine, food chemistry, etc. Their toxic characteristics and the consequent potential risk for the human health, even at low concentrations, make necessary the development of sensitive, robust, reliable and low-cost techniques for their fast detection. The emerging biosensors, although still at an early stage of development, could be the solution.

References

1. Kuiper-Goodman T. Food Safety: Mycotoxins and phycotoxins in perspective. In: Miraglia M, van Edmond H, Brera C et al, eds. Mycotoxins and phycotoxins-developments in chemistry, toxicology and food safety. Fort Collins: Alaken Inc, 1998:25-48.
2. Fink-Gremmels J. Mycotoxins: Their implications for human and animal health. Vet Q 1999; 21:115-120.
3. Richard JL, Payne GA, eds. Mycotoxins: risks in plant, animal and human systems. Raleigh: CAST Task Force Report 2003:R139.
4. Creppy EE. Update of survey, regulation and toxic effects of mycotoxins in Europe. Toxicol Lett 2002; 127:19-28.
5. Krogh P. Ochratoxins in food. In: Krogh P, ed. Mycotoxins in Food. London: Academic Press, 1987:97.
6. Commission directive 2002/26/EC of 13 March 2002 (p. 38). Amendments: - M1 Commission Directive 2004/43/EC of 13 April 2004 (p. 14)—amending Directive 98/53/EC and Directive 2002/26/EC as regards sampling methods and methods of analysis for the official control of the levels of aflatoxin and ochratoxin A in food for infants and young children. - M2—Commission Directive 2005/5/EC of 26 January 2005 (p. 38) amending Directive 2002/26/EC as regards sampling methods and methods of analysis for the official control of the levels of ochratoxin A in certain foodstuffs.
7. Pittet A. Modern methods and trends in mycotoxin analysis. Mitt Lebensm Hyg 2005; 96:424-444.
8. Valenta H. Chromatographic methods for the determination of ochratoxin A in animal and human tissues and fluids. J Chromatogr A 1998; 815:75-92.
9. Aresta A, Vatinno R, Palmesano F et al. Determination of ochratoxin A in wine at sub ng/mL levels by solid-phase microextraction coupled to liquid chromatography with fluorescence detection. J Chromatogr A 2006; 1115:196-201.
10. Aresta A, Cioffi N, Palmesano F et al. Simultaneous determination of ochratoxin A and cyclopiazonic, mycophenolic and tenuazonic acids in cornflakes by solid-phase microextraction coupled to high-performance liquid chromatography. J Agric Food Chem; 51:5232-5237.
11. Medina A, Valle-Algarra FM, Gimeno-Adelantado JV et al. New method for determination of ochratoxin A in beer using zinc acetate and solid-phase extraction silica cartridges. J Chromatogr A 2006; 1121:178-183.

12. Garcia-Villanova RJ, Cordón C, González Paramás AM et al. Simultaneous immunoaffinity column clean-up and HPLC analysis of aflatoxins and ochratoxin A in spanish bee pollen. J Agric Food Chem 2004; 52:7235-7239.
13. Hernández MJ, García-Moreno MV, Durán E et al. Validation of two analytical methods for the determination of ochratoxin A by reversed-phased high-performance liquid chromatography coupled to fluorescence detection in musts and sweet wines from Andalusia. Anal Chim Acta 2006; 566:117-121.
14. Timperio AM, Magrob P, Chilosi G et al. Assay of ochratoxin A in grape by high-pressure liquid chromatography coupled on line with an ESI-mass spectrometry. J Chromatogr B 2006; 832:127-133.
15. González-Peñas E, Leache C, López de Cerain A et al. Comparison between capillary electrophoresis and HPLC-FL for ochratoxin A quantification in wine. Food Chem 2006; 97:349-354.
16. Il'Ichev YV, Perry JL, Rüker F et al. Interaction of ochratoxin A with human serum albumin. Binding sites localized by competitive interactions with the native protein and its recombinant fragments. Chem-Biol Inter 2002; 141:275-293.
17. Yu F-Y, Chi T-F, Liu B-H et al. Development of a sensitive enzyme-linked immunosorbent assay for the determination of ochratoxin A. J Agric Food Chem 2005; 53:6947-6953.
18. Thirumala-Devi K, Mayo MA, Reddy G et al. Production of polyclonal antibodies against ochratoxin A and its detection in chillies by ELISA. J Agric Food Chem 2000; 48:5079-5082.
19. Clarke JR, Marquardt, Oosterveld A et al. Development of a quantitative and sensitive enzyme-linked immunosorbent assay for ochratoxin A using antibodies from the yolk of the laying hen. J Agric Food Chem 1993; 41:1784-1789.
20. Alarcón SH, Palleschi G, Compagnone D et al. Monoclonal antibody based electrochemical immunosensor for the determination of ochratoxin A in wheat. Talanta 2006; 69:1031-1037.
21. Adányi N, Levkovets IA, Rodriguez-Gil S et al. Development of immunosensor based on OWLS technique for determining Aflatoxin B1 and Ochratoxin A. Biosens Bioelectron 2007; 22:797-802.
22. Ngundi MM, Shriver-Lake LC, Moore MH et al. Array biosensor for detection of ochratoxin A in cereals and beverages. Anal Chem 2005; 77:148-154.
23. Sapsford KE, Ngundi MM, Moore MH et al. Rapid detection of foodborne contaminants using an array biosensor. Sens Actuat B 2006; 113:599-607.
24. Jodlbauer J, Maier NM, Lindner W. Towards ochratoxin A selective molecularly imprinted polymers for solid-phase extraction. J Chromatogr A 2002; 945:45-63.
25. Maier NM, Buttinger G, Welhartizki S et al. Molecularly imprinted polymer-assisted sample clean-up of ochratoxin A from red wine: merits and limitations. J Chromatogr B 2004; 804:103-111.
26. Turner NW, Piletska EV, Karim K et al. Effect of the solvent on recognition properties of molecularly imprinted polymer specific for ochratoxin A. Biosens Bioelectron 2004; 20:1060-1067.
27. Maier NM, Buttinger G, Welhartizki S et al. Molecularly imprinted polymer-assisted sample clean-up of ochratoxin A from red wine: merits and limitations. J Chromatogr B 2004; 804:103-111.
28. Navarro-Villoslada F, Urraca JL, Moreno-Bondi MC et al. Zearalenone sensing with molecularly imprinted polymers and tailored fluorescent probes. Sens Actuat B 2007; 121:67-73.
29. Pearson MJ, Ferguson AJD, Codd GA et al. Toxic blue-green algae. Report of the National Rivers Authority, Water Quality, Series no. 2, London: 1990.
30. Ressom R, Soong FS, Fitzgerald J et al. Health effects of toxic cyanobacteria (blue-green algae), National Health and Medical Council. Camberra: Australian Government Publishing Service 1994.
31. Carmichael WW, Beasley V, Bunner DL et al. Naming of cyclic heptapeptide toxins of cyanobacteria (blue-green algae). Toxicon 1988; 26:971-973.
32. Sivonen K, Jones G. Cyanobacterial toxins. In: Chorus I, Bartram J, eds. Toxic cyanobacteria in water—A guide to their public health consequences, monitoring and management, WHO. London: E and FP Spon, 1999:41-111.
33. Harada KI, Matsuura K, Suzuki M et al. Isolation and characterization of the minor components associated with microcystins LR and RR in the cyanobacterium (blue-green algae). Toxicon 1990; 28:55-64.
34. Eriksson JE, Toivela D, Meriluto JAO et al. Hepatocyte deformation induced by cyanobacterial toxins reflects inhibition of protein phosphatases. Biochem Biophys Res Commun 1990; 173:1347-1353.
35. Guidelines for drinking-water quality, 2on ed. Addendum to Recommendations, WHO. Geneva 1998.
36. Falconer IR. Measurement of toxins from blue-green algae in water and foodstuffs. In: Falconer IR, ed. Algal toxins in seafood and drinking water. London: Academic Press, 1993:165-175.
37. Lawton LA, Edwards C, Codd GA. Extraction and high-performance liquid chromatography method for the determination of microcystins in raw and treated waters. Analyst 1994; 119:1525-1530.
38. An JS, Carmichael WW. Use of a colorimetric protein phosphatase inhibition assay and enzyme-linked immunosorbent assay for the study of microcystins and nodularins, Toxicon 1994; 32:1495-1507.

39. Bouaïcha N, Maatouk I, Vincent G et al. A colorimetric and fluorometric microplate assay for the detection of microcystin-LR in drinking water without preconcentration. Food Chem Toxicol 2002; 40:1677-1683.
40. Rivasseau C, Racaud P, Deguin A et al. Development of a bioanalytical phosphatase inhibition test for the monitoring of microcystins in environmental samples. Anal Chim Acta 1999; 394:243-257.
41. Brooks WP, Codd GA. Immunoassay of hepatotoxic cultures and water blooms of cyanobacteria using Microcystis areuginosa peptide toxin polyclonal antibodies. Environ Technol Lett 1988; 9:1343-1348.
42. Chu FS, Huang X, Wei RD. Enzyme-linked immunosorbent assay for microcystins in blue-green algal blooms. J Assoc Off Analyt Chem 1990; 73:451-456.
43. Campàs M, Marty J-L. Highly sensitive amperometric immunosensors for microcystin detection in algae. Biosens Bioelectron 2007; 22:1034-1040.
44. Campàs M, Szydlowska D, Trojanowicz M et al. Towards the protein phosphatase-based biosensor for microcystin detection. Biosens Bioelectron 2005; 20:1520-1530.
45. Campàs M, Szydlowska D, Trojanowicz M et al. Enzyme inhibition-based biosensor for the electrochemical detection of microcystins in natural blooms of cyanobacteria. Talanta 2007; 72:179-186.

Chapter 22

Biosensors as Analytical Tools in Food Fermentation Industry

Lise Barthelmebs,* Carole Calas-Blanchard, Georges Istamboulie, Jean-Louis Marty and Thierry Noguer

Abstract

The food industries need rapid and affordable methods to assure the quality of products and process control. Biosensors, combining a biological recognition element and a sensitive transducer, are versatile analytical tools that offer advantages as classical analytical methods due to their inherent specificity, selectivity and simplicity. This paper reviews the recent trends in the development and applications of biosensors used in food fermentation industry, focusing on amperometric enzymatic and microbial sensors.

Introduction

Fermentation is one of the oldest and probably the most important method in food processing. Wine, bread, beer and cheese are produced since a very long time in order to preserve the quality of dairy products, cereals and milk. Fermented foods are food substrates that are modified by microorganisms such as bacteria, yeast and fungi, whose enzymes hydrolyse polysaccharides, proteins and lipids to nontoxic products with flavor, aromas and texture pleasant and attractive to the human consumer. Fermentation plays also other roles in food processing : (i) Preservation of substantial amounts of food through lactic acid, alcoholic, acetic acid, alkaline and high salt fermentations. (ii) Biological enrichment of food substrates with protein, essential amino acids, essential fatty acids and vitamins. (iii) Removal of the natural toxic component or prevention of the growth of disease-causing microorganisms.

Nowadays, food-fermentation is used worldwide to produce a wide diversity of beverages (wine, beer, cider and kefir) and food products (cheese, yogurt, bread, sausage, vinegar, pickled cucumbers, soy sauce and olives), which need to be control for the quality and for the freshness. Typically, the main analysis consist in the determination of food composition (sugars, amino-acids, vitamins, fermentation products) and the detection of contaminants (pesticides, heavy metals and nitrites), pathogenic microorganisms, toxins, antibiotics, allergens and hormones. Food product quality and safety are generally evaluated through periodic chemical and biological analysis. Except for microbiological assays, these procedures conventionally use chemical techniques like titration, chromatography, spectrophotometry, electrophoresis and in some cases biological methods such as immunoassays and enzymatic determination. These techniques are not very suitable for routine analysis since they require expensive instrumentation and highly trained operators, they are time consuming and sometimes require extraction

*Corresponding Author: Lise Barthelmebs—IMAGES EA 4218, Centre de Phytopharmacie, Université de Perpignan Via Domitia, 52 av. Paul Alduy, 66 860, Perpignan Cedex, France. Email: barthelm@univ-perp.fr

Bio-Farms for Nutraceuticals: Functional Food and Safety Control by Biosensors edited by Maria Teresa Giardi, Giuseppina Rea and Bruno Berra.

steps or sample pre-treatment which increase analysis time. There is an urgent need for food industries to find rapid and affordable tools to replace existing ones and, also to determine compounds that are not currently monitored. A reliable and promising alternative to classical methods is the development of biosensors. Due to their high selectivity and sensitivity, relatively low cost of construction, potential for miniaturization, facility of automation and portability, biosensors appear as efficient, rapid and simple tools for the control of food processing, quality and safety.

In this review, the role of biosensors for "off line" food fermentation compounds monitoring will be examined. We will focus mainly on enzymes and microbial amperometric sensors used for carbohydrates, alcohols and organic acids detection. In addition, the range of the different commercially available biosensors used in food analysis will be briefly surveyed.

Biosensors, General Aspects

Biosensors are analytical devices, which utilize the sensitivity and selectivity of a biological recognition element (such as enzyme, microorganisms, cells organelles, plant or animal tissue slice, antibody or DNA) closely connected to or integrated within a physical transducer (e.g., electrochemical, mass, optical, thermal) and coupled to a data acquisition and processing system.

The biological recognition element is usually immobilized in the close proximity of the transducer surface, thereby facilitating a direct or mediated signal transfer. The immobilization also plays an important function by stabilizing the biological material. The detection principle is based on the ability of the transducer to transform a biochemical and/or physico-chemical change into a measurable signal[1-3] as a result of a bio-recognition event between the biological recognition element and its target analyte. This interaction assures the selectivity of the biosensor, whereas the sensibility is determined by the transducer. The measured signal is generally proportional to the concentration of the analyte.

Chronologically, the first biosensor was an enzyme sensor developed by Clark and Lyons in 1962.[4] This sensor utilized glucose oxidase attached onto the surface of an amperometric oxygen electrode and was used to directly quantify the amount of glucose in a sample. Since then, a great number of biosensors in different configurations have been developed and many are used for the detection of compounds involved in food fermentation, as substrates or products.

Main Food Fermentation and Compounds of Interest

Food fermentation involves the cultivation of microorganisms. It is very important, for the management of the process, to monitor the concentrations of substrates and products. These compounds are mainly carbohydrates, alcohols and organic acids and specifically depend on the kind of fermentation.

Alcoholic fermentation is performed by the common yeast *Saccharomyces cerevisiae* which converts sugars in ethanol and carbon dioxide. This fermentation process is used for making bread, beer, and wine. In wine and beverages, glucose, fructose, sucrose and maltose found in grape must, juice and cereal are the substrates of the fermentation. Determination of glucose concentration, which often is not only the carbon source for the fermentation yeasts, but also the growth-limiting substrate, is very important during the fermentation process and for the final quality control. As ethanol is the main product of alcoholic fermentation, its determination is particularly important for the control of quality and conformity of wines and other alcoholic beverages. The monitoring of ethanol is also crucial for the control of fermentation processes because ethanol concentration has important effects on yeast growth and also on enzyme activities.[5] The ethanol content produced by fermentation, ranges in a concentration from a few percent up to about 14%.

Glycerol is the most important secondary product of alcoholic fermentation and contributes to the smoothness and viscosity of a wine with a favorable effect on the taste.[6] The amount of glycerol produced during the fermentation process is about 1:10 of the alcohol formed, with final concentrations between 1 to 10 g/L.[7] Monitoring of glycerol during alcoholic fermentation would allow detecting and preventing unwanted metabolic changes that affect the quality of the final product. Methanol can also be produced by yeast during wine fermentation at levels of 0.1 g/l (0.01% w/v). Most of the methanol in wine is derived from grape pectins with the highest level in red wines due to the extended maceration of the grape skins.

In winemaking, a secondary fermentation, called malo-lactic fermentation most currently occurs after completion of alcoholic fermentation. Malo-lactic fermentation is performed by the lactic bacteria *Leuconostoc* sp. and *Lactobacillus* sp. and results in the transformation of the L-malic acid to L-lactic acid and CO_2.[8-10] Malo-lactic fermentation is a crucial step conditioning the future taste of wine and inducing a decrease of total acidity. Depending on the type of wine, it should be avoided (sweet and syrupy wines), controlled (dry white wines), or encouraged (red wines). D-lactic acid detection is of interest for wine industries since this acid may be responsible of a dramatic increase of acidity.

Lactic acid fermentation, carried out by lactic acid bacteria (*Lactobacillus, Leuconostoc, Pediococcus, Streptococcus*) is used throughout the world to produce dairy products like yogurt, kefir with lactose as substrate and speciality foods like sourdough breads, sauerkraut, pickles vegetables and olives. The whole basis of lactic acid fermentation lays on the ability of the lactic bacteria to produce acid, which then inhibits the growth of other nondesirable organisms and improves the microbiological stability and safety of the food. The lactic acid bacteria belong to two main groups: the homofermenters and the heterofermenters. The pathways of lactic acid production differ for the two. Homofermenters produce mainly lactic acid, via the glycolytic (Embden–Meyerhof) pathway. Heterofermenters produce lactic acid plus appreciable amounts of ethanol, acetate and CO_2, via the 6-phosphoglucanate/phosphoketolase pathway. These other compounds are important as they impart particular tastes and aromas to the final product.[11]

Acetic fermentation is used for vinegar production. Ethanol present in cider, wine, beer, barley malt, is oxidized in acetic acid via acetic bacteria which belongs to the genus *Acetobacter*. For a good fermentation, it is essential to have an alcohol concentration of 10 to 13%. If the alcohol content is much higher, the alcohol is incompletely oxidised to acetic acid. If it is lower than 13%, there is a loss of vinegar because the esters and acetic acid are oxidised. The acetic acid strength of good vinegar should be approximately 6%.[12]

Many other compounds also produced by yeasts and bacteria in only minor quantities during these fermentations, are important since they can have a large impact on flavor. These are, for example several alcohols: 2,3 butanediol, a di-alcohol with a sweet flavor tending to bitterness at high concentrations, inositol a cyclic poly-alcohol found in grapes and metabolized from fructose and sorbitol, a poly-alcohol with a sweet flavor. Amino acids, arising from proteins degradation during fermentation, likewise contribute to a variety of tastes in food product. Glycine and alanine have a sweet taste, valine and leucine are bitter and aspartate and glutamate have a sour taste. Occurrence in fermented foods of amino acids with their respective taste influences the global taste of food. Some compounds have also to be detected due to their toxic effect as acetaldehyde founded in alcoholic beverages as the result of a decarboxylation of pyruvate by microbial pyruvate decarboxylase.

Potential Applications

Since the first description of an enzyme electrode by Clark and Lyons in 1962[4] and the work of Updike and Hicks,[13] for the determination of glucose, a huge number of biosensors have been described in the literature for the detection of food compounds. Tables 1 to 4 present some of the most important biosensors for food fermentation compounds, described in the literature these last 15 years, with a particular interest for those tested with real samples, such as wine, beer, cider and other beverages, grape fermentation, milk fermentation, yogurt, vinegar and soy sauce. The tables start with glucose and other carbohydrates and end with amino-acids biosensors. The type of biocomponent, the transducer and the detection limit and/or detection range for each biosensor are mentioned.

For the majority of the presented biosensors (Tables 1 to 4), the detection is performed by electrochemical transduction, via amperometry. This method is based on the measurement of the current generated through electro-oxidation/reduction of the products of an enzymatic reaction catalyzed by a free enzyme or a microorganism immobilized onto the transducer. The potential of the working electrode is maintained with respect to a reference electrode, usually Ag/AgCl, which is at the equilibrium. The most common working electrodes are noble metals, graphite, modified forms of carbon or conducting polymers. Amperometry provides linear concentration dependence over a defined range. These systems offer the advantage that they can be very sensitive, are usually small and robust, are rapid and economical, and are easily used outside the laboratory environment. Amperometric biosensors can suffer from poor selectivity, especially when applying potentials for oxygen/hydrogen peroxide detection, but this can usually be overcome by using mediators (e.g., ferrocyanide, quinones, phenoxazines etc.).

Two main biological recognition elements were used in these sensors: enzymes and microbial cells. The choice of the biological material depends on the analyte to detect and also on a number of factors such as the specificity, storage, stability. Amperometric biosensors are mainly based on oxidoreductase enzymes, which commonly use as electronic acceptor either molecular oxygen (oxidases) or pyridinic coenzymes such as NAD^+ and $NADP^+$ (dehydrogenases) (Table 5). Oxidase enzymes, such as glucose oxidase (EC 1.1.3.4) that is the most widely used enzyme for glucose determination (Table 1) generally have a good stability and also a high activity in absence of any additional cofactor. Oxidase-based sensors have been developed for the detection of ethanol, methanol and lactate using either alcohol oxidase (1.1.3.13) and L-lactate oxidase (EC 1.1.3.4) respectively (Tables 2 and 3). These sensors involve either the low sensitive measurement of oxygen consumption (using a Clark-type oxygen electrode), or the monitoring of hydrogen peroxide production, which requires high over-voltages responsible of many interferences due to the presence in real samples of readily oxidizable compounds such as ascorbic acid, uric acid, phenolic compounds, etc. The dehydrogenase enzymes offer a significant advantage when compared to oxidases : this enzyme family includes more than 250 $NAD(P)^+$ dependent enzymes, there is thus a great diversity of substrates which are potentially detectable using these specific enzymes. The main difficulties in the development of NAD^+-dependant dehydrogenase sensors are related to the necessary addition of the expensive $NAD(P)^+$ coenzyme in the reaction medium and to the oxidation of NADH, which requires high overpotentials. One strategy, based on chronoamperometric detection has been developed to overcome theses problems.[61]

Since 90% of the enzymes known to date are intracellular, the utilization of whole cells has been shown to be a good alternative to purified enzymes. It avoids the slow and expensive steps of enzyme purification, preserves the enzyme in its natural environment and protects it from external inhibitory compounds. The main limitation is the diffusion of the substrate and product through the cell wall resulting in a slow response as compared to enzyme-based sensors.[83,84] Amperometric microbial biosensors have been mainly developed for carbohydrates and alcohol detection in food analysis (Tables 1 and 2). Because of its importance in fermentation industry, microbial biosensors for ethanol have garnered a particular research attention. Different microorganisms metabolizing ethanol such as *Acetobacter aceti, Pichia methanolica, Gluconobacter oxydans, Saccharomyces ellipsoideus and Candida tropicalis* (Table 2) have been immobilized on oxygen electrode. While these sensors possess good stability and sensitivity, they usually have poor selectivity. A promising and simple approach to enhance selectivity is to use the size exclusion effect of a cellulose acetate membrane as performed with whole cells of *Gluconobacter oxydans*.[85]

Enzyme biosensors also constitute an interesting alternative to carry out multidetection, because they allow the analysis of samples containing analytes unable to be simultaneously detected at a conventional detector, by incorporation of different enzymes or coupling of several enzyme reactions, provided that each biological recognition element gives an individual response. Multianalyte systems have been used for the determination of analytes of interest in wine and fermented processes (Table 6).

Table 1. Carbohydrate biosensors for food analysis

Analyte	Food Matrix	Biocomponent(s)	Transducer	Detection Limit or/ and Linear Range	Reference
Glucose	Must and wine	GOX	Amp.	0.01-0.8 mM	14
Glucose	Beverages	GOX	Amp.	10 μM	15
Glucose	Beer	GOX	Amp.	40 mM	16
Glucose	Beverages	GOX	Optical	0.06-30 mM	17
Glucose	Wine	GOX	Amp.	1 μM	18
Glucose	Wine	GOX	Amp.	0.01-1.3 mM	19
Glucose	n.r.	Bacterial glucose-galactose binding protein	SPR	1-30 mM	20
Glucose	Grape must	GOX	Amp.	46 μM > 1.2 mM	21
Glucose	n.r.	PQQ-sGDH	Amp.	0.02 mM	22
Glucose	Milk fermentation	*Aspergillus niger*	Amp.	0-10 mM	23
Glucose	n.r.	*Gluconobacter oxidans*	Amp.	0-0.8 mM,	24
Lactose	n.r.	*G. oxydans* and *Kluyveromyces marxianus*	Amp.	0-4 mM	24
Maltose	Beer	Amyloglucosidase	Amp.	n.r.	16
Sucrose	n.r.	*G. oxydans* and *Saccharomyces cerevisiae*	Amp.	0-4 mM	24
Sucrose	Soft drinks	*Saccharomyces cerevisiae*	Pot.	3.2 μm 0.01-30 mM	25
Sucrose	n.r.	Invertase and GOX	Optical	0,0001-1 μm	26
Total sugars	n.r.	*Gluconobacter oxidans*	Amp.	1,1-2,2 g/L	27
Mono-and/ disaccha-rides	n.r.	*E. coli* K12 strains with defects in their carbohydrate transport systems	Amp.	0-2.5 mM m-saccharides 0-4 mM dissacharides	28

GOX: Glucose Oxidase; GDH: Glucose Dehydrogenase; PQQ-sGDH: Pyrroloquinoline Quinone-dependent soluble Glucose Dehydrogenase; Amp.: Amperometric; Pot.: Potentiometric; SPR: Surface Plasmon resonance; n.r.: not reported.

Table 2. Alcohol biosensors for food analysis

Analyte	Food Matrix	Biocomponent(s)	Transducer	Detection Limit or/and Linear Range	Reference
Ethanol	Alcoholic beverages	ADH + NOX	Amp.	0.3-200 μM	29
Ethanol	Cider, wine and whisky	ADH	Amp.	0.03-3 μM	30
Ethanol	Beer	AOX	Amp.	0-0.4 mM	31
Ethanol	Wine	ADH	Amp.	1 μM	32
Ethanol	Beer, wine and liquor	AOX + HRP	Amp.	0.2-20 μM	33
Ethanol	Alcoholic beverages	ADH	Amp.	0.1-10 mM	34
Ethanol	n.r.	ADH	Optical, ECL	10 μM 0.025-50 mM	35
Ethanol	Wine and beer	AOX	Amp.	0.5-25 μM	36
Ethanol	Wine	AOX + HRP	Amp.	0-2 mM	37
Ethanol	Wine	AOX	Amp.	n.r.	38
Ethanol	Wine, whisky and brandy	ADH	Amp.	0.05-10 mM	39
Ethanol	Wine, Beer and spirit	ADH	Amp.	LR > 1.5 mM	40
Ethanol	Beer and liquor	AOX	Amp.	30 μM 0.06-0.8 mM	41
Ethanol	n.r.	*Acetobacter aceti*	Amp.	<0.2 mM	42
Ethanol	n.r.	*Pichia methanolica*	Amp.	0.05 mM	43
Ethanol	Batch fermentation	*Gluconobacter oxydans*	Amp.	0.85 μM 2-270 μM	44
Ethanol	n.r.	*Saccharomyces ellipsoideus*	Pot.	0.05-50 mM	45
Ethanol	Alcoholic drink	*Candida tropicalis*	Amp.	0.5-7.5 mM	46
Glycerol	Alcoholic beverages	GlyDH	Amp.	0.02-0.2 mM	47
Glycerol	Grape must fermentation	GlyK + GlyPOX	Amp.	0.5 μM 2-1000 μM	7
Glycerol	Wine	PQQ-dependant GlyDH	Amp.	n.r.	48

continued on next page

Table 2. Continued

Analyte	Food Matrix	Biocomponent(s)	Transducer	Detection Limit or/and Linear Range	Reference
Glycerol	Products from anaerobic fermentation	GlyDH + DP GlyDH + GlyPOX + HRP	Amp.	mM 0.01-1.5 mM	49
Glycerol	Wine	NOX + GlyDH	Amp.	0.01-1 mM	50
Glycerol	n.r.	*Gluconobacter oxydans*	Amp.	20 μM	51
Methanol	Wine and liquor	AOX + HRP	Amp.	0.02-1.5 μM	33
Methanol	n.r.	MDH	Amp.	0.5 μM 0.5-200 μM	52
Sorbitol	n.r.	SDH + NAD(P) H:FMN Oxidoreductase + Luciferase	Optical	0.05-2 μM	53
Sorbitol	n.r.	SDH NAD^+ dependent	Amp.	40 μM	54
Sorbitol	n.r.	MIPS	QMB	1 mM 1-15 mM	55

ADH: Alcohol Dehydrogenase; AOX: Alcohol Oxidase; DP: Diaphorase; FMN: Flavin Mononucleotide; GlyDH: Glycerol Dehydrogenase; GlyK: Glycerokinase; GlyPOX: Glycerol-3-phosphate Oxidase; HRP: Horseradish Peroxidase; MDH: Methanol Dehydrogenase; MIPS: Molecularly Imprinted Polymers; NAD^+: Nicotinamide Adenine Dinucleotide; NADH: reduced form of NAD^+; $NAD(P)^+$: Nicotinamide Adenine Dinucleotide Phosphate; NAD(P) H: reduced form of $NAD(P)^+$; NOX: NADH Oxidase; PQQ-GlyDH: Pyrroloquinoline Quinone-dependent Glycerol Dehydrogenase; SDH: Sorbitol Dehydrogenase; Amp.: Amperometric; ECL: Electrogenerated Chemiluminescence; Pot.: Potentiometric; QMB: Quartz Micro Balance; n.r.: not reported.

Commercial Devices

In spite of the huge number of scientific publications on biosensors for food industries, only few devices are commercially available and essentially devoted to the determination of glucose and lactic acid. Drawbacks that have to be overcome are the limited lifetime of the biological components, mass production as well as practicability in handling. Thus, given the existing advances in biological and chemical sciences combined with advances in other various scientific and engineering fields, it is desirable that many biosensors be as competitive and reliable as many conventional instruments since biosensors offer unique solutions to food analysis in terms of specificity and time saving. Commercial biosensors are available in different forms, including autoanalyzers, manual laboratory instruments and portable systems.[86] These devices are mainly based on the use of oxidases and oxygen or hydrogen peroxide mediated detection (Table 7).

Table 3. Organic acid biosensors for food analysis

Analyte	Food Matrix	Biocomponent(s)	Transducer	Detection Limit or/ and Linear Range	Reference
Acetic acid	Wine, soy sauce, vinegar	AK + PK + PyOX	Amp.	5 μM	56
Acetic acid	Wine and vinegar	AK + PK + LDH	Amp.	0.2-8 mM 0.13 mM	57
Acetic acid	n.r.	*Fusarium solani*	Amp.	2-70 mg/L	58
Lactate	Wine	LDH + DP	Amp.	10 μM	59
Lactate	n.r.	DP + LDH	Amp.	0.005-1.5mM	60
Lactate	Wine	LDH	Chronoamp.	0.05 mM 0.1-1 mM	61
Lactate	n.r.	LDH	Optical, Fluorescence	0.2-1 mM	62
Lactate	Wine	LDH	Amp.	0.04 μM	63
Lactate	Wine	LOX	Amp.	3 μM 10-400 μM	64
Lactate	n.r.	Lactate-cytochrome c Oxidoreductase	Amp.	0-1 mM	65
Lactate	Wine and beer	LOX	Amp.	10 μM > 0.3 mM	66
Lactate	n.r.	LOX	Amp.	5 μM	67
Lactate	n.r.	LDH	Pot.	0,2 μM > 10 μM	68
Lactate	n.r.	*Escherichia coli* cytoplasmic membranes (LOX)	Fiber optic oxygen sensor	0.5-3 mM	69
Malate	Wine	MDH + Salicylate Hydroxylase	Amp.	0.01-1.2 mM	70
Malate	Wine	MDH + DP	Amp.	10 μM	59
Malate	Wine	Malic enzyme	Amp.	2 μM 5-1000 μM	63
Malate	Wine	Malic enzyme or MDH	Amp.	10 μM	18
Malate	Wine	MQO	Amp.	5 μM	71
Malate	Wine	MDH + NADH Oxidase	Amp.	4.5 μM	72

AK: Acetate Kinase; DP: Diaphorase; LDH: Lactate Dehydrogenase; LOX: Lactate Oxidase; MDH: Malate Dehydrogenase; MQO: Malate Quinone Oxidoreductase; PK: Pyruvate Kinase; PyOX: Pyruvate Oxidase; Amp.: Amperometric; Pot: Potentiometric; n.r.: not reported.

Table 4. Amino acid biosensors for food analysis

Analyte	Food matrix	Biocomponent	Transducer	Detection Limit or/ and Linear Range	Reference
Glutamate	Beverages	Glutamate oxidase	Amp.	3 μM	15
Glutamate	Soy sauce	Glutamate oxidase	Amp.	n.r.	73
Glutamate	n.r.	Glutamate oxidase	Amp.	2 μM	74
L. Amino acids	Muscatel grapes	L and D amino acid oxidase	Amp.	0.1-15 μM depending of AA	75
Lysine	Milk	Lysine oxidase	Amp.	2 μM 2-125 μM	76
Lysine	n.r.	Lysine oxidase	Pot.	n.r.	77
Lysine	n.r.	*S. cerevisiae*	Amp.	1-10 μM	78

Amp.: Amperometric; Pot.: Potentiometric; n.r.: not reported.

Table 5. Most frequently used enzymes in biosensors for food analysis

Enzyme(s)	Analyte	Enzymatic Reaction(s)
Oxidases		
AOX	Alcohol	Ethanol + $O_2 \rightarrow$ Ethanal + H_2O_2 Methanol + $O_2 \rightarrow$ Methanal + H_2O_2
GOX	Glucose	Glucose + $O_2 \rightarrow$ Gluconolactone + H_2O_2
GK and GPOX	Glycerol	Glycerol + ATP $\rightarrow$ Glycerol-3-Phosphate + ADP Glycerol-3-Phosphate + $O_2 \rightarrow$ Dihydroxyacetone-Phosphate + H_2O_2
LOX	Lactate	Lactate + $O_2 \rightarrow$ Pyruvate + H_2O_2
Dehydrogenases		
ADH	Alcohol	Ethanol + $NAD^+ \rightarrow$ Ethanal + NADH + H^+
GlyDH	Glycerol	Glycerol + $NAD^+ \rightarrow$ Dihydroxyacetone + NADH + H^+
LDH	Lactate	Lactate + $NAD^+ \rightarrow$ Pyruvate + NADH + H^+
MDH	Malate	Lactate + $NAD^+ \rightarrow$ Oxaloacetate + NADH + H^+
Others		
HRP	H_2O_2	$H_2O_2 \rightarrow O_2 + 2H^+ + 2e^-$
NOX	NADH	NADH + O_2 + $H^+ \rightarrow NAD^+ + H_2O_2$
DP	NADH	NADH + $2Fe(CN)_6^{3-} \rightarrow NAD^+ + 2Fe(CN)_6^{4-} + H^+$

Refer to Tables 1 to 4 for abbreviations.

Table 6. Multidetection system biosensors for food analysis

Food Matrix	Transducer	Biocomponent(s)	Analyte	Detection Limit ou Range	Reference
Wine	Amp.	GOX	Glucose	0.03-15 mM	79
		ADH + DP	Ethanol	0.014-10 mM	
		MDH + DP	Malate	0.015-1.5 mM	
		LDH + DP	Lactate	0.011-1.5 mM	
Wine	Amp.	PQQ-dependant GDH	Glucose	19 μM 20-800 μM	80
		PQQ-dependant GlyDH	Glycerol	1 μM 1-200 μM	
		PQQ-dependant ADH	Ethanol	1.2 μM 2.5-250 μM	
Wine	Amp.	GOX + HRP	Glucose	0.1-3.2 mM	81
		AOX + HRP	Ethanol	0.1-3.2 mM	
		LOX + HRP	Lactate	n.r.	
Wine,	Amp.	GOX	Glucose	1 μM 10-800 μM	82
		AOX	Ethanol	1 μM 10-700 μM	
Yogurt		LOX	Lactate	1 μM 10-500 μM	

Refer to Tables 1 to 4 for abbreviations.

Conclusion

Biosensor concepts are a vast area of research that continues to develop rapidly. Although the glucose sensor is still the largest area of research, mainly due to its medical applications but also as a model system, many biosensors have been developed for various fermentation compounds in food and beverage industry and related domains. The use of biosensors for food analysis can provide a route to a specific, sensitive, rapid, and an inexpensive method for monitoring a range of target analytes. This applies to monitoring carried out not only under laboratory conditions but also at on-site locations. These devices can be designed such that the non-specialist operator can use them effectively.

These sensors have, in great majority, been designed using oxidases and dehydrogenases enzymes, often used in combination with other enzymes in relatively complex systems. Achievement of stable biological sensing layers is one of the main prerequisites to the development of enzyme sensors. The main investigations have consisted in using thermophilic enzymes, mild and stable immobilization methods or biocompatible polymers. Enzyme stabilization will undoubtedly continue to be a key issue in biosensor technology. Additionally to the classical methods, current and future strategies will focus essentially in protein engineering, including directed evolution, stabilization by polypeptide chain extension and site-specific mutagenesis, and protein chemical modification.[87]

Microorganisms, due to their low cost, long lifetime and wide range of suitable pH and temperature, have been also widely employed as the biosensing element in the construction of biosensors. Some of the basic limitations of microbial biosensors as compared to enzyme sensors have been their long response time, low sensitivity and detection limits. Slow response of microbial sensors has been attributed to diffusional problems associated with the cell membranes. Possibility of genetically engineering the cell to express the enzyme of interest on cell surface can overcome this problem.[88]

The promise shown by biosensor technology is real. However, there are still technological obstacles to overcome. Advances in areas such as surface chemical analysis, protein stabilization

Table 7. Some commercially available sensors for food monitoring

Name Biosensor	Analyte(s)	Biocomponent(s)	Compagny and Reference
AM2 & AM3 AM5 GL6	Ethanol Methanol Glucose, ethanol, lactate, methanol and glycerol	AOX AOX GOX, AOX, LOX, GK + GPOX	Analox instruments (UK and USA) www.analox.com
LM5	Lactate	LOX	
Answer 8000	Glucose	GOX + HRP	Gwent sensors (UK) www.g-s-l.co.uk
Microzyme	Lactate	n.r.	Biosentec (France) www.biosentec.fr
OLGA	Glucose Lactate Sucrose Ethanol Glutamate	n.r.	Sensolytics (Germany) www.sensolytics.com
Per Bacco 2000 Per Bacco 2002	Glucose Lactate Malate	GOX LDH MDH	BioFutura s.r.l. (Italy) www.biofutura.com
Senzytech one	Glucose Lactate Malate	n.r.	Tectronik (Italy) www.tectronik.it
SIRE biosensor 101e	Glucose Sucrose Lactate Ethanol Methanol	n.r.	Chemel AB (Sweden) www.chemel.com
YSI 2700 select food analysis	Glucose Sucrose	GOX Invertase + Mutarotase + GOX	Yellow springs instruments (USA) www.ysilifesciences.com
	Lactose Lactate Ethanol Glutamate	Galactose oxidase LOX AOX Glutamate oxidase	

Refer to Tables 1 to 4 for abbreviations.

and engineering and automated manufacturing technologies will extend the market and allow biosensors to have a promising and bright future in the food industry sector.

Acknowledgement

This work was supported by the NUTRA-SNACKS project funded within the EU Framework Programme 6.

References

1. Turner APF, Karube I, WilsonGS. Editors, Biosensors: Fundamentals and Applications. Oxford University Press, Oxford 1987:770.
2. Thevenot DR, Toth K, Durst RA et al. Electrochemical biosensors: recommended definitions and classification. Biosens Bioelectron 2001; 16:121-131.
3. Patel PD. (Bio)sensors for measurement of analytes implicated in food safety. Trends Anal Chem 2002; 21:96-115.
4. Clark LC, Lyons C. Electrode systems for monitoring in cardiovascular surgery. Ann NY Acad Sci 1962; 102:29-45.
5. Kitagawa Y, Kitabatake K, Suda M et al. Amperometric detection of alcohol in beer using a flow cell and immobilized alcohol dehydrogenase. Anal Chem 1991; 63:2391-2393.
6. Ribéreau-Gayon J, Peynaud E, Sudraud P et al. Traité d'oenologie. Science et Technique du Vin 1972; 1. Dunod, Paris.
7. Compagnone D, Esti M, Messia MC et al. Development of a biosensor for monitoring of glycerol during alcoholic fermentation. Biosens Bioelectron 1998; 13:875-880.
8. Delcourt F, Taillandier P, Vidal F. Influence of pH, malic acid and glucose concentrations on malic acid consumption by Saccharomyces cerevisiae. Appl Microbiol Biotechnol 1995; 43:321-324.
9. Pretorius IS. Tailoring wine yeast for the new millennium: novel approaches to the ancient art of winemaking. Yeast 2000; 16:675-729.
10. Redzepovic S, Orlic S, Majdak A. Differential malic acid degradation by selected strains of Saccharomyces during alcoholic fermentation. Intern J Food Microbiol 2003; 83:49-61.
11. Axelsson L. Lactic acid bacteria: classification and physiology. In Lactic Acid Bacteria: Microbiology and Functional Aspects (2nd Edition ed.), Marcel Dekker Inc, New York, 1998; 1-72.
12. Tesfaye W, Morales ML, Garcia-Parrilla MC et al. Wine vinegar: technology, authenticity and quality evaluation. Trends Food Sci Tech 2002; 13:12-21.
13. Updike SJ, Hicks GP. The enzyme electrode. Nature 1967; 214:986-8.
14. Del Cerro MA, Cayuela G, Reviejo AJ et al. Graphite-teflon-peroxidase composite electrodes. Application to the direct determination of glucose in musts and wines. Electroanalysis 1997; 9:1113-1119.
15. Mizutani F, Sato Y, Hirata Y et al. High throughput flow-injection analysis of glucose and glutamate in food and biological samples by using enzyme/polyion complex bilayer membrane based electrodes as the detectors. Biosens Bioelectron 1998; 13:809-815.
16. Ge F, Zhang X-E, Zhang Z-P et al. Simultaneous determination of maltose and glucose using a screen-printed electrode system. Biosens Bioelectron 1998; 13:333-339.
17. Wu X, Choi MMF, Xiao D. A glucose biosensor with enzyme entrapped sol-gel and an oxygen sensitive optode membrane. Analyst 2000; 125:157-162.
18. Lupu A, Compagnone D, Palleschi G. Screen-printing enzyme electrodes for the detection of marker analytes during winemaking. Anal Chim Acta 2004; 513:67-72.
19. Wu B, Zhang G, Shuang S et al. Biosensors for determination of glucose with glucose oxidase immobilized on an eggshell membrane. Talanta 2004; 64:546-553.
20. Hsieh HV, Pfeiffer ZA, Amiss TJ et al. Direct detection of glucose by surface plasmon resonance with bacterial glucose/galactose-binding protein. Biosens Bioelectron 2004; 19:653-660.
21. Barsan MM, Klincar J, Batic M et al. Design and application of a flow cell for carbon-film based electrochemical enzyme biosensors. Talanta 2007; 71:1893-1900.
22. Lau C, Borgmann S, Maciejewska M et al. Improved specificity of reagentless amperometric PQQ-sGDH glucose biosensors by using indirectly heated electrodes. Biosens Bioelectron 2007; 22:3014-3020.
23. Kartlik J, Brandsteter R, Svoc J et al. Mediator type of glucose microbial biosensor based on Aspergillus niger. Anal Chim Acta 1997; 356:217-224.
24. Svitel J, Curilla O, Tkac J. Microbial cell-based biosensor for sensing glucose, sucrose or lactose. Biotechnol Appl Biochem 1998; 27:153-158.
25. Rotariu L, Bala C, Magearu V. Yeast cells sucrose biosensor based on a potentiometric oxygen electrode. Anal Chim Acta 2002; 458:215-222.
26. Bagal DS, Vijayan A, Aiyer RC et al. Fabrication of sucrose biosensor based on single mode planar optical waveguide using co-immobilized plant invertase and GOD. Biosens Bioelectron 2007; 22:3072-3079.
27. Tkac J, Gemeiner P, Svitel J et al. Determination of total sugars in lignocellulose hydrolysate by a mediated Gluconobacter oxydans biosensor. Anal Chim Acta 2000; 420:1-7.
28. Held M, Schuhmann W, Jahreis K et al. Microbial biosensor array with transport mutants of Escherichia coli K12 for the simultaneous determination of mono- and disaccharides. Biosens Bioelectron 2002; 17:1089-1094.
29. Leca B, Marty JL. Reusable ethanol sensor based on NAD(+)-dependent dehydrogenase without coenzyme addition. Anal Chim Acta 1997; 340:143-148.

30. Castañón MJL, Ordieres AJM, Blanco PT. Amperometric detection of ethanol with poly-(o-phenylenediamine)-modified enzyme electrodes. Biosens Bioelectron 1997; 12:511-520.
31. Boujtita M, Hart JP, Pittson R. Development of a disposable ethanol biosensor based on a chemically modified screen-printed electrode coated with alcohol oxidase for the analysis of beer. Biosens Bioelectron 2000; 15:257-263.
32. Niculescu M, Erichsen T, Sukharev V et al. Quinohemoprotein alcohol dehydrogenase-based reagentless amperometric biosensor for ethanol monitoring during wine fermentation. Anal Chim Acta 2002; 463:39-51.
33. Guzmán-Vázquez de Prada A, Pena N, Mena ML et al. Graphite-teflon composite bienzyme amperometric biosensors for monitoring of alcohols. Biosens Bioelectron 2003; 18:1279-1288.
34. Santos AS, freire RS, Kubota LT. Highly stable amperometric biosensor for ethanol based on Meldola's blue adsorbed on silica gel modified with niobium oxide. J Electrochem Chem 2003; 547:135-142.
35. Xu Z, Guo Z, Dong S. Electrogenerated chemiluminescence biosensor with alcohol dehydrogenase and tris(2,2'-bipyridyl)ruthenium (II) immobilized in sol–gel hybrid material. Biosens Bioelectron 2005; 21:455-461.
36. Kirgöz UA, Odaci D, Timur S et al. A biosensor based on graphite epoxy composite electrode for aspartame and ethanol detection. Anal Chim Acta 2006; 570:165-169.
37. Smutok O, Ngounou B, Pavlishko H et al. A reagentless bienzyme amperometric biosensor based on alcohol oxidase/peroxidase and an Os-complex modified electrodeposition paint. Sens Actuators B 2006; 113:590-598.
38. Shkotova LV, Soldatkin AP, Gonchar MV et al. Amperometric biosensor for ethanol detection based on alcohol oxidase immobilised within electrochemically deposited Resydrol film. Materials Science Engineering 2006; 26:411-414.
39. Santos AS, Pereira AC, Durán N et al. Amperometric biosensor for ethanol based on co-immobilization of alcohol dehydrogenase and Meldola's Blue on multi-wall carbon nanotube. Electrochimica Acta 2006; 52:215-220.
40. Tsai YC, Huang JD, ChiuCC. Amperometric ethanol biosensor based on poly(vinyl alcohol)–multiwalled carbon nanotube–alcohol dehydrogenase biocomposite. Biosens Bioelectron 2007; 22:3051-3056.
41. Wen G, Zhang Y, Shuang S et al. Application of a biosensor for monitoring of ethanol. Biosens Bioelectron 2007; 23:121-129.
42. Ikeda T, Kato K, Maeda M et al. Electrocatalytic properties of Acetobacter aceti cells immobilized on electrodes for the quinone-mediated oxidation of ethanol. J Electroanal Chem 1997; 430:197-204.
43. Reshetilov AN, Trotsenko JA, Morozova NO et al. Characteristics of Gluconobacter oxydans B-1280 and Pichia methanolica MN4 cell based biosensors for detection of ethanol. Process Biochem 2001; 36:1015-1020.
44. Tkac J, Vostiar I, Gemeiner P et al. Monitoring of ethanol during fermentation using a microbial biosensor with enhanced selectivity. Bioelectrochemistry 2002; 56:127-129.
45. Rotariu L, Bala C, Magearu V. New potentiometric microbial biosensor for ethanol determination in alcoholic beverages. Anal Chim Acta 2004; 513:119-123.
46. Akyilmaz E, Dinçkaya E. An amperometric microbial biosensor development based on Candida tropicalis yeast cells for sensitive determination of ethanol. Biosens Bioelectron 2005; 20:1263-1269.
47. Prodromidis MI, Stalikas CD, Tzouwara-Karayanni SM et al. Determination of glycerol in alcoholic beverages using packed bed reactors with immobilized glycerol dehydrogenase and an amperometric FIA system. Talanta 1996; 43:27-33.
48. Lapenaite I, Kurtinaitiene B, Razumiene J et al. Properties and analytical application of PQQ-dependent glycerol dehydrogenase from Gluconobacter sp. 33. Anal Chim Acta 2005; 549:140-150.
49. Katrlík J, Mastihuba V, Voštiar I et al. Amperometric biosensors based on two different enzyme systems and their use for glycerol determination in samples from biotechnological fermentation process. Anal Chim Acta 2006; 566:11-18.
50. Radoi A, Compagnone D, Devic E et al. Low potential detection of NADH with Prussian Blue bulk modified screen-printed electrodes and recombinant NADH oxidase from Thermus thermophilus. Sens Actuators B 2007; 121:501-506.
51. Tkac J, Svitel J, Novak R et al. Triglyceride assay by amperometric microbial biosensor: Sample hydrolysis and kinetic approach. Analytical Letters 2000; 33:2441-2452.
52. Liu Q, Kirchhoff JR. Amperometric detection of methanol with a methanol dehydrogenase modified electrode sensor. J Electroanal Chem 2007; 601:125-131.
53. Michel PE, Gautier SM, Blum LJ. A high-performance bioluminescent trienzymatic sensor for image-sorbitol based on a novel approach of the sensing layer design. Enz Microb Technol 1997; 21:108-116.

54. Saidman SB, Lobo-Castañón MJ, Miranda-Ordieres AJ et al. Amperometric detection of image-sorbitol with NAD+-image-sorbitol dehydrogenase modified carbon paste electrode. Anal Chim Acta 2000; 424:45-50.
55. Feng L, Liu Y, Tan Y et al. Biosensor for the determination of sorbitol based on molecularly imprinted electrosynthesized polymers. Biosens Bioelectron 2004; 19:1513-1519.
56. Mizutani F, Sawaguchi T, Sato Y et al. Amperometric determination of acetic acid with a trienzyme/poly(dimethylsiloxane)-bilayer-based sensor. Anal Chem 2001; 73:5738-5742.
57. Mieliauskiene R, Nistor M, Laurinavicius V et al. Amperometric determination of acetate with a tri-enzyme based sensor. Sens Actuators B 2006; 113:671-676.
58. Subrahmanyam S, Kodandapani N, Shanmugam K et al. Development of a sensor for acetic acid based on Fusarium solani. Electroanal 2001; 13:1275-1278.
59. Katrlík J, Pizzariello A, Mastihuba V et al. Biosensors for image-malate and image-lactate based on solid binding matrix. Anal Chim Acta 1999; 379:193-200.
60. Tap H, Gros P, Gué AM. An amperometric silicon-based biosensor for image-lactate. Sens Actu B 2000; 68:123-127.
61. Avramescu A, Noguer T, Magearu V et al. Chronoamperometric determination of image-lactate using screen-printed enzyme electrodes. Anal Chim Acta 2001; 433:81-88.
62. Li CL, Lin YH, Shih CL et al. Sol–gel encapsulation of lactate dehydrogenase for optical sensing of image-lactate. Biosens Bioelectron 2002; 17:323-330.
63. Avramescu A, Noguer T, Avramescu M et al. Screen-printed biosensors for the control of wine quality based on lactate and acetaldehyde determination. Anal Chim Acta 2002; 458:203-213.
64. Esti M, Volpe G, Micheli L et al. Electrochemical biosensors for monitoring malolactic fermentation in red wine using two strains of Oenococcus oeni. Anal Chim Acta 2004; 513:357-364.
65. Smutok O, Gayda G, Gonchar M et al. A novel image-lactate-selective biosensor based on flavocytochrome b2 from methylotrophic yeast Hansenula polymorpha. Biosens Bioelectron 2005; 20:1285-1290.
66. Parra A, Casero E, Vázquez L et al. Design and characterization of a lactate biosensor based on immobilized lactate oxidase onto gold surfaces. Anal Chim Acta 2006; 555:308-315.
67. Cui X, Li CM, Zang J et al. Highly sensitive lactate biosensor by engineering chitosan/PVI-Os/CNT/LOD network nanocomposite. Biosens Bioelectron 2007; 22:3288-3292.
68. Lupu A, Valsesia A, Colpo P et al. Development of a potentiometric biosensor based on nanostructured surface for lactate determination. Sens Actuators B 2007; doi:10.1016/j.snb.2007.05.020
69. Ignatov SG, Ferguson JA, Walt DR. A fiber-optic lactate sensor based on bacterial cytoplasmic membranes. Biosens Bioelectron 2001; 16:109-113.
70. Gajovic N, Warsinke A, Scheller FW. A bienzyme electrode for image-malate based on a novel and general design. J Biotechnol 1998; 61:129-133.
71. Bucur B, Mallat E, Gurban A-M et al. Strategies to develop malic acid biosensors based on malate quinone oxidoreductase (MQO). Biosens Bioelectron 2006; 21:2290-2297.
72. Gurban A-M, Prieto-Simon B, Marty J-L et al. Malate biosensors for the monitoring of malolactic fermentation: Different approaches. Anal Lett 2006; 39:1543-1558.
73. Kwong AWK, Grundig B, Hu J et al. Comparative study of hydrogel-immobilized L-glutamate oxidases for a novel thick-film biosensor and its application in food samples. Biotechnol Lett 2000; 22:267-272.
74. O'Neill RD, Chang R-C, Lowry JP et al. Comparisons of platinum, gold, palladium and glassy carbon as electrode materials in the design of biosensors for glutamate. Biosens Bioelectron 2004; 19:1521-1528.
75. Dominguez R, Serra B, Reviejo AJ et al. Chiral analysis of amino acids using electrochemical composite bienzyme biosensors. Anal Biochem 2001; 298:275-282.
76. Kelly SC, O'Connell PJ, O'Sullivan CK et al. Development of an interferent free amperometric biosensor for determination of image-lysine in food. Anal Chim Acta 2000; 412:111-119.
77. Garcia-Villar N, Saurina J, Hernandez-Cassou S. Potentiometric sensor array for the determination of lysine in feed samples using multivariate calibration methods. Anal Bioanal Chem 2001; 371:1001-1008.
78. Akyilmaz E, Erdogan A, Ozturk R et al. Sensitive determination of l-lysine with a new amperometric microbial biosensor based on Saccharomyces cerevisiae yeast cells. Biosens Bioelectron 2007; 22:1055-1060.
79. De Luca S, Florescu M, Ghica ME et al. Carbon film electrodes for oxidase-based enzyme sensors in food analysis. Talanta 2005; 68:171-178.
80. Niculescu M, Mieliauskiene R, Laurinavicius V et al. Simultaneous detection of ethanol, glucose and glycerol in wines using pyrroloquinoline quinone-dependent dehydrogenases based biosensors. Food Chem 2003; 82:481-489.

81. Miertus S, Katrlík J, Pizzariello A et al. Amperometric biosensors based on solid binding matrices applied in food quality monitoring. Biosens Bioelectron 1998; 13:911-923.
82. Guzman-Vázquez de Prada A, Peña N, Parrado C et al. Amperometric multidetection with composite enzyme electrodes. Talanta 2004; 62:896-903.
83. D'Souza SF. Microbial biosensors. Biosens Bioelectron 2001; 16:337-353.
84. Lei Y, Chen W, Mulchandani A. Microbial biosensors. Anal Chim Acta 2006; 568:200-210.
85. Tkac I, Vostiar L, Gorton P et al. Improved selectivity of microbial biosensor using membrane coating. Application to the analysis of ethanol during fermentation. Biosens Bioelectron 2003; 18:1125-1134.
86. Mello LD, Kubota LT. Review of the use of biosensors as analytical tools in the food and drink industries. Food Chem 2002; 77:237-256.
87. O'Fagain C. Enzyme stabilization—recent experimental progress. Enzyme Microb Technol 2003; 33:137-149.
88. Mulchandani A, Mulchandani P, Kaneva I et al. Biosensor for direct determination of organophosphate nerve agents using recombinant Escherichia coli with surface-expressed organophophorous hydrolase. Anal Chem 1998; 70:4140-4145.

Chapter 23

An Overview of the Functional Food Market:

From Marketing Issues and Commercial Players to Future Demand from Life in Space

Francesca Vergari,* Arianna Tibuzzi and Giovanni Basile

Abstract

Companies in the food industry have high expectations for food products that meet the consumers' demand for a healthy life style. In this context Functional Food plays a specific role. These foods are not intended only to satisfy hunger and provide the necessary human nutrients, but also to prevent nutrition-related diseases and increase the physical and mental well-being of their consumer. Among participants in space science and missions, recognition of nutraceuticals and dietary supplements is growing for their potential in reducing health risks and to improve health quality and eating habits during long-term flights and missions.

In 2008 the entire functional foods market was worth over an estimated US$80 billion, with the US holding a majority share in the nutraceuticals market (35%) followed by Japan (25%) and with the ever-growing European market, currently estimated at US$8 billion. India and China are the two major countries known for their production of traditional functional food products and nutraceuticals, but other South-East Asian countries and Gulf nations are developing potential markets.

Introduction

Historically the markets for Functional Food have developed only patchily in Europe. One indication in this respect is the fact that most "Functional" brands have been launched in only a limited number of countries in recent years. In addition, multinational food companies, as well as other companies, have mostly introduced single products rather than fully developed umbrella brands into the Functional Food market.

The development of the market in the future is influenced by the degree of familiarity and acceptance of Functional Food by the consumers. According to surveys in different European countries consumers, although they are often not aware of "Functional Food" or similar terms, express considerable interest in the concept. In the United Kingdom, France and Germany, up to 75% of the consumers have not heard the term "Functional Food", but more than 50% of them would agree to increased functional ingredients in specific food products.[1] Thereby the acceptance of a specific functional ingredient is linked to the consumers knowledge of its health effects. In the latter case consumers often do not even know the health benefits of the specific groups of ingredients and therefore are unable to assess their health effects.[2] In this sense, the health image of a Functional Food product

*Corresponding Author: Francesca Vergari—CNR-National Research Council, IC-Crystallography Institute, Via Salaria Km.29.300, 00015 Monterotondo Scalo (Roma) Italy. Email: francesca.vergari@ic.cnr.it

Bio-Farms for Nutraceuticals: Functional Food and Safety Control by Biosensors edited by Maria Teresa Giardi, Giuseppina Rea and Bruno Berra.

or a specific ingredient is a necessary prerequisite but cannot be regarded as being sufficient for a possible commercial success. In addition, consumers are not willing to change their daily lifestyle or eating patterns for the consumption of a specific Functional Food product.[2]

Marketing of Functional Food

Despite the general sociodemographic and behavioural trends which are in favour of Functional Food, there are specific challenges in the development and marketing of such products. While, in general, the total costs from the product idea to market introduction of new food products are estimated to be up to 1 or 2 million US$,[2] the development and marketing costs of Functional Food products may exceed this level by far.

Due to the limited consumers' awareness of the health effects of newly developed functional ingredients, there is a strong need for specific information and communication activities to eliminate these evident deficits. This relates in particular to pioneering companies opening a specific market segment, for which targeted information activities to consumers and opinion leaders (like e.g., medical doctors, nutritional advisers) are regarded as crucial success factor.

Another specific challenge represents the regulatory situation of Functional Food in Europe. From a legal point of view, Functional Food is positioned in a transitional zone between food and pharmaceuticals. In almost all European countries as well as in the European Union, these areas are traditionally regulated by separate institutions and are subject to different regulations, so that a kind of "grey zone" emerges with a high level of uncertainty. The classification of specific Functional Food products to one of the two categories is of high practical relevance since the factual prerequisites, authorities and procedures related to market entrance differ significantly between the two areas. Definition problems mainly exist for products with functions aiming to prevent nutrition-related diseases and/or to support health (so-called "health claims"). In the EU and related national legislation it is currently not permitted to use disease-related aspects in consumer information or product advertisements for Functional Food. Therefore, interested industry groups try to extend the type of claims allowed for Functional Food. In Europe, companies attempting to launch a functional food product have faced a variety of legislative frameworks regulating its approval, the type of nutrition information required on labels and the types of functional and health claims that were allowed in connection with a product, often in a way that was highly inconsistent among EU member states.[3] Recently, however, the "Regulation (EC) No. 1924/2006 of the European Parliament and of the Council of 20 December 2006 on nutrition and health claims made on foods has come into force. Legislative aspects have to be taken into account with particular care in this nutritional field because they can strongly affect product marketing.

Another important success factor for the marketing of Functional Food is the price for this type of food in comparison to "conventional" food products. Examples of recently launched Functional Food products indicate that consumers are only willing to accept limited price increments for such products. In general, price increments of 30-50% are observed in high-volume Functional Food segments like functional dairy products or ACE drinks.[2] It seems most likely that higher price increments will be accepted by consumers only for such Functional Food products that have a proven health effect related to a disease which directly influences them in the near future. However, as yet such products have been rarely launched in the European market. In this sense, relatively high price increments are probably the main reason for the limited market success of several Functional Food products introduced in Europe over recent years.

Consumer surveys and other market analysis studies in the USA and Europe indicate that the general success factors for the marketing of food are valid for Functional Food as well.[2,4,5] This relates in particular to tasty products, convenience attributes, a certain product variety as well as different packaging volumes. The "functional" component of a Functional Food product is mainly regarded as an added value but hardly determines the choice of the products by itself. In this sense, it appears that concepts for Functional Food products should be based on food products with a positive health image and avoid a distinct medical or clinical perspective.

For the market success of a Functional Food it is additionally required to serve high-volume distribution channels (supermarkets, general retail stores, discount retailers etc) that are the most important channels for food sales in most European countries. Therefore, consumers expect Functional Food in such retail outlets and most of them are not willing to go to specific shops just to buy Functional Food products. Such a strategy does not exclude serving specific distribution channels (pharmacies, health food shops) either with the same or a modified product. In addition, it should be Functional Food products can be made available for impulse buying (e.g., in specific convenience-oriented shops) as well.

One aspect not to be ignored is that Functional Food products help to ensure overall good health and/or to prevent/manage specific conditions in a convenient way (i.e., through daily diet).

Against the above-mentioned advantages, the development and commerce of these products is rather complex, expensive and risky as special requirements need to be matched.[6,7] This development and marketing requires significant research efforts. This involves identifying functional compounds and assessing their physiological effects; developing a suitable food matrix, taking into account bio-availability and potential changes during processing and food preparation, consumer education and clinical trials on a product's efficacy in order to gain approval for claims of health-enhancement.[8]

Businesses wishing to succeed in this market will need new methods of identifying critical technologies reliably and at a very stage in product development.

Effective scientific research alone does not make a product successful in the marketplace. The product must be in an adequate form so that the consumers can easily accept it. Therefore, as one of the first steps of product development, it is necessary to explore which diseases consumers are concerned about so that the product could be successful in the market. According to surveys, primary health concerns among consumers are cardiovascular diseases, stress, high blood pressure, malignant tumour diseases of the digestive system, arthritis and obesity.[6,7,9-11] It is more interesting, however, to identify those diseases that consumers consider as preventable/curable by way of nutrition.

From the consumer point of view, the success of functional foods relies on a number of inter-related factors, including the level of concern about general health and different medical conditions, the belief that it is possible to influence one's own health and, awareness and knowledge of foods/ingredients that are supposed to be beneficial. The marketing strategy can be focused on highlighting the positive effects ("life" marketing) of the product on human health or, on the other hand, the negative results ("death" marketing) occurring or subsisting in the absence of the functional food intake. The success of either marketing approach strongly depends on the health claims oriented by the product. For instance, in the case of a product playing a pivotal role in the reduction of the risk of cardiovascular diseases, death marketing is more effective, while life marketing is more expedient when advertising a product claiming to provide extra energy. This approach had been confirmed by Levin et al,[12] who stated that sometimes negative information is more informative, attracts more attention and stimulates deeper information processing than positive information.

The Nutraceutical and Functional Food Market

The world market for functional foods and beverages is highly dynamic, in many ways it may even be characterized as an experimental environment. The industry relies upon a network of supportive stakeholders with a vested interest, in one form or another, in providing consumers with alternative health products with potential to prevent diseases resulting from nutrient deficiencies or with products that have beneficial physiologic effects beyond those simply attributed to their nutrient content.[13]

Although the popularity of products marketed as Functional Foods and nutraceuticals is highly variable, it is often dependent upon historical and local allegiances to the industry as a whole. Popularity increases if it has an international presence and operates in many countries with a high growth potential.

The countries playing a major role in nutraceutical and Functional Food industries are identified in the map below (Fig. 1).

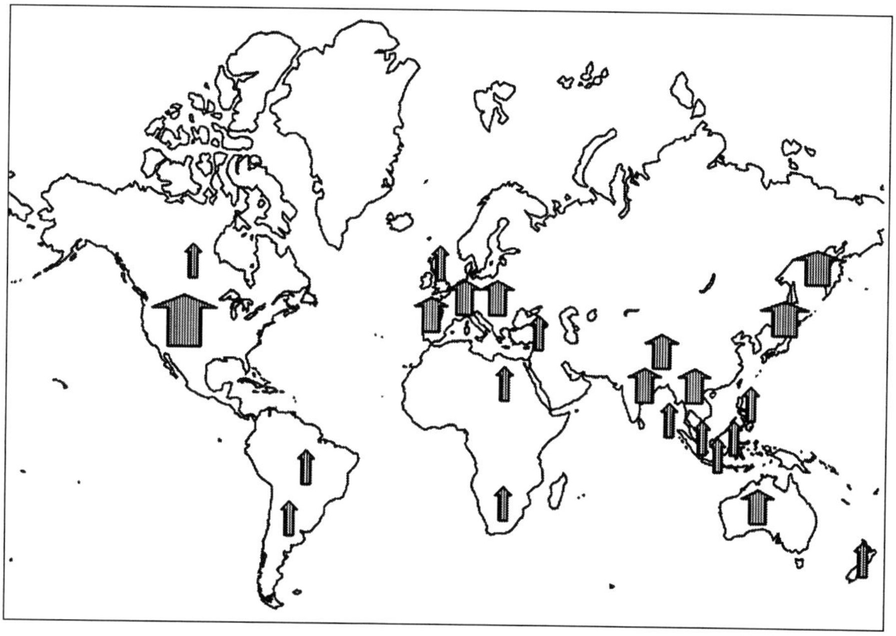

Figure 1. Key nutraceutical markets: arrows represent the areas with most rapid expansion and important niche markets. The size of the arrows indicates projected proportional expansion of the key nutraceutical markets within the next ten years.

Different factors are responsible for the rapid global growth of the nutraceutical and Functional Food industry:

1. An increase in public health consciousness: increased access to the information through education and an enquiring medium has resulted in a rapidly emerging self-care movement among consumers;
2. Change in perspective from food as sustenance to food as having nutritional benefit;
3. An aging population;
4. Escalating health care costs;
5. Recent advances in research and technology;
6. Change in Government Regulations and Accountability;
7. Expansion of the global marketplace.

Competition in the industry is driven by several factors such as price, safety, efficacy, packaging and brand loyalty, among others. With growing similarity among products and formulations, maintaining consumer brand loyalty is emerging as a critical yet complex issue. The global nutraceuticals market is characterized by intense competitive conditions, as a result of which consolidation has gathered momentum. Consequently, large pharmaceuticals companies are taking over smaller and regional players to boost their position in the intensely competitive market.

The global nutraceuticals market is projected to witness healthy growth to cross US$187 billion in sales by 2010. The United States, Europe and Japan dominate the global market, with the combined share estimated at about 86% for 2007.

The USA currently possesses the largest and most rapidly expanding Functional Food and nutraceutical market in the world.[14] In 2006 the value of the industry was $21.3 billion.[15] Its strong domestic market supports major imports from Japan, North and South Korea, China, India, Brazil, the European Union, Australia, New Zealand and other parts of the world.[14] For the USA it has been suggested that about 50% of its multi-billion dollar food market can be related to use of nutraceuticals and Functional Food products.[16]

Nutraceutical and Functional Food markets in the EU have grown over the past eight years, from about $1.8 billion out of a $5.7 billion global market in 1999[17] to $8 billion[15] out of a global market of $75.5 billion in 2006.[18] While growth of Functional Food markets within the EU was estimated to be only 2-5 fold in 2001,[17] it reached more than 10 fold in 2006.[14] Germany, France, the United Kingdom and the Netherlands represent the most important countries within the Functional Food market in Europe.

Nevertheless, compared to the world's largest regional markets, the Japanese and the US markets, the European markets are less developed. A number of factors have contributed to restricted growth within the industry throughout the EU. Strict regulations governing food labelling, product formulation, food processing, packaging, marketing, registration and licensing details are all strictly monitored in the EU and have been identified as restricting the size of the consumer market in these countries.[14,19,20] Major trading partners with the EU are the USA, Japan, south, south-east, far east and middle east Asia and Pacific regions.[14,17]

India has a large share of the international functional food and nutraceutical market and exports products to the Far East, south-east, west and middle east Asia as well as to parts of north Africa and the EU. However, India's major export destinations are USA and Japan.[20]

Both Functional Foods and nutraceuticals are part of the traditional Chinese diet and are also a large component in Traditional Chinese Medicine. The industry is represented by around 1000 small to medium sized enterprises located throughout China.[20,13] There is a steady demand for Functional Foods and nutraceuticals in the country and a friendly business environment.

Japan is the second largest market in the world for nutraceutical products after the USA, with a steady average growth rate of 9.6% per annum for the past decade. In 2006 its Functional Food industry was estimated to have a value of $27.1 billion.[13] The per capita consumption of nutraceuticals by the Japanese is actually higher ($166.00 per annum) than that observed in the USA ($136.00 per annum) and in the EU ($92.00 per annum). Two types of Functional Foods have been approved by the Japanese government, i.e., those with approved health claims or FOSHU (Foods for Specified Health Use) and foods that may provide health benefits (without any health claims).

Potential markets also can be found in oil rich middle-eastern or gulf countries like Saudi Arabia, the United Arab Emirates, Qatar, Oman and Kuwait.[14]

African markets are still not well organized, although functional food and nutraceuticals are part of the African diet and culture. Scattered opportunities also have been identified in parts of northern and sub-Saharan Africa and in some southern African nations.[13]

It is estimated that demand for Functional Food and nutraceutical products will grow internationally by about 6% per annum through 2010 and that China and India will be the fastest growing markets, while the USA will continue to be the largest, followed by Japan.

Functional Food Market Segments in Europe

From the nutraceutical and Functional Food market the consumers will expect in the future:

1. Products for heart, gut and bone health;
2. Products to enhance physical performance and general well-being;
3. Products with proven nutritional profile;
4. Products that not only taste good, but are also nutritional;
5. Guarantees of safety.

Functional Foods have been developed in virtually all food categories. According to the alternative classification some functional products: (1) "add good to your life", e.g., improve the regular stomach and colon functions (pre and probiotics) or "improve children's life" by supporting their learning capability and behaviour. It is difficult, however to find good biomarkers for cognitive, behavioural and psychological functions; (2) are designed for reducing an existing health risk problem, such as high cholesterol or high blood pressure; (3) "make your life easier" (e.g., lactose-free, gluten-free products).[21]

Functional Food products are not homogeneously scattered over all segments of the food and drink market and consumer health concerns and product preferences may vary among markets. Functional Food products have been mainly launched in the dairy-, confectionery-, soft-drinks-, bakery- and baby-food markets. The most prominent types of marketed functional products are presented briefly in the following.[21]

Special mention is also given to the newly emerging market segment in the field of the space industry. We describe a potentially profitable commercial niche, a relevant example of a successful technology transfer from biotechnology research laboratories and space science to market applications.

Probiotics and Prebiotics

Among probiotics, dairy products are the key product sector, accounting for sales of around 1.35 billion US$ in 1999 and about 56% of global sales of functional foods (31.1 billion US$ in 2004).[21,22] The main markets for dairy probiotics are Scandinavia, The Netherlands, Switzerland, Croatia, Estonia, while Greece, France and Spain can be considered as developing markets.[21]

Germany, France, The United Kingdom and The Netherlands account for around two thirds of all sales of functional dairy products in Europe.[22] Such products have shown an impressive growth during recent years, bringing the market volume in Germany from around 5 million US$ in 1995 to 419 million US$ in 2000, of which 301 million US$ account for pro-, prebiotic and other functional yoghurts and around 118 million US$ for functional dairy drinks.[2] In Central-Eastern Europe, e.g., in The Czech Republic, Hungary or Romania, the probiotics market is dominated by international companies such as Unilever or Danone and the majority of the national producers are only able to adopt technologies and product ingredients developed in other countries.[21] There is however extensive research and development activity concerning probiotics resulting in a large number of special new dairy products (e.g., Synbiofir drinking kefir, Synbioghurt drinking yoghurt, HunCult fermented drink, Milli Premium sour cream, Aktivit quark dessert, New Party butter cream, Probios cheese cream).[26] Some commercial examples of probiotic products are listed in Table 1 on the following page.[21,24]

Prebiotics are nondigestible food ingredients that beneficially affect the host by stimulating the growth and/or activity of one or a limited number of bacteria in the colon, thus improving host health.[25,26] The world demand for prebiotics is estimated to be around 167,000 tons and 390 million Euro.

Functional Spreads and Drinks

It can be assumed that cholesterol-lowering spreads will gain increasing relevance in the coming years due to the market introduction of e.g., a functional variety of Becel margarine of Unilever (named "Becel pro-activ"), containing phytostanol esters which are supposed to lower the cholesterol level. Doctors and dieticians have long recommended that a diet lower in saturated fat and cholesterol is an effective way to manage cholesterol. Since the 1950s, more than 600 studies have demonstrated that a diet rich in plant sterols can manage cholesterol.[27] Plant sterols are naturally occurring substances that have been in the human diet for thousands of years and are found in everyday foods such as vegetable oils, fruit, vegetables and grains. A diet rich in plant sterols can help consumers manage their cholesterol. Leading nutritionists describe the use of margarine with plant sterols as very significant development in dietary management for people concerned about their cholesterol. It is yet another food that can be so very easily incorporated into a healthy diet. Millions of people around the world are already enjoying Becel pro-activ as part of their dietary approach to cholesterol management. Unilever has launched this food product in 20 countries around the world, including the United States, Australia and European countries.

Another important product category within the Functional Food segment is non-alcoholic beverages fortified with vitamins A, C and E or other functional ingredients. The market is still small and fragmented in most European countries. Germany is the only country in Europe with a sizeable functional drink market, mainly due to the success of ACE drinks. In 1999 these beverages reached a market volume of 89 million US$ up from sales of around 15 million US$ in

Table 1. Commercial examples of probiotic products

Brand/Trade Name	Description	Producer
Aciforce	Freeze-dried product containing *Lactococcus lactis, Lactobacillus acidophilus, Enterococcus faecium, Bifidobacterium bifidum*	Biohorma, The Netherland
Actimel	Probiotic drinking yogurt with *L. casei Imunitass* cultures	Danone, France
Activia	Creamy yogurt containing *Bifidus ActiRegularis,*	Danone, France
Bacilac	Freeze-dried product containing *Lactobacillus acidophilus, Lactobacillus rhamnosus*	THT, Belgium
Bactisubtil	Freeze-dried product containing *Bacillus sp.strain IP5832*	Synthelabo, Belgium
Bififlor	Freeze-dried product containing *Lactobacillus acidophilus, Lactobacillus rhamnosus, Bifidobacterium bifidum*	Eko-Bio, The Netherlands
Gefilus	A wide range of LGG products	Valio, Finland
Hellus	Dairy products containing *Lactobacillus fermentum ME-3*	Tallinna Piimatööstuse AS, Estonia
Jovita Probiotisch	Blend of cereals, fruit and probiotic yogurt	H and J Bruggen, German
Pohadka	Yogurt milk with probiotic cultures	Valašske´ Meziřičıcı´ Dairy, Czech Republic
Proflora	Freeze-dried product containing *Lactobacillus acidophilus, Lactobacillus delbrueckii subsp bulgaricus, Steptococcus thermophilus, Bifidobacterium*	Chefaro, Belgium
Provie	Fruit drink containing *Lactobacillus plantarum*	Skane Mejerier, Sweden
ProViva	Refreshing natural fruit drink and yogurt in many different flavours containing *Lactobacillus plantarum*	Skane Mejerier, Sweden
Rela	Yogurts, cultured milks and juices with *L. reuteri*	Ingman Foods, Finland
Revital Active	Yogurt and drink yogurt with probiotics	Olma, Czech Republic
Snack Fibra	Snacks and bars with natural fibers and extra minerals and vitamins	Celigűeta, Spain
SOYosa	Range of products based on soy and oats and includes a refreshing drink and a probiotic yogurt-like soy—oat product	Bioferme, Finland
Soytreat	Kefir type product with six probiotics	Lifeway, USA
Yakult	Milk drink containing *Lactobacillus casei* Shirota	Yakult, Japan
Yosa	Yogurt-like oat product flavoured with natural fruits and berries containing probiotic bacteria (*Lactobacillus acidophilus, Bifidobacterium lactis*)	Bioferme, Finland

continued on next page

Table 1. Continued

Brand/Trade Name	Description	Producer
Vitality	Yogurt with pre- and probiotics and omega-3	Műller, Germany
Vifit	Drink yogurts with LGG, vitamins and minerals	Campina, the Netherlands
Vitamel	Dairy products containing *Lactobacillus casei GG,Bifodobacterium bifidum, Lactobacillus acidophilus*	Campina, the Netherlands

1996.[28] In 2000 more than 117 million of vitaminized non-alcoholic beverages were consumed in Germany, which equals to around 1% of the total consumption of these beverages.[2]

Other types of functional drinks are those of cholesterol-lowering drinks (with combination of omega-3 and soy), "eye health" drinks (with lutein) or "bone health" drinks (with calcium and inulin). The European functional drink market was estimated to be around 7% of the total soft drink market in 2004, with a further increase to 8% in 2005. According to the predictions the consumption will reach 5.1 billion by 2009, which corresponds to 23% increase compared to 2005.[10]

Functional Cereals and Bakery Products

Cereals, in particular oat and barley, offer another alternative for the production of functional foods.

Additionally, cereals can be applied as sources of nondigestible carbohydrates that, besides promoting several beneficial physiological effects, can also selectively stimulate the growth of lactobacilli and bifidobacteria present in the colon and act as prebiotics. Cereals contain water soluble fibre, such as beta-glucan and arabinoxylan, oligosaccharides, such as galacto- and fructo-oligosaccharides and resistant starch, which have been suggested to fulfil the prebiotic concept.

While Functional Foods are rapidly increasing in popularity in such sectors as dairy products or confectionery, in bakery products they are still relatively underdeveloped. These products however provide the ideal matrix by which functionality can be delivered to the consumer in a highly acceptable food. In late 2003, Unilever innovated the bakery sector by introducing a white bread called Blue Band Goede Start, which was the first white bread containing the nutritional elements normally available in brown bread including fibres, vitamins B1, B3 and B6; iron; zinc; inulin, a starch that comes from wheat.[21]

Functional Meat and Eggs

Meat and its derivatives may also be considered functional foods to the extent that they contain numerous compounds thought to be functional. The meat industry can explore various possibilities of implementing functionality, including the control of the composition of raw and processed materials via reformulation of fatty acid profiles or inclusion of antioxidants, dietary fibers or probiotics, etc.

Eggs are of particular interest from a functionality point of view, because they are relatively rich in fatty acids and the associated fat-soluble compounds. The type and ratio of fatty acids is an important determinant of human health. The idea of egg enrichment with omega-3 fatty acids simultaneously with antioxidants and other vitamins has recently been used to produce VITA Eggs by Freshlay Foods (Devon, UK).

Nutraceuticals in the Space Industry

The European (ESA) and National Space Agencies have been recently planning to extend the time of human life in Space by preparing long term missions for astronauts to Mars and the Moon and by promoting important scientific experiments to study and deliver solutions to the several health and living problems linked to the Space environment.[29,30]

The health related consequences of long term space flights include weight loss (e.g., lean and bone tissue), haematological changes, increased renal stone risk and central nervous system problems. Radiation exposure is a significant concern. The maintenance of crew health is closely related to the maintenance of a proper nutritional status and great attention to dietary intake for space travellers in both low-Earth orbiting spacecraft and in long-term missions to distant planets.

Space radiation is known to induce oxidative stress in astronauts after extended space flight.[31] Reactive oxygen species may contribute to radiation-induced cytotoxicity like chromosome aberrations, protein oxidation and muscle injury and to metabolic and morphologic changes like increased muscle proteolysis and changes in the central nervous system in animals and humans during space flight. All these phenomena accelerate the aging process, leading to arteriosclerosis and cognitive deficit which can be prevented and reduced through specific nutritional interventions based on antioxidants intake.[32] It has been reported that exposure to space radiation may compromise the capacity of the host antioxidant defence system, therefore dietary antioxidants may be useful radioprotectors to defend astronauts against radiation-induced tissue lethality and other deleterious effects, among which emotional instability, depression and stress feeling are able to negatively affect the space workers' efficiency and level of attention.[33]

Bioactive compounds extracted from plant cell cultures suitable for astronauts' health protection and for production of edible biomass during the period of segregation have already been identified and are listed in Table 2 below.

In particular, photosynthetic organisms such as higher plants and algae have demonstrated a special capacity to survive and adapt to the unfriendly space environment. Monitoring the physiology of such organisms on board BIOPAN facilities by fluorescence records during space flights (experiments PHOTO I and II in 2005 and 2007 sponsored by ESA and Italian Space Agency ASI) showed that radiation seems to be responsible for an even increased activity of the photosynthesis and of the synthesis of metabolites with antioxidant power, thanks to a stimulation phenomenon. These results are the basis of the approach recently supported by ASI to use such photosynthetic organisms as "Biological Farms" for production of antioxidant and anti-aging compounds in Space with direct clear implications for terrestrial application.

This interesting framework might soon have commercial implications by launching a market segment, strongly supported by scientific evidence, within the space sector, wich is rich in investments and careful to profitable application/technology transfers from Earth to space and viceversa.

Table 2. Botanical biologic compounds with proved healthy activity

Compound	Main Botanical Sources	Therapeutic Category
Quinones	higher plants (Spinacia oleracea, Salvia spp)	Antioxidants, Antimicrobial
Zeaxanthin and Lutein, total Xanthophylls	algae (C. reinhardtii)	Antioxidants, protector of vision
Lycopene	Tomatoes	Anticancer
Phenylpropanoids	Broad beans	Cardioprotective
Isoflavones- flavones	Soybeam, Glycine max	Cardioprotective, Anticancer
Phytic acid	Pisum sativum Spinacea oleracea	Antioxidant
Resveratrol	Vitis vinifera	Antioxidant, Cardioprotective

Suppliers of Functional Food

Six main types of actors can be identified in the EU market[2]:

1. Multinational food companies with a broad product range;
2. Pharmaceutical and/or dietary products producing companies;
3. National "category leaders";
4. Small and medium-sized companies (SMEs) of the food industry;
5. Retail companies;
6. Supplier of "functional ingredients".

Since the mid 90s several multinational food companies (e.g., Nestlé, Danone, Unilever, Kellogg, Quaker Oats) have introduced Functional Food products into the EU market. The first three companies still have a leading position in the functional dairy market in Europe. A household example is the multinational company Unilever which introduced the Becel-margarine mentioned above.

Those multinational food companies with established and well-known brands have the resources necessary for the product development and marketing of Functional Food. While, in general, the total costs from the product idea to market introduction of new food products are estimated to be several million US$, the development and marketing costs of Functional Food products may exceed this level by far. Some of these companies spend up to 2% of their turnover for R and D activities.[2] In addition to product development the proof of efficacy of Functional Food products in clinical studies also requires some time (several months up to more than one year) and relatively high financial investments.

A second type of Functional Food producers represent pharmaceutical or dietary products producing companies like e.g., Novartis Consumer Health, Glaxo SmithKline, Johnson and Johnson or Abbott Laboratories. In particular, Novartis Consumer Health launched a series of products including biscuits, cereal, cereal bars and beverages in different European countries under the "AVIVA" brand in 1999.

A third group of Functional Food producers are companies specialised in a particular product category which mostly belong to the market leaders on a national level. Examples for this type of company are Molkerei Alois Müller (with its functional "ProCult" dairy products), Ehrmann ("DailyFit" dairy products), Bauer (with several probiotic dairy products), Eckes (ACE drinks) or Becker Fruchtsäfte (ACE fruit juice) in Germany. In most EU countries as well as in Switzerland, in particular in the dairy industry, the leading national companies are often among the producers of pro- or prebiotic dairy products.[2]

There is a limited number of SMEs active in the Functional Food market as well. These companies mostly produce functional products for market niches or offer "me-too" products following the pioneering products of the multinational companies. These products often "survive" only for a rather short time period (e.g., up to two years). In general, SMEs lack the know-how and resources for conducting their own intensive R and D activities and cannot afford to spend high sums in specific information or advertising activities necessary to open a specific segment of the Functional Food market as pioneering companies. The same applies to long-lasting clinical trials (e.g., intervention studies with high number of patients) which may be necessary to prove the efficacy of a specific "functional ingredient".

Food retail companies are increasingly starting to introduce private label brands especially in the relatively "mature" markets of functional dairy products. In Germany this applies in particular to food discounters like Aldi, Lidl and Penny which launched pro- and prebiotic dairy products in recent years.

Suppliers of food ingredients play a significant role as innovation sources in the Functional Food segment. Nearly all main food ingredient producers have introduced "functional ingredients" or tried to acquire companies specialising in this field in recent years. This relates e.g., to the worldwide most important producers of vitamins (e.g., Roche Vitamins, BASF AG) who introduced specific "bioactive" ingredients to the market. In this sense, innovative suppliers of food ingredients are of high relevance, in particular for product innovations from SME food companies.

A Market Focus on Carotenoids

In the growing market of Functional Food, examples of high-demand ingredients are the carotenoids. For many years, the most prominent representative of the carotenoids, beta-carotene, was used as a food colouring agent. In the meantime, due to the anti-oxidative properties of carotenoids, the over-the-counter sector has become one of the fastest growing outlets for such products. In addition, the feed area still is a large sector, with a demand for the entire range of carotenoids to colour fish, broilers and eggs. The worldwide market value of all commercially-used carotenoids is expected to rise at an average annual growth rate of 2.9% to just over $1 billion and to reach $1.06 bn by 2010 as consumers continue to look for natural ingredients.

Figure 2 shows the world market trend of carotenoids in the years 2004 and 2009.

Astaxanthin (3,3′-dihydroxy-β-carotene-4,4′-dione) is now one of the fastest growing carotenoids in Europe with a world market standing poised to reach $219 m in the next three years. Today, essentially all commercial Astaxanthin for aquaculture is produced synthetically from petrochemical sources, with an annual turnover of over $200 million and a selling price of ~$2000 per kilo of pure Astaxanthin. In recent years, there has been a growing trend toward using natural ingredients in all forms of food nutrients, resulting from increasing concerns for consumer safety and regulatory issues over the introduction of synthetic chemicals into the human food chain. This is also true for the nutraceutical and cosmetic markets. A report from market analysts Frost and Sullivan claims that from 2000 to 2009 the €15 million total European antioxidant market is set to grow by a mere 1.7 per cent, while natural antioxidants will experience a compound annual growth rate of 35%.

Consumers are mainly responsible for driving this increase. There has been continued debate over possible health risks involved in consumption of certain synthetic antioxidants. Any potential risks in food drive demand for 'safe' alternatives. As economic conditions return to a more healthy state and, as supply and demand become more closely attuned, uptake of natural antioxidants is perceived as a key growth area for the antioxidants sector as a whole. In this fast-growing, multi-billion dollar nutraceutical market, the demand for natural Astaxanthin has recently emerged. On this market, this compound is sold at a price of about 8000 to 10000$/Kg.

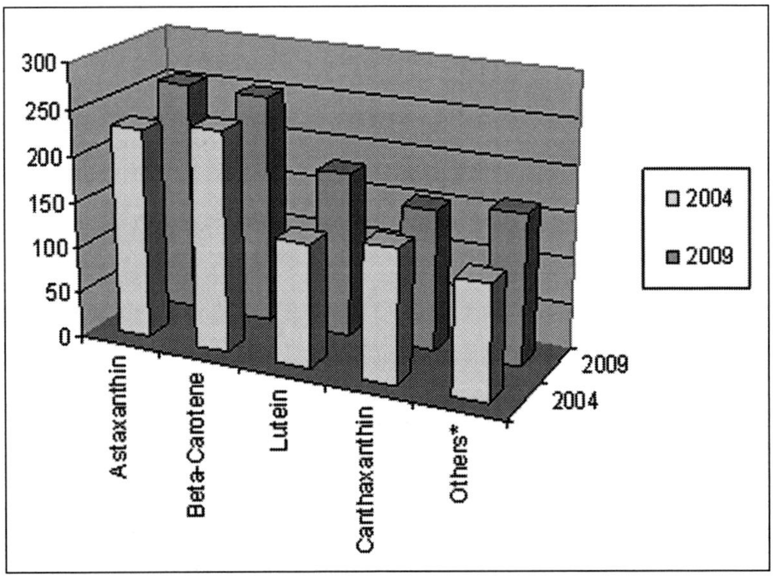

Figure 2. Global nutraceutical market in $ millions by product for 2004 and 2009. *Includes Lycopene, Annatto, Zeaxanthin, Apo-carotenal and Apo-carotenal-ester.

While only a negligible part of today's market, the demand for clinical and pharmaceutical applications of antioxidants is also expected to grow significantly as a result of promising reports from numerous medical studies performed during the last 5 years in the area of antioxidant applications.

Antioxidants represent our front line of defence against major health conditions such as heart disease, cancer and macular degeneration. 90% of diabetes, 2.80% of cardiovascular diseases and 30% of all cancers are preventable by diet and physical exercise.[34] Nonbalanced diets increase Chronic Inflammation (Metabolic Syndrome, MS) diagnosed by biomarkers of cardiovascular diseases (CVD), diabetes, some cancers and obesity.

Major chronic and diet related diseases are:

1. Ischaemic heart diseases (thrombosis, arteriosclerosis etc) and stroke;
2. Cancers (colon, lung, breast, prostate, uterus...);
3. Diabetes Mellitus;
4. Osteoporosis;
5. Gastrointestinal disorders;

In addition, further functional foods target:

6. Obesity (regulating satiety and metabolism);
7. Mental/cognitive, e.g., Alzheimer's, Parkinson's, learning, mood;
8. Immune function;
9. Physical performance;
10. Ageing.

La Société Française de Santé Publique reports (2000) a European expenditure for treatment and losses due to CVD of 180 billion €/year. The diet potential makes up 144 billion € savings each year, with even higher figures for cancers. The incidence of cancers in Europe is 2.6 million each year. The UK heart health market is 150 million £ (2007).

128 out of 156 epidemiological health studies found a protective role of fruits and vegetables on cancers, CVD and other chronic diseases. It is estimated that low intake of fruits and vegetables is responsible for 31% of ischemic heart diseases, 11% of strokes and 19% of cancers in the digestive tract. Fruits and vegetables are high in antioxidants and other beneficial components. Plant antioxidants are among the promising constituents which have biological activities and may influence biomarkers for chronic diseases and reduce the risk. They are potential anti-carcinogenic agents.

A health related antioxidant is characterized by these properties and actions:

1. Bio-availability;
2. Biological antioxidant reducing the oxidative stress in serum and tissues (redox balance);
3. Preventing/reducing oxidative degradation of major biological components, e.g., DNA, cholesterol and proteins;
4. Often influences the expression of genes related to major chronic diseases.

The most important Astaxanthin producer companies for the US, Europe and rest of the world are: Allied Industrial Corporation, Ltd., BASF AG, Chr. Hansen A/S, Carotech Bhd, Cognis Group, Cognis Nutrition and Health, Cyanotech Corporation, DSM Nutritional Products, GNT Europa GmbH, Kemin Industries, Inc., LycoRed Natural Products Industries, Ltd., Map Technologies Ltd., Phytone Ltd., Valensa International, BioReal (Sweden) AB, subsidiary of Hawaii, including the subsidiary Fuji Health Science, Inc., wholly owned by the pharmaceutical Group Fuji Chemical Industry Co., Ltd, Japan.

Conclusion

A wide overview of the functional food products and market has been presented. The most critical points related to functional ingredients commercialization and consumers' acceptance have been highlighted to warn on some aspects that sometimes are not considered by entrepreneurs and scientists working in this field. On the other hand, two success stories from an industrial and research point of view have been described, the wealthy carotenoids market with the Astaxanthin natural production from microalgae, and the use of nutraceuticals and functional food in Space missions to support the quality of life of astronauts in the harsh environment.

The wide range of functional products and their several formats allow to support various kinds of business, from the multinational company to the SME and retail shop. All stakeholders in the future will have to pay increasing attention to the health claims made for their functional products to meet the consumers' demand for beneficial effects while complying with the national, EU and USA regulations, which are especially dynamic in this field at the border between food and pharma.

Acknowledgments

The study reported here was conducted within the FP6 EU Project NUTRA-SNACKS N. FOOD-CT-2005-023044 and with the support of the national project MOMA "From MOlecules to MAn" sponsored by ASI.

References

1. Hilliam M. Functional foods. The World of Food Ingredients 1999; 3/4:46-49.
2. Menrad K. Market and marketing of functional food in Europe. J Food Process Eng 2003; 56:181-188.
3. Bech-Larsen T, Scholderer J. Functional foods in Europe: consumer research, market experiences and regulatory aspects. Trends in Food Science and Technology 2007; 18:231-234.
4. Bech-Larsen T, Grunert KG, Poulsen JB. The acceptance of functional foods in Denmark, Finland and the United States: A study of consumers' conjoint evaluations of the qualities of functional food and perceptions of general health factors and cultural values. 2001; Working Paper No. 73. Aarhus, Denmark: MAPP.
5. Childs NM. Functional foods and the food industry: consumer, economic and product development issues. Journal of Nutraceuticals, Functional and Medical Foods 1997; 1:25-43.
6. Van Kleef E, Van Trijp HCM, Luning P et al. Consumer-oriented functional food development: How well do functional disciplines reflect the 'voice of the consumer'? Trends in Food Science and Technology 2002; 13:93-101.
7. Van Kleef E, Van Trijp HCM, Luning P. Functional foods: health claim food product compatibility and the impact of health claim framing on consumer evaluation. Appetite 2005; 44:299-308.
8. Kotilainen L, Rajalahti R, Ragasa C et al. Health enhancing foods: opportunities for strengthening the sector in developing countries. Agriculture and Rural Development Discussion 2006; Paper 30.
9. Hilliam M. The market for functional foods. Int Dairy J 1998; 8:349-353.
10. Keller C. Trends in beverages and "Measurable Health". In Proceedings of the third functional food net meeting 2006.
11. Korzen-Bohr S, O'doherty Jensen K. Heart disease among postmenopausal women: Acceptability of functional foods as a preventive measure. Appetite 2006; 46:152-163.
12. Levin IP, Schneider SL, Gaeth GJ. All frames are not created equal: a typology and critical analysis of framing effects. Organ Behav Hum Decis Process 1998; 76:149-188.
13. Basu SK, Thomas JE, Acharya SN. Prospects for Growth in Global Nutraceutical and Functional Food markets: a Canadian Perspective. Aust J Basic Appl Sci 2007; 1(4):637-649.
14. World Nutraceuticals. Industry Study with Forecasts to 2010 and 2015. The Freedonia Group, Cleveland, OH USA 2006.
15. Functional food and drink consumption trends. Datamonitor 2007.
16. Belem MAF. Application of biotechnology in the product development of nutraceuticals in Canada. Trends Food Sci and Technol 1999; 10(3):101-106.
17. Kleter GA, WM van der Krieken, EJ Kok et al. Regulation and exploitation of the genetically modified crops. Nature Biotechnol 2001; 19:1105-1110.
18. Just-food. Global market review of functional foods—forecasts to 2012. Aroq Limited 2007.
19. Breithaupt H. GM plants for your health. EMBO Rep 2004; 5(11):1031-1033.
20. Moon W, Balasubramanian SK. Is there a market for genetically modified foods in Europe? Contingent valuation of GM and nonGM breakfast cereals in United Kingdom. Agri Bio Forum 2003; 6:128-133.
20. Patwardhan B, Warude D, Pushpangadan P et al. Ayurveda and Traditional Chinese Medicine: a comparative overview. eCAM 2005; 2(4):465-473.
21. Sirò I, Kàpolma E, Lugasi A. Functional food. Product development, marketing and consumer acceptance-A review, Appetite 2008; 51:456-467.

22. Hilliam M. Functional food—How big is the market? The World of Food Ingredients 2000; 12:50-52.
23. Szakaly S. Development and distribution of functional dairy products in Hungary. In Proceedings of the fourth international FFNet meeting on functional foods 2007.
24. Temmerman R, Scheirlinck I, Huys G et al. Culture-Independent Analysis of Probiotic Products by Denaturing Gradient Gel Electrophoresis. Appl Environ Microbiol 2003; 69(1):220-226.
25. Charalampopoulos D, Pandiella SS, Webb C. Evaluation of the effect of malt, wheat and barley extracts on the viability of potentially probiotic lactic acid bacteria under acidic conditions. Int J Food Microbiol 2003; 82:133-141.
26. Stanton C, Ross RP, Fitzgerald GF, et al. Fermented functional foods based on probiotics and their biogenic metabolites. Curr Opin Biotechnol 2005; 16:198-203.
27. Executive Summary of the Third Report of the National Cholesterol Education Program (NCEP). Expert Panel on Detection, Evaluation and Treatment of High Blood Cholesterol in Adults (Adult Treatment Panel III). JAMA 2001; 285:2486-2497.
28. Hilliam M. Fortified juice trends. The World of Food Ingredients 2000; 12:17-19.
29. Miroshnichenko LI. Radiation hazard in space. Series: astrophysics and Space Science Library Vol. 297. Dordrecht: Kluwer Academic Publishers, 2003.
30. George K, Durante M, Wu H et al. Chromosome aberrations in the blood lymphocytes of astronauts after space flight. Radiat Res 2001; 156:731.
31. Kennedy AR, Guan J, Ware JH. Countermeasures against space radiation induced oxidative stress in mice. Radiat Environ Biophys 2007; 46(2):201-203.
32. Wan XS, Ware JH, Zhou Z et al. Protection against radiation-induced oxidative stress in cultured human epithelial cells by treatment with antioxidant agents. Int J Radiat Oncol Biol Phys 2006; 64:1475-1481.
33. Rabin BM, Shukitt-Hale B, Joseph J et al. Diet as a factor in behavioral radiation protection following exposure to heavy particles. Pathol Int 2007; 57(8):461-73.
34. The World Health Report 2002: Reducing Risks, Promoting Healthy Life. World Health Organization (WHO), Geneva, 2002.

Chapter 24

Legislation on Nutraceuticals and Food Supplements:

A Comparison between Regulations in USA and EU

Roberto Giunta, Giovanni Basile and Arianna Tibuzzi*

Abstract

This chapter provides a short review of the main regulations concerning functional food and nutraceutical products set out by the US Food and Drug Administration and in the European Union by the European Commission. The directives are presented in chronological order with an overview on the reaction and feedback they received from consumers and manufacturing industry. A comparison between the US and EU regulations, their acceptance and impact on the market is presented, together with a final suggestion to enhance the future development and distribution of functional products, which belong to an equivocal area between food and drugs.

Introduction

Functional foods and nutraceutical products possess a high potential to improve the long-term health of populations through disease prevention in cooperation with healthcare professionals. Research on functional food and supplements could become an important section of health care for its properties and for its high level of acceptance among the general population. Complementary and alternative medicine have been taught for many years in some medical schools and are recognised by many biomedical practitioners as having a sufficient evidence base for recommendation by physicians (e.g., medical herbalism and enzyme therapy). Moreover, increased interest in health questions has drastically changed the factors dictating nutritional behaviour and enhancing the focus on the benefits of nutraceuticals. Consumers have started subscribing to diet regimens that reduce the risk of chronic diseases. People have begun to use dietary supplements merely as foodstuffs that will prevent diseases and enhance physiological performance rather than consuming them to meet the recommended dietary allowance. Fewer synthetic ingredients are being consumed as people opt more and more for natural and organic foods, functional beverages and natural supplements.

In addition, the media have drawn people's attention to scientific developments in the health and nutritional field and the nutraceutical market has benefited immensely from this greater consumer awareness. Moving from a launch phase to a faster market development the increased public interest required new certainties and a more accurate regulatory environment.

*Corresponding Author: Arianna Tibuzzi—CNR-National Research Council, IC-Crystallography Institute, Via Salaria Km.29.300,00015 Monterotondo Scalo (Roma) Italy. Email: arianna.tibuzzi@mlib.ic.cnr.it, a.tibuzzi@biosensor.it

Bio-Farms for Nutraceuticals: Functional Food and Safety Control by Biosensors edited by Maria Teresa Giardi, Giuseppina Rea and Bruno Berra.

In this context of multiple increasing interests, government regulation of functional foods is needed to provide consumer protection and market support. Two different regulatory approaches took place in Europe and in US trying to follow the market tendencies and to protect public health.

Regulations on Dietary Supplements in USA

In the United States, 6 out of 10 people use dietary supplements. For decades, the Food and Drug Administration (FDA) protected the public from mislabeled and unsafe products by regulating as foods those dietary supplements that included only those essential nutrients listed in The Nutrition Labelling and Education Act of 1990 which included herbs as dietary supplements.

The claims that a dietary supplement makes are essential to its classification. The FDA states that "a product sold as a dietary supplement and promoted on its label as a treatment, prevention or cure for a specific disease or condition would be considered an unapproved—and thus illegal—drug". Dietary supplements are permitted to make structure/function specifications. These are broad specifications clarifying how a product can support the structure or function of the body (e.g., "glucosamine helps support healthy joints", "the hormone melatonin helps establish normal sleep patterns").

In October 1994, the Dietary Supplement and Health and Education Act (DSHEA) was signed into law by President Clinton. Before this time, dietary supplements were subjected to the same regulatory requirements as other food were. This new law, which amended the Federal Food, Drug and Cosmetic Act, created a new regulatory framework for the safety and labelling of dietary supplements.

The DSHEA expanded the definition of dietary supplements beyond essential nutrients. Dietary supplements are no longer considered food additives, which makes them exempt from prescreening or from any safety and efficacy studies before they are released to the public. Unlike drug products that must be proven safe and effective for their intended use before marketing, there are no provisions for FDA to "approve" dietary supplements for safety or effectiveness before they reach the consumer.[1] Under DSHEA, the manufacturer is responsible for determining that the dietary supplements it manufactures or distributes are safe. Except in the case of a new dietary ingredient (see below), a company does not have to provide FDA with the evidence it relies on to substantiate safety or effectiveness before or after it markets its products. Also, unlike drug products, manufacturers and distributors of dietary supplements are not currently required by law to record, investigate or forward to FDA any reports they receive of injuries or illnesses that may be related to the use of their products. Moreover, manufacturers do not need to register themselves nor their dietary supplement products with FDA before producing or selling them. Until 2007, there were no FDA regulations specific to dietary supplements that established a minimum standard of practice for manufacturing dietary supplements.

Under DSHEA, once the product is marketed, FDA may take action if a product poses a direct health threat and only after adverse health effects have already occurred[2] and has the responsibility for showing that a dietary supplement is "unsafe," before it can take action to restrict the product's use or removal from the market.

As regards new dietary ingredients, the DSHEA requires that a manufacturer or distributor notify FDA if it intends to market a dietary supplement in the U.S. that contains a "new dietary ingredient". The manufacturer (and distributor) must demonstrate to FDA why the ingredient can be reasonably expected to be safe for use in a dietary supplement, unless it has been recognized as a food substance and is present in the food supply. There is no authoritative list of dietary ingredients that were marketed before October 15, 1994. Therefore, manufacturers and distributors are responsible for determining if a dietary ingredient is "new" and if it is not, for documenting that the dietary supplements they sell, containing the dietary ingredient, were marketed before October 15, 1994.

When the DSHEA was passed, the FDA lost its regulatory power. In USA sales of dietary supplements (vitamins, minerals, botanicals such as herbal products and other specialty products) have grown from \$8.8 billion in 1994 to \$15.7 billion in 2000. This growth has been attributed to a number of factors, including the DSHEA's enactment of 1994, which made it easier to market dietary supplements with attractive properties, and the growth of the self-care movement, which has resulted in an increased consumers' responsibility for their health and health care.

Public Reaction to the DSHEA and Final Rule

With broader use of a wide variety of supplements a greater concern arose about the safety and quality of these products. Some dietary supplement ingredients are unsafe for all consumers, while others ought not be taken by certain segments of the population (e.g., children, elderly persons, pregnant women and persons suffering from particular diseases). Of particular concern for older users are possible interactions between dietary supplements and prescription medicines. In addition, poor manufacturing practices can result in products being contaminated with toxins, metal, or pesticides and in products containing too little or too much of the key ingredients.

These considerations contrast the direction of deregulation of the supplement industry brought about by the DSHEA. In fact the popular support enjoyed by the new directive may have been based on a misunderstanding of the situation.

A 2001 study, published in Archives of Internal Medicine,[3] found broad public support for greater governmental regulation of dietary supplements than was currently permitted by DSHEA. The authors found that the majority of Americans supported premarketing approval by the FDA, increased oversight of harmful supplements, authority to remove them from sale and greater scrutiny of the truthfulness of supplement label claims.

A large survey by the AARP Public Policy Institute[4] found that 77% of respondents (including both users and non-users of supplements) believed that the federal government should review the safety of dietary supplements and approve them before they could be marketed.

In an October 2002 nationwide Harris poll, 59% of respondents believed that supplements had to be approved by a government agency before they could be marketed; 68% believed that supplements had to list potential side effects on their labels; and 55% believed that supplement labels could not make claims of safety without scientific evidence.

All of these beliefs were incorrect as a result of provisions of the DSHEA.

On the 22nd of June 2007 FDA issued a Dietary Supplements Final Rule[5] establishing regulations for current Good Manufacturing Practices (cGMP) for dietary supplements. Before this rule, the manufacturer was responsible for establishing its own manufacturing practice guidelines to ensure that the dietary supplements it produces are safe and contain the ingredients listed on the label. The new rule ensures that dietary supplements are produced in a quality manner, do not contain contaminants or impurities and are accurately labelled. The new regulations establish the cGMP needed to ensure the quality throughout the manufacturing, packaging, labelling and storing of dietary supplements. The final rule includes requirements for establishing quality control procedures, designing and constructing manufacturing plants and testing ingredients and the finished product. It also includes requirements for recordkeeping and handling consumer product complaints.

In addition, the industry is required to report to the FDA "all serious dietary supplement related adverse events".

Regulation on Food Supplements in the European Union

The Food Supplements Directive 2002/46/EC of the European Commission[6] sets the regulations for dietary/food supplements. The Directive states that "'food supplements' means foodstuffs the purpose of which is to supplement the normal diet and which are concentrated sources of nutrients or other substances with a nutritional or physiological effect, alone or in combination, marketed in dose form, namely forms such as capsules, pastilles, tablets, pills and other similar forms, sachets of powder, ampoules of liquids, drop dispensing bottles and other similar forms of liquids and powders designed to be taken in measured small unit quantities". There is a wide range of nutrients and other ingredients that might be present in food supplements including, but not limited to, vitamins, minerals, amino acids, essential fatty acids, fibre and various plants and herbal extracts. Vitamins and minerals are the only recognized nutrients listed in the Directive 2002/46/EC.

The Directive requires supplements to be demonstrated as safe, both in quantity and quality and also available to be used by the body. The Directive states that upper safe level of vitamin and mineral established by scientific risk assessment based on generally scientific data should be taken into account, together with recommended dietary allowances of different consumer groups. Some vitamins are essential in small quantities but dangerous in large quantities, notably Vitamin A. Consequently, only those supplements that have been proven to be safe may be sold without prescription. In practice, however, there appears to be little risk to supplement users experiencing adverse side effects due to excessive intakes of micronutrients.[7]

In order to implement the directive, a so-called "positive list" of vitamins and minerals usable in the manufacture of food supplements was established (Annex I and II of the Directive) in 2002 by the Standing Committee on the Food Chain and Animal Health (instituted by Regulation (EC) No 178/2002).

The EU Directive concerns food supplements marketed as foodstuffs and does not define them as drugs, as does the US counterpart. In Europe, the established view is that food supplements should not be labelled with drug claims: "The labelling, presentation and advertising must not attribute to food supplements the property of preventing, treating or curing a human disease" but can bear health properties. However, such claims shall be adopted after consultation with the European Food Safety Authority.

In order to ensure a high level of consumer protection and to facilitate their choices, products to be placed on the market must be safe and bear adequate and appropriate labelling. The labelling shall include the following particulars:

a. the names of the categories of nutrients or substances that characterise the product or an indication of the nature of those nutrients or substances;
b. the dose of the product recommended for daily consumption;
c. a warning not to exceed the recommended daily dose;
d. a statement explaining that food supplements should not be used as a substitute for a varied diet.

The amount of the nutrients or substances with a nutritional or physiological effect present in the product shall be declared on the labelling in numerical form.

In case of plant/herbal products, the label must contain the following information:

- Botanic name
- Plant origin
- Material origin
- Plant seed
- Preparation model
- Activity
- Biological marker
- Toxicology
- Pollution
- Side effect(s)

Public Reaction to the Directive 2002/46/EC and Directive EC 1924/2006

The dietary supplements industry in the UK, which is the biggest market for functional foods and beverages (US $2.6 billion) in the European Union, strongly opposed the EU Directive. In addition, a large number of consumers throughout Europe (over one million in the UK) and many doctors and scientists signed petitions against what were viewed by the petitioners as unjustified restrictions of consumer choice.[8] In August 2004, the British Health Food Manufacturers Association, the National Association of Health Stores and Alliance for Natural Health (ANH) made a legal challenge to the European Union's Food Supplements Directive referred to the European Court of Justice by the High Court in London.[9] The industry argued the legislation threatened 5000 products, containing more than 200 nutrients, including some vitamin C products. They also contended that it was unfair for health food manufacturers to bear the cost of applying for approval for products they had been selling for many years.

Although the European Court of Justice's Advocate General subsequently ruled that the EU's plan to tighten rules on the sale of vitamins and food supplements should be cancelled or reviewed,[10] he was eventually overruled by the European Court, which decided that the measures in question were necessary and appropriate for the purpose of protecting public health.[11]

ANH, however, interpreted the ban as applying only to synthetically produced supplements.[12]

Nevertheless, the European judges did acknowledge the Advocate General's concerns, stating that an application to have a substance included on a list may be refused only on the basis of a full risk assessment, established on the basis of the most reliable scientific data available and the most recent results of international research. The judges promised to simplify the application process for new ingredients and ruled that the burden of proof should shift from those seeking approval to those seeking to ban a product: "In order to keep up with scientific and technological developments it is important to revise the lists promptly, when necessary. Such revisions would be implementing measures of a technical nature and their adoption should be entrusted to the Commission in order to simplify and expedite the procedure".[6] They also declared that any refusal to add a product to the list must be open to challenge in the courts.[13]

The successive Regulation (EC) No. 1924/2006 of the 20th December 2006[14] (entered into force on the 1st July 2007) lays down specific provisions concerning the use of nutrition and health claims concerning foods to be delivered as such to the consumer and provides special support to scientific research and small and medium-sized enterprises (SMEs) operating in this field.

The Regulation aims at protecting all consumers from misleading claims and requires that the substances for which a claim is made have been shown to have a beneficial nutritional or physiological effect. In order to ensure that the claims made are truthful, it is necessary that the substance that is the subject of the claim is present in the final product in quantities that are sufficient, or that the substance is absent or present in suitably reduced quantities, to produce the nutritional or physiological effect claimed. The substance should also be available to be used by the body.

Health claims other than those referring to the reduction of disease risk and to children's development and health, based on generally accepted scientific evidence, will undergo a different type of assessment and authorisation. It is therefore necessary to adopt a Community list of such permitted claims after consulting the European Food Safety Authority. Furthermore, in order to stimulate innovation, those health claims which are based on newly developed scientific evidence should undergo an accelerated type of authorisation. In order to keep up with scientific and technological developments, the list mentioned above should be revised promptly whenever necessary. Such revisions are implementing measures of a technical nature and will be supervised by the Commission in order to simplify and expedite the procedure.

In order to stimulate research and development within the agro-food industry, it is appropriate to protect the investment made by innovators in gathering the information and data supporting an application under this Regulation.

This Regulation grants a facilitated access to claims by SMEs, which rarely have the financial capacity to carry out research activities. SMEs represent an important added value to the European food industry in terms of quality and preservation of different dietary habits. In order to facilitate the implementation of this Regulation, the European Food Safety Authority should make available appropriate technical guidance and tools, in due time, especially for SMEs.

Conclusion

The regulatory pictures in USA and EU describe almost opposite governmental attitudes and decisions with regards to the practical implementation of the food/dietary supplements regulations. While the DSHEA in USA at the time of its entry into force seemed to over-support the supplements industry resulting in disappointment among consumers and a strong request for tighter control and concern for health protection, the EU Directive 2002/46/EC is stricter and sets out several rules that, at the time of its issue, seemed to spell instant damage to the food supplements industry, which reacted, predictably, by legal means.

The consequent replies to the industry and consumers' reactions from the responsible institutional bodies in USA and EU improved the adversarial situations by respectively increasing the desired control on the dietary supplements marketed in USA and by relieving the EU industry of the burden of long and costly safety procedures.

The general behaviour of the EC towards the increasing commercialization and intake of food supplements demonstrates the importance the EU gives to nutraceuticals, as popular foodstuffs able to support public health and consumers, in terms of health protection and food safety.

The EC recognizes the high potential impact that nutraceuticals will have in EU business, market and health. Nevertheless, additional research is required to disseminate this standpoint. At the moment no specific rules concerning nutrients other than vitamins and minerals or other substances with a nutritional or physiological effect used as ingredients of food supplements exist. The EC plans to introduce such rules at a later stage, provided that adequate and appropriate scientific data become available.

The number of career scientists attracted to nutraceutical research should increase to provide scientific expertise where it is presently underdeveloped. Moreover, adequate research funds must be made available to fuel this process.

In order to bring the nutraceutical field a step further, the research should target validation in the medical field to be able to claim healthy properties for some food supplements supported by scientific evidence. Rigorous research is certainly not confined to randomised clinical trials. For defining the safety of nutraceuticals other methodologies could be much more adequate, for example post marketing surveillance studies. Also many students could be invited to participate as human subjects on a voluntary basis in clinical test projects. Generally speaking, the research question determines the optimal research methodology, especially in such an innovative field as food supplements.

Today research ethics committees have the remit to monitor medical research. Some of the core points they must consider include: scientific quality of proposal, getting and applying evidence, risks for study participants, informed consent, indemnity cover, data protection.

The responsible investigator has to ensure that his study is conducted in agreement with either the Declaration of Helsinki or the laws and regulations of the country, whichever provides the greatest protection for the patient.

A protocol must be written and the study must be conducted according to the Guidelines for Good Clinical Practice and subjected to review by the competent Ethical Committee.

The central policy question is whether the requirements that healthy food must meet should be the same or higher than those that ordinary food does.

To answer this question the community will need to better assess the risk of side effects of these new foods and the medium/long-term effects on eating habits and public health in general.

Acknowledgements

This work was supported by the NUTRA-SNACKS project funded within the EU Framework Programme 6 and by the MOMA project sponsored by the Italian Space Agency (ASI).

References

1. FDA. Dietary Supplements. http://www.cfsan.fda.gov/~DMS/supplmnt.html.
2. Larsen LL, Berry JA. The regulation of dietary supplements, J Am Acad Nurse Pract 2003; 15(9):410-4.
3. Blendon RJ, DesRoches CM, Benson JM et al. Americans' views on the use and regulation of dietary supplements, Arch Intern Med 2001; 161(6):805-10.
4. AARP Public Policy Institute. Dietary supplements and older consumers. Data Digest 2001; 66:1-8.
5. 72 Fed. Reg. 34752. FDA Issues Dietary Supplements Final Rule. 2007.
6. Directive 2002/46/EC of the European Parliament and of the Council of 2002 on the approximation of the laws of the Member States relating to food supplements.

7. Kiely M, Flynn A, Harrington KE et al. The efficacy and safety of nutritional supplement use in a representative sample of adults in the North/South Ireland Food Consumption Survey. Public Health Nutrition 2001; 4(5a):1089-1097(9).
8. Knight, Sam. "Controversial EU Vitamins Ban To Go Ahead." The Times, July 12, 2005. Europe news section, Online edition.
9. "Court Victory for Vitamin Firms." BBC News, January 30, 2004. BBC Health section, Online edition.
10. "EU Health Foods Crackdown Wrong." BBC News, April 5, 2005. BBC Health section, Online edition.
11. Commission reply PE 374.206 to Petition 0694/2005 by Graham Allen Green (British), on the Food Supplements Directive. May 12, 2006.
12. "Vitamin Controls Cacked by Europe." BBC News, 2005.
13. "EU Court Backs Health Supplements Ban." The Guardian, July 12, 2005. Society health section, Online edition.
14. Regulation (EC) No 1924/2006 of the European Parliament and of the Council of December 20, 2006 on nutrition and health claims made on foods.

Index

A

Adaptogen 99, 103
Adventitious regeneration 187
Age-related macular degeneration and cataracts (AMD) 4, 21, 77, 80, 319
Ajoene 111-113, 115
Alfa-lipoic-acid (LA) 52, 55, 57, 63, 176
Alginate 9, 116-118, 227
Alliance for Natural Health (ANH) 325, 326
Alliin 5, 8, 110, 111, 113-116, 118
Alliinase (Alliin lyase) 5, 93, 110, 111, 115, 116, 118
Alzheimer's disease (AD) 2, 18, 27-30, 62, 63, 93, 136, 319
Antibody 11, 148, 236, 269-271, 283, 284, 286-288, 294
Antioxidant 1-6, 8, 10-12, 17-21, 23, 28-32, 42, 43, 45, 52-63, 71, 75-77, 81-86, 88, 89, 92-94, 99-101, 105, 106, 113, 115, 123, 126, 127, 135-137, 145, 147, 153, 175, 176, 204, 213, 214, 216, 217, 222-226, 228-230, 234-236, 241-248, 255, 256, 267, 272, 273, 275-278, 315, 316, 318, 319
Antioxidant capacity 12, 59, 99, 101, 105, 106, 126, 127, 222-226, 228-230, 236, 241-248, 277
Antiproliferative activity 126, 127
AP-1 18, 20, 36, 45-47
Apoptosis 3, 5, 31, 36, 43, 44, 46, 47, 52, 56, 58, 62, 63, 75, 83-86, 94, 106, 115, 148
Astaxanthin 10, 318, 319
Augmentation 36, 44
Autotrophy 185, 196, 197, 200

B

Basil 5, 175, 176, 216
Batch culture 206, 210, 213
Beta-carotene 4, 20, 22, 77-80, 122, 123, 126-128, 130, 131, 135, 136, 225, 318
Bilberry 82, 99, 104-106
Bilirubin 29, 52, 55, 59, 60, 63
Bioactive 6, 8, 10, 32, 75, 85, 99, 100, 101, 102, 103, 138, 144, 147-149, 175, 179, 185, 186, 200, 203, 206, 215, 217, 218, 251, 253, 258, 278, 316, 317
Biolistic 145
Bioreactor 1, 10, 12, 103, 144, 145, 149, 167, 169, 189, 203, 204, 206-218
Biosensor 1, 12, 13, 222-227, 229, 230, 234-239, 267-278, 282, 285-287, 289, 290, 293-303
Brain 27-30, 32, 60-62, 69, 89, 94, 125, 226, 274

C

Caffeic acid phenethyl ester (CAPE) 30, 31
Callus 8, 9, 103, 129, 171, 172, 176, 187, 188, 205, 206, 216, 252, 258, 260
Cancer 2, 4-6, 8, 17-21, 27, 29, 31, 36-38, 41-49, 55, 59, 63, 68, 75-78, 83, 84, 86-89, 92-95, 99, 105, 106, 111, 113, 115, 122, 123, 126, 127, 135, 136, 186, 187, 222, 228, 268, 274, 275, 283, 319
Cancer prevention 37, 63, 94
Canola oil 71, 89
Capsaicinoid 74, 76, 95, 127, 273, 275
Carbohydrates 3, 4, 18, 38, 52, 53, 68, 69, 70, 72, 74, 115, 125, 126, 145, 154, 170, 187, 192, 196, 197, 203, 206, 208, 216, 217, 222, 271, 276, 294-297, 315
Carotenoid 4, 7, 10, 17, 20, 22, 23, 74, 75, 77-80, 100, 122, 123, 128, 130-132, 135, 136, 234, 244, 254, 273, 274, 318, 320
Cataract 4, 21, 80, 111, 124
Catechin 5, 31, 38-40, 44, 47, 75, 81, 83, 84, 105, 229, 235, 237, 238, 255, 273
Catechol oxidase (Laccase) 12, 223, 234, 237-239, 270, 275, 276
Celiac disease 122-124
Cell culture 6-12, 31, 102, 103, 152, 153, 155, 157-159, 161, 173, 177-179, 186, 187, 189, 203-206, 209, 211, 213, 217, 250-253, 259, 261, 262, 316

Cell factory 144
Cell line 6, 9, 21, 31, 75, 84, 90, 94, 186, 187, 204, 253
Cell suspension 9, 12, 89, 161, 177, 187, 188, 204, 205, 207-209, 213, 216, 258, 259
Cellular stress response 30, 32, 61
Chemotherapy 6, 38, 44, 47, 93
Chemotherapy-sensitization 6, 38, 44, 47, 93
Chlamydomonas reinhardtii 7, 8, 10, 144-149, 273, 316
Chloroplast 4, 10, 131, 144-149, 216, 251, 252
Chronic disease 4, 11, 17, 18, 20, 23, 77, 113, 126, 319, 322
Citric acid 71
Clone 176, 185-187, 189, 204
Coenzyme Q (CoQ) 52, 55, 57, 58, 63
Commercial device 299
Compound 1-13, 18, 21, 22, 27-32, 36, 38-44, 46, 48, 54, 55, 57, 60, 61, 63, 71, 74-78, 81, 83-90, 93-95, 99-106, 110-118, 124, 136, 144-147, 149, 152-155, 159, 160, 168, 173, 175, 176, 178, 179, 185-187, 189, 199, 200, 203-207, 214, 216, 218, 222, 223, 225, 226, 228-230, 234, 236-238, 239, 241-248, 251, 253, 254, 256-261, 268, 273-278, 282, 283, 287, 289, 290, 294-296, 302, 310, 315, 316, 318
Coneflower 214
Conjugated linoleic acid (CLA) 36, 49, 74, 76, 90-92
Consumer 2, 3, 11, 13, 19, 72, 84, 145, 267, 268, 278, 287, 293, 308-313, 315, 317-320, 322-327
Continuous culture 206, 211, 218
Cow's milk protein allergy (CMPA) 123-125, 139
Culture 1, 6-12, 31, 37, 61, 76, 89, 102, 103, 114, 146, 152-155, 157-159, 161, 167, 168, 171-173, 175-179, 185-190, 192, 193, 196-200, 203-218, 250-254, 256-262, 312, 314, 316
Curcuma zedoaria 213
Curcumin 6, 8, 27-32, 38, 44-47, 86, 213, 214, 273
Current good manufacturing practices (cGMP) 324
Cyanobacteria 7, 60, 272, 273, 282, 286, 290
Cyanotoxin 282, 286, 287
Cytochrome *c* (Cyt *c*) 56, 58, 223, 275, 277, 278, 300

D

Dietary Supplement and Health and Education Act (DSHEA) 323, 324, 326
Diet-gene interaction 17
Diet-related disease 122-125, 139
Disease 1-5, 7, 8, 11, 13, 17-21, 23, 27-29, 31, 36-38, 43, 53-55, 57-63, 68, 71, 72, 74, 75, 77, 78, 80, 83, 85, 87-90, 92-95, 99, 101, 104, 105, 111, 113, 122-127, 129, 135, 136, 137-139, 148, 153, 167, 203, 214, 217, 222, 228, 252, 267, 274, 278, 282, 293, 308-310, 319, 322-326
DNA 2, 5, 11, 20, 28, 37, 45, 47, 52, 55-58, 60, 61, 63, 71, 83, 84, 86, 94, 95, 129, 136, 145, 146, 167, 168, 170, 177, 189, 213, 222, 228-230, 236, 239, 243, 247, 248, 252, 269, 270, 272, 274-278, 294, 319
DNA-based biosensors 228-230, 272, 277
Docosahexaenoic acid (DHA) 71, 89, 90
Drosera rotundifolia 99, 103, 104
Drug delivery 117

E

Echinacea purpurea 177, 178, 214, 215
Eicosanoid 68, 69, 71
Eicosapentaenoic acid 11, 69, 71, 89, 90
Electrochemical biosensor 222, 230, 274-276, 289, 290
Electrochemical detection 275, 287
Electrochemical transduction 270, 284, 289, 296
Elicitor 8, 9, 103, 152-161, 173, 175, 176, 179, 216, 217, 218
elicitation 7-10, 153, 156-158, 161, 167, 175, 176, 179, 217, 218, 259
Encapsulation 110, 111, 116
Endangered plant 186

Endogenous antioxidant 52, 54, 55
Enzyme 2, 3, 5, 11, 12, 22, 27, 29, 30, 32, 42, 45-47, 52, 56, 59-61, 63, 76, 78, 83, 85, 86, 88, 89, 93, 94, 101-103, 110-113, 115-118, 123-127, 129, 135, 138, 145, 148, 155-157, 160, 167, 172, 199, 214, 217, 222, 223, 225-227, 229, 234, 236-239, 252, 269-271, 277, 284, 287-289, 293-296, 300-302, 322
Enzyme immobilisation 238
Enzyme-linked immunosorbent assay (ELISA) 269, 270, 284, 285, 287, 288
Enzyme sensor 226, 289, 294, 302
Ethyl ferulate (EFE) 30, 31

F

Fatty acid 2, 4, 11, 17-23, 28, 36, 45, 49, 53, 54, 60, 69-71, 74, 75, 89, 92, 113, 146, 154, 257, 267, 273, 274, 277, 278, 293, 315, 324
Fed batch culture 206, 210
Fenton reaction 58, 227, 228, 230, 247
Ferritin 30, 52, 55, 58, 63, 138
Ferulic acid 8, 31, 83
Fish oil 17, 71, 75, 89, 90
Flavenoid 5, 7, 38-41, 44, 45, 47, 48, 54, 81-83, 87, 99, 101, 102, 104-106, 113, 122, 123, 126, 127, 133, 137, 138, 153, 160, 175, 204, 224, 234-237, 242, 250, 254, 255, 259, 260, 273
Flexible 199, 234, 258
Folic acid 122, 126, 127, 129, 132, 136, 137, 267, 273, 274, 276
Food and Drug Administration (FDA) 10, 71, 322-324
Food quality 1, 101
Food safety 145, 267, 325-327
Free radical 2-6, 11, 12, 28, 29, 32, 52, 53, 55, 56, 58-63, 71, 75, 76, 81, 82, 101, 116, 158, 213, 215, 222, 223, 230, 234, 241-245, 247, 248, 273, 277, 278
Functional food 2, 8, 11-13, 36, 71, 95, 100, 123, 136, 139, 148, 267, 268, 273, 278, 308-313, 315, 317-320, 322, 323, 325
Fungus 9, 39, 60, 61, 77, 88, 92, 94, 104, 115, 149, 154, 192, 236, 237, 259, 268, 272, 282, 290, 293

G

Galactosemia 123-125, 139
Garlic 5, 38, 75, 76, 93, 94, 110-116, 118
Genetic engineering 10, 17, 122, 127, 135, 137, 144-147, 149, 171, 177, 260
Genetic enhancement 136
Genetics 2, 9, 10, 17, 18, 37, 57, 101, 103, 122-125, 127, 135-139, 144-147, 149, 160, 167-169, 171, 177, 179, 187, 189, 205, 218, 251, 252, 260
Ginseng 99, 103, 158, 170, 178, 179, 204, 205, 212, 216, 217, 258, 277
Global market 278, 311, 312
Glucosinolate 74, 76, 94, 95, 126, 153, 158, 159, 161, 273, 275-277
Glycogen storage disorder 123-126, 139
Green tea 31, 39, 40, 45-47, 75, 81-84, 86, 225
Growth regulator 173, 187, 188, 189, 192, 197, 203, 206

H

Hairy root (HR) 9, 10, 12, 159, 167-180, 189, 207, 209, 211-213, 216-218, 252, 255, 259, 260
Hairy root culture 167, 175, 207, 209, 212, 213, 217, 218, 260
Health 1-5, 8, 10, 11, 13, 17, 19, 20, 22, 23, 31, 36, 38-40, 42, 43, 45, 69, 71, 72, 74-78, 80-82, 86-88, 90, 93-95, 99, 101, 105, 110, 111, 122-124, 127, 135, 136, 138, 145, 147, 152, 160, 191, 203, 222, 267, 268, 273, 276, 278, 282, 287, 290, 308-313, 315-320, 322, 323-327
Heart disease prevention 4, 17, 21, 23, 59, 68, 71, 75, 83, 85, 90, 111, 123, 126, 319
Heat shock protein 11, 61
Heme oxygenase 27, 29, 32, 59
Heterodimers Of NF-e2-Related Factors 2 (Nrf2) 30-32, 60
Heterotrophy 196
Hormone balance 89
Horseradish peroxidase (HRP) 225, 234, 239, 276, 278, 284, 288, 298, 299, 301-303

Hydrogen peroxide 2, 53, 59, 60, 158, 241, 245, 276, 277, 296, 299
Hydroxyl radical 2, 29, 31, 52, 227, 243, 247, 277

I

Immobilization 9, 12, 118, 225, 226, 228, 237-239, 289, 294, 302
Immune response 17, 18, 23, 47, 48, 78, 83, 90, 147, 148
Immunity 47, 48, 135
Immunosensor 270, 271, 284-288
Inflammation 2, 6, 17, 19, 20, 28, 30, 44, 83, 86, 217, 319
Isoflavone 5, 39-41, 75, 81, 86-88, 122, 133, 137, 235, 242, 256, 259, 273, 316

L

Lavandula vera 216, 217
Lavender 216
Layered double hydroxide (LDH) 116-118
L-carnitine 52, 55, 60, 61, 63
Light 2, 4, 7, 8, 10, 21, 22, 40, 58, 63, 78, 83, 85, 99, 100, 103, 104, 106, 136, 147, 155, 161, 178, 186, 187, 191, 196, 197, 206, 208, 209, 228, 235, 243, 246, 256, 268, 270, 272, 273, 275, 276, 286, 289
Lignan 75, 86, 88, 89, 254, 273
Limonene 74, 76, 93, 273, 275
Lipid peroxidation 11, 21, 28-31, 53, 54, 58, 63, 71, 85, 93, 158, 213, 217, 247
Lycopene 20, 22, 75, 77-80, 101, 122, 123, 126-128, 130, 131, 135, 136, 316, 318
Lycopene-cyclase 22

M

Major crop 128, 129, 135-138
MAP-kinase 157, 158
Marketing 3, 95, 116, 118, 145, 185, 267, 268, 270, 277, 278, 303, 308-313, 315-325, 327
Master control gene 17
Medicinal and aromatic plant 203, 204, 213, 218
Melatonin 52, 55, 61-63, 225, 323
Metabolite 1, 3, 4, 6-13, 38, 74, 81, 86, 95, 99-103, 122, 124, 126, 135, 144, 146, 152-161, 167, 169-173, 175-180, 185-187, 189, 199, 203-206, 208, 210, 211, 213-218, 234-236, 250-254, 256, 259-261, 267, 282, 290, 316
Metabolomic 127, 152, 159, 161, 250, 251, 253, 257, 260, 261
Metallothioneine 52, 63
Microalgae 10, 144-146, 149, 320
Microbial sensor 293, 302
Microcystin 282, 286, 287
Micropropagation 173, 185, 187, 189, 191, 192, 196, 198-200, 216
Minor crop 122, 129, 135-138
Molecularly 286, 289, 299
Molecularly imprinted polymer (MIP) 286, 299
Monitoring 11, 160, 186, 213, 214, 223, 227, 229, 238, 243, 244, 248, 251, 254, 268, 269, 271, 273, 277, 278, 283, 284, 294-296, 302, 303, 316
Monogenic disease 123, 124, 126
Monophenol monooxygenase 12, 223, 234, 237-239, 270, 275-277
Multidetection system 302
Mycotoxin 112, 282, 283, 285, 286

N

Natural compound 4, 154, 200, 204, 251, 254
Natural product 74, 77, 152-155, 159, 160, 255, 319
Neurodegenerative disorder 2, 27-30, 61, 90
Neuron 27, 29-32, 59, 60, 274
Neuropsychiatric and neurodevelopmental disorders 19
NF-κB (nuclear factor kappa B) 18, 20, 29, 31, 36, 45-47, 55-57
Nitric oxide radical 229, 277
Nutraceutical 1, 5, 11, 17, 20, 22, 71, 96, 110, 118, 122, 123, 126, 127, 136, 138, 147, 159, 173, 185-187, 189, 200, 203, 271, 273, 274, 276, 308, 310-312, 315, 318, 320, 322, 327
Nutraceuticals in space 315
Nutraceutics 1, 3, 11, 13, 72, 99-102, 152, 274-276

Nutrition 3, 4, 8, 9, 11, 32, 36, 37, 39, 43, 68-72, 74, 77, 92, 122, 123, 145, 152, 268, 278, 308-310, 319, 323, 326
Nutritional strategy 31, 68

O

Obesity 2, 31, 58, 68, 69, 90, 95, 111, 310, 319
Ochratoxin A 282-284
Ocimum basilicum 175, 190, 216
Omega-3 fatty acid 11, 18-20, 36, 49, 71, 75, 89, 90, 267, 273, 274, 315
Omega-6 fatty acid 18-20, 71, 89, 90
Omega-6 to omega-3 fatty acid ratio 19
Optical biosensor 274-277
Organic acid 38, 204, 294, 300
Organogenesis 187, 199, 200
Organosulphate 110
Overweight prevention 68
Oxidation 2, 4-6, 20, 21, 28-32, 45, 52, 53, 56, 58, 60, 63, 71, 83, 84, 123, 127, 158, 223, 225-230, 234, 236-238, 242-244, 247, 248, 261, 272, 276, 277, 289, 296, 316
Oxidative stress 1-5, 7, 11, 28-32, 52-54, 60-62, 105, 127, 160, 178, 214, 217, 222, 235, 241, 275, 277, 278, 316, 319

P

Panax ginseng 178, 204, 205, 216, 217, 258
Pathogen-associated molecular pattern 154, 159
Pharmaceutical 6, 29, 37, 71, 94, 101, 139, 147, 148, 152, 159, 175-177, 203, 250, 251, 254, 267, 268, 309, 311, 317, 319
Phenolic 7, 20, 22, 27-31, 36, 38-48, 54, 74, 81, 83, 86, 99-106, 123, 127, 134, 137, 138, 155, 168, 175-178, 216, 229, 234-238, 242, 254-256, 259, 273, 276, 277, 296
Phenylketonuria 124-126, 139
Photobioreactor 146
Photosystem II 11, 145, 272, 273
Phytochemical 3, 8, 9, 12, 17, 18, 20, 23, 27, 36-39, 43, 44, 74, 95, 111, 116, 153, 167, 177, 200, 203, 204, 242, 243, 250-253, 257-262, 267, 273, 276-278
Phytoestrogene 74
Phytonutrient 17, 122-124, 126, 127, 136, 138
Piezoelectric biosensor 269, 285
Plant 1, 3-10, 12, 17-23, 29-32, 36-45, 48, 54, 60, 61, 63, 71, 74, 77, 81, 83, 85-89, 92-95, 99-106, 110, 115, 122, 127, 136-139, 144-149, 152-161, 167-173, 175-180, 185-192, 196, 197, 199, 200, 203-218, 225, 228, 234-236, 239, 242, 250-262, 269, 272, 277, 278, 290, 294, 313, 316, 319, 324, 325
Plant cell culture 6-9, 12, 102, 152, 153, 158, 159, 186, 203-206, 209, 217, 250, 261, 262, 316
Plant secondary metabolism 100, 154, 155, 159, 161, 171, 172, 261
Polyphenol 5, 6, 27, 28, 30-32, 42, 46-48, 54, 71, 81, 84-86, 126, 127, 137, 223, 226, 228, 234-239, 242, 243, 255, 259, 270, 275-277
Polyunsaturated fatty acid 4, 11, 17-20, 22, 45, 49, 74
Prebiotic 76, 94, 313, 315, 317
Prevention 5, 7, 11, 13, 18, 29, 36-38, 46-48, 58, 59, 61, 63, 72, 83, 86-88, 90, 93, 94, 111, 123, 126, 217, 293, 322, 323
Probiotic 312-315, 317
Programmed cell death 17, 21-23, 36, 46
Protein expression 31, 146, 147
Protein phosphatase 287

R

Reactive oxygen species (ROS) 2, 3, 6, 11, 21, 29, 45, 52-57, 61-63, 78, 85, 93, 105, 123, 145, 153, 156, 158, 172, 216, 222-225, 241-244, 248, 277, 316
Reduced glutathione 55
Resveratol 5, 13, 30, 38, 42, 44-48, 75, 85, 86, 122, 127, 134, 137, 159, 172, 180, 237, 316
Rhodiola 99, 102, 103, 260
Rooting 170, 171, 191, 192, 196-198
Rose root 99
Rosmarinic acid 5, 13, 171, 173, 175, 176, 205, 216, 229, 256, 259
Round-leaved sundew 99, 103, 104

S

Sage 5, 8, 173, 175, 217
Salvia miltiorrhiza 173, 175, 217, 218
Screening tool 282, 284, 288, 289
Secondary 1, 3, 4, 6, 8-10, 12, 13, 38, 74, 81, 86, 95, 99-103, 146, 152-161, 167, 169-173, 175, 176, 178-180, 185-187, 189, 196, 199, 203-206, 208, 213, 214, 217, 218, 229, 235, 246, 250-254, 256, 259-261, 282, 290, 294, 295
Secondary metabolite 1, 3, 4, 6, 8-10, 12, 13, 38, 74, 81, 86, 95, 99-103, 146, 152-161, 167, 169-173, 175, 176, 178-180, 185-187, 189, 199, 203-206, 208, 213, 214, 217, 218, 235, 250-254, 256, 259-261, 282
Secondary metabolite production 9, 101, 156, 169, 172, 203, 204, 206, 208, 213, 217, 250, 252, 254
Selection 1, 6, 8, 9, 127, 146, 186, 192, 204, 253
Signalling pathway 83, 86, 155-159
Signal transduction 17, 30, 44, 60, 84, 86, 93, 153-155, 157, 161, 171, 236, 269, 276
Somaclonal variation 9, 170, 187
Somatic embryogenesis 187, 189
Soybean oil 71
Specific and accurate analytical tool 234
Stilbene 41, 42, 75, 85, 86, 127, 134, 137, 158, 172, 254, 273
Superoxide dismutase 2, 11, 55, 214, 225, 275, 277, 278
Superoxide radical 223, 277
Supplement 3, 5, 20, 23, 32, 71, 88, 89, 95, 100, 101, 106, 110, 111, 115, 116, 118, 126, 127, 136, 138, 149, 308, 322-327

T

Temperature 5, 8, 12, 22, 62, 70, 99, 100, 101, 112, 113, 125, 178, 186, 191, 197, 206, 208, 209, 214, 235, 245, 253-256, 283, 286, 289, 302
Tocopherol 6, 11, 21, 28, 54, 56, 60, 61, 71, 78, 81, 92, 93, 122, 127, 132,-134, 138, 273, 274, 276
Tocotrienol 75, 92, 93
Transcription factor 3, 18, 20, 29, 30, 31, 36, 44, 45, 55, 60, 63, 83, 86, 138, 153, 154, 157, 158, 160, 172, 173, 179
Transformation 9, 63, 103, 113, 115, 135, 136, 144-147, 149, 160, 168, 170, 172, 173, 175, 177, 179, 180, 189, 218, 238, 260, 295
Transgenic plant 146, 168, 170
Treatment 5, 6, 11, 28, 29, 30, 36, 37, 47, 48, 61, 85, 86, 89, 90, 106, 110-113, 123, 124, 126, 139, 152, 155, 156, 158-161, 177, 179, 188, 189, 196, 206, 217, 256, 276, 278, 286, 294, 319, 323

U

Uric acid 52, 55, 59, 63, 223, 224, 229, 296

V

Vaccinium 85, 99, 104-106
Vegetable crop 124, 127, 129
Vitamin 4, 17, 19-21, 29, 38, 57, 58, 62, 70-72, 74, 77, 78, 81, 88, 92, 93, 111, 122-127, 129, 132, 135, 137, 146, 192, 204, 226, 244, 267, 273, 276, 278, 293, 313-315, 317, 323-327
Vitamin E 19, 20, 21, 29, 58, 92, 93, 127, 132, 273, 276

W

Whole-cell biosensor 270, 271

X

Xanthine oxidase 45, 223, 277
Xanthophylls and xanthophyll cycle 8, 17, 20-23, 77, 316

Z

Zeaxanthin and lutein 4, 7, 17, 20-23, 78, 80, 122, 128, 135, 316
Zeaxanthin epoxidase 22
Zedoary 213, 214
Zone diet 3, 68-71